Musculoskeletal Imaging THE REQUISITES

SERIES EDITOR **James H. Thrall,** M.D.
Radiologist-in-Chief
Department of Radiology
Massachusetts General Hospital
Boston, Massachusetts

OTHER VOLUMES IN THE REQUISITES SERIES

Gastrointestinal Radiology

Pediatric Radiology

Neuroradiology

Nuclear Medicine

Ultrasound

Cardiac Radiology

Genitourinary Radiology

Thoracic Radiology

Mammography

Musculoskeletal Imaging

THE REQUISITES

DAVID J. SARTORIS, M.D.
Professor of Radiology
Chief, Quantitative Bone Densitometry
University of California School of Medicine
University of California Medical Center
San Diego, California

With 663 illustrations

St. Louis Baltimore Boston Carlsbad Chicago Naples New York Philadelphia Portland
London Madrid Mexico City Singapore Sydney Tokyo Toronto Wiesbaden

Dedicated to Publishing Excellence

A Times Mirror
Company

Editor-in-Chief: Susan Gay
Managing Editor: Elizabeth Corra
Project Manager: Mark Spann
Production Editor: Elizabeth Fathman
Production and Editing: Carlisle Publisher Services
Designer: David Zielinski
Manufacturing Supervisor: Tony McAllister

Printed in the United States of America

Composition by Carlisle Communications, Ltd.
Printing/Binding by Maple Vail Book Manufacturing Group

Mosby–Year Book, Inc.
11830 Westline Industrial Drive
St. Louis, Missouri 63146

Library of Congress Cataloging-in-Publication Data

Sartoris, David J.
    Musculoskeletal imaging : the requisites / David J. Sartoris.
        p.   cm. — (Requisites series)
    Includes bibliographical references and index.
    ISBN 0-8151-8002-0
    1. Musculoskeletal system—Imaging.   I. Title. II. Series.
    [DNLM: 1. Musculoskeletal System—radiography.   2. Musculoskeletal
    Diseases—physiopathology.   WE 141 S251m 1996]
    RC925.7.S27 1996
    616.7′0754—dc20
    DNLM/DLC
    for Library of Congress   95-25625
                                                        CIP

95 96 97 98 99 / 9 8 7 6 5 4 3 2 1

To God's greatest gifts to me:
*my dear parents, **Helen** and **Cornelius**,*
*and to my loving wife, **Cyd**.*
Their hard work also resides within this volume.

To those who might learn from this book,
who have been my inspiration and purpose.
Be not selfish with the knowledge derived from it,
but having light, forever pass it on to others.
For to teach is to truly understand
and to experience life's greatest reward.

D.S.

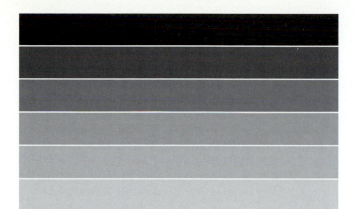

Foreword

Musculoskeletal Imaging: The Requisites is the sixth book in an ongoing series designed to provide core material in major subspecialty areas of radiology for residents in their initial training programs, for practicing radiologists seeking a concise overview, and for physicians in related disciplines who require knowledge of imaging in their area of specialization.

David Sartoris, M.D., has done an outstanding job in capturing the philosophy of *The Requisites* in this book. Musculoskeletal examinations represent one of the largest areas of radiology practice. At the chapter level, Dr. Sartoris has structured his text in a logical format based on major categories of disease that cover the essential components of the practice of musculoskeletal radiology.

Within the broad chapter headings, Dr. Sartoris has designed the presentation of material in a number of ways that should prove very beneficial to the reader. First, Dr. Sartoris has taken advantage of his enormous experience as a musculoskeletal radiologist and as an author to structure each chapter to emphasize more common and therefore more frequently encountered conditions first, thereby alerting the reader to their importance in radiology practice. Second, Dr. Sartoris has selected clinical "pearls" to reinforce important material in each area. These highlighted statements provide an emphasis that is hard to come by individually and is again based on the author's vast experience as an expert in musculoskeletal radiology.

A too common approach in textbooks is to present a discussion of each disease or condition within a category and include an exhaustive list of the radiographic findings associated with the condition. This is, however, not the way disease presents itself to us clinically or radiographically; rather, we are typically presented a limited number of signs and symptoms from which an initial differential diagnosis must be formulated. Where

appropriate, Dr. Sartoris has structured his book in a way that corresponds to our actual practice of radiology and not from the standpoint of creating an encyclopedia of disease descriptions. This is best illustrated in Chapters 3 and 4, where Dr. Sartoris first presents an in-depth discussion of radiographic signs—the building blocks of the interpretive process—followed by a discussion of specific diseases.

One of the additional challenges in musculoskeletal radiology is that the specialty area now encompasses every imaging modality. Dr. Sartoris has done an excellent job in sorting out the strengths, weaknesses, and indications for the different imaging methods available for the study of a particular disorder. With the increasing emphasis on cost-effective care, these considerations are particularly relevant, even to the beginning student of radiology.

I believe that the neophyte resident in radiology will find *Musculoskeletal Imaging: The Requisites* to be of great value in engaging the subject, and more senior residents, fellows, and practicing radiologists should find this an eminently readable text for a concise review. Physicians and trainees in other disciplines including rheumatology, orthopedic surgery, endocrinology, and oncology will also find this to be a user friendly text to forward their understanding of musculoskeletal radiology in the areas of interest to their practices.

In a sense, people entering training in radiology are forced to "start over" in terms of their medical knowledge base. The fundamental principles of basic science and clinical medicine that are learned in medical school are necessary background for training in radiology; however, the actual knowledge and skills required for correctly analyzing radiologic images and formulating diagnoses are left almost entirely to residency and fellowship training. Thus, a resident entering a subspecialty rotation in a radiology training program has the formidable task of going from a minimal knowledge of subspecialty specifics to a working knowledge in a very short period of time.

This observation was the basis for creating *The Requisites* in radiology series. One book specifically written for each of the major subspecialty areas is planned. The length of each book is dictated by the material requiring coverage, but the goal is to provide the resident with a text that might by reasonably read within several days at the beginning of each subspecialty rotation and perhaps reread several times during both initial and subsequent rotations. The books are not intended to be exhaustive but should provide the basic conceptual, factual, and interpretive material required for clinical practice. Each book is written by an internationally recognized authority in the respective subspecialty areas and, because this is a completely new series, each author has had the opportunity to present material in the context of today's contemporary practice of radiology rather than grafting information about new imaging modalities and strategic approaches on to text material originally developed for conventional radiography.

I congratulate Dr. Sartoris for this outstanding addition to *The Requisites in Radiology* series.

James H. Thrall, M.D.

Radiologist-in-Chief
Department of Radiology
Massachusetts General Hospital
Boston, Massachusetts

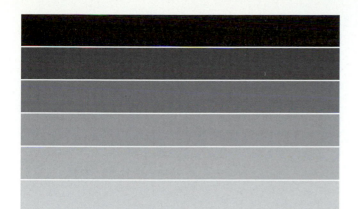

Preface

The mission of Mosby's *Requisites in Radiology* series is to fill an important gap in the diagnostic imaging literature by identifying and presenting the core material that is absolutely fundamental to each subspecialty area. Its target audiences include radiology residents/fellows and nonradiologists who require an overview of the field for primary learning and/or examination preparation. I have carefully prepared this volume to achieve this goal, based upon the premise that repetition is the key to learning.

Information is presented simultaneously in a broad spectrum of formats: text, lists, outlines, tables, figures with legends, and "pearls" of knowledge emphasizing critical concepts. Each of these components has been designed specifically to reinforce the others, resulting in an integrated and comprehensive coverage of the material.

Throughout the volume, I have placed emphasis upon differential diagnosis. This is particularly true for the figure legends, where the question "what else could it be?" is addressed in detail for all nonpathognomonic images.

In preparing this book, I have attempted to provide adequate coverage of the complete spectrum of diagnostic imaging methods applicable to musculoskeletal disease. Conventional radiography and tomography, contrast arthrography, computed tomography, skeletal scintigraphy, ultrasonography, and quantitative bone densitometry are discussed appropriately when indicated in the management of specific osseous, articular, and soft-tissue disorders.

Organization of the chapters is based upon presentation of disease categories in order from most to least common. This approach parallels that of *Musculoskeletal Imaging Workbook* (Mosby, 1993). This textbook and the workbook are intended for use in tandem by the reader.

Chapter 1 includes relevant fundamental concepts of musculoskeletal trauma and specific injuries in both the axial and appendicular skeleton. Chapter 2 discusses and illustrates the complete spectrum of musculoskeletal infections, including pyogenic, tuberculous, fungal, viral, and parasitic varieties. Chapter 3 is devoted to rheumatologic disorders and includes clinical concepts, general imaging features, and target site–specific radiographic findings in each disease. Chapter 4 is a comprehensive overview of benign and malignant categories of osseous and soft-tissue neoplasia, with emphasis on magnetic resonance imaging as the preferred method of staging these conditions.

Chapter 5 discusses metabolic and endocrine disorders of the musculoskeletal system and includes information on the pathophysiology of these conditions and techniques for noninvasive bone densitometry. Chapter 6 provides a comprehensive review of congenital bone and joint disease, including important syndrome complexes and short-limbed dysplasia. Chapter 7 addresses hematologic disorders such as hemophilia, sickle cell anemia, thalassemia, mastocytosis, and myelofibrosis, with emphasis on the importance of magnetic resonance in bone marrow imaging. The final chapter is devoted to miscellaneous conditions and imaging techniques, including scoliosis, orthopaedic hardware, environmental bone disease, soft-tissue calcification and ossification, periostitis in children, sarcoidosis, skeletal scintigraphy, and magnetic resonance imaging.

Musculoskeletal disease and diagnostic imaging thereof encompass one of the most complex and extensive areas in all of medicine. Consequently, the definition of "requisite" material within this subspecialty may not be precisely agreed upon by all authorities. In preparing this volume, I have endeavored to cover all of the specific topics emphasized by the American College of Radiology in its teaching file and self-assessment syllabi, which are used extensively by radiology trainees in preparation for board examinations. I have covered the requisite material within the framework of a concise, succinct yet comprehensive presentation that is amenable to completion by the average trainee, for whom individual study time is generally limited. George Savile, the Marquess of Halifax, would thus have been pleased with this book: "There is no stronger evidence of a crazy understanding than the making too large a catalogue of things necessary."

In conclusion, I sincerely hope that this *Requisites* volume has accomplished its stated objectives and will prove beneficial to all who use it. I would like to emphasize that many concepts in musculoskeletal imaging are controversial; the information conveyed in this book reflects the most widely accepted dogma. In his last public address on April 11, 1865, Abraham Lincoln eloquently summarized this universal point: "Important principles *may* and *must* be inflexible."

David J. Sartoris, M.D.

Acknowledgments

This book could not have been completed without the knowledge and dedication of the many outstanding instructors with whom I have been blessed during my career in radiology. With regard to musculoskeletal imaging in particular, my mentors, Henry Jones, Fred Silverman, Howard Steinbach, and Donald Resnick deserve my eternal gratitude for their unparalleled commitment to teaching.

I wish to sincerely thank the staff at Mosby for their invitation to undertake this project and for their support and encouragement during its preparation. In particular, Elizabeth Corra, Maura Lieb, and Anne Patterson have led an outstanding team of publishing experts whom any author would find a pleasure to work with. I would also like to thank Cindy Trickel of Carlisle Publishers Services for coordinating the editing and production of the book.

I would like to extend my deep appreciation to the radiology residents at the University of California–San Diego with whom I have personally worked since becoming a faculty member in 1985. The opportunity to teach and interact with such a dedicated group of talented individuals has been the most rewarding aspect of my career in academic medicine. It is their unrivaled enthusiasm and hunger for knowledge that has inspired me to accept the challenge of authoring this volume.

Finally, I wish to thank in earnest my dear and irreplaceable secretary, Gale Hurley, for her invaluable assistance with this volume. The one and only reason that I was able to take on this project while simultaneously managing a full-time academic schedule is that I have been blessed with the world's greatest administrative assistant. Despite her other overwhelming responsibilities, she dedicated hours of hard work to convert the manuscript to a suitable form for the publisher.

D.S.

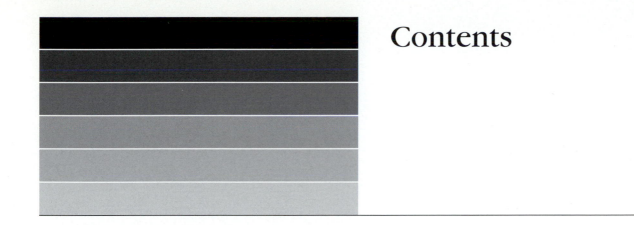

Contents

Musculoskeletal Imaging   THE REQUISITES

CHAPTER 1

Trauma

FUNDAMENTAL CONCEPTS

1. The tubular bones consist of an epiphysis, physis, zone of provisional calcification, metaphysis, and diaphysis, in this order, starting at the adjacent joint. The periosteum is attached at the metaphysis and, in adults, along the diaphysis.

2. Secondary ossification centers present at birth are the proximal part of the humerus (38 weeks), proximal part of the tibia (35 weeks), calcaneus, and distal part of the femur (33 weeks). The epiphysis of the proximal femur usually begins to ossify between the second and fifth months.

3. Bone growth is divided into interstitial growth (which causes lengthening) and appositional growth (which produces widening). Interstitial bone growth at the ends of long bones occurs at the physis by a process known as enchondral ossification. The periosteum forms bone by appositional growth producing widening of the bone.

4. Facial bones arise from membranous ossification. The carpal and tarsal bones are enchondral bones. The occipital bone is both membranous and enchondral bone. The clavicle is one of the first bones to ossify and consists of both enchondral and membranous bone.

5. A fracture is a complete or incomplete disruption in the normal continuity of bone or cartilage.

6. Bone is anisotropic because it has different mechanical properties when it is loaded in various directions.

7. The three basic types of loading that can be applied to a bone are compressive, tensile, and shear.

8. Cortical bone withstands the greatest stress with compressive loading and the least stress with shear loading.

9. Concerning magnetic resonance imaging (MRI), cortical bone has an extremely long T1 value. As the T1 relaxation time increases, the signal intensity decreases. Cortical bone has a low signal intensity because its relaxation time is long. Cortical bone also lacks mobile hydrogen atoms. Proton density represents the number of MR-visible protons in a unit volume of tissue and generally increases with H_2O content. Protons in solid material, such as cortical bone or dense calcifications, produce no signal on MR. Solids of this nature produce no signal because the protons in these materials are essentially frozen in position. Each proton is therefore constantly exposed to potent dephasing influences by nearby protons that are also fixed in position. Other musculoskeletal tissues exhibit the following MR relaxation times:
   - Fat: short T1, intermediate T2
   - Muscle: intermediate T1, intermediate T2
   - Fluid: long T1, long T2

10. A stress riser is a mechanical (frequently surgical) defect that concentrates stress and significantly decreases bone strength.

11. An open fracture is more likely to become infected than a closed fracture.

12. Incomplete fractures occur most commonly in children and are of four major types: buckle, torus, greenstick, and plastic bowing.

13. The buckle fracture is an indentation in the cortex on the compressive side of the bone.

14. The torus fracture is a circumferential buckling of the cortex induced by compressive loading.

15. The greenstick fracture is a partial cortical disruption occurring on the tensile side of the bone.

16. The plastic bowing fracture is a bending of the bone without angular deformity or subsequent remodeling.

PEARL

Pediatric fractures:
- Common sites of greenstick fractures include the proximal part of the metaphysis or diaphysis of the tibia and the middle third of the radius and ulna.
- Torus fractures are common in metaphyseal regions and in patients with osteoporosis.
- Greenstick fractures commonly convert to complete fractures because the traumatic deformity is exaggerated as the bone continues to grow.
- Bowing fractures constitute a plastic response, usually resulting from longitudinal stress in a bone. Bowing fractures most commonly involve the radius and ulna, although the clavicle, tibia, humerus, fibula, and femur can also be affected.
- The deforming force responsible for a bowing fracture must exceed the maximum strength of the bone (100% to 150% of body weight), and its duration must be shorter than the time necessary to reach the point of fracture.
- Torus fractures constitute a form of impacted fracture.
- The toddler's fracture is a spiral hairline fracture of the tibia occurring in children 9 months to 3 years of age, associated with sudden refusal to bear weight on the leg.

17. Complete long-bone fractures are usually transverse, oblique, or spiral in orientation.

18. A comminuted fracture yields more than two fragments.

19. A segmental comminuted fracture is characterized by two fracture lines that isolate a discrete fragment of bone.

20. A butterfly fragment is a wedge-shaped component of a comminuted fracture created at the apex of the applied force (Fig. 1-1).

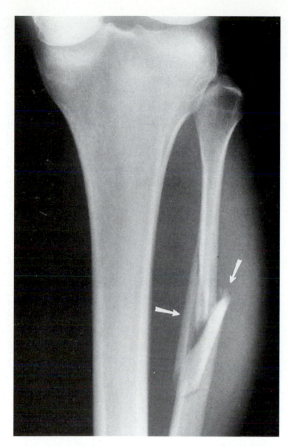

Fig. 1-1 Comminuted fracture of midfibular diaphysis with several butterfly fragments (*arrows*).

21. An intraarticular fracture predisposes the patient to premature degenerative joint disease.
22. Apposition refers to contact between the ends of two fracture fragments.
23. Anatomic alignment refers to the absence of significant displacement and angulation between two fracture fragments.
24. Bayonet apposition refers to overlap between two fracture fragments such that their shafts rather than their ends are in contact.
25. Distraction is a complete lack of apposition and may occur secondary to soft-tissue interposition, excessive traction, or bone resorption during early healing.
26. Angulation may be described as either (1) the direction of displacement of the distal fragment relative to the proximal fragment or (2) the direction of pointing of the fracture apex.
27. Valgus angulation is relative deviation of the distal fragment away from the midline of the body.
28. Varus angulation is relative deviation of the distal fragment toward the midline of the body.
29. The degree of rotation between two long-bone fracture fragments is best determined by computed tomography (CT) (Fig. 1-2).

30. An avulsion fracture is a fragment that is separated from the parent bone by traction from an attached ligament or tendon.
31. A chip fracture is a small isolated fragment induced by direct impact on the surface of a bone.

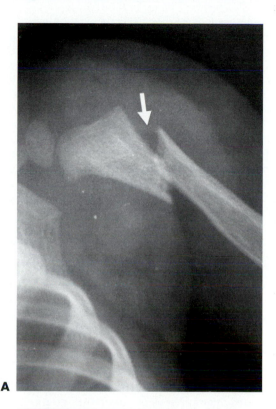

A

B

Fig. 1-2 **A,** Oblique or spiral fractures of long bones such as the humerus (*arrow*) are frequently characterized by torsion at the fracture site, which may be difficult to quantify by conventional radiography. **B,** CT is a useful technique for accurately measuring the degree of rotation between the major fragments (*arrows*) in such injuries (*T = tibia, F = fibula*). Typically, slices are taken at reproducible landmarks above and below the fracture, with comparative reference made to the contralateral uninjured extremity. This approach can also be used to measure femoral anteversion and tibial torsion in atraumatic situations.

32. Dislocation is complete disruption of contact between the articular surfaces of two bones at a joint.

PEARL

Joint dislocations and subluxations:
- Posterior dislocation of the glenohumeral joint constitutes only about 2% to 4% of all shoulder dislocations.
- Complete coracoclavicular ligament disruption in suspected acromioclavicular joint dislocation is suggested by an increase of the coracoclavicular distance by 40% to 50%.
- Sternoclavicular joint dislocations most commonly occur anteriorly.
- In children, traumatic dislocation of the radial head must be distinguished from congenital dislocation (as in the Larsen or Ehlers-Danlos syndromes, or that associated with hereditary osteo-onychodysplasia [nail-patella syndrome]).
- Dorsiflexion carpal instability is a dorsal tilting of the lunate bone with a scapholunate angle exceeding 60 degrees.
- Palmar flexion instability is indicated by palmar tilting of the lunate bone with a scapholunate angle less than 30 degrees.
- Anterior dislocation of the hip is due to forced abduction and constitutes only 5% to 10% of all hip dislocations.

---

33. Subluxation is partial disruption of contact between the articular surfaces of two bones at a joint.
34. Diastasis is an abnormal separation between two bones that form an articulation where motion is normally minimal.
35. A stress fracture is caused by repetitive application of force to a bone, the magnitude of which is not sufficient to cause an acute fracture; the two types are fatigue and insufficiency fractures.

PEARL

Characteristic sites of stress fracture:
- Prolonged standing: metatarsal sesamoid bones
- Trapshooting: coracoid process of scapula
- Clay shoveling: lower cervical, upper thoracic spinous processes
- Hurdling: patella
- Postoperative radical neck dissection: clavicle
- Carrying heavy pack: ribs
- Baseball pitching: coronoid process of ulna
- Propelling wheelchair: ulnar shaft
- Scrubbing floors: patella
- Baseball batting: hook of hamate bone

PEARL

Stress fractures:
- Fatigue fracture results from excessive stress placed upon a bone with normal elastic resistance.
- Insufficiency fracture results from application of normal stress or torque to a bone with deficient elastic resistance.
- Insufficiency fractures occur in rheumatoid arthritis, osteoporosis, Paget disease, rickets or osteomalacia, hyperparathyroidism, renal osteodystrophy, osteogenesis imperfecta, osteopetrosis, fibrous dysplasia, and irradiated bone.
- Metaphyseal stress fractures or those occurring in cancellous areas manifest focal sclerosis representing trabecular condensation, without prominent periostitis. Diaphyseal stress fractures exhibit linear or band-like radiolucency, frequently associated with periosteal and endosteal cortical thickening.
- Spondylolysis occurs most commonly at the L5 level (67% of cases), followed by the L4 (15% to 30%) and L3 (1% to 2%) levels. Increased uptake of bone-seeking radiopharmaceuticals indicates increased stress in the pars interarticularis, with or without a definite fracture.

---

36. A fatigue fracture occurs when abnormal repetitive force is applied to a bone of normal strength.
37. An insufficiency fracture occurs when normal repetitive force is applied to a bone with subnormal load-bearing capacity.
38. A pathologic fracture occurs when normal force is exerted on a bone that has been weakened by an underlying disease process, such as neoplasia or infection.
39. Systems of injury classification are commonly employed for the shoulder (Neer), ankle (Lauge-Hansen), subcapital portion of the proximal femur (Garden), and calcaneus (Essex-Lopresti).
40. Delayed union is an abnormally slow rate of fracture healing, given the clinical circumstances (patient age, percentage of bone loss, fracture location, degree of immobilization, etc.).

PEARL

The terms *primary healing* and *secondary healing* are used in reference to fracture healing by an analogy to postoperative healing of soft-tissue wounds. In a bone, primary healing is the process that occurs after internal fixation where there is minimal external callus and callus forms primarily intracortically. Callus formation is

Continued

PEARL—cont'd

much more prominent in secondary healing, and this is thought to be due to greater motion at a fracture site when external rather than internal fixation is used. Healing rates are dependent on a patient's age, the vascular supply to the bone, and the degree of bony- and soft-tissue injury, among other factors. Delayed healing may progress to healing or nonunion, where healing has stopped. Inadequate immobilization of fractures leads to excessive callus formation. Nonunion is most common in the tibia or femur and less common in the humerus, radius, ulna, and clavicle. It is least common in short or irregular bones such as the carpal and tarsal bones.

41. Nonunion is an absence of bridging callus with rounding and sclerosis of the apposed fragments at a site of prior fracture; the two types are atrophic and hypertrophic.
42. Ischemic necrosis is a complication that may occur following fractures of the femoral neck, scaphoid waist, and talar neck.
43. Osteomyelitis of a residual pin tract is manifested radiographically as an annular sequestrum.
44. Severe regional osteopenia with soft-tissue swelling as a complication of fracture is known as sympathetic reflex dystrophy or Sudeck's atrophy.

PEARL

Fracture complications:
- Sudeck's atrophy (sympathetic reflex dystrophy) has diagnostic criteria including pain and tenderness, soft-tissue swelling, decreased motor function, trophic skin changes, vasomotor instability, and patchy osteoporosis.
- Gas gangrene is caused by several species of clostridia, the most common being *Clostridium perfringens.*
- The average onset of symptoms from fat emboli occurs at 48 hours following fracture, with a range of 1 to 5 days. It is uncommon at less than 24 hours or more than 7 days.
- Lucency adjacent to an internal fixation screw may indicate movement of the screw or infection.

45. Malunion is healing of a fracture with the fragments not in anatomic alignment.
46. Limb length discrepancy following fracture is considered clinically significant when it exceeds 2 cm.

47. Angulation is acceptable at a healed fracture site if it occurs in the plane of motion of an adjacent hinge joint.
48. The Salter-Harris classification has relevance to the extent, prognosis, and management of fractures involving the open physis in skeletally immature patients.

PEARL

An apophysis represents an accessory ossification center that develops late and forms a bony prominence that serves as an attachment for a tendon or ligament. Examples of these are the humeral greater tuberosity and femoral greater trochanter. Apophyses do not contribute significantly to longitudinal growth of long bones, although the Salter-Harris classification also can be applied to injuries in these areas.

49. A Salter-Harris type V fracture is a crush injury to an open physis.
50. A Salter-Harris type II fracture is the most commonly encountered growth plate injury.
51. A Salter-Harris type I fracture extends through an open physis without violating epiphyseal or metaphyseal bone.
52. A Salter-Harris type IV fracture involves an open physis, as well as the adjacent epiphysis and metaphysis.
53. Salter-Harris type IV and V fractures are relatively uncommon but often result in partial premature physeal closure with secondary deformity.
54. The triplane injury of the ankle is an example of a Salter-Harris type IV fracture.
55. Slipped capital femoral epiphysis is an example of a Salter-Harris type I fracture.
56. A Salter-Harris type III fracture involves an open physis, as well as the adjacent epiphysis.
57. A Salter-Harris type II fracture involves an open physis, as well as the adjacent metaphysis.

PEARL

Salter-Harris type V fractures can lead to focal areas of diminution or loss of bone growth resulting in an angular deformity. Premature osseous fusion of the injured portion of the plate may be identified. Such physeal trauma may cause metaphyseal cupping. Other causes of this appearance include scurvy, infection, vitamin A poisoning, sickle cell anemia, numerous hereditary bone disorders, and rickets. Type IV Salter-Harris fractures can lead to localized osseous bridging between the epiphysis and metaphysis. A Salter-Harris type I injury is a slip of the epiphysis.

58. Acute osteochondral fractures and osteochondritis dissecans are examples of type VII pediatric injuries.
59. A type VI pediatric injury involves the perichondrium of the physis, resulting in an external osseous bridge and subsequent angular growth deformity.
60. Lipohemarthrosis is nearly always indicative of an intraarticular fracture, although false positive results may occur from soft-tissue injury.

PEARL

Fractures:
- Slipped capital femoral epiphysis is most typically observed in boys between the ages of 10 and 17 years and is bilateral in 25% of cases.
- Approximately 25% to 30% of patients with growth plate injuries develop some degree of growth deformity, which is very significant in 10% of cases.
- The type VII Salter-Harris fracture involves an epiphyseal alteration in the absence of physeal or metaphyseal abnormalities, and it includes transchondral fractures or osteochondritis dissecans.
- The Type VI Salter-Harris fracture involves an injury to the perichondrium that is caused by a glancing blow or burn and produces reactive bone formation external to the growth plate, which may result in angulation.
- The triplane fracture of the ankle represents a form of Salter-Harris type IV injury.
- Salter-Harris type II injuries represent the most common type of growth plate injury (75%).

61. Skeletal scintigraphy is the imaging procedure of choice for the detection of radiographically occult fractures because of its high sensitivity.
62. Identification of occult fractures or complete characterization of recognized osseous injuries can be accomplished using conventional tomography or computed tomography.

PEARL

Facial trauma:
- Mandibular fractures occur frequently in at least two sites.
- A pyramidal fracture of the facial bones is similar to a Le Fort type II injury but may also represent a Le Fort type III fracture.
- In children, one nasal bone may be fractured independently of the other.
- Emphysema of the orbit may be the only radiologic manifestation of an ethmoid fracture.

SPINAL TRAUMA (Boxes 1-1, 1-2, 1-3, 1-4)

Box 1-1   Spinal Anatomy

ANTERIOR COLUMN

Vertebral bodies
Intervertebral disks
Anterior and posterior longitudinal ligaments

POSTERIOR COLUMN

Facets
Apophyseal joints
Pedicles
Laminae
Spinous processes
Intervening ligaments

Note: The posterior longitudinal ligament and posterior vertebral body margins are frequently referred to as the middle column.

Box 1-2   Patterns of Spinal Injury

- Peak incidence of injury

    C1-2
    C5 to C7
    T12 to L2

- Approximately 20% of spinal fractures are associated with fractures elsewhere.
- Multiple discontiguous injuries:
    Present in 4% of patients with spinal injury
    Uncommon in neurologically intact patients
    Important in explaining pain, deficits
    Determinants of definitive treatment
- Patterns of multilevel injuries:

1	2
C5 to C7	T12 or L
T2 to T4	C
T-L junction	L4-5
Mid-T	T-L junction or C

Box 1-3   Spinal Cord Injuries

INTRINSIC INJURIES

Edema
Hematomyelia
Partial or complete transection

Box 1-3—cont'd

EXTRINSIC INJURIES

Disk or bone fragments
Intradural and/or epidural hematomas

INCIDENCE OF CERVICAL INJURY

Fracture-dislocation: 100%
Bilateral facet lock: 85%
Teardrop or severe crush: 75%
Unilateral facet lock: 30%

PEARLS

- In 10% of patients with traumatic cord injury, no overt radiographic evidence of vertebral injury is present
- Approximately 60% of patients sustaining vertebral body and posterior element fractures with malalignment have neurologic deficit
- Spinal cord injuries occur in 10% to 14% of spinal fractures and dislocations

ONSET OF INJURY

Immediate: 85%
Early posttrauma period: 5% to 10%
Late: 5% to 10%

INCIDENCE OF NEUROLOGIC DAMAGE BY SITE OF INJURY

Cervical: 40%
Thoracolumbar junction: 4%
Thoracic: 10%

Cervical Region

Fundamental concepts

1. Adequate clearance of a lateral radiograph in the cervical region of the spine requires that the upper portion of the T1 vertebral body be visualized.

PEARL

The upper cervical region of the spine may be optimally radiographed by using an anteroposterior (AP) view with moving-jaw technique or an open-mouth view. The swimmer's lateral view is useful for the lower cervical region of the spine. Paired oblique projections can also evaluate cervicothoracic alignment.

2. The prevertebral soft tissues in adults should not exceed 5 mm in width at the C3-4 level.
3. The prevertebral soft tissues in adults should not exceed 22 mm in width at the C6 level.

4. The prevertebral soft tissues in children should not exceed two thirds of the width of the C2 vertebral body at the C3-4 level.
5. The prevertebral soft tissues in children should not exceed 14 mm in width at the C6 level.
6. Loss of the normal cervical lordosis may occur secondary to normal variation, flexion, or muscle spasm.
7. The four continuous curves that describe the normal position of the cervical vertebrae on a lateral radiograph are the anterior vertebral body line, posterior vertebral body line, spinal laminar line, and posterior spinous process line.
8. Widening between the spinous processes at the C3 to C5 levels indicates a posterior ligamentous injury, but this sign is less reliable at other levels because these levels may normally manifest wide interspinous distances.
9. The odontoid process is normally tilted posteriorly with respect to the body of C2.
10. In adults, the normal atlantoaxial distance does not exceed 2.5 mm and does not change with flexion (Fig. 1-3).
11. In children, the normal atlantoaxial distance may be up to 5 mm and may increase by 1 to 2 mm with flexion.
12. The double spinous process sign on a frontal radiograph is indicative of a clay-shoveler fracture.
13. Bilateral lateral offset of the C1 pillars with respect to the C2 pillars, as seen on an open-mouth odontoid view, may be a normal variant in children; however, in adults, it indicates a Jefferson fracture.
14. Flexion-extension radiographs are indicated in the setting of a suspected ligamentous injury.
15. The pillar view is a frontal radiograph obtained with 20 to 30 degrees of caudal angulation to provide an

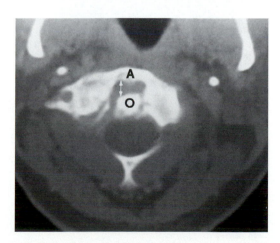

Fig. 1-3 Atlantoaxial dislocation (*double-headed arrow*) demonstrated by CT (*O = odontoid process, A = anterior C1 arch*). Possible causes include trauma, rheumatoid arthritis, seronegative spondyloarthropathies, infection, Down syndrome, and Griesl syndrome of infection-induced hyperemia in childhood.

optimal depiction of the articular pillars and apophyseal joints.

16. Computed tomography may not detect fractures oriented perpendicularly to the long axis of the cervical region of the spine, such as the type II fracture of the odontoid process.

Normal and congenital variation that may simulate injury

1. Occipitalization of the atlas is characterized by lack of segmentation between the atlas and occiput, with high positioning of the former, widening of the C1-2 interspinous distance, and deformity of the odontoid process, and may simulate traumatic atlantoaxial subluxation.

2. The most common congenital defect in C1 is located in the midline posteriorly at the laminar arch synchondrosis.

3. The pillar-arch synchondroses of C1 may fuse asymmetrically and simulate a fracture.

4. The body–odontoid process synchondrosis and body-arch synchondroses of C2 normally fuse between the ages of 3 and 6 years, with the latter fusion frequently occurring asymmetrically.

5. A persistent body–odontoid process synchondrosis in C2 is readily distinguished from a type II odontoid fracture because it occurs below the typical site of the latter.

6. The os terminale normally unites with the remainder of the odontoid process by the age of 12 years, but it may remain isolated and simulate a type I odontoid fracture.

7. The os odontoideum may represent an old ununited odontoid fracture or a congenital anomaly consisting of odontoid hypoplasia with overgrowth of the os terminale.

8. Osteophytosis involving an uncinate process may simulate a vertebral body fracture on a lateral radiograph.

9. On a lateral radiograph, an elongated transverse process may simulate a posttraumatic bone fragment.

10. In children, the annular apophysis, as seen on a lateral radiograph, may simulate a chip fracture, avulsion fracture, or anterior wedging of a vertebral body.

Traumatic injuries

1. The most common form of injury involving C1 is a pattern of bilateral neural arch fractures.

2. The Jefferson fracture is caused by vertical compression that produces a burst type of injury involving the anterior and posterior arches of C1; the fracture is stable, and neurologic deficit does not occur unless the transverse ligament is also disrupted (Fig. 1-4).

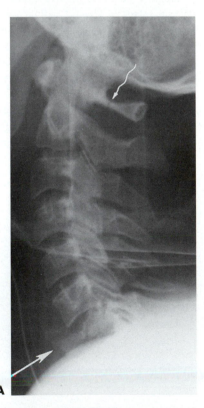

A

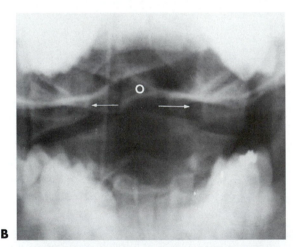

B

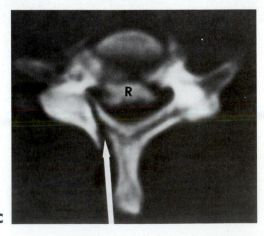

C

Fig. 1-4   A, Jefferson fracture of C1 (*wavy arrow*) associated with burst fracture of C7 (*straight arrow*). **B,** Open-mouth view reveals characteristic lateral displacement of C1 pillars (*arrows*) with respect to the odontoid process (*O*). **C,** CT at the C7 level documents a large retropulsed fragment (*R*) and a laminar fracture (*arrow*).

3. The atlantoaxial rotary fixation injury is a rotational facet lock between C1 and C2 that usually occurs in children and has a presenting symptom of fixed torticollis; the condition may be diagnosed by an open-mouth odontoid series including lateral bending and rotation views or by computed tomography.

4. The most common injury of the odontoid process is the type II fracture, which extends through the base of the dens.

5. Oblique extension of an odontoid fracture into the C2 vertebral body is characteristic of a type III injury.

6. The hangman injury is a C2-3 fracture-dislocation with bilateral neural arch fractures of C2; the mechanism of injury is hyperextension, and neurologic deficit is uncommon (Fig. 1-5).

7. Anterior wedging of a vertebral body is usually a relatively minor injury induced by hyperflexion force.

8. The hyperflexion sprain injury is characterized by disruption of the posterior ligamentous complex, anterior subluxation, interspinous widening, localized disk space widening, and focal kyphotic angulation accentuated by flexion; delayed instability occurs in up to 20% of cases.

9. The unilateral facet lock injury is caused by hyperflexion, distraction, and rotation forces; the condition occurs most commonly at the C4-5 and C5-6 levels, and approximately 35% of cases are associated with fracture, usually involving the facets.

10. The bilateral facet lock injury is caused by hyperflexion and distraction forces, and it manifests a high incidence of spinal cord damage. Radiographically, the dislocated vertebra is anteriorly displaced by approximately 50% of the vertebral body width.

11. The clay-shoveler fracture is a spinous process avulsion induced by hyperflexion force; the C7 and C6 vertebrae are most commonly affected.

12. The most severe hyperflexion injury compatible with life is the burst fracture, characterized by comminution of the vertebral body with a triangular teardrop fragment arising from its anteroinferior margin; neurologic deficit is likely, resulting from retropulsion of the posterior aspect of the vertebral body (see Fig. 1-4).

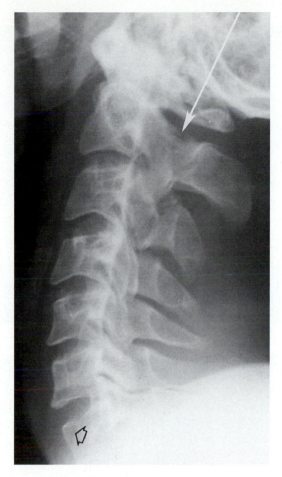

Fig. 1-5 Hangman's fracture of C2 lamina (*white arrow*) associated with teardrop fracture of anterosuperior corner of C7 (*black arrow*).

13. Cervical injuries in children under 12 years of age most commonly involve the atlantooccipital and atlantoaxial regions.

14. The distribution of cervical injuries in adolescents is similar to that observed in adults.

Hyperextension injuries

1. Spinal cord damage may occur in the absence of fracture or dislocation.

2. Radiographic findings may be subtle or occult (Fig. 1-6).

3. Compression fractures of the articular pillars tend to occur unilaterally.

4. Avulsion fractures of the anteroinferior vertebral body margins are most common at levels C2 and C3.

5. Prevertebral soft-tissue swelling is a frequent radiographic finding.

6. Nerve root compression may occur as a complication.

7. Intervertebral disk space widening may occur, especially anteriorly.

8. Posterior displacement of vertebral bodies may occur.

A

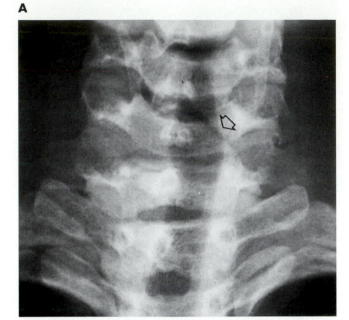

B

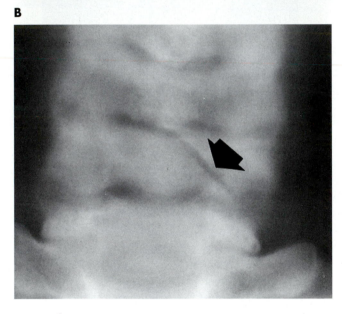

Fig. 1-6   Laminar fracture. **A,** The abnormality (*arrow*) is subtle on a frontal radiograph. **B,** Such injuries (*arrow*) are more confidently visualized by conventional tomography than by CT because of the normal close approximation and parallelism of the cervical laminae.

9. In patients with acute trauma, vacuum phenomenon within the anulus fibrosus strongly suggests anterior soft-tissue injury only in the absence of degenerative disk disease, because it is also a feature of the latter.

Radiographic signs of instability following trauma (Fig. 1-7)
1. Widening of apophyseal joints
2. Intervertebral disk space widening seen on non-stress radiographs

3. Widening of the interspinous distance
4. Intervertebral disk space widening seen on flexion or extension radiographs
5. Radiographic evidence of injuries involving both the anterior and posterior columns of the spine
6. Intervertebral disk space widening with the patient in traction

PEARL

Avulsion of nerve roots occurs most commonly in the lower cervical region as a result of severe force separating the neck and shoulder. Motorcycle accidents account for a great number of these injuries. Newborns can suffer this injury during delivery. If the damage is limited to the brachial plexus beyond the central canal, it is potentially repairable. Avulsed nerve roots are generally considered permanently lost. Lumbar nerve root avulsions are rare and usually are seen only in conjunction with severe pelvic trauma.

Thoracic Region

PEARL

Breathing technique is used to blur out the ribs and lungs on radiographs of the sternum and the thoracic region of the spine (autotomography).

1. Thoracic spinal injuries are usually stable because of apophyseal joint orientation in this region and support provided by the thoracic cage.
2. Widening of the interpediculate distance as seen on a frontal radiograph is strongly suggestive of a burst fracture.
3. Slight anterior wedging of the T12 vertebral body is a common normal variant that is readily distinguished from an acute fracture by the absence of cortical disruption in the former.
4. The upper thoracic region of the spine is poorly visualized on routine lateral radiographs because of the obscuring influence of thick soft tissue; hence, a swimmer's view must also be obtained for a comprehensive examination.
5. Aside from the thoracolumbar junction, the thoracic region of the spine is a relatively unusual site of traumatic fracture.

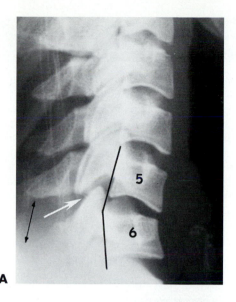

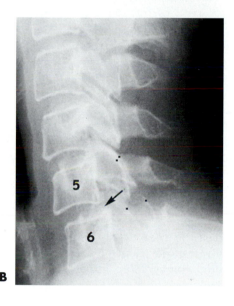

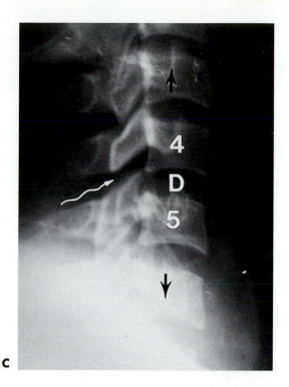

Fig. 1-7 Variations of cervical apophyseal joint injury. **A,** Widening of the C5-6 apophyseal joints bilaterally (*arrow*) with acute angular kyphosis (*lines*) and interspinous widening (*double-headed arrow*) indicating capsular disruption. **B,** Rotary facet subluxation with C5-6 spondylolisthesis (*arrow*). An abrupt change in paired alignment of pillars occurs between C5 and C6 (*dots = posterior margins of pillars*). **C,** Hyperextension sprain injury (*black arrows = direction of force*) with widening of the disk spaces (*D*) and distraction of the apophyseal joints (*wavy arrow*) at C4-5.

6. Thoracic spinal injuries are more likely to be unstable when associated multiple rib fractures or a sternal fracture are present.
7. The normal thoracic kyphosis as measured from T4 to T12 ranges from 20 to 40 degrees.
8. An important clue to thoracic spinal injuries on chest radiographs is the presence of a paraspinous soft-tissue mass representing hematoma.

Lumbar Region

Fundamental concepts

1. On frontal radiographs, the interpediculate distance normally increases gradually from L1 to L5.
2. On lateral radiographs, the intervertebral disk space height normally increases gradually from L1-2 to L4-5.

3. The L5-S1 intervertebral disk is normally slightly narrower than the L4-5 intervertebral disk.
4. A limbus vertebra represents an annular apophysis that has failed to fuse to the remainder of the vertebral body because of anterior or posterior displacement of disk material; anteriorly, the finding may simulate a fracture, and posterior lesions may be symptomatic.
5. On computed tomographic studies, the superior facets of the apophyseal joints are directed posteromedially, and their articular surfaces are concave in shape.
6. On computed tomographic studies, the inferior facets of the apophyseal joints are directed anterolaterally, and their articular surfaces are flat or convex in shape.

Traumatic injuries (Box 1-4)

Box 1-4 Thoracolumbar Spine Fractures

- Two thirds occur at levels T12 to L2
- 90% occur between levels T11 and L4
  Uncommon in mid or upper thoracic region in adults, except those with convulsive disorders
  Peak incidence in children: T4, T5, L2
  Compression fractures: 75%
  Fracture-dislocations: 20%
  Isolated to posterior elements: 5%

1. Approximately 60% of thoracolumbar spinal fractures occur at the T12 to L2 levels.
2. Approximately 90% of thoracolumbar spinal fractures involve the T11 to L4 levels.
3. Approximately 75% of thoracolumbar fractures are of the compression type with intact posterior elements.
4. Approximately 20% of thoracolumbar fractures are fracture-dislocations with involvement of the vertebral bodies and posterior elements.
5. Computed tomography is the procedure of choice for the evaluation of burst fractures, which may be associated with retropulsion of bone fragments into the spinal canal and may require both anterior and posterior stabilization (Fig. 1-8).
6. The naked facet sign on computed tomography indicates apophyseal joint subluxation or dislocation, which can be confirmed on image reformations in the sagittal plane.
7. The Chance fracture is a transverse fracture through the posterior and anterior columns of the spine that is commonly associated with abdominal injury; the mechanism involves hyperflexion in the setting of tensile stress imposed by a lap-type seat belt, which acts as a fulcrum.

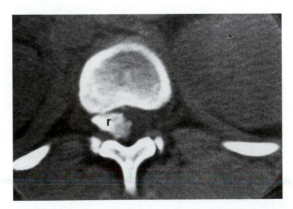

Fig. 1-8 CT image in the lower thoracic region demonstrates a large retropulsed fragment (*r*) within the spinal canal, resulting from a burst fracture.

PEARL

The horizontal Chance fracture of a vertebral body or tearing of an intervertebral disk is frequently combined with laminar and spinous process fractures or ligamentous tear. Chance fracture occurs as a result of flexion around a lap-type seat belt or other fulcrum, most commonly at L1. There is a horizontal fracture of the vertebral body or disk and neural arch or posterior ligaments. In about 15% of cases there are associated intraabdominal injuries, including tears of mesentery, rupture of the small bowel, and/or laceration of solid organs.

8. A defect in the pars interarticularis of a lumbar vertebra is termed *spondylolysis* and represents an insufficiency type of fracture superimposed upon dysplastic or hypoplastic bone.
9. Pars interarticularis defects are best visualized on oblique conventional radiographs.
10. Pars interarticularis defects exhibit a predilection for white male patients.
11. Approximately two thirds of all pars interarticularis defects involve the L5 vertebra.
12. Spondylolysis is usually diagnosed during the second or third decade of life.
13. Bilateral pars interarticularis defects may be associated with spondylolisthesis.
14. Unilateral defects or congenital absence of the pars interarticularis may be associated with sclerosis and hyperplasia of the contralateral pars interarticularis and adjacent posterior elements.
15. On computed tomographic images, bilateral pars interarticularis defects can be distinguished from apophyseal joints by horizontal versus oblique orientation, by irregular versus smooth osseous margins, and by disruption of the complete osseous ring normally formed by the vertebral body and neural arch.

PELVIC TRAUMA

Fundamental Concepts

1. The four bones of the pelvis are the ilium, ischium, pubis, and sacrum.
2. The iliopsoas muscle inserts on the lesser trochanter of the femur; the long head of the rectus femoris muscle arises from the anterior inferior iliac spine.

3. The innominate bone is formed by fusion of the ilium, ischium, and pubis with closure of the triradiate or Y cartilage of the acetabulum, which may be mistaken for a fracture in skeletally immature patients.

4. The three rings of the pelvis are the pelvic inlet-outlet and the paired obturator foramina; although exceptions occur, identification of one fracture or joint separation in any ring generally implies disruption of at least one additional site in the same ring.

5. Disruption of the iliopubic line on a frontal radiograph is usually associated with a fracture extending into the obturator foramen.

6. Fractures of the acetabulum may disrupt the ilioischial line, teardrop shadow, or anterior or posterior acetabular rims on a frontal radiograph.

7. Occult fractures of the sacrum may be manifested on frontal radiographs as disruption of the sacral foraminal lines.

8. Identification of a fracture involving the L5 transverse process should suggest the presence of an occult sacral fracture.

9. The sacroiliac joints are normally wide in children and adolescents but should not exceed 4 mm in width in adults.

10. The width of the symphysis pubis may be up to 5 mm in adults and 10 mm in adolescents.

11. Malalignment of the iliopubic lines at the symphysis pubis of up to 2 mm is acceptable.

12. For adequate visualization of the acetabular columns, 45-degree oblique or Judet radiographic views are required; the posterior oblique film depicts the posterior column and the anterior acetabular rim, whereas the anterior oblique film depicts the anterior column and the posterior acetabular rim.

13. The most common direction of femoral head subluxation or dislocation is posterolateral.

14. Medial migration of the femoral head as seen on a frontal radiograph implies a fracture of the medial acetabular wall.

15. Computed tomography is the imaging procedure of choice for complete characterization of pelvic injuries (Fig. 1-9).

16. Intraarticular osteochondral bodies are a common complication of acetabular fractures and hip dislocation, and they are best detected by computed air arthrotomography of the hip joint.

17. Computed tomography depicts more extensive injury than conventional radiography in approximately 30% of unstable pelvic fractures.

18. Conventional radiography is most useful for detecting injuries in the anterior portion of the pelvis, whereas computed tomography is best suited to the identification of injuries in the acetabula and posterior portion of the pelvis.

Traumatic Injuries

1. Approximately two thirds of all pelvic fractures are stable and do not warrant routine evaluation by computed tomography.

2. Stable pelvic fractures are characterized by single or peripheral breaks, and they include avulsions as well as fractures of the iliac wings, sacrum, or ischiopubic rami.

3. Unstable fracture-dislocations constitute approximately one third of all pelvic injuries and are characterized by disruption at two or more sites.

4. The most common unstable pelvic injury is the vertical shear or Malgaigne type, which usually consists of fractures involving the sacrum, superior ischiopubic ramus, and inferior ischiopubic ramus on

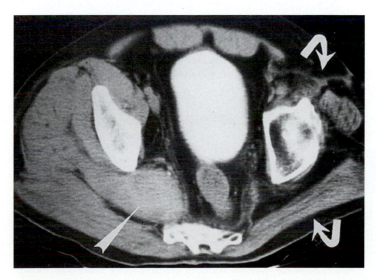

Fig. 1-9 Gluteal hematoma depicted by CT. A high-density mass (*arrowhead*) occupies the space between the ilium and sacrum on the left. Diffuse muscle atrophy (*arrows*) is also evident on the right.

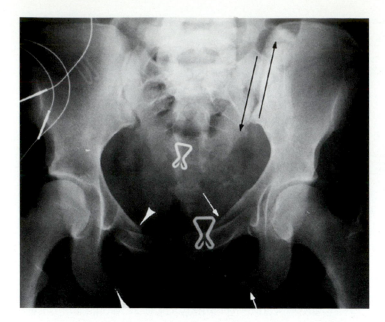

Fig. 1-10 Malgaigne injury of the pelvis, consisting of ipsilateral sacroiliac joint disruption (*black arrows*) and ischiopubic ramus fractures (*white arrows*). The latter are also present on the contralateral side (*arrowheads*).

the same side of the body; variations may include fracture of the iliac wing or diastasis of the sacroiliac joint or symphysis pubis (Fig. 1-10).

5. The straddle injury is unstable and includes bilateral fractures of the superior and inferior ischiopubic rami (Fig. 1-10).

6. Unstable fractures should be managed by internal or external fixation; they may be complicated by internal hemorrhage and visceral injury.

7. The open-book injury is unstable and involves diastasis of the sacroiliac joint and symphysis pubis with associated increase in the angle between the sacrum and iliac wing; frontal radiographs demonstrate an asymmetric appearance of the two hemipelves, and computed tomography is the procedure of choice for determining the severity of the injury.

8. Four pelvic apophyses that are vulnerable to avulsion injuries between puberty and age 25, along with their respective muscular attachments are as follows: (1) inferior pubic ramus, adductor magnus muscle; (2) ischial tuberosity, hamstring muscle group; (3) iliac crest (anterior superior iliac spine), sartorius muscle; and (4) anterior inferior iliac spine, rectus femoris muscle. The importance of these avulsions lies in the fact that they may simulate neoplasia or infection radiographically during the healing phase.

PEARL

Pelvic avulsion fractures occur at the anterior superior iliac spine (sartorius muscle origin), anterior inferior iliac spine (rectus femoris muscle origin), pubic ramus (adductor magnus muscle origin), and ischial tuberosity (hamstring muscle origin).

Sacral Fractures

1. Sacral stress fractures are usually of the insufficiency type, occurring in the setting of severe osteoporosis, often secondary to long-term corticosteroid therapy (Fig. 1-11).

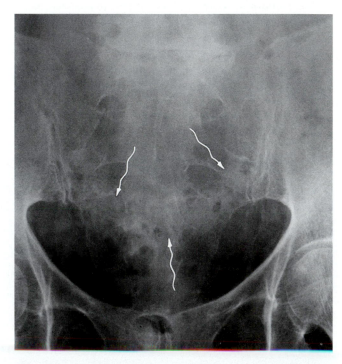

Fig. 1-11 Healing sacral insufficiency fracture (*arrows*) in steroid-induced osteoporosis. Such injuries are amenable to early detection by skeletal scintigraphy or magnetic resonance imaging, although CT may be useful in distinguishing them from other pathologic conditions.

2. Skeletal scintigraphy and computed tomography are superior to conventional radiography for detecting sacral stress fractures.
3. Sacral stress fractures often are shaped like the letter *H,* with a horizontal component connecting vertical limbs involving each ala of the bone.
4. Isolated traumatic sacral fractures are usually transverse in orientation.
5. In the setting of complex pelvic trauma, sacral fractures tend to be vertically oriented, disrupting the foraminal lines.
6. Fracture or dislocation of the coccygeal segments may be difficult to diagnose with certainty because of significant normal variation in this region.

Pubic Fractures

1. Fatigue-type stress fractures tend to occur at the junction between the ischium and pubis, and they are more common in women than in men.
2. In the setting of severe osteoporosis, insufficiency-type stress fractures frequently occur in the superior and inferior ischiopubic rami.
3. Pseudofractures of the ischiopubic rami are common in osteomalacia.
4. The healing phase of insufficiency-type stress fractures of the pubis is frequently associated with significant bone resorption that can simulate infection or neoplasia radiographically.
5. The ischiopubic synchondroses normally fuse by 12 to 13 years.
6. In children, the ischiopubic synchondroses tend to be asymmetrical, irregular, and bulbous radiographically, and they can therefore be confused with bone-forming tumors or fracture callus.

HIP TRAUMA

Fundamental Concepts

1. The three fat planes adjacent to the hip joint are the gluteal, iliopsoas, and obturator internus.
2. The frog-leg lateral radiograph of the hip is the best projection for demonstrating slipped capital femoral epiphysis, subcapital fracture, or ischemic necrosis of the femoral head.
3. The normal fat planes around the hip joints do not always appear symmetric on frontal radiographs.
4. Apparent bulging of the fat planes around the hip joint, which can lead to the false positive diagnosis of effusion, can be caused by suboptimal positioning of the patient, with the hip slightly flexed or externally rotated.

5. Lateral or inferior displacement of the femoral head with respect to the acetabulum is a reliable radiographic sign of hip joint effusion.
6. The neck-shaft angle of the femur as measured on a true lateral radiograph of the hip closely approximates the angle of femoral anteversion.
7. Cross-sectional imaging is the most reliable method of measuring the true angle of femoral anteversion.
8. The calcar femorale is a plate of cortical bone that extends perpendicularly into the medullary cavity from the endosteal surface of the femoral cortex medially in the vicinity of the lesser trochanter; consequently it is evident only on cross-sectional imaging studies and is best seen by computed tomography.

Epidemiology of Hip Fractures

1. The occurrence of femoral neck fractures is twice as common as that of intertrochanteric fractures.
2. By the age of 80 years, approximately 20% of white women and 10% of white men will sustain a hip fracture.
3. In the elderly, hip fractures are highly correlated with fractures in the spine, distal region of the radius, and proximal region of the humerus.
4. Hip fractures are relatively uncommon among young and middle-aged individuals.
5. By the age of 90 years, approximately 30% of white women and 20% of white men will sustain a hip fracture.
6. In approximately 5% of cases, the hip fracture is actually the cause of the fall. Osteopenia is so severe in such patients that the femur is unable to sustain the stress of normal ambulation and fails.
7. Postmenopausal and senile osteoporosis are the most common predisposing causes of hip fractures in the elderly.

Femoral Neck Fractures

1. Ischemic necrosis is a complication in up to 30% of subcapital fractures.
2. Subcapital fractures are far more common than basicervical or transcervical fractures.
3. Annular osteophytes secondary to osteoarthritis may simulate the appearance of an impacted or subacute fracture.
4. Nonunion is a relatively common complication of basicervical fractures.
5. Incomplete or impacted subcapital fractures may be treated conservatively with bed rest.
6. A single Knowles pin is inadequate orthopedic management for a complete or displaced subcapital fracture, owing to the potential for rotation at the

fracture site. Three- or four-point fixation is thus preferred.

7. Depending upon the configuration of the fracture, as well as patient condition and age, an Austin-Moore endoprosthesis may be the preferred management strategy.

8. Skeletal scintigraphy and magnetic resonance imaging (MRI) offer high sensitivity for the detection of stress and radiographically occult fractures that have not yet begun to heal, whereas conventional tomography is well suited to the identification of early callus formation; because such injuries in the femoral neck tend to be horizontally oriented, they are readily overlooked (even if healing) by the transverse scanning plane of computed tomography.

Intertrochanteric Fractures

1. Patients with intertrochanteric fractures tend to be older than those with femoral neck fractures.

2. Comminution is greatest in the posteromedial proximal portion of the femur, around the lesser trochanter and calcar femorale.

3. Intertrochanteric fractures may have two, three, or four major fragments, with the latter type including greater and lesser trochanteric segments.

4. The major fracture line is usually obliquely oriented and extends from the greater trochanter superiorly to the lesser trochanter inferiorly. Fracture orientations deviating from this are uncommon and tend to be more unstable.

5. Optimal internal fixation of intertrochanteric fractures involves the use of a dynamic or sliding hip screw because of the tendency for the proximal fragment to settle.

6. For optimal results the dynamic screw head is placed in a slightly posterior and inferior position in the femoral head, with the tip approximately 6 mm from the articular surface.

7. Because of the tendency for intertrochanteric fractures to be unstable and settle, rigid nail fixation results in the potential for the hardware to cut out through the femoral head or neck.

8. Anatomic reduction of intertrochanteric fractures is acceptable only if posteromedial comminution is not severe.

9. Valgus reduction of intertrochanteric fractures involves an osteotomy to reduce shear forces and provide stability.

10. Medial impaction reduction of intertrochanteric fractures is associated with a slight decrease in leg length.

11. Ischemic necrosis rarely occurs following intertrochanteric fractures.

12. Major complications of intertrochanteric fractures include instability and cutting out of the hardware

as excessive varus alignment develops between the proximal and distal fragments.

Miscellaneous Femoral Fractures (Fig. 1-12)

1. Avulsion fractures of the lesser trochanter are common in children and adolescents, and these fractures represent type I injuries in the Salter-Harris classification.

2. Isolated avulsion of the lesser trochanter in an adult should suggest the diagnosis of underlying malignancy.

3. Highly comminuted femoral shaft fractures are optimally managed by the use of an intramedullary rod with circumferential wiring.

4. Femoral shaft fractures that do not manifest significant comminution or large butterfly fragments (Fig. 1-13) are optimally managed by the use of a tight-fitting intramedullary rod.

5. An interlocking screw should be used in conjunction with an intramedullary rod for fractures in the proximal third or distal third of the femur to ensure rotational stability because of widening of the medullary canal in these regions.

6. Complications of intramedullary rod fixation for femoral shaft fractures include distraction and rota-

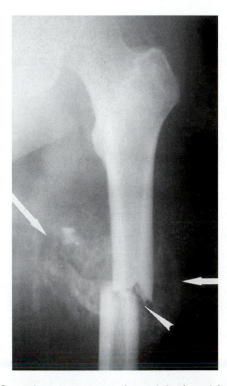

Fig. 1-12   Subacute transverse femoral diaphyseal fracture (*arrowhead*) with exuberant immature callus formation (*arrows*). Fracture healing with abundant callus occurs in the setting of extensive hemorrhage, corticosteroid therapy, neuropathic disease, osteogenesis imperfecta, and osteopetrosis.

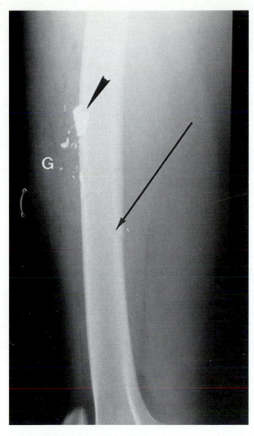

Fig. 1-13 Gunshot wound (*arrowhead*) to the thigh with femoral fracture (*arrow*) and soft-tissue gas (*G*). Angiography is frequently indicated in such situations to exclude associated vascular injury.

tion of the major fracture fragments, as well as migration and breakage of the rod.

7. Stress fractures of the proximal portion or mid-diaphysis of the femur usually involve the medial portion of the cortex.
8. Stress fractures in the distal third of the femoral diaphysis usually involve the posterior portion of the cortex.
9. Pseudofractures affecting the proximal part of the femur in osteomalacia usually involve the compressive (medial) side of the bone.
10. Pseudofractures affecting the proximal part of the femur in Paget disease tend to involve the tensile (lateral) side of the bone.

Hip Dislocation

1. Hip dislocation is frequently associated with femoral diaphyseal fracture.
2. Posterior dislocation is the most common pattern and occurs in approximately 90% of cases.
3. Posterior dislocation of the hip is frequently associated with a fracture of the posterior acetabular margin.
4. On frontal radiographs of the pelvis, a posteriorly dislocated femoral head will appear smaller than

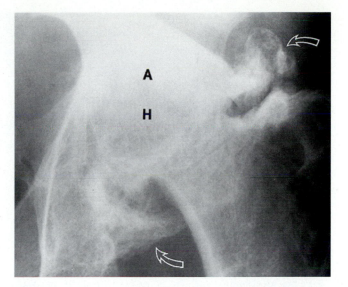

Fig. 1-14 Chronic unreduced posterior hip dislocation following trauma. On a frontal radiograph, the femoral head (*H*) appears small (because of posterior displacement) and overlaps the acetabular dome (*A*), which exhibits marked sclerosis; exuberant osteophytosis and osteochondral bodies are also evident (*arrows*). Additional potential causes for this appearance include neuroarthropathy and osteonecrosis with secondary osteoarthritis (stage 5).

the contralateral normally positioned femoral head (Fig. 1-14).
5. In posterior dislocation of the hip, the femoral head is usually located superiorly and laterally, with the femur held in internal rotation.
6. In the rare anterior hip dislocation, the femoral head usually projects over the obturator foramen on frontal radiographs.
7. Delayed reduction significantly increases the probability of ischemic necrosis, which approaches 50% if the hip is left dislocated for more than 24 hours.

PEARL

Ischemic necrosis of the femoral head following hip dislocation and femoral neck fracture is usually confined to the anterolateral weight-bearing surface.

8. Computed air arthrotomography is the preferred method for detecting intraarticular fracture fragments following reduction of a dislocated hip.
9. Lateralization of the relocated femoral head is a radiographic sign suggesting the presence of retained intraarticular fracture fragments.

Slipped Capital Femoral Epiphysis

1. The idiopathic variety tends to occur between the ages of 10 and 17 years.

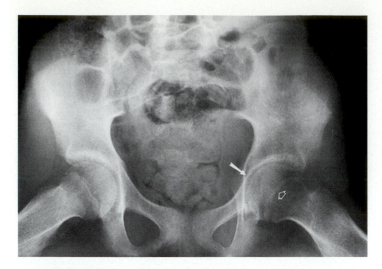

Fig. 1-15 Chondrolysis following slipped capital femoral epiphysis. Frontal radiograph of the pelvis demonstrates characteristic periarticular demineralization and diffuse joint space narrowing (*closed arrow*) affecting the left hip. The previous epiphyseal slip is manifested as apparent narrowing of the growth plate (*open arrow*) on the affected side, made particularly evident by comparison to the normal right hip.

2. The idiopathic variety is more common in male children, particularly black individuals.
3. The idiopathic variety is rarely symmetric but occurs bilaterally in approximately 20% of cases.
4. Obesity and shear stress across the physis of the proximal femur (secondary to the change of femoral neck orientation from valgus to varus during the years of rapid growth) are factors implicated in the pathogenesis of the idiopathic variety.
5. The direction of femoral head slippage in the disorder is nearly always posteromedial and is best demonstrated on a frog-leg lateral radiograph.
6. On frontal radiographs, the physis appears widened with less distinct margins, the epiphysis appears shortened, and a line drawn along the lateral portion of the femoral neck may barely intersect or miss the femoral head.
7. Other potential causes of slipped capital femoral epiphysis, particularly in children under 10 years of age, include developmental hip dysplasia, osteomyelitis, rickets, hyperparathyroidism, radiation therapy, and severe trauma.
8. Potential complications of the condition include ischemic necrosis, chondrolysis, and degenerative joint disease.
9. Ischemic necrosis of the femoral head occurs in approximately 10% of cases and is more likely in the setting of acute severe slippage, open reduction, and attempted repositioning (Box 1-5).

Box 1-5 Traumatic Causes of Ischemic Necrosis of the Femoral Head

Developmental dysplasia of the hip
Legg-Calvé-Perthes disease
Slipped capital femoral epiphysis
Hip dislocation
Femoral neck fracture

10. Chondrolysis, which simulates septic arthritis radiographically, occurs both idiopathically and as a complication of slipped capital femoral epiphysis; in the latter situation, it is more common in girls and following penetration of the subchondral cortex by an internal fixation pin (Fig. 1-15).
11. Degenerative joint disease occurs as a delayed complication of slipped capital femoral epiphysis, occurring up to 30 years following presentation.

PEARL

Idiopathic slipped capital femoral epiphysis occurs in boys 10 to 17 years of age and girls 8 to 15 years of age. Occurrence is more common in black and male individuals. Most of the children are overweight. It is more common on the left, especially in male patients. Bilateralism occurs about 20% of the time and is more common in girls. Trauma appears to be an important precipitating event in infants and young children but only plays a minor contributing role in older individuals. Fewer than 50% of affected children have a history of significant injury. The sequelae of slipped capital femoral epiphysis are varus deformity with osteonecrosis and early degenerative joint disease.

KNEE TRAUMA

Normal Anatomy

1. The distal posterior aspect of the femur gives rise to the gastrocnemius muscle.
2. The soleus muscle arises from both the fibular head and adjacent tibia.
3. The tensor fascia lata arises from the ileum and inserts along the lateral aspect of the tibia.

4. The lateral collateral ligament extends from the lateral femoral condyle to the fibular head.
5. The gracilis muscle inserts on the upper part of the medial surface of the tibia, below the plateau.
6. The main portion of the biceps femoris muscle inserts into the head of the fibula, whereas the remainder attaches to the lateral tibial plateau.
7. The tensor fascia lata also inserts on the medial and lateral intermuscular septa.
8. The quadriceps femoris muscle inserts on the patella.
9. The posterior cruciate ligament inserts on the posterior intercondylar fossa of the tibia and the posterior extremity of the lateral meniscus.
10. The medial collateral ligament inserts on the medial condyle and medial surface of the tibia, with intimate adherence to the medial meniscus.
11. The popliteus tendon is seen on MRI of the knee as a vertical structure adjacent to the posterior aspect of the lateral meniscus.
12. The cruciate ligaments are intraarticular but extra-synovial.
13. The anterior cruciate ligament extends from the lateral femoral condyle to the anterior medial tibial plateau.
14. Lateral support of the knee is provided by the iliotibial band, fibular collateral ligament, biceps femoris tendon, popliteus muscle and tendon, lateral patellar retinaculum, lateral capsular ligament, arcuate ligament, and fabellofibular ligament.
15. The lateral meniscus is a round, ring-shaped, or more discoid structure. The medial meniscus is C-shaped.
16. The deep portions of the collateral ligaments attach to the menisci forming the coronary ligaments; however, the popliteus tendon separates the lateral meniscus from the lateral collateral ligament.
17. The posterior cruciate ligament extends from the posterior intercondylar depression of the tibia to the lateral aspect of the medial femoral condyle.

Knee Joint Effusions

1. On lateral radiographs, the suprapatellar bursa separates the suprapatellar and prefemoral fat pads.
2. In the absence of a knee joint effusion, the fluid-density suprapatellar bursa is less than 5 mm in width.
3. On a cross-table lateral radiograph, the presence of a fat-fluid level indicates lipohemarthrosis, which is nearly always secondary to an intraarticular fracture. The osseous injury may itself be occult (Fig. 1-16).

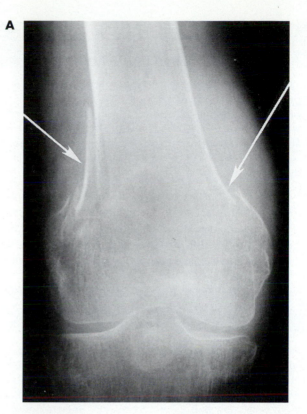

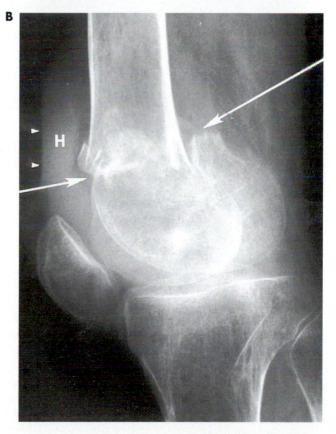

Fig. 1-16 A and B, Frontal and lateral radiographs demonstrate a comminuted and impacted fracture (*arrows*) of the distal femoral metaphysis in severe postmenopausal osteoporosis, resulting from minor trauma; a large hemarthrosis (*H*) with fat-fluid level (*arrowheads*) is also present.

---PEARL---

The cross-table lateral view of the knee is obtained primarily for identifying effusions and establishing the presence or absence of a fat-fluid level in the posttraumatic situation. Tumoral calcinosis around the knee is also well documented on the cross-table view because semiliquid calcific deposits may tend to layer out in a dependent position.

4. Posterior displacement of the fabella, when present, has also been described as a radiographic sign of knee joint effusion; this structure is a sesamoid bone located in the lateral head of the gastrocnemius tendon.

---PEARL---

Sesamoid bones are covered with cartilage on all surfaces, so they have perichondrium, not periosteum. The patella is the largest and lies in the quadriceps tendon. The fabella lies in the lateral head of the gastrocnemius tendon. Sesamoid bones function to protect tendons from wear in areas where they contact osseous surfaces.

Femoral Condylar Fractures (Fig. 1-17)

1. Osteochondritis dissecans most commonly involves the lateral (non–weight-bearing) aspect of the medial femoral condyle (Fig. 1-18).
2. Osteochondritis dissecans is caused by repetitive minor trauma and occurs most commonly in adolescents and young adults.
3. Intraarticular osteochondral bodies are a potential complication of osteochondritis dissecans that may give rise to joint locking (Fig. 1-18).
4. The stability of an osteochondral fragment in osteochondritis dissecans can be predicted by the use of arthrography or magnetic resonance imaging, particularly following intraarticular injection of gadolinium-DTPA (Figs. 1-19, 1-20).
5. Osteochondritis dissecans must be distinguished from normal irregularity of the distal femoral epiphysis, which is most prominent medially and occurs between the ages of 3 and 6 years.
6. Osteochondritis dissecans must be distinguished from normal posterior irregularity of the femoral condyles, which is bilateral, is best visualized on a notch view, and occurs between the ages of 10 and 13 years (Fig. 1-21).
7. Ischemic necrosis is a potential complication of acute intraarticular fractures of the femoral condyles.

8. Spontaneous osteonecrosis tends to affect older women, who have the presenting symptom of sudden onset of pain; the condition exhibits a predilection for the weight-bearing aspect of the medial femoral condyle (Fig. 1-22).
9. Approximately 75% of patients with spontaneous osteonecrosis have an associated medial meniscal tear; magnetic resonance imaging is an ideal technique for demonstrating the latter, as well as for establishing an early diagnosis of the bone marrow abnormality (Fig. 1-23).

---PEARL---

Osteochondritis dissecans:
- Most patients are adolescents at the time of onset.
- Clinical manifestations include pain aggravated by movement, limited motion, clicking, locking, and swelling.
- The most common site of involvement is the lateral aspect of the medial femoral condyle.
- Patellar lesions have a predilection for the medial facet, and involvement of the most medial or "odd" facet is rare.
- Talar dome involvement is most likely in the middle third of its lateral border and the posterior third of its medial border.
- Spontaneous osteonecrosis of the knee can be distinguished by its older age of onset, involvement of the weight-bearing aspect of the femoral condyle, and sudden clinical onset.

Proximal Tibial and Fibular Fractures

1. Approximately 80% of tibial plateau fractures are limited to the lateral side because most result from valgus stress on the knee joint.
2. Internal fixation is required for tibial plateau fractures with depression exceeding 10 mm or fragment separation greater than 5 mm.
3. Conventional tomography is the best method for establishing the extent of depression in tibial plateau fractures (Fig. 1-24).
4. Tibial plateau fractures commonly result when pedestrians are struck by automobiles, because of the level of impact from bumpers.
5. Oblique and cross-table lateral radiographs of the knee are indicated in the setting of suspected tibial plateau fractures.
6. Because the tibial plateau slopes anteroposteriorly by 10 to 20 degrees, it is not profiled by frontal radiography; anterior depressed fragments may thus be overlooked because the anterior margin is projected above the posterior margin, whereas

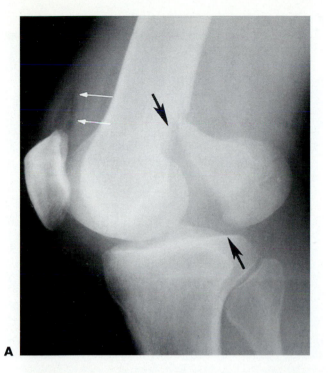

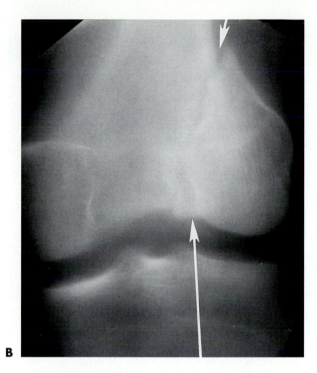

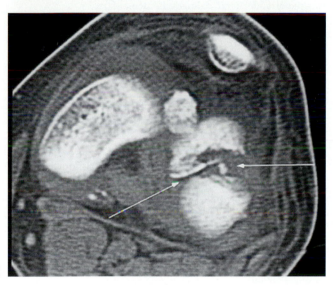

Fig. 1-17 **A,** Lateral femoral condylar fracture (*black arrows*) with lipohemarthrosis and fat-fluid level (*white arrows*) in the suprapatellar bursa. Intraarticular extension of such fractures (*arrows*) can be optimally characterized by either conventional tomography (**B**) or CT (**C**).

depression of posterior fragments tends to be exaggerated.

7. Anterior tibial spinal fractures are more common in children and adolescents than in adults.

8. Posterior cruciate ligament avulsions occur from the posterior aspect of the tibia, slightly below the articular surface of the plateau.

9. Avulsion of the lateral capsular attachment to the tibia, known as the Segond injury, is associated with disruption of the anterior cruciate ligament (Fig. 1-25).

10. Stress fractures and contusions occur in the cancellous bone of the tibial plateau and can be diagnosed early by skeletal scintigraphy or magnetic resonance imaging, particularly using inversion recovery sequences.

11. Cortical stress fractures tend to involve the posterior cortex of the proximal tibial diaphysis.

12. Proximal tibiofibular joint dislocation is a rare injury.

13. Peroneal nerve damage is a potential complication of proximal fibular injuries.

14. Fibular head fractures frequently result from avulsion of the conjoined tendon (lateral collateral ligament and biceps femoris insertion).

Patellar Injuries

1. Approximately 60% of patellar fractures are transversely oriented through the midportion of the

A

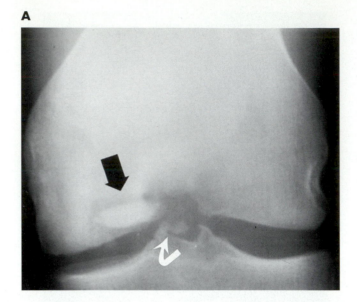

B

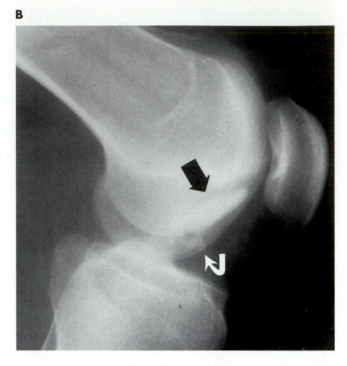

Fig. 1-18   Osteochondritis dissecans. **A** and **B,** Frontal and lateral radiographs demonstrate fragmentation (*straight arrows*) of the lateral aspect of the medial femoral condyle, in association with multiple intraarticular osteochondral bodies (*curved arrows*).

bone and result from sudden traction exerted by the quadriceps tendon.

2. Approximately 25% of patellar fractures are stellate and result from direct trauma to the bone.

3. Vertical fractures of the patella are unusual.

4. The accessory centers of bipartite or multipartite patellae occur on the superolateral margin of the bone.

5. Bipartite and multipartite patellae are often but not always bilateral; therefore, distinction from fracture is most reliably made on the basis of location and configuration of the fragments.

6. The dorsal defect of the patella is a normal variant that may simulate chondroblastoma or osteochondritis dissecans radiographically.

7. Osteochondral fractures of the patella usually arise from the medial facet.

8. Osteochondral fractures of the patella are associated with lateral subluxation or dislocation of the bone.

9. The axial view of the patella should be obtained with the knees flexed 20 degrees, because greater degrees of flexion force the patella into the patellar groove of the femur and prevent the detection of subtle alignment abnormalities. The merchant view is thus preferred because it is performed with less knee flexion than the sunrise projection.

10. Patellar subluxation or dislocation usually occurs in a lateral direction (Fig. 1-26).

11. The lateral articular facet of the patella is normally longer and more horizontally oriented than the medial articular facet.

12. Lateral subluxation of the patella occurs in approximately 30% of patients with chondromalacia patellae.

13. Abnormal patellar tilt is diagnosed on a sunrise or merchant view when the angle formed by the dicondylar line and a tangent to the lateral facet opens medially.

14. On a sunrise or merchant view, the most medial point of the patella should lie within 1 mm of a line drawn perpendicular to the dicondylar line through the highest point of the medial femoral condyle.

15. On a lateral view with the knee in 20 to 30 degrees of flexion, the length ratio of patellar tendon/patella is normally 1.0 plus or minus 0.2.

16. Patella alta is diagnosed when the length ratio of patellar tendon/patella exceeds 1.2 (Fig. 1-27).

17. Patella baja is diagnosed when the length ratio of patellar tendon/patella is less than 0.8.

18. Patella alta is associated with recurrent patellar subluxation.

A

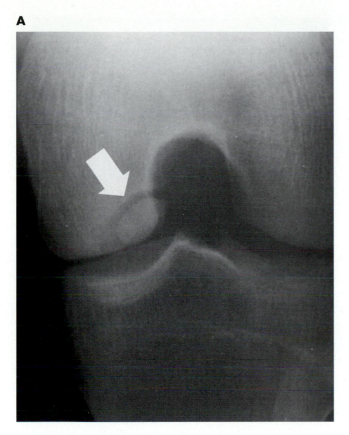

C

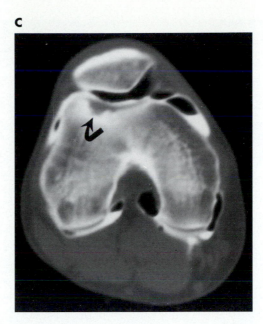

Fig. 1-19 Visualization of osteochondritis dissecans (*arrows*) involving the distal region of the femur can be enhanced by use of a notch view (**A**), CT (**B**), or computed arthrotomography (**C**), as well as MRI.

B

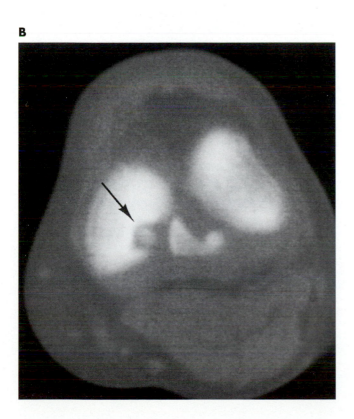

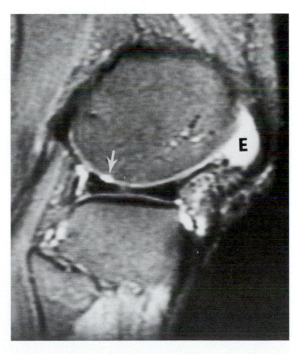

Fig. 1-20 Femoral cartilage defect following acute trauma. Sagittal T2-weighted MR image demonstrates high–signal intensity joint fluid within the area of disrupted hyaline cartilage (*arrow*). Presence of a joint effusion (*E*) facilitates detection of subtle intraarticular pathologic conditions such as this, which has given rise to the technique of MR-arthrography using gadolinium–diethylenetriamine pentaacetic acid and other agents.

A

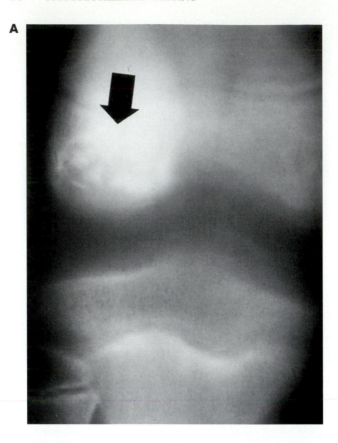

B

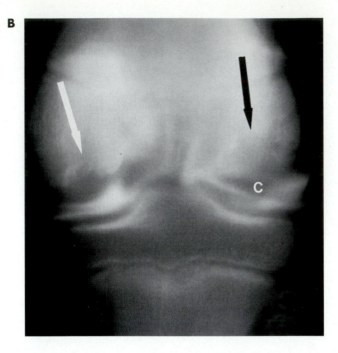

C

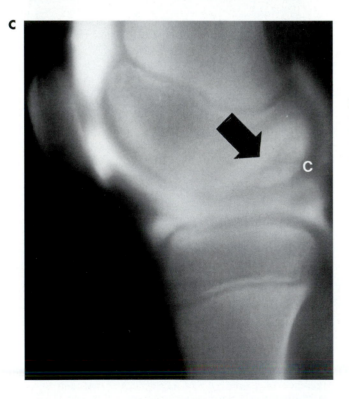

Fig. 1-21 Normal fragmentation and irregularity of the femoral condyles (*arrows*) during childhood is a developmental variation that should not be misinterpreted as osteochondritis dissecans or osteonecrosis. **A,** Frontal tomogram. **B** and **C,** Frontal and lateral arthrotomograms demonstrating intact cartilage (*c*) over the areas.

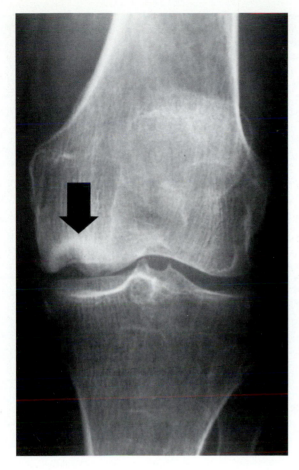

Fig. 1-22 Spontaneous osteonecrosis of the knee in a middle-aged woman. Frontal radiograph reveals typical subchondral radiolucency with a sclerotic margin (*arrow*) on the weight-bearing aspect of the medial femoral condyle. Differential diagnosis should include consideration of other causes of osteonecrosis in adults, such as corticosteroid therapy, alcoholism, radiation therapy, and caisson disease.

Knee Dislocation

1. Anterior dislocation of the knee joint is more frequent than posterior, medial, or lateral displacement of the tibia relative to the femur.
2. Arteriography is an indicated procedure following knee dislocation because of significant potential for vascular damage.
3. Approximately 30% of patients with dislocation of the knee joint have an associated injury to the popliteal artery.

Knee Injuries in the Immature Skeleton

1. Physeal injuries rarely involve the knee.
2. Approximately 70% of physeal injuries involving the knee are Salter-Harris type II.
3. Approximately 15% of physeal injuries at the knee are Salter-Harris type III.

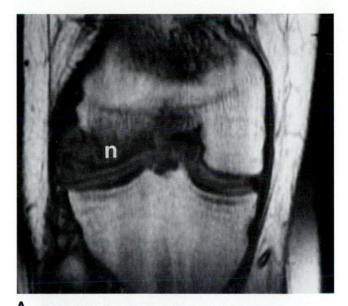

A

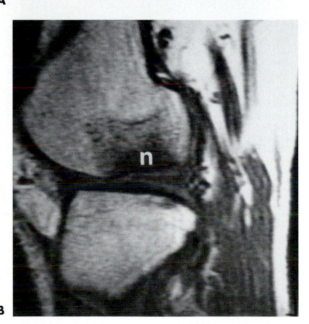

B

Fig. 1-23 Osteonecrosis of the lateral femoral condyle. **A** and **B,** Coronal and sagittal T1-weighted MR images demonstrate abnormally low signal intensity (*n*) in the weight-bearing subchondral bone. Involvement of the medial femoral condyle is more common in both this condition and osteochondritis dissecans, which tends to be centered closer to the intercondylar notch.

4. The medial femoral condyle is the most common site for Salter-Harris type III fractures involving the knee.
5. Salter-Harris type III fractures of the medial femoral condyle result from valgus stress on the knee and are commonly undisplaced and occult. Oblique or valgus stress radiographs may be necessary for their detection.
6. The knee region is the most common site in the skeleton for Salter-Harris type V fractures.

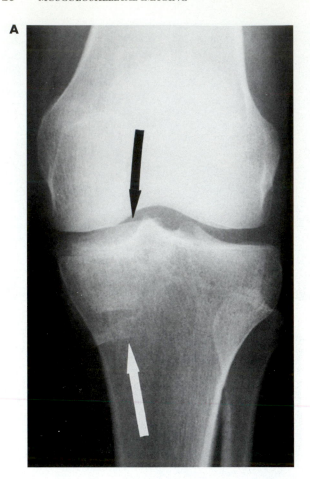

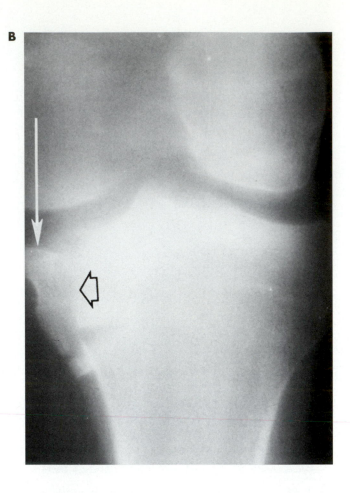

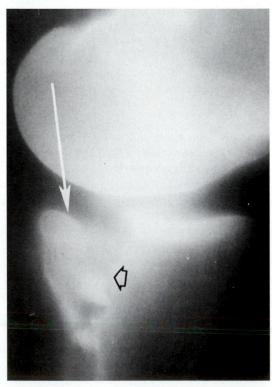

Fig. 1-24   Tibial plateau fracture. **A,** Frontal radiograph reveals abnormal radiodensities (*arrows*) overlying the medial tibial plateau. **B** and **C,** Frontal and lateral tomography documents a comminuted posteromedial fracture (*black arrows*), with significant depression of the articular surface (*white arrows*).

A

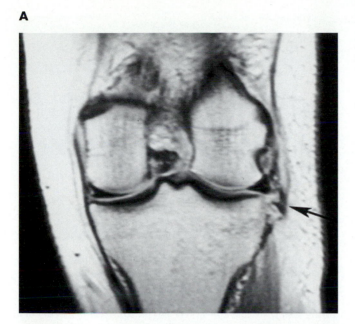

B

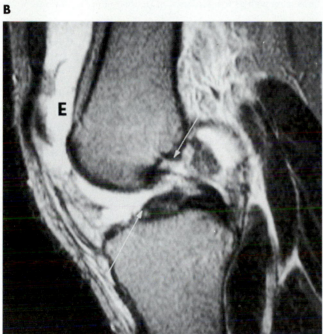

C

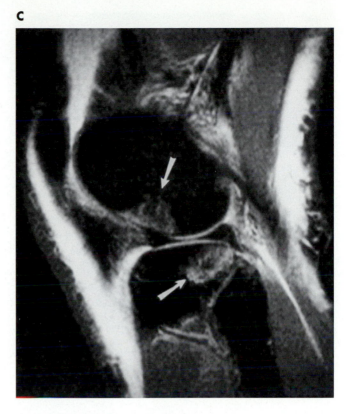

Fig. 1-25   Segond fracture of lateral tibial plateau. **A,** Coronal T1-weighted MR image demonstrates the lateral collateral ligament avulsion (*arrow*). **B,** Sagittal T2-weighted image reveals the associated anterior cruciate ligament tear (*between arrows*) along with a joint effusion (*E*). **C,** Sagittal inversion recovery image documents marrow contusions in the lateral femoral condyle and posterolateral tibial plateau (*arrows*). This constellation of findings represents a specific pattern of knee injury induced by varus stress on the knee joint.

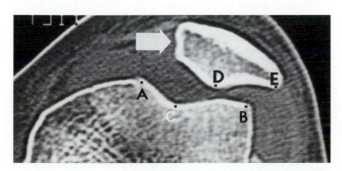

Fig. 1-26   Lateral patellar subluxation (*arrow*) demonstrated by CT. Normally, a line bisecting angle ACB should form an angle of less than 10 degrees with line CD. Normal patellar tilt is present when lines AB and DE form an angle that opens laterally. These measurements are usually applied to merchant or sunrise radiographic projections of the knee.

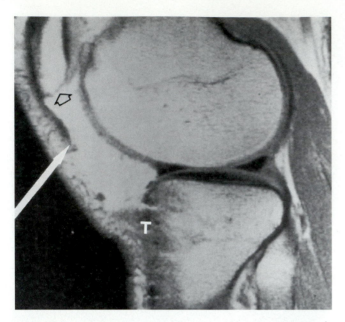

Fig. 1-27   Patellar tendon rupture documented by MRI. Sagittal T1-weighted image reveals free end of tendon (*closed arrow*) that has become detached from the tibial tubercle (*T*), resulting in patella alta (*open arrow*).

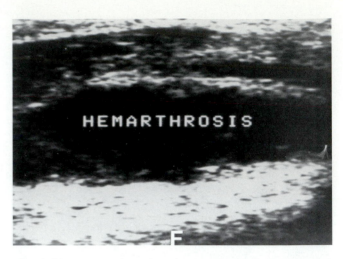

Fig. 1-28   Ultrasonography of a large hemarthrosis (*labeled*) within the suprapatellar bursa. Typical findings include a homogeneously anechoic region with enhanced through-transmission of sound (*F = femoral shaft*). Sonography of joints is useful in identifying abnormal fluid collections and intraarticular osteochondral bodies, as well as for guidance of percutaneous aspiration.

7. The proximal tibial physis is the most common site for Salter-Harris type V fractures involving the knee.
8. Salter-Harris type V fractures of the proximal tibial physis are frequently associated with fractures of the tibial diaphysis.
9. Growth disturbances frequently complicate physeal injuries involving the knee.

Internal Derangements of the Knee Joint (Fig. 1-28)

1. Meniscal tears occur most commonly in the posterior horn and body of the medial meniscus (Fig. 1-29).

PEARL

Medial meniscal tears are more common than lateral meniscal tears because of the firm attachment of the medial collateral ligament. Discoid menisci, usually lateral, are commonly torn and present during childhood. For this reason, in children, tears of the lateral meniscus are more common. Most tears of the medial meniscus involve the posterior horn. Most tears of the lateral meniscus involve the anterior horn.

2. The O'Donoghue triad includes tears of the medial meniscus, anterior cruciate ligament, and medial collateral ligament (Fig. 1-30).
3. Anterior cruciate ligament tears are common injuries. Posterior cruciate ligament tears seldom occur as isolated injuries.
4. Fluoroscopic or magnetic resonance arthrography may be preferred over magnetic resonance imaging for the detection of meniscal tears involving meniscal remnants following partial meniscectomy.
5. Children under 10 years of age who have clinical signs of an internal derangement of the knee are likely to harbor a torn discoid meniscus.
6. The discoid meniscus is characterized by widening and/or thickening of the fibrocartilage with a predisposition to tearing, and it is most commonly encountered laterally (Fig. 1-31).
7. Magnetic resonance imaging is the imaging procedure of choice for the evaluation of internal derangements of the knee joint. (Figs. 1-32, 1-33, 1-34, 1-35).
8. Unequivocal clinical evidence of an internal derangement of the knee joint is an indication for arthroscopic surgery without cross-sectional imaging.
9. Computed air arthrotomography is the imaging procedure of choice for the detection of intraarticular osteochondral bodies in the knee joint.
10. Meniscal tears may be associated with either meniscal or popliteal (Baker's) cysts (Fig. 1-36).

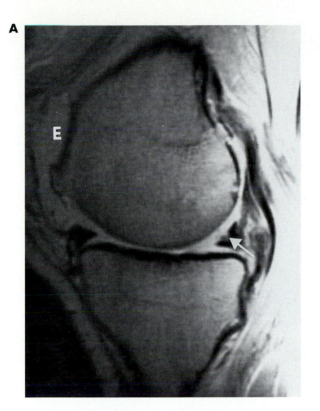

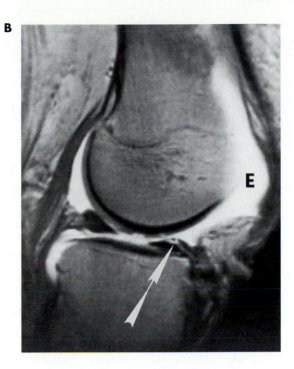

Fig. 1-29   **A,** Medial meniscal abrasion or tear limited to the superior surface of the posterior horn (*arrow*) on a sagittal proton density–weighted MR image. A joint effusion (*E*) is also present. **B,** Superiorly surfacing tear (*arrow*) in the anterior horn of the medial meniscus on a T2-weighted sagittal image from another patient. A large joint effusion (*E*) is also evident.

PEARL

The popliteal (Baker's) cyst typically occurs between the semimembranosus muscle and medial head of the gastrocnemius muscle and may be associated with a meniscal tear. The meniscal cyst communicates with a meniscal tear. Often multiple, ganglion cysts are attached to an underlying joint capsule or tendon sheath.

ANKLE TRAUMA

Normal Anatomy

1. The lateral malleolus lies approximately 1 cm distal and 1 cm posterior to the medial malleolus.
2. The horizontal distal tibial articular surface is termed the *plafond.*
3. The mortise view of the ankle is obtained with 15 to 20 degrees of internal obliquity and normally demonstrates a uniform joint space of 3 to 4 mm around the entire talar articular surface.
4. Joint space widening that exceeds 2 mm in any area on a mortise radiograph is abnormal.
5. The articular surface of the talus narrows posteriorly and acts to prevent posterior dislocation of the bone.
6. Stress radiographs of the ankle include varus, valgus, and anterior drawer views (Fig. 1-37).
7. Although talar tilt with stress is normally 10 to 12 degrees, it may be up to 20 degrees in patients with ligamentous laxity; hence, comparison views of the contralateral side are recommended.
8. The os trigonum is a normal sesamoid bone located posterior to the talus.
9. An ankle joint effusion is best visualized on a lateral radiograph as convex soft-tissue densities located anterior and posterior to the tibiotalar joint space.
10. The pre-Achilles fat pad is located between the Achilles tendon and the deep muscles of the leg and is well defined in the absence of ankle joint effusion or Achilles tendon inflammation (Figs. 1-38, 1-39, 1-40).

A

B

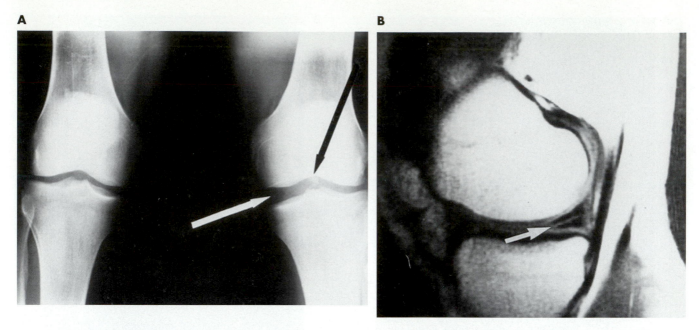

Fig. 1-30   O'Donoghue triad of injuries to the knee following valgus stress. **A,** Frontal radiograph demonstrates an anterior cruciate ligament avulsion from the tibial plateau (*black arrow*) and widening of the medial joint space (*white arrow*) secondary to medial collateral ligament disruption. **B,** Sagittal T1-weighted MR image reveals a horizontal tear (*arrow*) of the posterior horn of the medial meniscus.

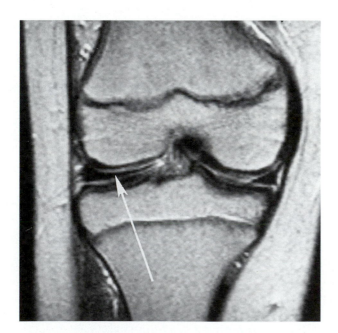

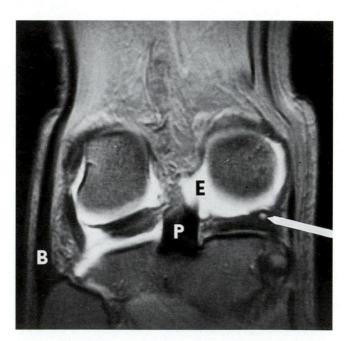

Fig. 1-31   Discoid lateral meniscus (*arrow*) in a child, demonstrated by use of MRI. The meniscal fibrocartilage may be abnormally broad and/or thick and is associated with aberrant ligamentous attachments that predispose to tearing.

Fig. 1-32   Meniscal ossicle. Coronal T2-weighted MR image reveals a rounded marrow-containing structure (*arrow*) within the posterior horn of the medial meniscus, along with a joint effusion (*E*). This uncommon finding may be posttraumatic or vestigial in origin. (*P = posterior cruciate ligament, B = biceps femoris tendon*).

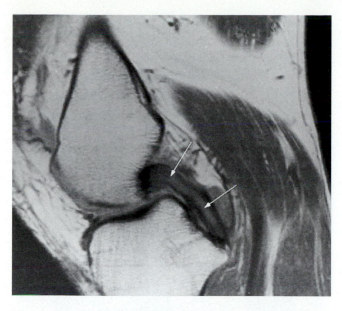

Fig. 1-33 Partial tear of posterior cruciate ligament. Sagittal proton density–weighted MR image reveals diffuse high signal intensity within the substance of the ligament (*arrows*) that normally is homogeneously dark.

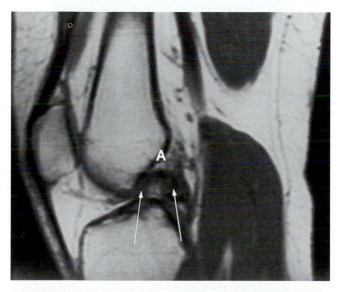

Fig. 1-34 Complete tear of posterior cruciate ligament. Sagittal T1-weighted MR image demonstrates abnormal orientation of the ligament (*arrows*), indicating complete detachment from the femur or tibia (*A = Blumensadt line*). Bowing of the posterior cruciate ligament has also been described as an indirect sign of anterior cruciate ligament tear.

Injuries (Fig. 1-41)

1. Soft-tissue swelling distal to the medial or lateral malleolus suggests ligamentous injury.
2. Oblique malleolar fractures usually result from impaction forces.
3. Transverse malleolar fractures usually result from avulsion forces.
4. Inversion and eversion injuries are usually complicated by rotational forces.

PEARL

The inversion ankle injury results in transverse fracture of the distal fibula (lateral malleolus) and oblique fracture of the medial malleolus. In inversion-adduction injury, the tensile force generated by the lateral collateral ligament causes either lateral ligamentous rupture or a transverse fracture of the lateral malleolus. The angular force generated by movement of the talus against the medial malleolus yields an oblique fracture of the medial malleolus. If inversion occurs with external rotation of the foot, then the fracture of the lateral malleolus is oblique. A pseudo-Jones fracture, or transverse fracture of the base of the fifth metatarsal bone, is caused by avulsion by the peroneus brevis tendon (this occurs with inversion and plantar flexion).

PEARL

Standard eversion injuries of the ankle usually cause a transverse fracture of the medial malleolus and an oblique fracture of the distal fibula at or below the tibiofibular ligaments, with or without deltoid and tibiofibular ligament tears. Pott's fracture disrupts the deltoid ligament and fractures the fibula above the tibiofibular ligaments with an intact interosseous membrane. The Maisonneuve fracture tears the tibiofibular ligaments and interosseous membrane and causes fracture of the fibula (midshaft or fibular neck).

5. The Lauge-Hansen classification of ankle injuries is based on mechanism (Box 1-6).

A

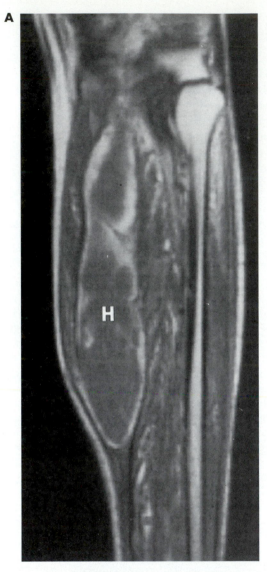

B

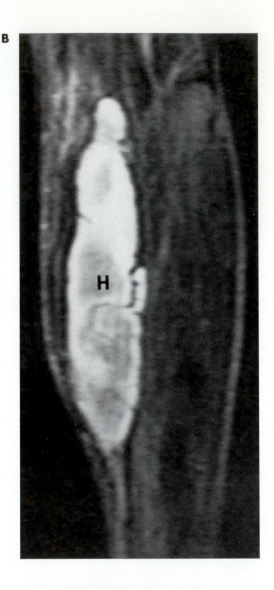

Fig. 1-35   Soleus hematoma following acute trauma. **A,** Coronal proton density–weighted MR image of the calf demonstrates a well-defined mass (*H*) with central low signal intensity and peripheral high signal intensity. **B,** Corresponding T2-weighted image reveals predominantly high signal intensity within the mass (*H*), which maintains some areas of low signal intensity centrally. The findings are characteristic of evolving hemorrhage.

6. The Weber or AO classification of ankle injuries uses the level of fibular fracture to predict the degree of tibiofibular ligament disruption, and it indicates both management and prognosis.

7. The Weber stage A injury involves a transverse fracture of the lateral malleolus at or distal to the distal tibiofibular joint that spares the tibiofibular ligaments and may be associated with an oblique fracture of the medial malleolus.

8. The Weber stage B injury involves a spiral fracture of the lateral malleolus beginning at the level of the plafond, with partial disruption of the tibiofibular ligaments and variable diastasis of the ankle mortise.

9. The Weber stage C injury involves a fibular fracture proximal to the plafond, with complete tearing of the tibiofibular ligaments and lateral talar instability.

10. The Weber stage B injury may be associated with a transverse fracture of the medial malleolus below the ankle joint or a deltoid ligament rupture.

11. The Weber stage C injury may be associated with avulsion of the anterior (Tillaux-Chaput) or posterior (Volkmann) tubercles of the distal tibia or the medial aspect of the distal fibula (Fig. 1-42).

12. The Weber stage C injury is associated with avulsion of the medial malleolus below the ankle joint or a tear of the deltoid ligament.

A

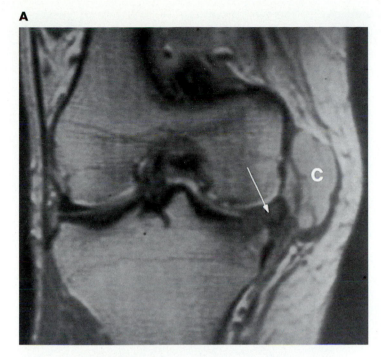

B

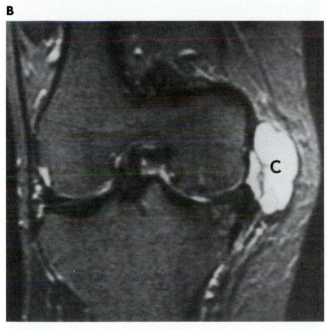

Fig. 1-36   Meniscal cyst. **A** and **B,** Coronal proton density– and T2-weighted MR images demonstrate a typical septated fluid collection (*c*) located at the medial joint line of the knee. Altered signal intensity within the medial meniscus (*arrow*) represents the associated horizontal tear, which allowed fluid to enter the cyst from the joint.

13. Occult injuries of the ankle include impaction of the plafond and osteochondral fractures of the medial or lateral talar dome.
14. The Maisonneuve injury involves a transverse medial malleolar fracture or medial mortise widening, tibiofibular ligament tears, and fracture of the proximal third of the fibula, caused by eversion force.
15. Closure of the distal tibial physis begins at 12 to 13 years of age and initially involves its central portion, followed respectively by involvement of its medial and lateral portions.

16. The juvenile Tillaux injury is a Salter-Harris type III fracture of the lateral portion of the distal tibial physis that occurs subsequent to closure of the medial portion of the distal tibial physis (Fig. 1-43).
17. The triplane injury to the distal tibia involves a vertical fracture through the lateral portion of the epiphysis, a horizontal fracture through the physis, and an oblique fracture through the metaphysis that begins at the anterior aspect of the physis and extends posteriorly and superiorly; it is a Salter-Harris type IV fracture.
18. A two-fragment triplane injury occurs after the medial portion of the distal tibial physis has closed, and it leaves the medial malleolus intact.
19. A three-fragment triplane injury occurs before the distal tibial physis begins to close, and it isolates the medial malleolus as a separate fragment.

A

B

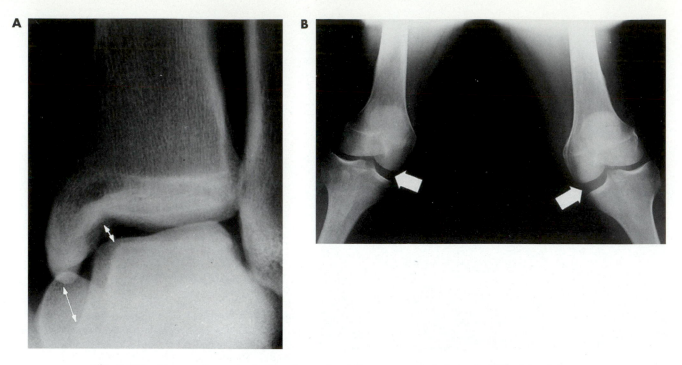

Fig. 1-37   Stress radiography in the diagnosis of ligamentous pathology. **A,** Widening of the ankle joint space medially (*double-headed arrows*) with eversion stress indicates a deltoid ligament tear. **B,** Symmetric laxity of the medial collateral ligaments of the knees (*arrows*) demonstrated by use of valgus stress in a patient with Ehlers-Danlos syndrome.

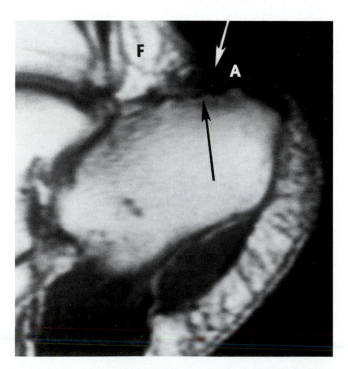

Fig. 1-38   Pre-Achilles bursitis. Sagittal T1-weighted MR image demonstrates edema (*arrows*) in the posterior aspect of the pre-Achilles fat (*F*), immediately anterior to the Achilles tendon (*A*).

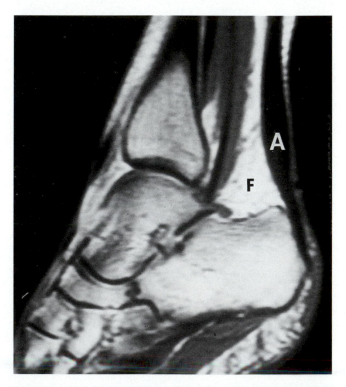

Fig. 1-39   Chronic Achilles tendinitis. Sagittal T1-weighted MR image demonstrates abnormal thickening of the tendon (*A*), which bulges convexly into the pre-Achilles fat (*F*).

Box 1-6 Lauge-Hansen Classification System

SUPINATION-EXTERNAL ROTATION (SER) FRACTURE—STAGES I, II, III, IV

Mechanism

This fracture is caused by an external rotation of the supinated foot. In supination, the deltoid ligament is relaxed and therefore an outward rotation of the talus produces pressure on the lateral malleolus, resulting in rupture of the anterior tibiofibular ligament (stage I). As the mechanism of injury continues, a spiral oblique fracture of the lateral malleolus (stage II) occurs, followed by a fracture of varying size of the posterior aspect of the tibia (stage III). Finally, as the mechanism of injury continues, rupture of the deltoid ligament or fracture of the medial malleolus occurs (stage IV).

Stages

I. Rupture of anterior tibiofibular ligament
II. Spiral or oblique fracture of lateral malleolus
III. Rupture of posterior tibiofibular ligament or fracture of posterior tibial margin
IV. Rupture of deltoid ligament or fracture of medial malleolus

Characteristic Radiographic Feature

Spiral oblique fracture of the lower aspect of the fibula

Rule

If there is a spiral or oblique fracture of the lateral malleolus (SER II), suspect possible deltoid rupture (SER IV), especially if there is medial soft-tissue swelling.

PRONATION-EXTERNAL ROTATION (PER) FRACTURE—STAGES I, II, III, IV

Mechanism

With pronation, the deltoid ligament tightens. Forceful outward rotation of the foot results in a medial malleolus fracture or deltoid ligament rupture (stage I). Continued forces result in anterior tibiofibular ligament and interosseous membrane rupture (stage II), a short spiral oblique fracture of the fibula commonly 7 to 8 cm above the ankle joint (stage III), and finally, posterior tibiofibular ligament rupture or a posterior malleolus fracture (stage IV).

Stages

I. Rupture of deltoid ligament or avulsion of medial malleolus
II. Rupture of anterior tibiofibular ligament and interosseous ligament
III. Short spiral fracture of the fibula (typically 7 to 8 cm above ankle joint but not infrequently more proximal)
IV. Fracture of posterior tibial margin or rupture of posterior tibiofibular ligament

Characteristic Radiographic Feature

Spiral oblique fracture of the fibula at a relatively high position relative to the ankle joint

Rules

A short spiral fracture of the fibula (PER III) should have a deltoid ligament tear if no fracture of the medial malleolus is seen.

If there is a tibiofibular diastasis (from rupture of the anterior tibiofibular ligament and interosseous membrane), a high fibular fracture (or dislocation) must be excluded. The entire fibula must be visualized radiographically.

PRONATION-ABDUCTION (PA) FRACTURE—STAGES I, II, III

Mechanism

This fracture is produced by subjecting the pronated foot to a laterally directed force. Initial forces lead to tightening of the deltoid ligament with ligament rupture or a transverse fracture of the medial malleolus (stage I). Continued forces lead to rupture of the anterior and posterior tibiofibular ligaments (stage II). Finally (stage III), a short oblique (bending) fracture of the fibula occurs, just above the level of the ankle joint.

Stages

I. Fracture of the medial malleolus or rupture of the deltoid ligament
II. Rupture of both the anterior and posterior tibiofibular ligaments
III. Bending fracture of the fibula (generally just above the ankle joint, often with displacement of a triangular fragment from the lateral fibular surface)

Characteristic Radiographic Feature

Short oblique fracture of the fibula that appears horizontally oriented on the lateral film

Rule

Visualization of a characteristic horizontal or bending fracture of the lateral malleolus (PA III) should be associated with a deltoid rupture if a fracture of the medial malleolus is not seen.

SUPINATION-ADDUCTION (SA) FRACTURE—STAGES I, II

Mechanism

This fracture is produced by medially directed force exerted on the supinated foot. Supination results in tightening of the calcaneofibular ligament resulting in a rupture of the talofibular ligaments or a transverse fracture of the lateral malleolus (stage I). Continued pressure from the medially directed talus results in a fracture of the medial malleolus (stage II) that is usually vertical in orientation.

Stages

I. Traction fracture of lateral malleolus at or below level of ankle joint, or rupture of talofibular ligaments
II. Near-vertical fracture of medial malleolus

Characteristic Radiographic Features

Transverse fracture of the distal fibula and relatively vertical fracture of medial malleolus

Rule

Visualization of a vertical fracture of the medial malleolus (SA II) should be associated with talofibular ligament rupture if fracture of the lateral malleolus is not seen.

A

B

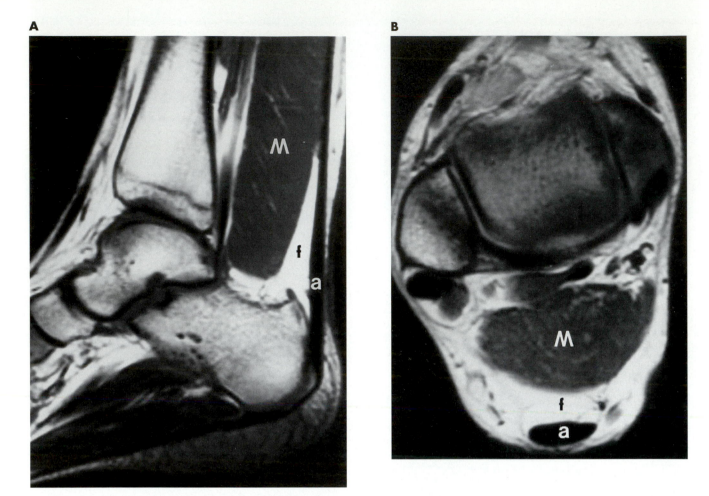

Fig. 1-40   The accessory soleus muscle is a normal anatomic variant that may be encountered on MRI studies of the ankle and foot. Sagittal (**A**) and transaxial (**B**) T1-weighted images demonstrate a tubular structure (*M*), with signal behavior and texture identical to that of normal muscle, occupying the pre-Achilles space (*a* = normal Achilles tendon, *f* = pre-Achilles fat).

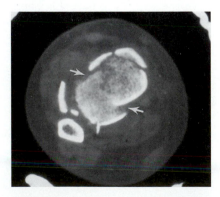

Fig. 1-41   Pillion fracture of the distal tibia resulting from axial loading. CT demonstrates marked comminution and malalignment of fracture fragments in the metaphyseal region (*arrows*).

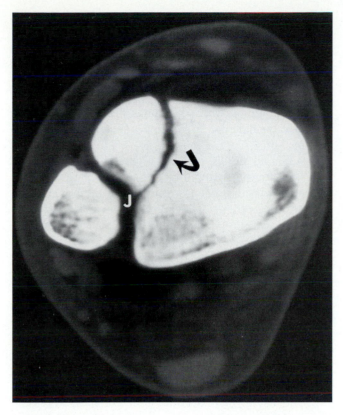

Fig. 1-42 Tillaux fracture of distal tibia. CT demonstrates characteristic disruption (*arrow*) of the anterolateral aspect of the bone with extension into the distal tibiofibular joint (*J*).

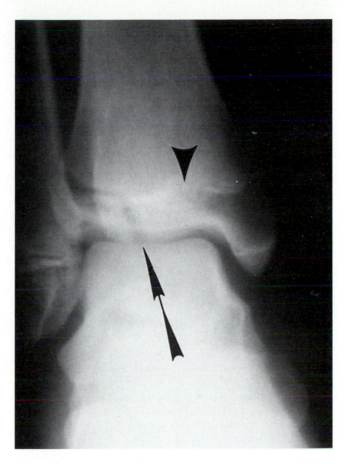

Fig. 1-43 Juvenile Tillaux or Salter-Harris type III fracture (*arrow*) of distal tibia in a child. Note partial closure of the medial portion of the distal tibial growth plate (*arrowhead*).

20. Computed tomography with multiplanar reformation is the procedure of choice for complete characterization of the extent and degree of displacement of triplane injuries.
21. Fatigue-type stress fractures in runners tend to occur approximately 3 to 7 cm from the tip of the lateral malleolus, or in the tibial diaphysis (Figs. 1-44, 1-45).
22. Insufficiency-type stress fractures may involve both the distal tibia and fibula, usually within 4 cm of the plafond.
23. Following reduction of complex ankle injuries, signs of persistent joint subluxation include posterior displacement of either a posterior malleolar or distal fibular fragment that exceeds 2 mm.

FOOT TRAUMA

Normal Anatomy

1. The plantaris muscle arises from the femur just above the origin of the lateral head of the gastrocnemius muscle. Its tendon usually inserts into the calcaneus.

2. The peroneus longus muscle arises from the lateral surface of the fibula and from the adjacent deep fascia. It courses across the foot within a tendon sheath and inserts into the lateral side of the medial cuneiform and base of the first metatarsal bone.
3. On lateral radiographs, the Bohler angle of the calcaneus measures the severity of fragment depression in the setting of fracture and normally measures 20 to 40 degrees.
4. On frontal radiographs, the lateral margin of the first metatarsal base normally aligns with the lateral margin of the medial cuneiform, and the medial margin of the second metatarsal base normally aligns with the medial margin of the intermediate cuneiform.
5. On oblique radiographs, the lateral margin of the third metatarsal base normally aligns with the lateral margin of the lateral cuneiform, and the medial margin of the fourth metatarsal base normally aligns with the medial margin of the cuboid bone.
6. The keystone of the arch of the foot is the second metatarsal base, which articulates with five other

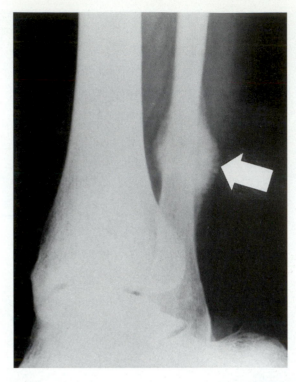

Fig. 1-44 Healing stress fracture (*arrow*) of the distal fibula in a military recruit. Other possible causes for this appearance would include bone-forming neoplasm (particularly parosteal osteosarcoma) in the appropriate clinical setting.

bones, including the first and third metatarsal bases and all three cuneiforms.

7. The apophysis of the peroneal tubercle of the fifth metatarsal is parallel to the shaft of the bone and may be bipartite; it is distinguished from a fracture by the transverse orientation of the latter.

8. The epiphysis of the proximal phalanx of the hallux may be bipartite, simulating a Salter-Harris type III fracture.

9. Sesamoid bones may be bipartite or multipartite, simulating fracture.

10. Accessory ossicles are common in the foot and may be unilateral or bilateral; their location and well-marginated appearance distinguish them from acute avulsion fractures (Fig. 1-46).

11. The os peroneum is a common sesamoid bone that lies in the peroneus longus tendon adjacent to the cuboid bone.

12. The paired sesamoid bones that lie beneath the first metatarsal head are located within the paired flexor hallucis brevis tendons; the flexor hallucis longus tendon courses between them.

13. The os trigonum is a common accessory ossicle that lies immediately posterior to the talus.

14. The os tibiale externum is a common accessory ossicle that lies immediately proximal to the navicular tuberosity and is the site of attachment of the tibialis posterior tendon.

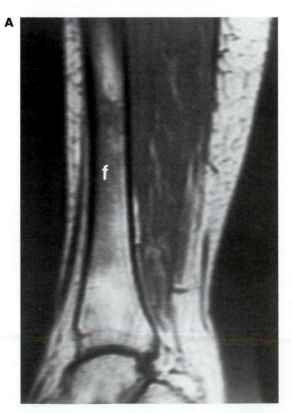

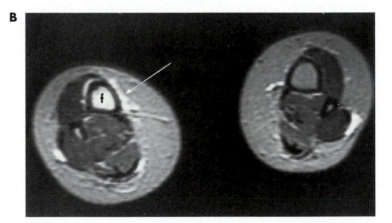

Fig. 1-45 Stress fracture of distal tibial diaphysis. **A,** Sagittal T1-weighted MR image demonstrates diffuse low signal intensity within the tibial marrow (*f*), representing edema in the vicinity of the fracture. **B,** Transaxial T2-weighted image reveals abnormally high signal intensity within both the marrow cavity (*f*) and the adjacent soft tissues (*arrow*). Other possible causes for this appearance include osteomyelitis, neoplastic disease, and acute intramedullary bone infarction.

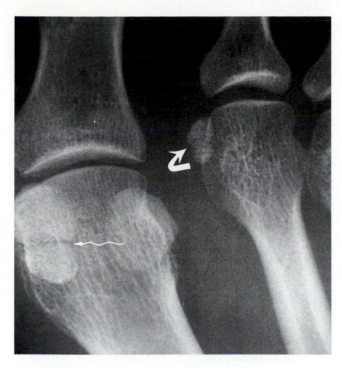

Fig. 1-46   Bipartite (*wavy arrow*) and accessory (*curved arrow*) sesamoid bones are common normal variants, particularly in the foot.

15. There are two subtalar joints; the anterior subtalar joint includes the talonavicular joint and the articulation between the talus and the sustentaculum tali of the calcaneus, as well as a synovial space supported inferiorly by the plantar calcaneonavicular (spring) ligament as it passes beneath the inferior aspect of the talar head.

16. The posterior subtalar joint lies posterior and lateral to the anterior subtalar joint.

PEARL

The posterior subtalar joint is located posterolaterally between the talus and calcaneus. Subtalar stability is provided by the interosseous ligaments within the sinus tarsi and the medial and lateral talocalcaneal ligaments. The anterior subtalar joint is also known as the talocalcaneonavicular joint.

17. The most common site of subtalar coalition is the anterior subtalar joint at the level of the sustentaculum tali.

Traumatic Injuries

1. Radiographic assessment of suspected calcaneal fractures should include lateral and axial (Harris-Beath) views.

PEARL

Calcaneal fractures are commonly sustained in falls from heights. Bohler's angle, normally 20 to 40 degrees, is of value in the detection and characterization of fractures of the calcaneus. When there is impaction, this angle is reduced and may approach zero or become negative.

2. Calcaneal fractures may or may not extend into the subtalar and calcaneocuboid joints, and they occur bilaterally in approximately 10% of cases (Fig. 1-47).

3. Approximately 10% of calcaneal fractures are associated with compression or burst fractures of the thoracolumbar spine, because both calcaneal and vertebral body fractures are induced by axial loading.

4. Computed tomography is the procedure of choice for complete characterization of calcaneal fractures; views in the transaxial plane of the foot (coronal plane of the ankle) are best for demonstrating subtalar joint extension, whereas views in the plantar plane of the foot (transaxial plane of the ankle) optimally depict calcaneocuboid joint involvement.

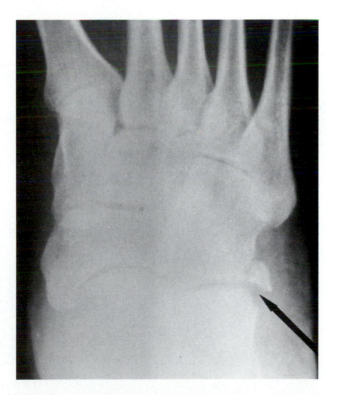

Fig. 1-47   Anterolateral process avulsion (*arrow*) of the calcaneus. Fractures in this area may result from traction on the extensor digitorum brevis muscle origin or the bifurcate ligament.

5. Stress fractures of the calcaneus usually involve cancellous bone because the calcaneus contains only 5% cortical bone; during the healing phase, the poorly defined bands of sclerosis are oriented perpendicularly to the major trabecular lines (Fig. 1-48).

PEARL

Subtalar dislocation usually occurs as a result of the patient's landing on the inverted foot after a fall, and therefore it is more common medially. Both the talonavicular and talocalcaneal joints are dislocated, with rupture of the ligaments between them. The tibiotalar and calcaneocuboid joints are not affected. Prognosis is poor in the presence of talar fracture or significant tendinous injury, and it is not related to direction of dislocation.

6. The most common fracture of the talus (aside from chips and avulsions) involves the talar neck and may be associated with talar dislocation.
7. Ischemic necrosis is a common complication of talar injuries, including fractures and dislocations.

PEARL

Osteonecrosis of the talus following injury involves the body more commonly than the head and neck. Hawkins' sign is an early radiographic indicator of an intact blood supply to the talar dome that effectively excludes osteonecrosis.

8. Acute osteochondral fractures or osteochondritis dissecans may involve the medial or lateral aspects of the talar dome, particularly in patients with ligamentous laxity; osteochondritis dissecans frequently occurs bilaterally (Figs. 1-49, 1-50).
9. Conventional or computed tomography and magnetic resonance imaging are useful radiographic methods for demonstrating stress fractures of the navicular bone, which tend to occur in athletes and usually manifest horizontal orientation.
10. The Lisfranc fracture-dislocation involves dorsal and lateral displacement of the metatarsals relative to the bones of the midfoot; the fractures in this injury are of the avulsion or chip type.
11. A homolateral Lisfranc injury involves lateral dislocation of the first through fifth or second through fifth metatarsals.

A

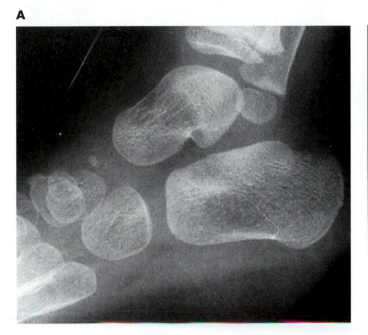

B

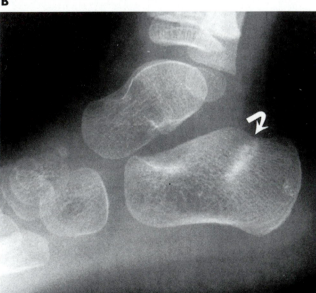

Fig. 1-48 Calcaneal stress fracture in a child. **A,** Radiograph obtained at the onset of symptoms is normal. **B,** Follow-up film taken 3 weeks later reveals a poorly defined band of sclerosis (*arrow*) in the superior aspect of the tuberosity.

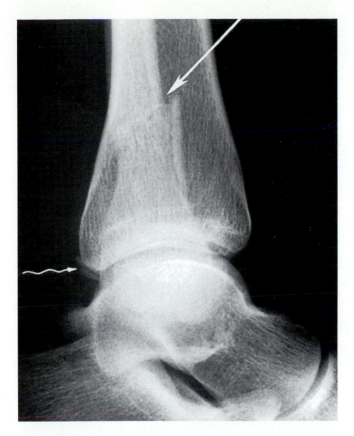

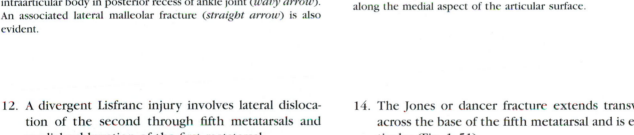

Fig. 1-49   Acute osteochondral fracture of talar dome with intraarticular body in posterior recess of ankle joint (*wavy arrow*). An associated lateral malleolar fracture (*straight arrow*) is also evident.

Fig. 1-50   Osteochondritis dissecans affecting the talar dome. A focal well-defined subchondral radiolucency (*arrows*) is present along the medial aspect of the articular surface.

12. A divergent Lisfranc injury involves lateral dislocation of the second through fifth metatarsals and medial subluxation of the first metatarsal.

13. The Lisfranc fracture-dislocation occurs more commonly as a consequence of diabetic neuroarthropathy than as a consequence of trauma.

PEARL

The Lisfranc fracture-dislocation involves displacement at the tarsometatarsal joints, usually involving two or more metatarsals. Most commonly there is lateral displacement of all five metatarsals (homolateral) or lateral displacement of metatarsals two through five with medial displacement of the first metatarsal (divergent). The injury is usually associated with a fracture at the base of the second metatarsal and often smaller fractures at the joint margins. Trauma or neuropathic disease (as in diabetes mellitus) can cause this phenomenon.

14. The Jones or dancer fracture extends transversely across the base of the fifth metatarsal and is extraarticular (Fig. 1-51).

15. The most common sites of stress fracture in the skeleton are the second metatarsal and third metatarsal.

16. Metatarsal stress fractures are also known as march fractures and usually involve predominantly cortical as opposed to cancellous bone (Fig. 1-52).

17. Because the nail bed of the hallux is attached to the periosteum of the distal phalanx at the level of the proximal metaphysis, a stubbed great toe injury in a child may result in an open Salter-Harris type I or II fracture, which can be complicated by osteomyelitis.

18. Bunk-bed injury occurs in children younger than 10 years, as a result of jumping and landing on the first metatarsal, which fractures its lateral proximal aspect and the medial cuneiform.

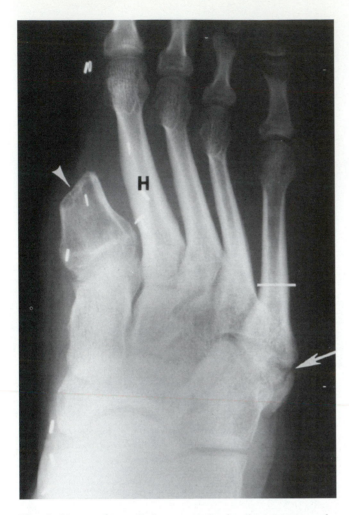

Fig. 1-51 Avulsion of the personal tubercle (*arrow*) at the peroneus brevis insertion. This injury should be distinguished from a true Jones or dancer fracture, which occurs at the proximal portion of the fifth metatarsal diaphysis (*line*) and is extraarticular. The patient has undergone first-ray resection (*arrowhead*) as part of a toe-hand transplant procedure and exhibits compensatory stress hypertrophy (*H*) of the second metatarsal.

A

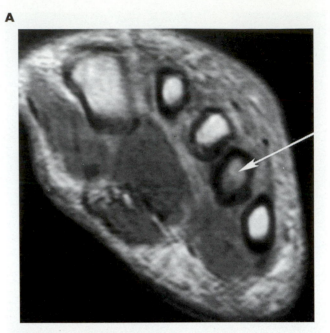

B

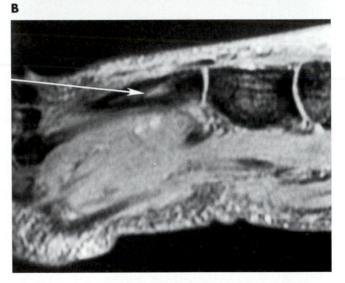

Fig. 1-52 Fatigue-type stress fracture of the fourth metatarsal, diagnosed by use of MRI in the setting of normal radiographs. **A,** Transaxial protein density–weighted image demonstrates low signal intensity within the marrow of the affected bone (*arrow*). **B,** Sagittal gradient-echo image reveals high signal intensity in the same area (*arrow*), indicating bone marrow edema. Other possible causes of these findings include acute hematogenous osteomyelitis, neoplastic disease, and acute infarction.

SHOULDER TRAUMA

Sternoclavicular Joint Injuries

1. Approximately 50% of the medial clavicular articular surface extends above the superior aspect of the manubrium.
2. Computed tomography is the procedure of choice for the evaluation of sternoclavicular joint injuries because it can demonstrate both osseous and soft-tissue abnormalities.
3. The most useful conventional radiographic view of the sternoclavicular joints is a frontal projection with 40 degrees of cephalad angulation.
4. The medial clavicular epiphysis ossifies at 18 to 20 years of age, and the medial clavicular physis closes at approximately 25 years of age.
5. Many sternoclavicular joint dislocations that occur between the ages of 18 and 25 years are actually Salter-Harris type I or II fractures.
6. Anterior dislocations are more common than posterior dislocations, and in both types, the medial portion of the clavicle also moves superiorly.

7. Posterior dislocation of the sternoclavicular joint may cause secondary injury to the trachea or great vessels, and surgical reduction is frequently associated with complications.

PEARL

Fractures of the first rib may occur secondary to blunt trauma, stress, heavy lifting, or sternotomy.

Clavicular Injuries

1. The clavicle is the first bone to ossify, beginning between the fifth and sixth fetal weeks.
2. The rhomboid fossa is a normal concavity on the inferior aspect of the medial clavicle at the costoclavicular ligament attachment, and it may be asymmetric in appearance.
3. Approximately 80% of clavicular fractures involve the middle third of the bone, and they are often associated with exuberant callus formation.
4. Approximately 15% of clavicular fractures involve the distal third of the bone and may be associated with disruption of the coracoclavicular ligament.
5. Posttraumatic osteolysis is a condition that may follow major or repetitive minor trauma and is characterized by distal clavicular resorption that stabilizes approximately 18 months after onset; many affected patients have a history of weight lifting.
6. Some possible causes of distal clavicular resorption are posttraumatic osteolysis, hyperparathyroidism, rheumatoid arthritis, and infection.
7. Condensing osteitis is a stress-related condition characterized by unilateral sclerosis of the medial portion of the clavicle (Fig. 1-53).

Acromioclavicular Joint Injuries (Box 1-7)

Box 1-7 Acromioclavicular Joint Separations

Type I: sprain of the acromioclavicular (AC) ligaments.
Type II: complete disruption of the ligaments of the acromioclavicular joint.
Type III: AC separation plus disruption of the coracoclavicular (CC) ligaments (or coracoid fracture).

Box 1-7 —cont'd

Type IV: type 3 plus superior and posterior displacement of the lateral clavicle.
Type V: type 3 plus marked superior displacement of the lateral clavicle.
Type VI: AC ligament disruption with lateral clavicle displaced inferiorly below the acromion or the coracoid; the CC ligaments may be intact.

1. In the setting of trauma, stress radiographs of both joints are indicated and should be performed with weights suspended from the wrists.
2. False negative findings may result from stress radiographs if the examination is performed with handheld weights.
3. The coracoclavicular ligament complex includes the trapezoid and conoid components.

PEARL

Acromioclavicular separations are divided into three common types. Two main ligaments stabilize the joint: the acromioclavicular ligaments and the coracoclavicular ligament, of which the trapezoid and conoid are components. In a type I acromioclavicular joint separation, the acromioclavicular ligaments are stretched. In a type II separation, the acromioclavicular ligaments are torn with stretching of the coracoclavicular ligament. In a type III separation, both the acromioclavicular and the coracoclavicular ligaments are torn.

4. The type I acromioclavicular joint separation represents a sprain and is associated with normal nonstress but abnormal stress radiographs.
5. The type IV acromioclavicular joint separation involves posterior dislocation of the clavicle relative to the acromion; the diagnosis is best made on the basis of the axillary view, and the injury may be occult on the frontal projection.
6. The type III acromioclavicular joint separation includes disruption of both the acromioclavicular and coracoclavicular ligaments, which allows the scapula to move inferiorly relative to the clavicle.
7. The type II acromioclavicular separation results in widening of the joint space on nonstress radiographs secondary to disruption of the acromioclavicular ligaments but not the coracoclavicular ligaments.
8. The width of the acromioclavicular joint space is normally 3 to 5 mm, but occasionally may be as much as 8 mm.
9. Asymmetry between the acromioclavicular joint space widths should not exceed 3 mm.

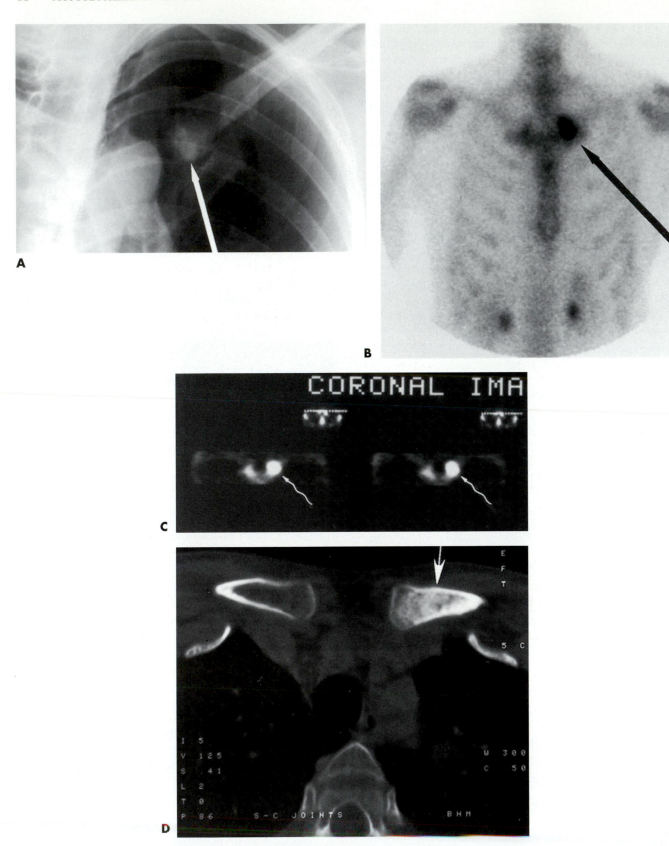

Fig. 1-53   Condensing osteitis of the medial clavicle. **A,** Oblique radiograph reveals increased density of the clavicular head (*arrow*). **B** and **C,** The lesion exhibits increased activity (*arrows*) on both planar and tomographic bone scan images. **D,** The extent of osteosclerosis (*arrow*) is optimally depicted by CT. Additional possible causes of this appearance include chronic osteomyelitis (particularly the recurrent multifocal type), fibrous dysplasia, and metastatic disease.

10. The coracoclavicular distance normally measures 11 to 13 mm; asymmetry between the two sides of the body should not exceed 5 mm.
11. As part of a type III acromioclavicular joint separation in a skeletally immature patient, the apophysis of the coracoid process may be avulsed, representing a Salter-Harris type I fracture.

Scapular Fractures

1. Scapular fractures are relatively uncommon.
2. Because of extensive muscular coverage of the scapula, scapular fractures are usually not significantly displaced and therefore tend to be radiographically occult (Fig. 1-54).
3. Fractures involving the articular surface of the glenoid are likely to result in premature osteoarthritis of the glenohumeral joint.
4. Normal scapular foramina and secondary ossification centers (including the coracoid process, acromion process, glenoid rim, inferior angle, and medial marginal apophyses) may simulate traumatic injuries.
5. Fractures at the base of the coracoid process are best detected on tangential or Y views of the bone.
6. Fatigue-type stress fractures in the coracoid process of the scapula occur in trapshooters.

Proximal Humeral Fractures

1. The proximal humerus is a relatively common site of fracture, particularly in patients with osteoporosis.

2. The four bone fragments considered in the Neer classification of proximal humeral fractures are the humeral head, humeral diaphysis, greater tuberosity, and lesser tuberosity.
3. In the Neer classification, significant malposition is defined as fragment displacement exceeding 1 cm or angulation exceeding 45 degrees.
4. According to the Neer classification, an n-part fracture is defined as one in which n−1 fragments are malpositioned.

PEARL

The Neer classification uses four principal components of proximal humeral fractures: the humeral head, the greater tuberosity, the lesser tuberosity, and the humeral diaphysis. The fractures are classified as one-part (80%), two-part (13%), three-part (3%), and four-part (4%), according to the number of fragments and their degree of angulation or displacement. Minimal amounts of either angulation or displacement are ignored, and there must be greater than 45 degrees of angulation or 1 cm of displacement from normal position in order for the angulation or displacement to be counted.

5. Approximately 80% of all proximal humeral fractures manifest fragment malposition that is insignificant according to the criteria of the Neer classification because of the stabilizing influence of the rotator cuff, joint capsule, and periosteum.

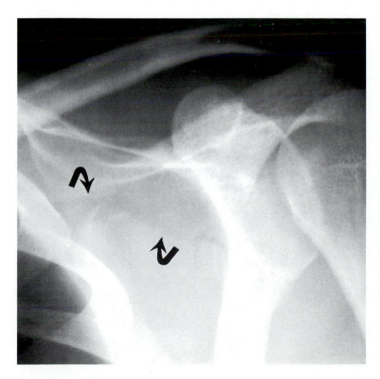

Fig. 1-54 Fractures of the scapular body (*arrows*) are often difficult to confidently diagnose because of the thin structure of the bone in this area, as well as the normal foramina and vascular grooves that may occur.

6. Axillary views are mandatory in the setting of proximal humeral fracture, because posterior angulation of the humeral shaft relative to the head often occurs and may be occult on other projections.

7. Although it may persist for months, pseudosubluxation of the glenohumeral joint following proximal humeral fracture is a self-limited phenomenon that will disappear with resolution of hemarthrosis and return of normal muscle tone.

8. Humeral shaft fractures generally heal readily and seldom require internal fixation.

9. Malunion of humeral shaft fractures is usually not important because the ball-and-socket nature of the glenohumeral joint can compensate for all but severe angular or rotational malalignment.

Glenohumeral Joint Dislocation

1. The glenohumeral joint space is depicted tangentially on 45-degree posterior oblique and axillary radiographs.

2. Approximately 95% of all glenohumeral joint dislocations involve anterior, medial, and inferior displacement of the humeral head relative to the glenoid, resulting in subcoracoid positioning of the former with capsular stripping from the latter (Fig. 1-55).

3. The Hill-Sachs defect occurs on the posterolateral aspect of the humeral head where it impacts on the anteroinferior aspect of the glenoid with anterior dislocation, and it is best visualized on the internal rotation view (Fig. 1-56).

4. The Bankart lesion is a chip fracture of the anteroinferior aspect of the glenoid, secondary to anterior

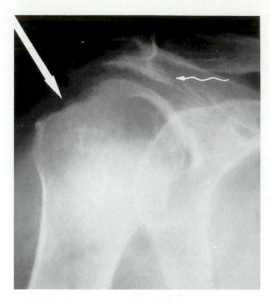

Fig. 1-56   Internal rotation view of the shoulder optimally demonstrates flattening of the posterolateral aspect of the greater tuberosity, or Hill-Sachs lesion (*straight arrow*), related to prior anterior glenohumeral dislocation. A large subacromial spur (*wavy arrow*) with potential for shoulder impingement syndrome is also present.

dislocation, that may be visualized on frontal or axillary radiographs (Fig. 1-57).

5. Injuries to the fibrocartilaginous glenoid labrum (including the nonosseous Bankart lesion) that occur in association with glenohumeral joint dislocation can be detected by computed air arthrotomography, magnetic resonance imaging, or arthroscopy.

6. A Hill-Sachs or Bankart lesion occurs after a single episode of dislocation in approximately 35% of cases.

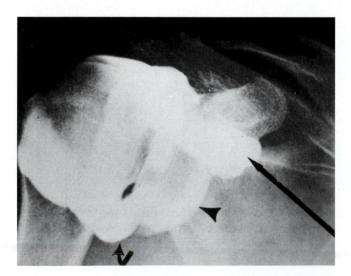

Fig. 1-55   Arthrography in anterior glenohumeral joint instability. Capsular stripping from the glenoid (*arrowhead*) is associated with large axillary (*curved arrow*) and subscapular (*straight arrow*) recesses.

PEARL

The Hill-Sachs deformity can occur after a single dislocation. It is caused by the humeral head impacting on the inferior lip of the glenoid. The defect occurs on the posterolateral aspect of the humeral head and is best visualized on the internal rotation view. With a posterior dislocation, if an associated trough fracture occurs, it is located on the anteromedial aspect of the humeral head.

7. A Hill-Sachs or Bankart lesion may predispose to instability and recurrent dislocation.

8. The incidence of recurrent glenohumeral joint dislocation is inversely related to patient age at the time of initial dislocation.

9. Posterior dislocation of the glenohumeral joint accounts for approximately 3% of all cases.

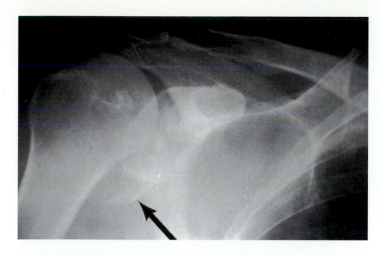

Fig. 1-57   Osseous Bankart lesion (*arrow*) resulting from previous anterior glenohumeral joint dislocation. Avulsed fragments of the lesser tuberosity related to the subscapularis tendon insertion may occur in a similar location following trauma.

10. Seizure disorders and electroconvulsive therapy are responsible for most cases of posterior glenohumeral joint dislocation.

11. Posterior glenohumeral joint dislocation may be overlooked on frontal radiographs because the humeral head moves directly posteriorly with respect to the glenoid, although it does tend to be locked in internal rotation.

12. The trough sign of posterior dislocation consists of impaction on the anteromedial aspect of the humeral head, and it is also known as the reverse Hill-Sachs lesion.

13. The reverse Bankart lesion of posterior dislocation is a chip fracture of the posterior aspect of the glenoid.

14. The rim sign of posterior dislocation consists of apparent lateral displacement of the humeral head relative to the glenoid, which simulates joint space widening.

15. Luxatio erecta is an inferior dislocation of the humeral head relative to the glenoid and accounts for less than 1% of all glenohumeral joint dislocations; in this injury, the humeral shaft is maintained in a position of fixed abduction.

16. The drooping shoulder, or pseudosubluxation of the glenohumeral joint, is characterized by mild inferior displacement of the humeral head relative to the glenoid, caused by decreased muscle tone and/or joint effusion; the condition frequently occurs secondary to proximal humeral fracture (Fig. 1-58).

Rotator Cuff Tears

1. The muscles of the rotator cuff are the supraspinatus, infraspinatus, teres minor, and subscapularis; the majority of tears involve the supraspinatus tendon.

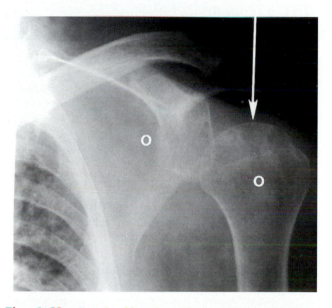

Fig. 1-58   Pseudosubluxation of the glenohumeral joint (drooping shoulder) following brachial plexus injury in adulthood. The humeral head is positioned inferiorly (*arrow*) with respect to the glenoid, and disuse osteopenia (*O*) is present. The condition most commonly occurs in the setting of surgical neck fracture with hemarthrosis, but it may be induced by any cause of muscular hypotonia or tense effusion.

2. The supraspinatus, infraspinatus, and teres minor muscles insert on the greater tuberosity of the humerus and are responsible for external rotational motion.

3. The subscapularis muscle inserts on the lesser tuberosity of the humerus and is responsible for internal rotational motion.

4. The deltoid muscle inserts on the deltoid tubercle of the proximal humeral diaphysis.

5. The biceps brachii muscle originates at the superior glenoid labrum and coracoid process and inserts on the bicipital tuberosity of the radius.

6. Rotator cuff tears may be an acute traumatic or chronic degenerative phenomenon, and they are common among patients with rheumatoid arthritis.

7. In the setting of rotator cuff tear, conventional radiography may demonstrate elevation of the humerus and sclerosis of the greater tuberosity and inferior aspect of the acromion secondary to mechanical apposition.

8. In patients with rheumatoid arthritis, mechanical erosion and occasionally fracture of the surgical neck of the humerus may occur secondary to impaction on the inferior aspect of the glenoid.

9. Contrast arthrography of the normal glenohumeral joint will opacify the axillary and subscapularis recesses (Fig. 1-59).

PEARL

Normal structures visualized on the shoulder arthrogram are the glenohumeral joint, axillary recess, subscapularis or subcoracoid recess, and bicipital groove containing the long head of the biceps tendon. If contrast material is noted in the subacromial or subdeltoid bursa, this is indicative of a full-thickness rotator cuff tear.

---

10. In the setting of a full-thickness rotator cuff tear, contrast arthrography will opacify the axillary, subscapularis, subacromial, and subdeltoid bursae.

11. In the setting of a partial tear on the inferior aspect of the rotator cuff, arthrography will opacify the axillary and subscapularis bursae, and contrast material will accumulate within the supraspinatus tendon.

PEARL

Rotator cuff tears:
- Partial undersurface tears are reliably detected by arthrography. Visualization of the subacromial and subdeltoid bursae allows diagnosis of a full-thickness tear.
- The humerus migrates toward the acromion in a chronic tear.
- The muscles of the rotator cuff are the supraspinatus, infraspinatus, teres minor, and subscapularis.
- The most common tendon to rupture is the supraspinatus.
- The subdeltoid bursa is continuous with the subacromial bursa.
- If the tear is incomplete, contrast material will enter the tendon substance but will not opacify the subacromial and subdeltoid bursae.
- Shoulder impingement syndrome and degenerative calcific tendinitis of the rotator cuff predispose it to tearing.

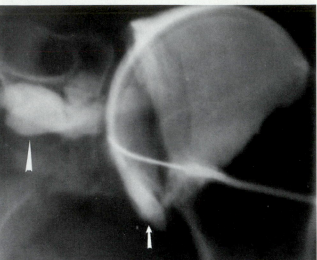

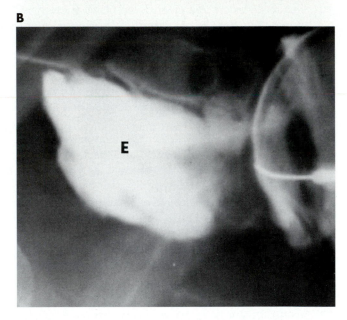

Fig. 1-59 Adhesive capsulitis of the glenohumeral joint. **A,** Arthrography demonstrates a low-capacity synovial space with small axillary (*arrow*) and subscapular (*arrowhead*) recesses. **B,** Following a brisement procedure, rupture of the capsule with extravasation (*E*) beneath the pectoralis musculature signifies termination of the therapeutic distension effect. The patient is subsequently sent for immediate and follow-up physical therapy.

12. In the setting of a partial tear on the superior aspect of the rotator cuff, contrast arthrography of the glenohumeral joint will be normal; direct injection of the subacromial-subdeltoid bursae will demonstrate the lesion.

13. The geyser sign consists of opacification of the acromioclavicular joint during contrast arthrography of the glenohumeral joint, and it indicates tears of both the rotator cuff and the acromioclavicular joint capsule inferiorly (Fig. 1-60).

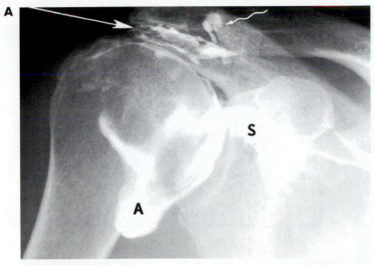

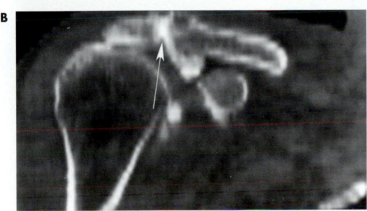

Fig. 1-60   Full-thickness rotator cuff tear with associated disruption of the inferior capsule of the acromioclavicular joint. **A,** Single-contrast shoulder arthrogram demonstrates abnormal opacification of the subacromial-subdeltoid bursae (*straight arrow*), as well as the acromioclavicular joint space (*wavy arrow*). Opacification of the latter constitutes the geyser sign (*A = normal axillary recess, S = normal subcoracoid or subscapularis recess*). **B,** Coronal re-formation following CT (performed to evaluate glenoid labrum, glenohumeral joint capsule, and biceps tendon) confirms presence of contrast material within the acromioclavicular joint (*arrow*).

14. Ultrasonography is a useful screening procedure for rotator cuff tear, but it is operator dependent and requires significant experience.
15. Magnetic resonance imaging can readily identify full-thickness rotator cuff tears, if imaging is performed in the oblique coronal and sagittal planes using T2 weighting; other soft-tissue injuries in the shoulder region are also readily demonstrated (Figs. 1-61, 1-62).
16. It is difficult to make the distinction among small complete rotator cuff tear, partial rotator cuff tear, and tendinitis by the use of magnetic resonance imaging unless intraarticular injection of gadolinium–diethylenetriamine pentaacetic acid (DTPA) is first performed (Fig. 1-63).

Shoulder Impingement Syndrome

1. Symptoms occur during abduction, external rotation, or elevation of the arm.
2. Radiographic findings include sclerosis and flattening of the greater humeral tuberosity and a bony prominence arising from the anteroinferior aspect of the acromion (see Fig. 1-56).
3. The condition is encountered in young as well as older patients, particularly athletes.
4. Pain is caused by entrapment of the rotator cuff between the greater tuberosity and the coracoacromial ligamentous arch, undersurface of the acromion process, or osteophytes arising from degenerative disease of the acromioclavicular joint.
5. Fluoroscopy and magnetic resonance imaging are useful in establishing the diagnosis.
6. The shoulder impingement syndrome may progress to supraspinatus tendinitis or rotator cuff tear.
7. Surgical acromioplasty is indicated in severe cases.

ELBOW TRAUMA

Normal Anatomy and Variation

1. The radial head articulates with the capitellum of the humerus, and the coronoid process of the ulna articulates with the trochlea of the humerus.

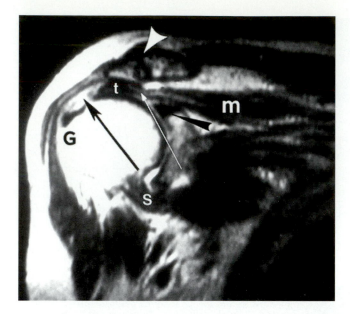

Fig. 1-61   Full-thickness rotator cuff tear. Oblique coronal T2-weighted MR image demonstrates abnormally high signal intensity (*black arrow*) within the supraspinatus tendon (*t*) at its insertion on the greater tuberosity (*G*). Osteoarthritis of the acromioclavicular joint (*white arrowhead*) has also resulted in impingement on the tendon more proximally (*white arrow*) (*m = supraspinatus muscle; black arrowhead = superior glenoid labrum; s = subscapularis tendon*)

2. The brachialis muscle inserts on the ulnar tuberosity and anterior coronoid process of the ulna.

3. The normal carrying angle of the elbow is 165 degrees, with valgus angulation of the forearm with respect to the upper arm.

4. The Bowman angle describes the normal cubitus valgus in children, and it is used to detect abnormal varus or valgus angulation in the setting of supracondylar fracture; it is formed by two lines constructed on frontal radiographs: the long axis of the humeral shaft, and the tangent to the medial epicondylar metaphysis.

5. A discrepancy between the Bowman angle of the normal elbow and that of the traumatized elbow is considered significant if it exceeds 5 degrees; the angle decreases with varus angulation and increases with valgus angulation.

6. The anterior humeral line is constructed along the anterior humeral cortex on true lateral radiographs and should intersect the middle third of the capitellum; this line is used to detect abnormal dorsal or ventral angulation in the setting of supracondylar fracture.

7. The radiocapitellar line is constructed along the long axis of the proximal radial shaft on any radiographic projection and should intersect the capitellum; this line is used to detect dislocation of the radial head.

A

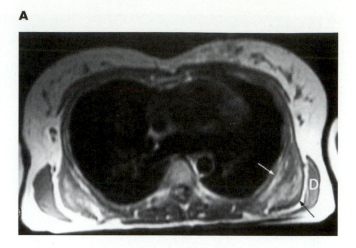

B

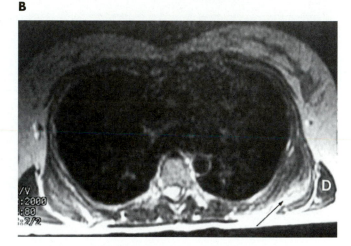

C

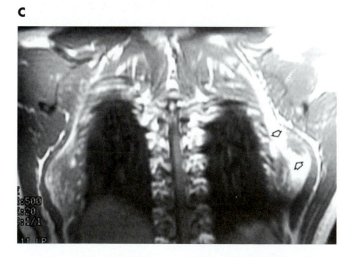

Fig. 1-62   Serratus anterior muscle tear. A, Transaxial T1-weighted MR image reveals enlargement and inhomogeneity of the affected muscle on the right (*arrows*). **B,** High signal intensity (*arrow*) is evident within the muscle on a corresponding T2-weighted image. **C,** Coronal T1-weighted image reveals retraction of the torn muscle away from the chest wall (*arrows*) (*D = latissimus dorsi muscle*).

A

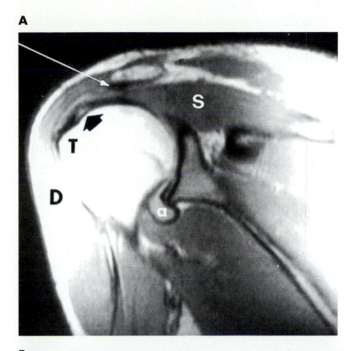

B

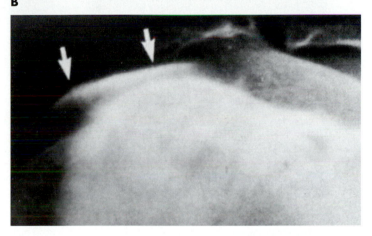

Fig. 1-63 Tendinitis or partial tear involving the inferior aspect of the supraspinatus tendon. **A,** Oblique coronal proton density–weighted MR image reveals abnormally high signal intensity (*black arrow*) within the tendon near its insertion on the greater tuberosity (*T*). A subacromial spur (*white arrow*) with potential for impingement is also evident (*S = supraspinatus muscle, D = deltoid muscle, a = axillary recess*). **B,** Glenohumeral arthrography is a more reliable method for distinguishing a partial tear on the undersurface of the tendon from tendinitis. In this case, contrast material intravasates into the tendon substance (*arrows*) but does not opacify the subacromial-subdeltoid bursa.

8. On lateral radiographs with 90 degrees of flexion, the anterior fat-pad is normally visualized in the coronoid fossa as a radiolucency with a straight margin, whereas the posterior fat-pad is not seen because it is hidden within the olecranon fossa; in the setting of elbow joint effusion, the margins of the anterior fat-pad become convex and the posterior fat-pad is also visualized as a radiolucency with convex margins.

9. The chevron refers to the normal trabecular pattern in the supracondylar region of the humerus on frontal radiographs.

10. The teardrop refers to the normal appearance of the distal humerus on lateral radiographs, caused by the olecranon and coronoid fossae; disruption of the neck of the teardrop can be a useful indicator of supracondylar fracture.

11. The radial tuberosity may simulate a lytic lesion when viewed en face, because of its normally thin cortex as compared with that of the surrounding bone.

12. The supratrochlear ossicle is a normal variant that may occur in a supratrochlear foramen; neither of these should be mistaken for a traumatic lesion.

13. The supracondylar process is a hooklike osseous excrescence arising from the anterior humeral cortex in the supracondylar region that may occasionally fracture, causing secondary injury to the adjacent median nerve.

14. There are six secondary ossification centers in the normal pediatric elbow; the order of appearance of these centers is as follows: capitellum, radial head, medial epicondyle, trochlea, olecranon, and lateral epicondyle.

15. The olecranon normally ossifies between the ages of 9 and 10 years and may be bipartite.

16. The capitellum normally ossifies at the age of 1 year.

17. The medial epicondyle normally ossifies between the ages of 5 and 7 years and is the last center to fuse.

18. The lateral epicondyle normally ossifies between the ages of 9 and 13 years and may simulate an avulsion fracture.
19. The radial head normally ossifies between the ages of 3 and 6 years.
20. The trochlea normally ossifies between the ages of 9 and 10 years and may appear fragmented.

PEARL

Order of appearance of secondary ossification centers of the elbow:

Center	Approximate age of ossification
Capitellum	1 year
Radial head	5 years
Medial epicondyle	7 years
Trochlea	9 years
Olecranon	10 years
Lateral epicondyle	11 years

The medial epicondyle is the last to fuse.

21. Because the trochlea never ossifies before the medial epicondyle, avulsion of the latter by the forearm flexor muscles with entrapment of the fragment within the elbow joint, simulating an ossified trochlea, should not be overlooked; the apparent presence of an ossified trochlea in the absence of an ossified medial epicondyle is thus diagnostic of avulsion and entrapment of the medial epicondyle.
22. The normal maturation sequence of the secondary ossification centers in the pediatric elbow can be remembered using the mnemonic CRITOE: capitellum, radial head, internal (medial) epicondyle, trochlea, olecranon, external (lateral) epicondyle.

PEARL

The mnemonic for remembering the order of ossification for the secondary centers of the elbow is CRITOE, which stands for capitellum, radial head, internal or medial epicondyle, trochlea, olecranon, and external or lateral epicondyle.

Fractures

1. Radial head fractures account for approximately 50% of all elbow fractures in adults and are the most common cause of a positive fat-pad sign in this age group.

2. Because approximately 50% of all radial head fractures are minimally displaced, specialized radiographic projections or fluoroscopy may be necessary for diagnosis (Figs. 1-64, 1-65).
3. The Essex-Lopresti injury involves a severely comminuted fracture of the radial head associated with subluxation of the distal radioulnar joint.
4. Fractures of the olecranon that occur distal to the insertion of the triceps tendon tend to be widely displaced.
5. The transcondylar fracture of the humerus tends to occur in older individuals with significant osteoporosis.

A

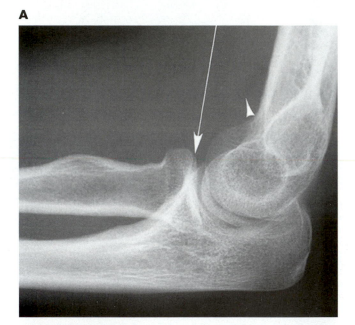

B

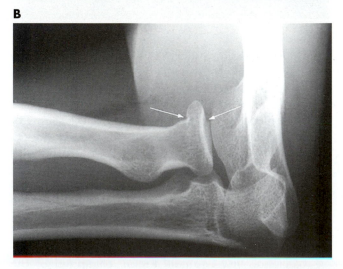

Fig. 1-64   Radial head fracture. **A,** Lateral radiograph demonstrates elevation of the anterior fat-pad (*arrowhead*), indicating joint effusion. A hairline fracture is suggested (*arrow*). **B,** A radial head–capitellar view is more definitive, clearly documenting a minimally displaced fracture (*arrows*).

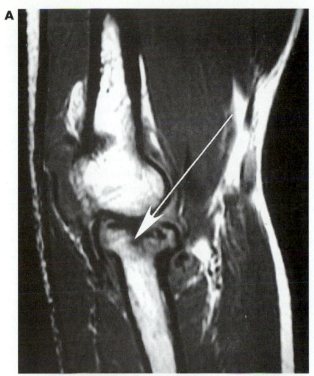

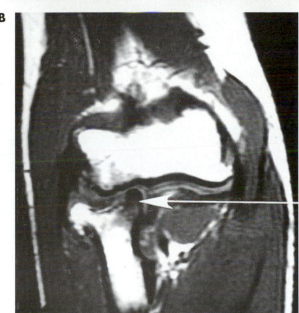

Fig. 1-65 Impacted radial head fracture with secondary osteonecrosis. **A,** Sagittal T1-weighted MR image demonstrates marked incongruity of the articular surface at the fracture site (*arrow*). **B,** Coronal T1-weighted image documents abnormally low signal intensity in the medial aspect of the radial head (*arrow*), indicating disruption of its blood supply.

6. The most common distal humeral fracture of adulthood is the intracondylar T or Y variety, with the vertical component of the injury usually involving the trochlear ridge; an important complication of this injury is diastasis of the humeral condyles.

7. Approximately 60% of all elbow fractures in children are of the supracondylar type.
8. Most supracondylar fractures of the humerus manifest dorsal angulation of the distal fragment, resulting in an anterior humeral line that does not cross the capitellum or intersects its anterior third.
9. Supracondylar fracture is the most common cause of an elbow joint effusion (positive fat-pad sign) in children.
10. Approximately 15% of all elbow fractures in children involve the lateral condyle.
11. Fractures of the lateral humeral condyle in children usually involve the capitellum and are type IV injuries in the Salter-Harris classification; the metaphyseal fragment is displaced posteriorly and distally by the extensor musculature, and the injury is best managed by internal fixation.
12. Approximately 10% of all elbow fractures in children are medial epicondylar avulsions, which may be complicated by entrapment of the avulsed fragment in the elbow joint.
13. Separation of the distal humeral physis in children is usually associated with medial displacement of the epiphyseal fragment and may resemble elbow joint dislocation, except that the radiocapitellar line is normal and displacement tends to be posterolateral with true dislocation.

PEARL

In widely displaced supracondylar fractures or in those in which treatment is delayed, there is a possibility of ischemia due to impingement or compression of the brachial artery and subsequent development of a Volkmann contracture. This is defined as contraction of the fingers and sometimes the wrist with loss of power after a severe injury to the elbow or improper use of a tourniquet. Fractures of the lateral epicondyle are the second most common elbow injury in childhood. When complete, the fractures are Salter-Harris type IV epiphyseal injuries. Entrapment of the medial epicondyle occurs when the epicondyle is avulsed and then drawn into the joint by traction. This can simulate a normal trochlea.

Dislocation

1. Approximately 85% of all elbow dislocations occur posterolaterally.
2. Elbow dislocation is frequently associated with radial head or coronoid process fracture, which may be complicated by entrapment of fragments within the joint.

3. Isolated radial head dislocation is rare in adults; more commonly, it is associated with fracture of the ulna in the Monteggia injury.
4. A frequent complication of elbow fracture-dislocation is myositis ossificans, which usually occurs on the ventral aspect of the joint in the brachialis muscle; the condition is more likely to occur in the setting of delayed reduction.
5. Volkmann's contracture involves ischemia of the forearm muscles with secondary contracture, which may be due to brachial artery occlusion associated with elbow dislocation or fracture of the distal humerus.
6. Congenital dislocation of the elbow is associated with radial head overgrowth and deformity of the radial head articular surface and is thus readily distinguished from traumatic dislocation.
7. The nursemaid elbow refers to subluxation of the radial head from the annular ligament, caused by sudden extension of the joint; radiographs are normal with the forearm pronated, and spontaneous reduction often occurs with supination of the forearm.

WRIST TRAUMA

Distal Forearm Anatomy

1. The tip of the radial styloid process lies more distal than the tip of the ulnar styloid process.
2. The distal radial articular surface normally tilts approximately 17 degrees toward the ulna.
3. The ulnar notch of the distal radius articulates with the head of the ulna at the distal radioulnar joint.
4. On lateral radiographs, the soft-tissue bulge of the pronator quadratus muscle may be an indirect sign of injury.
5. On the lateral projection, the distal radial articular surface is normally angled 10 to 15 degrees toward the volar aspect of the wrist.
6. On a true lateral radiograph, the dorsal surface of the distal ulna normally lies up to 3 mm posterior to the dorsal surface of the distal radius.
7. The ulna-plus variance is a relatively long ulna that bears increased axial loading and is associated with tears of the triangular fibrocartilage complex and impaction upon the proximal row of carpal bones.
8. The ulna-minus variance is a relatively short ulna that bears decreased axial loading and is associated with Kienböck's disease, owing to increased axial loading across the radioulnate articulation.

Carpal Anatomy

1. The scaphoid, lunate, triquetrum, and pisiform bones constitute the proximal carpal row of bones.
2. The trapezium, trapezoid, capitate, and hamate bones constitute the distal carpal row of bones.
3. On frontal radiographs, the parallel arcs described by Gilula are useful in excluding carpal bone malalignment and are constructed along the proximal margins of the bones of the proximal carpal row (scaphoid, lunate, triquetrum), along the distal margins of the bones of the proximal carpal row (scaphoid, lunate, triquetrum), and along the proximal margins of the bones of the distal carpal row (capitate, hamate).
4. Approximately 50% to 70% of the lunate width on a frontal projection normally articulates with the distal radius; this decreases to 25% in radial deviation and increases to 100% in ulnar deviation.
5. Inability to visualize the hook of the hamate bone (hamulus) in profile on a frontal radiograph indicates hypoplasia, fracture with malalignment, or ununited os hamuli proprium.
6. Scaphoid waist fractures are more readily visualized on films obtained with ulnar deviation than on those with radial deviation.
7. The most common accessory ossicle in the wrist lies adjacent to the ulnar styloid process and resembles an old ununited fracture.
8. The most common coalition in the wrist occurs between the lunate and triquetrum bones; when partial, it may simulate fracture.
9. On a true lateral radiograph, the most dorsal carpal surface is formed by the triquetrum bone.

PEARL

Wrist anatomy:
- The radius and ulna are separated distally by 1 to 2 mm.
- The ulnar styloid process is longer than the radial styloid process.
- The ulna and triquetrum bone are aligned dorsally, which is important in the evaluation of carpal or ulnar fractures and dislocations.
- The radial articular surface is tilted volarly and ulnarly.
- The proximal row of carpal bones is much more tightly bound to the radius by the radiocarpal ligaments than is the distal row; consequently, dislocation is more likely to occur between the distal row and the proximal row than between the proximal row and the radius.

Distal Forearm Fractures

1. Distal forearm fractures are the most common injuries in the entire musculoskeletal system and are usually caused by falling on an outstretched hand.

2. The most common distal forearm injuries among patients between the ages of 4 and 10 years are transverse fractures of the distal radial and ulnar metaphyses; these may be complete or incomplete (buckle, torus, or greenstick).

3. The most common distal forearm injury among patients between the ages of 11 and 16 years is a Salter-Harris fracture of the distal radial physis; these tend to be type II fractures with dorsal displacement of the distal fragment.

4. The most common injury to the wrist among patients between the ages of 17 and 40 years is the scaphoid fracture.

5. The most common distal forearm injury among patients over 40 years of age is the Colles fracture; it is more common among women because of the high prevalence of postmenopausal osteoporosis, and it may be associated with fractures of the vertebral bodies, proximal femur, or proximal humerus.

6. The Colles fracture is characterized by dorsal angulation and impaction of the distal fragment; variable features include intraarticular extension and associated ulnar styloid process fracture.

7. As seen on postreduction radiographs, potential complications of Colles fracture include loss of normal volar tilt of the distal radial articular surface, shortening of the radius with secondary ulna-plus variance, and intraarticular diastasis with loss of normal ulnar tilt of the distal radial articular surface.

8. The Smith or reverse Colles fracture of the distal radius is associated with volar angulation of the distal fragment relative to the proximal fragment.

9. The Barton fracture is an intraarticular fracture of the dorsal margin of the distal radius, with the carpus maintaining normal relationship to the dorsal fragment; the injury is unstable and requires either external or internal fixation.

10. The reverse Barton fracture is an intraarticular fracture of the volar margin of the distal radius.

11. The Hutchinson or chauffer fracture is an intraarticular fracture of the radial styloid process.

PEARL

Distal forearm fractures:
- Smith fracture is a distal radius fracture that tends to collapse into palmar flexion (volar angulation of the distal fragment).
- Colles fracture is a fracture of the distal radius with dorsal displacement (dorsal angulation of the distal fragment).
- Chauffer's fracture or Hutchinson's fracture is a radial styloid process fracture.
- Barton fracture is a fracture of the dorsal rim of the distal radius with dorsal displacement of the carpus.

Dislocation of the Distal Radioulnar Joint

1. The condition may occur as an isolated injury.
2. The Galeazzi fracture-dislocation consists of distal radioulnar joint dislocation and fracture of the radial diaphysis.

PEARL

The Galeazzi fracture-dislocation of the forearm consists of a mid- to distal radial diaphyseal fracture and dislocation of the ulnar head. Monteggia fracture-dislocation is a fracture of the middle to proximal third of the ulnar diaphysis and radial head dislocation.

3. The condition may be associated with Colles fracture.
4. On lateral radiographs, the injury is suggested by dorsal displacement of the ulnar head with respect to the ulnar notch of the distal radius.
5. Computed tomography of both the injured and normal sides is the best procedure for establishing the diagnosis, and the examination requires only a small number of transaxial images in both pronation and supination.

Carpal Injuries

1. Carpal injuries occur approximately 10 times less frequently than forearm injuries.
2. Carpal injuries are rare in patients under 12 years of age.
3. Approximately 65% of all carpal injuries are scaphoid fractures.
4. Approximately 10% of all carpal injuries are dislocations or fracture-dislocations.
5. Dorsal chip fractures of the carpus usually involve the triquetrum.
6. Conventional or computed tomography is useful for complete characterization of carpal injuries.

Carpal Fractures

1. Approximately 70% of all scaphoid fractures involve the waist of the bone and are displaced.
2. The early diagnosis of a suspected radiographically occult scaphoid fracture is best accomplished by skeletal scintigraphy or conventional or computed tomography, although it is more cost effective to obtain a follow-up radiograph.
3. Fractures of the distal pole of the scaphoid tend to heal rapidly.

4. Potential complications of scaphoid waist fractures include delayed union, nonunion, ischemic necrosis of the proximal pole, and premature degenerative joint disease (Fig. 1-66).

PEARL

Proximal scaphoid fractures are more likely to cause osteonecrosis of the proximal pole than are distal fractures. Of all scaphoid injuries, 10% to 15% result in osteonecrosis.

5. Union of scaphoid waist fractures may be delayed for up to 2 years, and approximately 90% of all scaphoid fractures eventually unite.
6. Acute fracture of the lunate bone is rare, although the bone is vulnerable to osteonecrosis (Kienböck disease) secondary to trauma, which is characterized radiographically by sclerosis and collapse and is associated with ulna-minus variance (Fig. 1-67).
7. Fractures of the triquetrum are usually of the chip variety and are best visualized on lateral radiographs because they tend to affect the dorsal aspect of the bone.

A

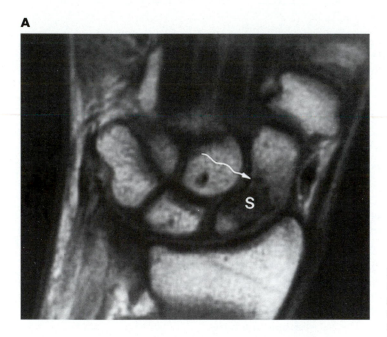

B

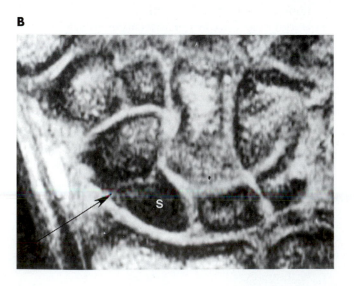

Fig. 1-66 Ischemia of the proximal pole following scaphoid waist fracture. Coronal T1-weighted (**A**) and gradient-echo (**B**) MR images from two different patients demonstrate abnormally low signal intensity (*s*) involving most of the bone proximal to the fractures (*arrows*). These findings may be transient and resolve following fracture healing with revascularization, or they may progress to osteonecrosis with nonunion and collapse.

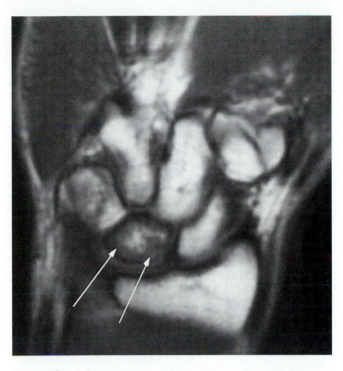

Fig. 1-67 Early diagnosis of Kienböck disease. Coronal T1-weighted MR image demonstrates mottled low signal intensity within the marrow of the lunate bone (*arrows*), compatible with osteonecrosis.

---PEARL---

The triquetrum is the second most common site of fracture in the carpus. The most common fracture is an avulsion from the dorsal surface at the attachment site of the radiotriquetral and ulnotriquetral ligaments. Fracture of the triquetrum is thus associated with dorsal carpal instability. The dorsal fracture is usually visible only on the lateral view of the wrist. Vertical and horizontal fractures of the triquetrum can occur from compression between the hamate bone and distal ulna. Horizontal fractures may also occur in perilunate dislocations, in which case there may also be a capitate and/or scaphoid fracture.

8. Isolated fractures of the capitate bone are rare.
9. Complex carpal fracture-dislocations may be associated with transverse capitate fractures, and the proximal fragment of the bone tends to rotate up to 180 degrees.
10. Visualization of the base of the hook of the hamate (hamulus), where most fractures of this structure occur, is best accomplished by use of a supinated oblique radiograph, conventional lateral tomography, or computed tomography (rather than by use of the carpal tunnel radiograph, which is difficult to obtain in injured patients).

---PEARL---

The hook of the hamate bone (hamulus) protrudes into the hypothenar eminence; injury is often due to playing racquet sports. The ulnar nerve lies outside the carpal tunnel, coursing between the hamulus and the pisiform bone, so it is frequently injured with fracture of the former. The median nerve lies on the radial side of the carpal tunnel, so an isolated median nerve palsy will not occur with hamulus fracture, but a generalized carpal tunnel syndrome may develop. Complex perilunate dislocation with capitate dislocation may fracture the body of the hamate bone, but not the hamulus.

11. Fracture of the proximal pole of the hamate bone may occur as part of a lunate or perilunate fracture-dislocation complex.

Carpal Dislocation

1. Carpal dislocations on the ulnar side of the wrist tend to be more severe than those on the radial side.
2. Carpal dislocations on the radial side of the wrist tend to be more common than those on the ulnar side.

---PEARL---

Keys to understanding carpal injuries:
- Knowing that the articular surface of the distal radius is the anatomic point of reference in both diagnosing and classifying carpal injuries
- Understanding the three arcs of the proximal and distal rows of carpal bones on the frontal projections

Three arcs can be drawn on the frontal projection along the contour of the carpal bones. A break in the continuation of one of these arcs strongly suggests a traumatic lesion at the site of the break.

Carpal dislocations are basically of two types: lunate and perilunate. With a perilunate dislocation, the lunate bone maintains a normal relationship with the distal radius. The remainder of the carpus does not maintain a normal relationship with the distal radius and is generally dorsally dislocated. With a lunate dislocation, the lunate bone does not maintain a normal relationship with the distal radius although the remainder of the carpus does.

When there is a fracture as well as dislocation, the terminology becomes more confusing. A navicular fracture accompanying a perilunate dislocation is termed a transscaphoid perilunate dislocation.

3. Scapholunate dissociation involves less severe trauma than perilunate dislocation.

4. Perilunate dislocation involves less severe trauma than lunate dislocation.

5. Scapholunate dissociation is more common than perilunate dislocation.

6. Perilunate dislocation is more common than lunate dislocation.

7. Fracture-dislocations of the wrist are more common than pure dislocations.

8. Disruption of the first and second carpal bone alignment arcs on a frontal radiograph indicates carpal dislocation.

9. Disruption of the normal coaxial arrangement between the distal radius, lunate bone, and capitate bone on a lateral radiograph indicates carpal dislocation.

PEARL

On a lateral radiograph of a normal wrist, the capitate bone, lunate bone, and radius lie along the same axis. These relationships are disrupted in lunate and perilunate dislocations. In perilunate dislocation, there is a dislocation of the capitate bone (usually dorsal) relative to the lunate bone, which maintains normal alignment with the radius. In lunate dislocation, the lunate bone is displaced in a volar direction, while the central axis of the capitate bone remains in line with the radius. Rotary scaphoid subluxation is caused by disruption of the dorsal radiocarpal ligaments. Increased distance between the lunate bone and scaphoid constitutes the "Terry Thomas" sign, and the scapholunate angle on lateral radiographs is increased.

10. In a perilunate dislocation, the capitate bone is usually dislocated dorsally with respect to the lunate, which maintains its normal articulation with the distal radius.

PEARL

Fracture of the navicular waist is commonly associated with perilunate dislocation. In perilunate dislocation, the distal row of carpal bones is displaced dorsally. Because the scaphoid bridges the proximal and distal rows of carpal bones, the force of a perilunate dislocation is transmitted to the waist of the scaphoid. Seventy-five percent of perilunate dislocations are associated with navicular fracture.

11. In a lunate dislocation, the lunate bone is dislocated in a volar direction and rotated approximately 90 degrees with respect to both the distal radius and capitate bone; the capitate bone remains aligned with the distal radius but migrates proximally.

12. In a midcarpal dislocation, the lunate bone tilts in a volar direction but is not dislocated with respect to the distal radius, whereas the capitate bone is slightly dislocated dorsally with respect to the lunate bone; the overall appearance is intermediate between those of the perilunate and lunate dislocations.

Carpal Instability Patterns

1. Carpal instability patterns result from ligamentous injury.

2. The normal space between the scaphoid and lunate bone measures 2 mm.

3. On lateral radiographs, the scapholunate angle normally measures 30 to 60 degrees whereas the normal lunate-capitate angle measures less than 20 degrees.

4. Videofluoroscopy is indicated in patients with normal conventional radiographs of the wrist who have midcarpal pain and/or audible carpal clicks; the procedure should include radial deviation, ulnar deviation, dorsiflexion, palmar flexion, and fist-clenching maneuvers.

5. Scapholunate dissociation is best visualized on the frontal projection with a clenched fist and radial or ulnar deviation, which drives the capitate bone proximally.

6. Instability between the lunate and capitate bones is best visualized on either frontal or lateral projections with longitudinal traction, hand flexion, and pressure applied to the scaphoid tuberosity.

7. Instability between the hamate and triquetrum bones is indicated by shifting of the proximal carpal row from palmar flexion to dorsiflexion as the carpus is moved from radial to ulnar deviation.

8. Radiographic signs of scapholunate dissociation include a scapholunate gap exceeding 2 mm (frontal view), foreshortening and annular appearance of the scaphoid (frontal view), and increased scapholunate angle without abnormal lunate volar flexion (lateral view).

9. Dorsiflexion carpal instability or dorsal intercalated segmental instability (DISI) is characterized by lunate dorsiflexion with increased lunate-capitate angle, palmar flexion of the scaphoid with increased scapholunate angle, and variable scapholunate dissociation.

PEARL

Dorsal intercalated segmental instability (DISI) of the wrist is due to disruption of the dorsal radiocarpal ligaments. Sixty percent of cases are associated with scaphoid fracture. Ten percent are associated with dorsal chip fractures of the triquetrum bone. DISI may also occur secondary to purely ligamentous injury, such as a rotary subluxation of the scaphoid or a perilunate dislocation, or to other carpal fractures. DISI is diagnosed on the lateral radiograph where there is a change in the relative orientation of the scaphoid and lunate bones. The vertical axes of the lunate bone and scaphoid usually form a 30- to 60-degree angle. In DISI the angle is greater than 60 degrees.

10. Volar flexion carpal instability or volar intercalated segmental instability (VISI) is characterized by capitate dorsiflexion and lunate volar flexion with abnormal lunate-capitate angle and a decreased scapholunate angle (less than 30 degrees).
11. VISI is less common than DISI in the setting of trauma, but the former is a relatively frequent manifestation of rheumatoid arthritis.
12. In ulnar translocation, the entire carpus migrates toward the ulna and away from the radial styloid process; the condition is usually secondary to inflammatory disease such as rheumatoid arthritis.
13. Dorsal carpal subluxation is dorsal subluxation of the entire carpus relative to the distal radius; the condition is usually secondary to residual fracture deformity, such as Colles fracture with relative dorsal angulation of the healed distal fragment.

Carpal Arthrography and Magnetic Resonance Imaging

1. Comprehensive evaluation of the wrist by contrast arthrography should involve up to three separate injections; the three sites (in the usual sequential order) are the midcarpal compartment, radiocarpal compartment, and distal radioulnar joint.
2. The pisiform-triquetral joint normally communicates with the radiocarpal compartment.
3. Opacification of the radiocarpal compartment resulting from a midcarpal injection indicates disruption of the trapezioscaphoid, scapholunate, lunate-triquetral, and/or hamate-triquetral ligaments; identification of the exact site of tear is best accomplished using digital subtraction arthrography or videofluoroscopy.

4. Opacification of the distal radioulnar joint resulting from a radiocarpal injection indicates disruption of the triangular fibrocartilage complex.
5. Potential causes of disruption of the interosseous ligaments and triangular fibrocartilage complex include acute or repetitive trauma, age-related degeneration, and inflammatory arthropathy.
6. For midcarpal arthrography the needle tip is optimally positioned between the scaphoid and capitate using an angled dorsal approach.
7. For radiocarpal arthrography the needle tip is optimally positioned between the radius and scaphoid using an angled dorsal approach; the scapholunate ligament should specifically be avoided.
8. Magnetic resonance imaging of the wrist in patients with carpal tunnel syndrome may demonstrate tendon sheath thickening, median nerve enlargement, thickening of the flexor retinaculum, or carpal tunnel encroachment by soft-tissue masses (including ganglia) or bone (Fig. 1-68).
9. Magnetic resonance imaging is the procedure of choice for the early diagnosis of Kienböck disease and ischemic necrosis of the proximal pole of the scaphoid following waist fracture (see Figs. 1-66, 1-67).
10. Although magnetic resonance imaging can identify tears of the interosseous ligaments and triangular fibrocartilage complex, three-compartment contrast arthrography remains the most definitive examination in this regard.

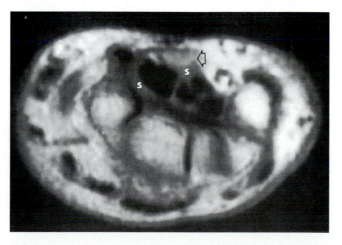

Fig. 1-68 Carpal tunnel syndrome. Transaxial T1-weighted MR image demonstrates thickening of the flexor tendon sheaths (*s*), resulting in compression of the median nerve (*arrow*).

HAND TRAUMA

Baseball or Mallet Finger

1. Forced flexion of an extended finger is the mechanism of injury.
2. An avulsion fracture may occur at the insertion site of the common extensor tendon at the dorsal aspect of the distal phalangeal base.
3. The fracture is an intraarticular avulsion fracture.
4. The injury may be limited to the extensor tendon, with no associated fracture.
5. The patient demonstrates flexion and volar subluxation of the distal interphalangeal joint, sometimes with hyperextension of the proximal interphalangeal joint.

PEARL

Mallet (baseball) finger occurs when a blow to the extended finger flexes the distal interphalangeal joint. It causes a partial or complete tear of the extensor tendon or avulsion fracture of the dorsal base of the distal phalanx, resulting in a flexed distal interphalangeal joint without ability to extend (fixed flexion or "mallet" appearance). Avulsion fracture is present in only 25% of cases.

Injury to the Flexor Digitorum Profundus Insertion

1. Forced hyperextension of a flexed finger is the mechanism of injury.
2. Clinically, the patient has the presenting symptom of inability to flex the distal interphalangeal joint.
3. The injury may be limited to the flexor digitorum profundus tendon, with no associated fracture, in which case a stress radiograph is required for diagnosis.
4. The fracture line, if present, extends into the distal interphalangeal joint.
5. The fracture fragment may be retracted proximally to the volar aspect of the middle phalanx.

Boutonnière or Buttonhole Deformity

1. The mechanism of injury involves flexion of the proximal interphalangeal joint and extension of the distal interphalangeal joint.
2. The middle fibers of the extensor tendon mechanism normally insert at the dorsal aspect of the middle phalangeal base.

3. With rupture of the middle fibers of the extensor tendon mechanism, the lateral fibers migrate toward the palm, forming a buttonhole through which the proximal interphalangeal joint flexes.
4. The distal interphalangeal joint is pulled into hyperextension by the lateral fibers of the extensor tendon mechanism.
5. Avulsion of the dorsal aspect of the middle phalangeal base is rare because the injury is usually limited to the middle fibers of the extensor tendon mechanism.
6. The buttonhole or boutonnière deformity is a slowly evolving consequence of the acute injury and can be prevented by early diagnosis and splinting.

Volar Plate Fracture

1. The volar plate is a fibrocartilaginous structure that crosses the metacarpophalangeal and proximal interphalangeal joints.
2. The distal attachment of the volar plate is the strongest, which explains the tendency for avulsion fracture, as opposed to pure soft-tissue injury, at this site.
3. Hyperextension of the proximal interphalangeal joint is the mechanism of injury.
4. The fracture line, if present, extends into the proximal interphalangeal joint.
5. The injury may occur in association with the mallet or baseball finger.

PEARL

A volar plate injury is usually a forced hyperextension injury at the proximal interphalangeal joint, with avulsion or ligamentous separation at the base of the middle phalanx, usually accompanied by dorsal subluxation or dislocation of the middle phalanx. There may be an associated hyperflexion deformity of the distal interphalangeal joint. Volar plate fractures may also occur at metacarpophalangeal joints. Dorsal avulsions (baseball finger) occur most commonly at the distal interphalangeal joint. Occasionally these fractures may involve more than 25% of the articular surface, in which case dislocation is common.

Gamekeeper Thumb

1. The injury affects the ulnar collateral ligament of the first metacarpophalangeal articulation.
2. Similar injuries may affect any of the metacarpophalangeal or interphalangeal joints of the hand but are less common.

3. Valgus stress across the thumb is responsible for the injury.
4. Modern examples of the injury are commonly encountered in skiers.
5. The fracture line, if present, extends into the first metacarpophalangeal joint.
6. Frontal stress radiographs are necessary for diagnosis if the injury is limited to the collateral ligament with no associated avulsion fracture, and comparison with the uninjured side is essential.

PEARL

The Gamekeeper thumb involves injury of the ulnar collateral ligament of the first metacarpophalangeal joint; an avulsion fracture is often associated. A stress view showing angulation greater than 10 degrees indicates that the injury is unstable. Less than 10 degrees of angulation can be a normal variation resulting from ligamentous laxity.

Metacarpal Fractures

1. The Bennett injury accounts for approximately one third of all first metacarpal fractures.
2. The Bennett fracture-dislocation includes an oblique intraarticular fracture of the first metacarpal base and dorsal subluxation or dislocation of the first metacarpal bone.
3. The smaller fragment of a Bennett fracture retains its normal articulation with the trapezium.
4. The Bennett injury of the first metacarpal bone is usually treated with open reduction and internal fixation.
5. The Rolando injury consists of a comminuted Bennett type of fracture, a separate dorsal fracture fragment, and dorsal subluxation of the first metacarpal bone.
6. The Rolando injury of the first metacarpal bone is usually treated by closed reduction with casting, because pin fixation is not successful in the setting of significant comminution.
7. The boxer fracture is characterized by relative volar angulation of the distal fragment that is best appreciated on the lateral view.
8. The boxer fracture usually involves the fifth metacarpal bone.
9. Stable reduction of the boxer fracture is difficult because of comminution along the volar aspect; hence, healing seldom occurs in anatomic alignment.
10. Extraarticular metacarpal fractures including the boxer injury are usually treated with closed reduction.

PEARL

The Bennett injury is an oblique, noncomminuted fracture of the base of the first metacarpal bone. It is usually treated with open reduction and internal fixation. The Rolando injury is a comminuted version of the Bennett fracture and is much less common. Open reduction and internal fixation are performed if only a few fragments are present; otherwise it requires closed reduction. Both Bennett and Rolando fractures are intraarticular.

Dislocations

1. Occult dislocation is likely in the setting of fracture involving the distal carpal row or metacarpal bases.
2. The carpometacarpal joints are an uncommon site of dislocation because of an interlocking zigzag pattern of articulation that provides stability.
3. Approximately half of all carpometacarpal dislocations involve the fifth metacarpal bone.
4. Dorsal metacarpal dislocation occurs in approximately two thirds of cases.
5. Despite the interlocking articulation of the second metacarpal base with the trapezoid, capitate, and third metacarpal bone, approximately 25% of all carpometacarpal dislocations involve this bone.
6. Approximately 80% of carpometacarpal dislocations involve multiple metacarpal bones, usually the fifth plus one other.
7. Hyperextension usually results in dorsal dislocation of phalanges.
8. Volar plate or joint capsule interposition complicating phalangeal dislocation prevents reduction and is evident radiographically only when a sesamoid is interposed between the articular surfaces.
9. Phalangeal dislocation generally produces volar plate disruption and/or avulsion.
10. Phalangeal shaft fractures may be associated with dislocations in the hand and usually manifest angulation and rotational malalignment that is better evaluated clinically than radiographically.

BATTERED CHILD SYNDROME (Fig. 1-69)

Strongly Suggestive Radiographic Signs

1. Multiple fractures in different stages of healing, in the absence of dysplastic or metabolic bone disease (Fig. 1-70).

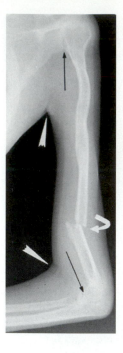

Fig. 1-69   Brachial plexus injury at birth. In a skeletally mature patient, typical findings include generalized hypoplasia of the osseous structures in the upper extremity, along with muscle atrophy and flexion contractures (*arrowheads*). The gracile nature of the long bone diaphyses and disuse osteopenia render them susceptible to fracture (*curved arrow*), and joint space narrowing (*straight arrows*) is indicative of cartilage atrophy. Brachial plexus injuries can also occur in child abuse.

2. Metaphyseal avulsion or bucket-handle fractures secondary to periosteal traction in children under 5 years of age (Fig. 1-71).

PEARL

Looseness of the periosteum in children is due to the relative paucity of and shortness of Sharpey fibers. The tighter attachment of the periosteum at the bone ends is the reason for typical bucket-handle metaphyseal fractures in child abuse. The periosteum has two well-defined layers, with the inner being osteogenic.

3. Fractured ribs, despite the normal plasticity of the pediatric thoracic cage, particularly at the costovertebral and costochondral junctions (Figs. 1-70, 1-72).

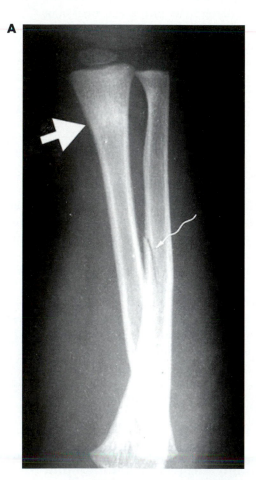

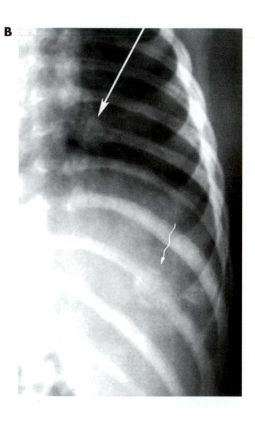

Fig. 1-70   Battered child syndrome. Multiple fractures in different stages of healing are characteristic. **A,** Subacute distal radius fracture (*straight arrow*) and acute ulnar diaphyseal fracture (*wavy arrow*). **B,** Healed rib fracture (*straight arrow*) and subacute rib fracture (*wavy arrow*) in the chest wall.

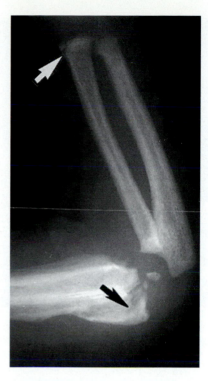

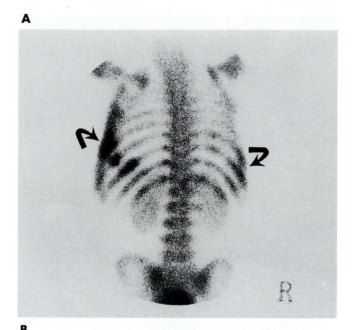

A

B

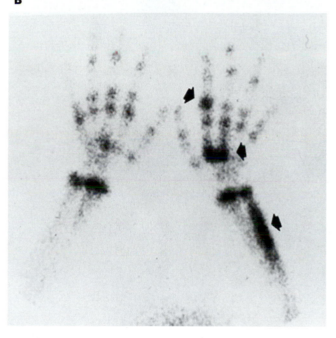

Fig. 1-71   Metaphyseal corner fractures (*arrows*) of differing ages in battered child syndrome. Similar findings may occur in Menkes' kinky-hair syndrome and in scurvy.

Syndrome Features

1. Skeletal trauma occurs in only 20% to 40% of child abuse cases, and strongly suggestive radiographic findings are unusual.
2. The long-bone diaphyses and skull are the most common sites of fracture. Spinal fractures (usually thoracic and lumbar) are rare, as are pelvic fractures (usually of the pubic rami) (Figs. 1-73, 1-74).

PEARL

Along with nonaccidental trauma, causes of vertebral fractures in children include Cushing syndrome, leukemia, metastatic neuroblastoma, eosinophilic granuloma, osteogenesis imperfecta, and idiopathic juvenile osteoporosis.

3. Skeletal trauma in battered children usually occurs during the first 2 years of life.
4. Approximately 90% of skull fractures in child abuse occur before the age of 2 years.
5. Skeletal trauma is usually associated with clinical evidence of physical injury as opposed to sexual abuse or neglect.

Fig. 1-72   Multiple rib and upper extremity fractures (*arrows*) in battered child syndrome. Bone scintigraphy is a sensitive technique for detecting occult injuries in affected patients, although physeal fractures may be difficult to identify because of the normally increased activity of the growth plates.

6. Skeletal scintigraphy in the battered child syndrome is particularly useful for the detection of acute occult fractures of the diaphyses, ribs, and spine (Fig. 1-72).
7. Radiographic skeletal survey should be the initial imaging strategy in clinically suspected cases of child abuse.
8. Metaphyseal avulsion fractures are difficult to detect by skeletal scintigraphy because of the normally increased activity of open physes.

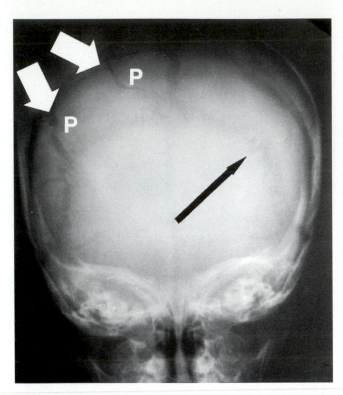

Fig. 1-73 Acute skull fractures (*white arrows*) of the right parietal bone (*P*) in child abuse. Sharp margins and linear orientation distinguish these abnormalities from sutures, which have an interdigitated appearance (*black arrow = coronal suture*).

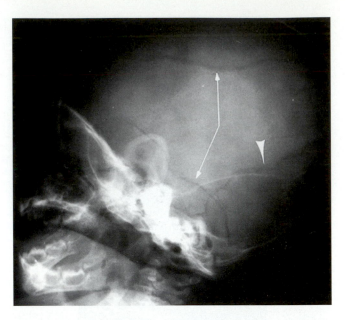

Fig. 1-75 Battered child syndrome. Multiple linear skull fractures (*arrows*) are readily distinguished from the lambdoid suture with its interdigitated appearance (*arrowhead*).

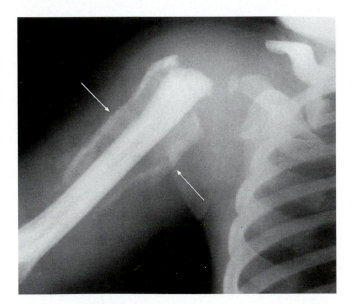

Fig. 1-74 Battered child syndrome. Extensive subperiosteal hemorrhage has resulted in prominent circumferential ossification (*arrows*) involving the proximal humerus. Additional possible causes of this appearance include scurvy and hemophilia.

9. Fractures in abused children are far more common in long bones than in the hands or feet.
10. Skull fractures in battered children tend to be linear rather than depressed (Fig. 1-75).

11. Nonspecific periosteal reaction may be the only radiographic manifestation of child abuse (Fig. 1-76).
12. Traumatic metaphyseal cupping and growth disturbances are nonspecific manifestations of battered child syndrome.
13. Visceral injury or subdural hematoma may accompany skeletal trauma in child abuse.
14. Fracture yield from skeletal survey of affected children is extremely low.

PEARL

Birth trauma to the skeleton:
- The bone most commonly fractured during delivery is the clavicle.
- Fractures of the femur and humerus at birth are usually associated with breech delivery.
- Diaphyseal fractures of the femur and humerus are usually spiral and involve the middle third of the diaphysis.
- Metaphyseal and apophyseal fractures of the long bones at birth are relatively common.

POSTTRAUMATIC MYOSITIS OSSIFICANS

1. Myositis ossificans is the heterotopic formation of nonneoplastic bone and cartilage in muscle and other soft tissues.

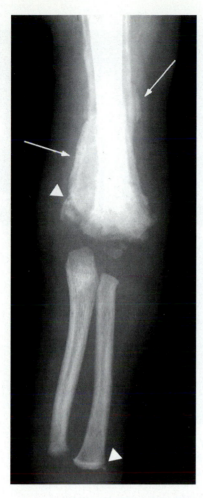

Fig. 1-76 Battered child syndrome. Extensive subperiosteal hemorrhage (*arrows*) with metaphyseal corner fractures (*arrowheads*) is virtually pathognomonic of this condition, although scurvy could produce a similar combination of findings.

PEARL

Posttraumatic calcification or ossification:

- Myositis ossificans is a condition in which heterotopic bone, and sometimes cartilage, forms in muscles, tendons, and fascia.
- Myositis ossificans may simulate parosteal osteosarcoma radiologically and histologically.
- Bone erosions, although unusual, may be observed.
- A thin soft-tissue cleavage plane is commonly observed between the calcific or ossific deposits and the adjacent bone.
- The thigh is a common site of involvement.

---

2. The thighs and elbows are common sites of involvement.
3. During the first 4 weeks of its evolution, the histologic appearance of the central zone in myositis ossificans simulates that of malignancy; during weeks 4 through 8, a characteristic zonal pattern is observed and features a cellular center surrounded by immature osteoid, which progressively organizes into mature bone.
4. During the initial 2 weeks following trauma, a warm and painful soft-tissue mass may be observed clinically (Fig. 1-77).
5. During the third and fourth weeks in the evolution of myositis ossificans, the mass contains flocculent amorphous densities and may be associated with periosteal reaction and/or apparent attachment to the underlying bone, simulating early parosteal osteosarcoma (Fig. 1-78).
6. The maturation of posttraumatic myositis ossificans progresses from the periphery to the center of the mass.
7. Maturation of myositis ossificans is associated with progressive reduction in mass size.
8. A radiolucent zone frequently separates mature myositis ossificans from the cortex of the underlying bone.
9. Serial skeletal scintigraphy or magnetic resonance imaging may be used to determine the state of maturation of the process.
10. Surgical resection of myositis ossificans should be undertaken only after maturation of the process; otherwise, a high rate of recurrence can be expected.
11. Posttraumatic periostitis or the blocker exostosis is broad-based myositis ossificans attached to a bone, resulting from subperiosteal hemorrhage. The underlying cortex is normal, unlike the situation with true exostosis (osteochondroma).
12. Up to 50% of patients with neurologic deficit following trauma exhibit ossification in atrophied muscles, tendons, and ligaments around paralyzed joints (Figs. 1-79, 1-80).
13. Myositis ossificans associated with burns can occur at sites remote from the areas of thermal trauma.
14. The radiographic differential diagnosis for posttraumatic myositis ossificans should include parosteal osteosarcoma, osteochondroma, tumoral calcinosis, periosteal osteosarcoma, and juxtacortical chondroid lesions.
15. The condition most likely to simulate posttraumatic myositis ossificans radiographically is parosteal osteosarcoma, and three important features (best demonstrated by computed tomography) distinguish these two lesions: (1) myositis ossificans matures (ossifies) from the periphery to the center, whereas the opposite is true for parosteal osteosarcoma; (2) myositis ossificans gradually decreases in size with maturation, whereas the opposite is true for parosteal osteosarcoma, and (3) parosteal osteosarcoma is always attached to the underlying

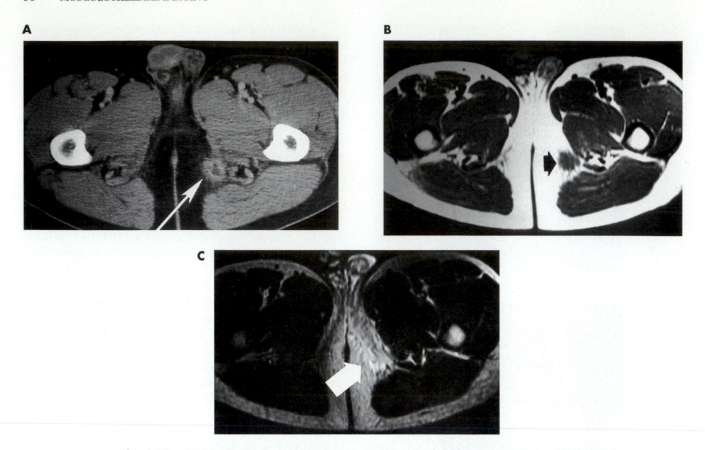

Fig. 1-77   Evolving posttraumatic myositis ossificans. **A,** CT demonstrates a poorly defined high-density soft-tissue mass (*arrow*) in the groin region. **B,** Transaxial T1-weighted MR image reveals a low–signal intensity mass (*arrow*) in the subcutaneous fat. **C,** Corresponding T2-weighted image exhibits diffuse high signal intensity in the area (*arrow*). The findings are nonspecific and simulate those of cellulitis, soft-tissue neoplasia, panniculitis, and other inflammatory processes.

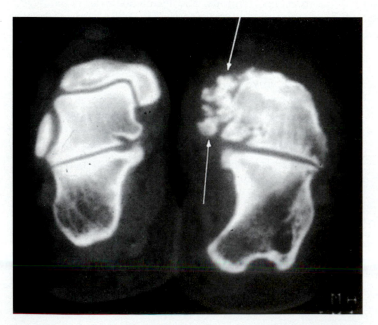

Fig. 1-78   Posttraumatic myositis ossificans. CT demonstrates soft-tissue ossification (*arrows*) medial to the talus. Additional potential causes of periarticular ossification include neurologic deficit, burns, previous surgery, and fibrodysplasia ossificans progressiva.

A

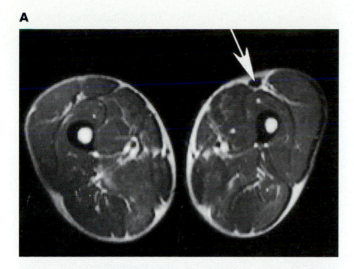

B

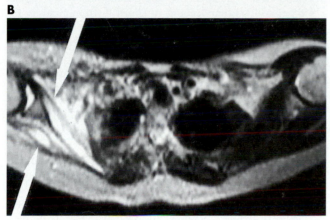

Fig. 1-79 Muscle atrophy documented by transaxial MRI. **A,** Rectus femoris (*arrow*), following trauma. **B,** Upper extremity musculature (*arrows*), following brachial plexus injury.

bone, whereas myositis ossificans is separated from the underlying bone by a radiolucent zone in most cases.

PEARL

Soft-tissue alterations following trauma:
- Dystrophic calcification may occur.
- Pellegrini-Stieda syndrome involves bone formation in the medial collateral ligament of the knee.
- Myositis ossificans is distinguished from ossifying neoplasms by its usual physical separation from adjacent bones and its relatively radiolucent center.
- Bone formation is well recognized following joint replacement and in surgical scars.
- Calcification of the pinna of the ear, also observed in endocrine disorders and relapsing polychondritis, can occur.

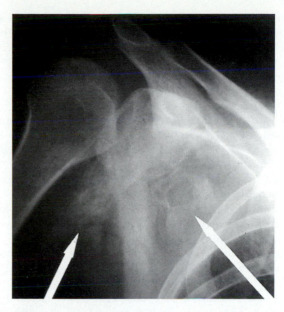

Fig. 1-80 Paraarticular ossification in quadriplegia. Extensive immature bone formation (*arrows*) is evident around the glenohumeral joint. Differential diagnosis should take into consideration other causes of neurologic deficit, burns, posttraumatic myositis ossificans, previous surgery, and fibrodysplasia ossificans progressiva.

SUGGESTED READINGS

1. Bellah RD et al: Low back pain in adolescent athletes: detection of stress injury to the pars interarticularis with SPECT, Radiology 180:509, 1991.
2. Berquist TH, editor: Imaging of orthopedic trauma, ed 2, New York, 1992, Raven Press.
3. Bluemke DA, Fishman EK, and Scott WW Jr: Skeletal complications of radiation therapy, Radiographics 14:111, 1994.
4. Cope R: Dislocations of the sternoclavicular joint, Skeletal Radiol 22:233, 1993.
5. Daffner RH and Pavlov H: Stress fractures: current concepts, AJR 159:145, 1992.
6. DeSmet AA and Graf DK: Meniscal tears missed on MR imaging: relationship to meniscal tear patterns and anterior cruciate ligament tears, AJR 162:905, 1994.
7. Egund N and Wingstrand H: Legg-Clavé-Perthes disease: imaging with MR, Radiology 179:89, 1991.
8. Falchook FS et al: Rupture of the distal biceps tendon: evaluation with MR imaging, Radiology 190:659, 1994.
9. Fitzgerald SW et al: Distal biceps tendon injury: MR imaging diagnosis, Radiology 191:203, 1994.
10. Friedman L, Yong-Hing K, and Johnston GH: Use of coronal computed tomography in the evaluation of Kienböck's disease, Clin Radiol 44:56, 1991.
11. Harris JH Jr and Edeiken-Monroe B: The radiology of acute cervical spine trauma, ed 2, Baltimore, 1987, Williams & Wilkins.
12. Harris JR Jr et al: Radiologic diagnosis of traumatic occipitovertebral dissociation, AJR 162:881, 1994.

13. Hayes CW, Conway WF, and Dainel WW: MR imaging of bone marrow edema pattern: transient osteoporosis, transient bone marrow edema syndrome, or osteonecrosis, Radiographics 13:1001, 1993.

14. Hayes CW et al: Seat belt injuries: radiologic findings and clinical correlation, Radiographics 11:23, 1991.

15. Hendrix RW et al: Fracture of the spine in patients with ankylosis due to diffuse skeletal hyperostosis: clinical and imaging findings, AJR 162:899, 1994.

16. Holder LE: Bone scintigraphy in skeletal trauma, Radiol Clin North Am 31:739, 1993.

17. Hu SS and Pashman RS: Spinal instrumentation: evolution and state of the art, Invest Radiol 27:632, 1992.

18. Kaplan PA et al: Occult fracture patterns of the knee associated with anterior cruciate ligament tears: assessment with MR imaging, Radiology 183:835, 1992.

19. Kaste SC, Hopkins KP, and Jenkins JJ III: Abnormal odontogenesis in children treated with radiation and chemotherapy: imaging findings, AJR 162:1407, 1994.

20. Kiss ZS, Kahn KM, and Fuller PJ: Stress fractures of the tarsal navicular bone: CT findings in 55 cases, AJR 160:111, 1993.

21. Klein MA: MR imaging of the ankle: normal and abnormal findings in the medial collateral ligament, AJR 162:377, 1994.

22. Kode L et al: Evaluation of tibial plateau fractures: efficacy of MR imaging compared with CT, AJR 163:141, 1994.

23. Lewis SL et al: Pitfalls of bone scintigraphy in suspected hip fractures, BR J Radiol 64:403, 1991.

24. Martinez CR et al: Evaluation of acetabular fractures with two- and three-dimensional CT, Radiographics 12:227, 1992.

25. Meaney JFM and Carty H: Femoral stress fractures in children, Skeletal Radiol 21:173, 1992.

26. Meyers SP and Wiener SN: Magnetic resonance imaging features of fractures using the short tau inversion recovery (STIR) sequence: correlation with radiographic findings, Skeletal Radiol 20:499, 1991.

27. Musculoskeletal trauma, Radiol Clin North Am 27(5):839, 1989.

28. Oshima M et al: Initial stage of Legg-Calvé-Perthes disease: comparison of three-phase bone scintigraphy and SPECT with MR imaging, Eur J Radiol 15:107, 1992.

29. Palmer WE, Brown JH, and Rosenthal DI: Labral-ligamentous arthrography, Radiology 190:645, 1994.

30. Rang M: Children's fractures, ed 2, Philadelphia, 1983, JB Lippincott Co.

31. Riddervold HO: Easily missed fractures, Radiol Clin North Am 30:475, 1992.

32. Rogers LF and Hendrix RW: Radiology of skeletal trauma, ed 2, New York, 1992, Churchill Livingstone, Inc.

33. Rogers LF and Poznanski AK: Imaging of epiphyseal injuries, Radiology 191:297, 1994.

34. Rubin DA et al: MR diagnosis of meniscal tears of the knee: value of fast spin-echo versus conventional spin-echo pulse sequences, AJR 162:1131, 1994.

35. Scott WW Jr et al: Subtle orthopedic fractures: teleradiology work station versus film interpretation, Radiology 187:811, 1993.

36. Slone RM, MacMillan M, and Montgomery WJ: Spinal fixation. I. Principles, basic hardware, and fixation techniques for the cervical spine, Radiographics 13:341, 1993.

37. Slone RM, MacMillan M, and Montgomery WJ: Spinal fixation. III. Complications of spinal instrumentation, Radiographics 13:797, 1993.

38. Slone RM et al: Orthopedic fixation devices, Radiographics 11:823, 1991.

39. Slone RM et al: Spinal fixation. II. Fixation techniques and hardware for the thoracic and lumbosacral spine, Radiographics 13:521, 1993.

40. Sonin AH et al: Posterior cruciate ligament injury: MR imaging diagnosis and patterns of injury, Radiology 190:455, 1994.

41. Tigges S, Stiles RH, and Roberson JR: Complications of hip arthroplasty causing periprosthetic radiolucency on plain radiographs, AJR 162:1387, 1994.

42. Tirman PFJ et al: Association of glenoid labral cysts with labral tears and glenohumeral instability: radiologic findings and clinical significance, Radiology 190:653, 1994.

43. Tuckman GA et al: Radial tears of the menisci: MR findings, AJR 163:395, 1994.

44. Weber WN et al: Lateral tibial rim (segond) fractures: MR imaging characteristics, Radiology 180:731, 1991.

45. Weissman BMW and Sledge CB: Orthopedic radiology, Philadelphia, 1986, WB Saunders Co.

46. Zobel MS et al: Pediatric knee MR imaging: pattern of injuries in the immature skeleton, Radiology 190:397, 1994.

Infection

PYOGENIC ORGANISMS (Fig. 2-1)

1. The most common organism encountered in infections of the musculoskeletal system (both in children and adults) is *Staphylococcus aureus*.

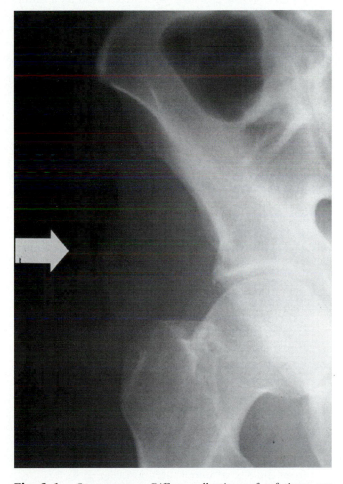

Fig. 2-1 Gas gangrene. Diffuse collections of soft-tissue gas (*arrow*) are evident in the gluteal region. Gas-forming organisms include clostridia, streptococci, *Bacteroides* organisms, and coliform bacteria.

PEARL

Staphylococcal arthritis is characterized by rapid destruction of bone and cartilage; loss of articular cortex is a distinctive finding. However, 8 to 10 days are required for the development of osseous alterations that can be seen on conventional radiographs. Sequestration occurs in the presence of secondary osteomyelitis. Tuberculous arthritis is a less aggressive process. Neonatal hip dislocation secondary to joint effusion and increased intraarticular pressure is a complication of staphylococcal arthritis. *S. aureus* occurs at all ages, but younger patients (under 4 years) may be affected by *Hemophilus influenzae* and older children (over 10 years) may acquire *Neisseria* strains as well.

2. In intravenous drug abusers, *S. aureus* is most commonly responsible for musculoskeletal infections, although *Serratia* and *Pseudomonas* organisms are encountered with greater frequency than in the general population.
3. In patients with sickle cell anemia, *S. aureus* is most commonly responsible for musculoskeletal infections, although *Salmonella* organisms are encountered with greater frequency than in the general population.
4. The most common organism encountered in neonatal infections of the musculoskeletal system is streptococcus (group B).

MECHANISMS OF MUSCULOSKELETAL INFECTION

1. The three basic mechanisms responsible for osteomyelitis and septic arthritis are hematogenous seeding, spread from adjacent soft-tissue infection, and direct inoculation secondary to penetrating trauma (Fig. 2-2; Box 2-1).
2. Human bite injuries incurred during fistfights are most commonly responsible for osteomyelitis involving metacarpals and phalanges and/or septic arthritis involving metacarpophalangeal and proximal interphalangeal joints (Fig. 2-3).
3. In the hand or foot, soft-tissue infection may spread along tendon sheaths and fascial planes, resulting in osteomyelitis and/or septic arthritis that is distant from the site of initial infection (Fig. 2-4).
4. Soft-tissue infection involving the terminal pulp of a digit is known as a felon and may progress to osteomyelitis of the distal phalangeal tuft.
5. Stubbed toe injury in children may result in an open physeal fracture and osteomyelitis of the distal phalanx of the hallux, because of close proximity between the toenail bed and periosteum.

6. Decubitus ulcers in paraplegic or bedridden patients may lead to osteomyelitis, particularly of the ischial tuberosities and sacrum (Fig. 2-5).
7. Staging of soft-tissue infections is best accomplished using magnetic resonance imaging (Fig. 2-6).

Box 2-1   Osteomyelitis from an Adjacent Source of Infection

- Soft-tissue swelling
- Mass formation
- Obliteration of tissue planes
- Periostitis
- Cortical erosion
- Cortical lucency and destruction
- Bone lysis

HEMATOGENOUS OSTEOMYELITIS (Table 2-1)

1. Typical localization of infection occurs according to division of age groups as follows: infants under 12 months of age, children over 12 months of age, and adults.

PEARL

Osteomyelitis:
- In hematogenous osteomyelitis, sequestration is less frequent in infants than among children or adults.
- *Salmonella* species are common offending organisms in bone infections among patients with sickle cell anemia; however, *S. aureus* remains the most common agent.
- The bony changes of osteomyelitis are radiographically apparent 10 to 14 days following clinical onset.
- The metaphysis is the most frequently involved region.
- In children under 1 year of age and adults, osteomyelitis may extend into the epiphysis because of vascular bridging of the growth zone.

2. Anastomosis between terminal metaphyseal and epiphyseal vessels is seen in adults.
3. Slipped epiphysis is a relatively frequent complication in infants under 12 months of age.
4. Septic arthritis commonly occurs secondary to osteomyelitis in infants under 12 months of age and adults.
5. Terminal metaphyseal vessels manifest sluggish flow and do not cross into epiphyses in children over 12 months of age.
6. Multifocal sites of involvement are relatively common in infants under 12 months of age.
7. The spine and small bones are more commonly affected than tubular bones in adults.

A

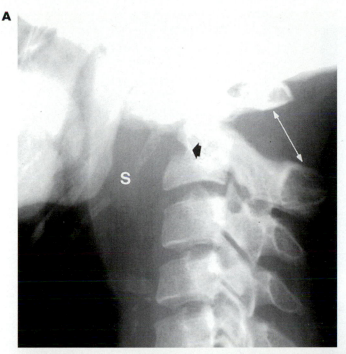

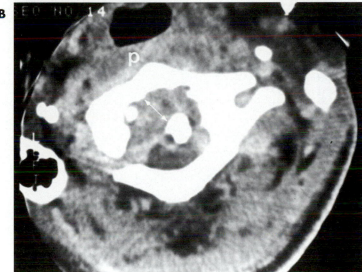

B

Fig. 2-2   Retropharyngeal tuberculosis with spinal involvement. **A,** Lateral radiograph demonstrates marked prevertebral soft-tissue swelling (*s*), along with osseous erosion (*black arrow*) and atlantoaxial subluxation (*white double-headed arrow*). **B,** CT confirms prevertebral soft-tissue swelling (*p*) and quantifies the severity of atlantoaxial dislocation (*double-headed arrow*). Additional possible causes of this appearance are other retropharyngeal infections, rheumatoid arthritis, seronegative spondyloarthropathies, and crystal deposition disease.

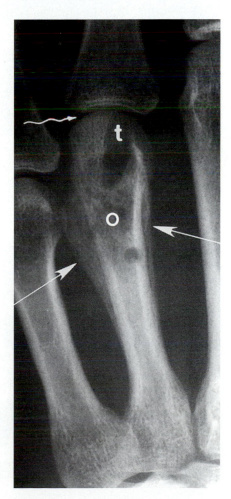

Fig. 2-3   Human bite infection. Typical findings of osteomyelitis include diffuse osteolysis (*o*) and periosteal reaction (*straight arrows*). Metacarpophalangeal joint space narrowing (*wavy arrow*) indicates associated septic arthritis; the two processes communicate via an intraosseous tract (*t*).

A

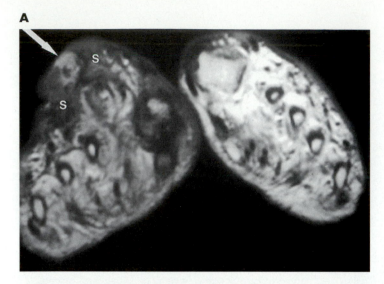

B

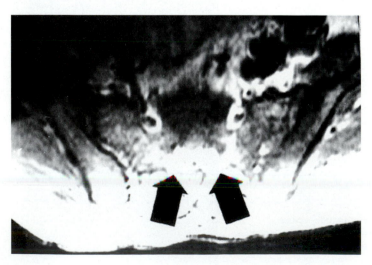

Fig. 2-4   Diabetic foot abscess. **A** and **B,** Proton density–and T2-weighted transaxial MR images demonstrate a focal collection of pus (*arrows*) in an area of extensive soft-tissue swelling (*S*).

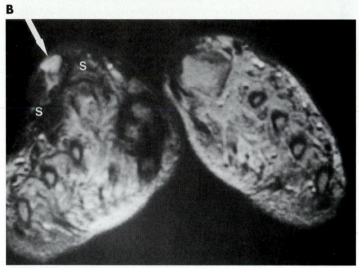

Fig. 2-5   Osteomyelitis of the sacrum. Transaxial T2-weighted MR image reveals diffuse high signal intensity within the sacral marrow (*arrows*). Other possible causes of this appearance include neoplastic disease, acute infarction, and insufficiency fracture; hence clinical correlation is extremely important in the interpretation of musculoskeletal MRI studies.

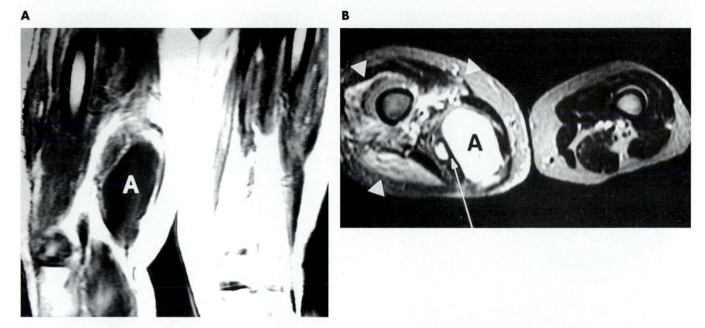

Fig. 2-6   Staging of thigh abscess by use of MRI prior to percutaneous drainage. **A,** Coronal T1-weighted image demonstrates a well-encapsulated mass of low signal intensity (*A*) in the medial aspect of the right thigh. **B,** Transaxial T2-weighted image reveals high signal intensity and septation (*arrow*) within the mass (*A*), as well as inflammation in the adjacent soft tissues (*arrowheads*).

8. Approximately 70% of cases involve the femoral or tibial metaphyses in children over 12 months of age.

Table 2-1	Hematogenous osteomyelitis of tubular bones		
	Infant	**Child**	**Adult**
Localization	Metaphyseal with epiphyseal extension	Metaphyseal	Epiphyseal
Involucrum formation	Common	Common	Not common
Sequestration	Common	Common	Not common
Joint involvement	Common	Not common	Common
Soft-tissue abscess	Common	Common	Not common
Pathologic fracture	Not common	Not common	Common*
Fistulae	Not common	Variable	Common

*In neglected cases

9. Growth deformity is a relatively frequent complication in infants under 12 months of age.
10. The physeal plate is violated by communicating vessels that link the metaphyseal and epiphyseal vessels in infants under 12 months of age.
11. Phagocytosis is relatively sparse in the metaphyses in children over 12 months of age.
12. Epiphyseal involvement is rare in children over 12 months of age (Fig. 2-7).
13. Langerhans cell histiocytosis or metastatic neuroblastoma may be simulated radiographically in infants under 12 months of age.
14. Tubular bones of the lower extremities are most commonly affected in children over 12 months of age.
15. Epiphyseal involvement is relatively common in infants under 12 months of age and adults (Fig. 2-7).
16. Septic arthritis is a rare complication in children over 12 months of age.
17. Infection limited to the metaphyses is the most common pattern of involvement in children over 12 months of age.

PEARL

Multiple foci of osteomyelitis can be seen in children with associated sickle cell disease, diabetes, or chronic granulomatous disease. Sickle cell disease is the most common cause of multiple foci of osteomyelitis in children. Chronic granulomatous disease of childhood is characterized by normal leukocytic phagocytosis, with inability to destroy the ingested organisms. In this disease, hematogenous osteomyelitis tends to involve small bones of the hands and feet, with minimal soft-tissue swelling and atypical destructive bony changes without sequestra.

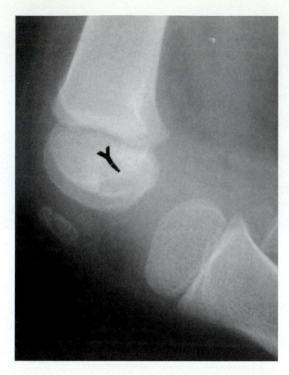

Fig. 2-7   Epiphyseal seeding of hematogenous infection in a young child. A well-defined lytic lesion (*arrow*) is identified in the femoral condyle. Differential diagnosis should include consideration of chondroblastoma, osteochondritis dissecans, and eosinophilic granuloma, although epiphyseal localization of the last is uncommon.

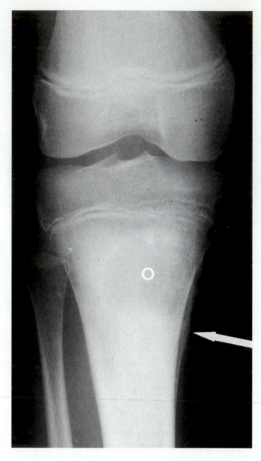

Fig. 2-8   Acute hematogenous osteomyelitis. Characteristic findings include metaphyseal localization, poorly defined osteolysis (*O*), and periosteal reaction (*arrow*).

ACUTE OSTEOMYELITIS (Box 2-2)

1. Radiographic findings do not become evident until 1 to 2 weeks following the onset of infection.
2. The initial radiographic manifestation of acute osteomyelitis is indistinctness or obscuring of normal fat planes in adjacent soft tissue.
3. Destruction of cancellous bone precedes that of cortical bone in hematogenous infections.
4. Periosteal reaction and endosteal scalloping are usually evident by 2 weeks after the onset of infection (Fig. 2-8).

PEARL

Inflammatory cellular infiltration is responsible for the periosteal elevation seen radiographically in osteomyelitis. This elevation can sometimes appear as a Codman triangle. Although this is one of the earliest signs, it may be delayed for 1 to 2 weeks after the infection has been seeded hematogenously.

5. A serpiginous pattern of osteolysis is virtually pathognomonic of osteomyelitis.
6. Acute osteomyelitis is best distinguished from malignant neoplasia by the more rapid rate of progression of the former.
7. A sequestrum is a necrotic bone fragment separated from viable bone by radiolucent granulation tissue and appears radiodense as compared to normal bone; it serves as a nidus for continuing infection and the development of chronic osteomyelitis.
8. An involucrum is caused by periosteal elevation and reactive new bone formation in the viable tissue around a sequestrum; it occurs more commonly in children than in adults because of the relatively loose attachment of the periosteum in the former.
9. The early diagnosis of acute osteomyelitis is best made by the use of skeletal scintigraphy or magnetic resonance imaging (Figs. 2-9, 2-10, 2-11).

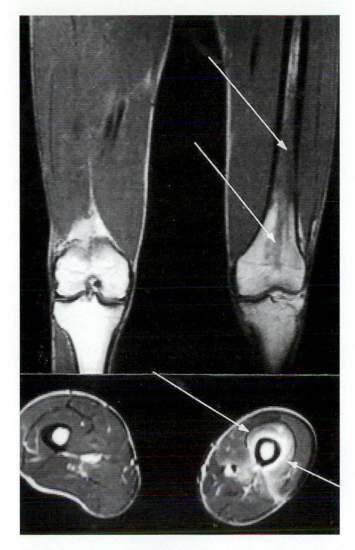

Fig. 2-9 Acute hematogenous osteomyelitis of the femur. A T1-weighted coronal MR image (*top*) demonstrates marrow with abnormally low signal intensity (*arrows*) within the femoral diaphysis and metaphysis. A T2-weighted transaxial image (*bottom*) documents abnormally high signal intensity (*arrows*) surrounding the affected bone.

Box 2-2 Radiographic Features of Acute Osteomyelitis

Soft-tissue swelling	Fistulae
Mass formation	Osteoporosis
Obliteration of tissue planes	Bone lysis
Single or multiple radiolucencies	Cortical radiolucency
	Periostitis
Cortical or medullary involvement	Sequestration
Lesions with surrounding sclerosis	Involucrum formation

SUBACUTE OSTEOMYELITIS

1. The Brodie abscess is a radiolucent focus of subacute osteomyelitis that is surrounded by a well-defined sclerotic margin (Fig. 2-12).
2. The radiographic appearance of subacute osteomyelitis is generally less aggressive than that of acute osteomyelitis.
3. The Brodie abscess most commonly occurs in the metaphysis of a child, and the most frequently affected bone is the tibia (Figs. 2-13, 2-14).
4. In the setting of subacute osteomyelitis, body temperature and erythrocyte sedimentation rate are frequently normal, in contrast to the situation with acute osteomyelitis.
5. The differential diagnosis of a typical-appearing Brodie abscess should include consideration of eosinophilic granuloma.
6. The Brodie abscess is occasionally located in the cortex of a bone, in which case periosteal reaction and prominent reactive sclerosis are observed radiographically (Fig. 2-15).
7. The differential diagnosis of an atypical-appearing Brodie abscess should include consideration of osteoid osteoma and healing stress fracture.

CHRONIC OSTEOMYELITIS (Box 2-3)

1. Radiographically, chronic osteomyelitis is characterized by prominent cortical thickening and a mixed pattern of osteosclerosis and osteolysis.
2. Radiographic signs of activity in chronic osteomyelitis include immature periostitis, draining sinus tracts, sequestra, and change in overall radiographic appearance over time (Figs. 2-16, 2-17; Box 2-4).
3. Sclerosing osteomyelitis of Garré is a variant of chronic osteomyelitis in which the radiographic appearance is dominated by osteosclerosis.
4. Technetium 99m–methylenediphosphonate bone scanning, gallium-67 citrate scanning, and indium-111 leukocyte scanning are frequently employed in combination for the imaging evaluation of chronic osteomyelitis.
5. Tropical ulcer is a form of chronic osteomyelitis that begins as an adjacent soft-tissue infection and may undergo malignant degeneration to squamous cell carcinoma; the benign lesion resembles osteoid osteoma radiographically, and the most commonly affected bones are the tibia and fibula (Box 2-5).

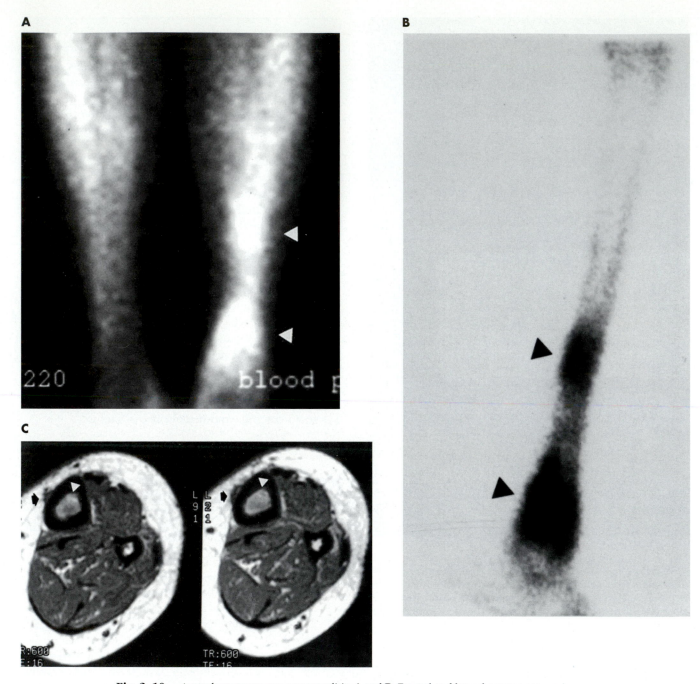

Fig. 2-10 Acute hematogenous osteomyelitis. **A** and **B,** Frontal and lateral gamma camera images reveal abnormal increased activity in the distal tibia (*arrowheads*). **C** and **D,** Sequential transaxial T1- and T2-weighted MR images demonstrate abnormal signal behavior within the tibial marrow (*arrowheads*) and adjacent soft tissues (*arrows*). **E,** Coronal inversion-recovery image exhibits high signal intensity within the tibial marrow (*m*) and adjacent soft tissues (*arrows*). **F,** Following intravenous administration of gadolinium-DTPA, sequential transaxial T1-weighted images with fat suppression reveal abnormal marrow (*arrowheads*) and soft-tissue (*arrows*) enhancement. The findings are nonspecific and could also occur in neoplastic disease and stress fracture.

6. Chronic recurrent multifocal osteomyelitis is a predominantly sclerotic form of chronic osteomyelitis that affects children and exhibits a predilection for involvement of the clavicle (Fig. 2-18).

7. Conventional tomography and computed tomography are the preferred techniques for the identification of sequestra in chronic osteomyelitis.

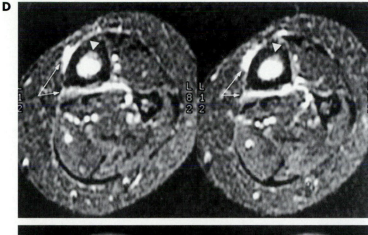

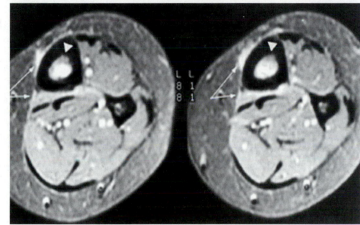

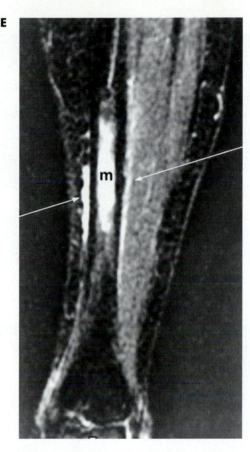

Fig. 2-10, cont'd.   For legend see opposite page.

Box 2-3   Chronic Osteomyelitis

Mixed osteosclerosis and osteolysis
Cortical thickening
Sequestrum (necrotic bone) within cloaca
Involucrum (thickened new bone surrounding cloaca)
Draining sinuses
Brodie abscess (circumscribed lytic area surrounded by
   sclerotic bone; most common in tibia)
Sclerosing osteomyelitis of Garré; little or no osteolysis
Chronic recurrent multifocal osteomyelitis; sclerotic
   lesions in children
Tropical ulcer (most common in tibia or fibula; cortical
   thickening with eventual cancellization;
   potential degeneration to squamous carcinoma)

**Box 2-4   Radiographic Signs of Activity
   in Chronic Osteomyelitis**

Change from previous film
Ill-defined areas of osteolysis
Thin, linear periostitis
Sequestration

Box 2-5   Tropical Ulcer

Characterized by acute infection involving *Bacillus
   fusiformis, Borrelia vincentii;* superinfection by
   many other organisms occurs in long-standing cases
Causes over 50% of skin cancer cases in Sri Lanka,
   Indonesia, Java, Papua New Guinea, Madagascar,
   West Africa, and Uganda
90% of cases occur below knee
Malignant degeneration occurs in 10% of cases, but is
   rare prior to age 20
Complications are osseous and soft-tissue deformity,
   flexion contractures, lymphedema, articular damage,
   osteoporosis, and gas gangrene
"Ivory osteoma" stage of disease resembles diaphyseal
   osteoid osteoma radiographically
Begins as minor cutaneous wound, progressing to
   bone involvement as secondary infection develops

A

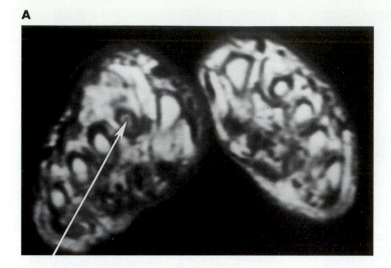

B

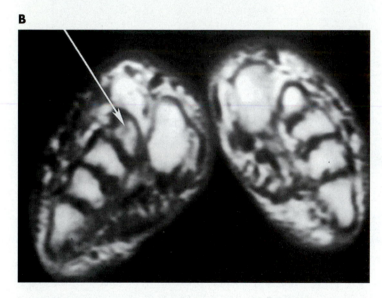

Fig. 2-11 MRI use for the early diagnosis of osteomyelitis secondary to soft-tissue infection in the diabetic foot.
A, T1-weighted transaxial image reveals low signal intensity within the marrow of the second metatarsal shaft (*arrow*).
B and **C,** T1- and T2-weighted images demonstrate extension of the process to the level of the metatarsal base (*arrows*). Conventional radiographs demonstrated no cortical bone erosion in this instance.

C

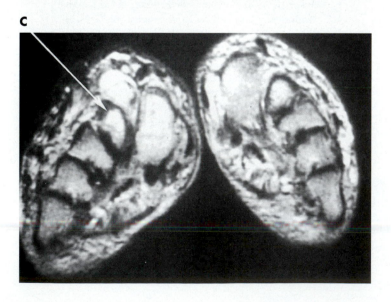

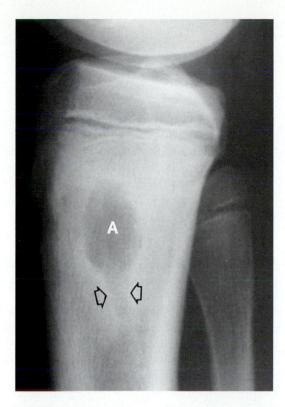

Fig. 2-12   A well-defined osteolytic process (*A*) in the tibial metaphysis with extension via an intraosseous tract (*arrows*) is virtually diagnostic of Brodie abscess.

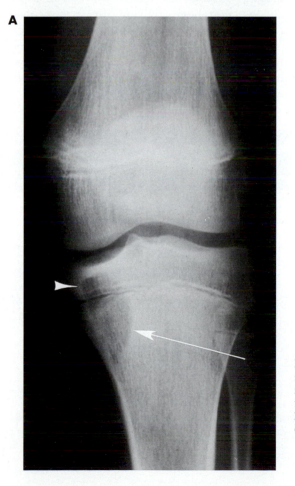

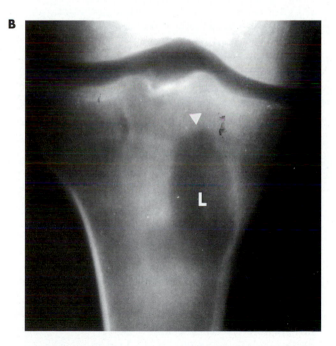

Fig. 2-13   Brodie abscess with epiphyseal extension. **A,** Frontal radiograph demonstrates a well-defined osteolytic lesion (*arrow*) centered in the proximal tibial metaphysis (most common site) that has transgressed the open growth plate to involve the epiphysis (*arrowhead*). **B,** Frontal tomogram of a similar lesion (*L*) with epiphyseal extension (*arrowhead*) in an adult.

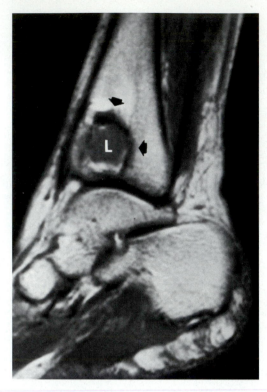

Fig. 2-14   Brodie abscess in the distal tibia. Sagittal T1-weighted MR image demonstrates a low–signal intensity lesion (*L*) with a sclerotic margin (*arrows*) adjacent to the tibiotalar articulation.

OSTEOMYELITIS: DIAGNOSIS AND COMPLICATIONS (Fig. 2-19; Table 2-2)

1. Increasing lesion–to–background activity ratios on 3- or 4-phase bone scans are more indicative of osteomyelitis than of cellulitis (Fig. 2-20).
2. Contrast enhancement within soft-tissue masses associated with osteomyelitis is usually nonuniform and serpiginous on computed tomographic or magnetic resonance images.
3. Delayed appearance of radiographic abnormalities may occur among patients with osteomyelitis who have had prior antibiotic therapy.
4. The addition of gallium-67 citrate scanning or indium-111 leukocyte imaging to skeletal scintigraphy improves specificity for osteomyelitis to approximately 80% (Fig. 2-21).
5. Because of the normally increased radiopharmaceutical uptake associated with open physes, skeletal scintigraphy is less sensitive for the detection of osteomyelitis in children than in adults; symmetric positioning for side-to-side comparison and pinhole collimation are essential to optimize sensitivity in skeletally immature patients.
6. Early radiographic abnormalities are seldom evident in the setting of infected total joint replacement, and therefore aspiration arthrography or combined

radionuclide studies are required for diagnosis; large bursal cavities and sinus tracts are characteristically observed with the former, although culture of aspirated fluid is the definitive test.

7. The final aspect of a 4-phase skeletal scintigram is a 24-hour delayed image.
8. The sensitivity of skeletal scintigraphy for the diagnosis of osteomyelitis exceeds its specificity.
9. Computed tomography affords optimal sensitivity for the detection of gas within bone or soft tissue; magnetic resonance imaging cannot distinguish gas from calcium because neither exhibits significant signal intensity.
10. Amyloid deposition may occur in the setting of chronic active osteomyelitis.
11. Potential complications of osteomyelitis include pathologic fracture and septic arthritis with secondary osteoarthritis.
12. The development of squamous cell carcinoma in a draining sinus tract usually occurs 10 to 20 years after the onset of chronic osteomyelitis.
13. Osteomyelitis in children may be complicated by slipped epiphyses and growth deformity.
14. Despite numerous major advances in diagnostic imaging, reliable distinction between neuropathic and infectious bone and joint disease in the diabetic foot remains a difficult problem; the two processes may give similar findings with conventional radiography, skeletal scintigraphy, gallium-67 citrate imaging, indium-111 leukocyte scanning, and magnetic resonance imaging (Figs. 2-22, 2-23).

Table 2-2 Radionuclide evaluation of osseous and soft-tissue infection			
Agent	Cellulitis	Acute osteomyelitis	Chronic osteomyelitis
Technetium phosphates	Early scans show increased uptake; later scans are normal	Early and late scans show increased uptake (scans in early acute osteomyelitis may reveal "cold" areas)	Scans may remain positive even in inactive disease
Gallium-67 citrate or indium-11 leukocytes	Increased uptake	Increased uptake	Increased uptake in areas of active disease

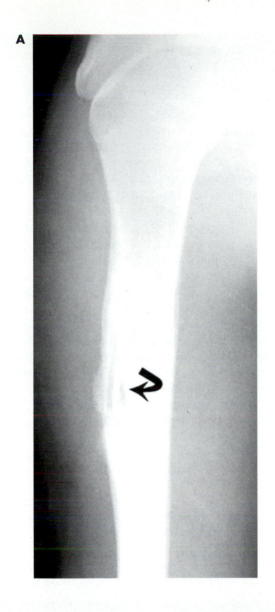

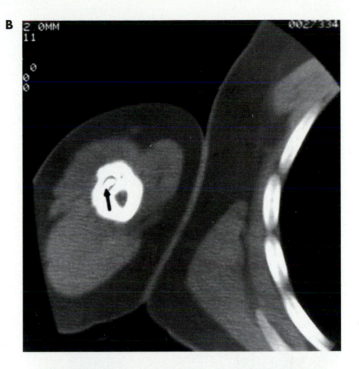

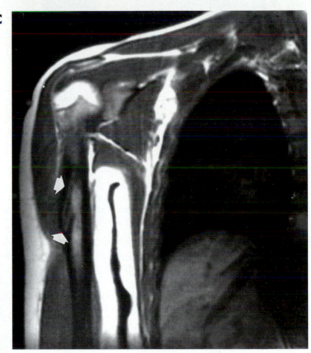

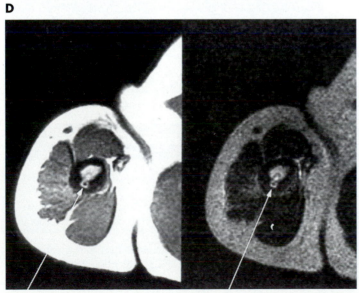

Fig. 2-15   Cortical abscess with sequestration.
A, Radiography demonstrates areas of cortical lucency
(*arrow*) subjacent to the deltoid tubercle. **B,** CT documents
a sequestrum (*arrow*) within one of the radiolucent areas.
C, Coronal T1-weighted MR image demonstrates the
intracortical sequestrum within an area of relatively low
signal intensity (*arrows*). **D,** Proton density– and
T2-weighted transaxial images confirm the cortical abscess
with central sequestration (*arrows*). Other possible causes of
this appearance include osteoid osteoma and juxtacortical
chondroma.

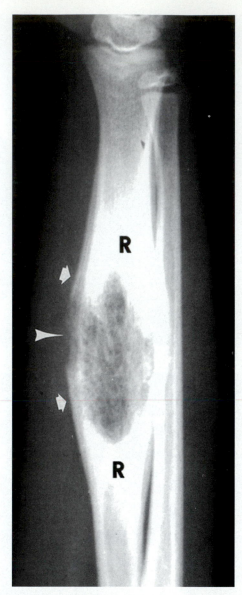

Fig. 2-16   Active chronic osteomyelitis. An expansile lytic process involving the midradial diaphysis is associated with reactive sclerosis (*R*), immature "onion skin" periostitis, and cortical breakthrough (*arrowhead*) resulting in Codman triangles (*arrows*). Other possible causes of this appearance include Ewing sarcoma and other round cell neoplasms of childhood.

15. Osteomyelitis complicating external or internal fixation pin placement is characterized by periostitis and an enlarging poorly defined radiolucent zone around the pin. Sterile motion of a pin is characterized by a uniform, well-defined radiolucent zone around the pin.
16. The annular sequestrum sign is a circular fragment of sclerotic bone within a poorly marginated and widened pin tract, and it is diagnostic of osteomyelitis.

TUBERCULOUS AND FUNGAL OSTEOMYELITIS (Figs. 2-24, 2-25; Boxes 2-6, 2-7, 2-8)

1. As compared to pyogenic osteomyelitis, tuberculous osteomyelitis tends to have a slower rate of progression and less reaction by the host bone (Figs. 2-26, 2-27).
2. Radiographically, the lesions of tuberculous osteomyelitis may be indistinguishable from those of fungal infection (Figs. 2-28, 2-29).
3. Tuberculous dactylitis or spina ventosa is a rare form of osteomyelitis that affects the tubular bones of the hands and feet; it is most commonly encountered in children and may be multifocal (Fig. 2-30).
4. The radiographic features of tuberculous osteomyelitis in the hands and feet include soft-tissue swelling, periostitis, and osseous expansion; the radiographic differential diagnosis of this condition should include consideration of sickle cell dactylitis, other infections, and juvenile chronic arthritis.
5. The neurovascular changes of leprosy result in development of osseous findings essentially identical to those seen in the neuropathic joints of diabetes and other conditions characterized by ischemia and loss of sensation (Box 2-7); acro-osteolysis is also a prominent feature.
6. Involvement of bone by coccidioidomycosis and cryptococcosis or torulosis tends to include bony prominences, such as the tibial tuberosity, ulnar olecranon, and calcaneal apophysis (Figs. 2-31, 2-32; Box 2-9).
7. Osteomyelitis involving the mandible or maxilla should suggest the diagnosis of actinomycosis.
8. Osteomyelitis in the hand of a farmer or florist should suggest the diagnosis of sporotrichosis.

Box 2-6   Tuberculous Osteomyelitis

- Poorly marginated osteolysis
- Minimal periosteal reaction
- Well-circumscribed cystic lesions
- Dactylitis (spina ventosa)
- Relatively slow rate of progression

Box 2-7   Leprosy (Hansen Disease)

- Neuropathic alterations are more common than osteomyelitis
- Acro-osteolysis occurs in the hands; arthropathy involves the tarsometatarsal and metatarsophalangeal joints in the feet
- Osteomyelitic involvement includes periosteal and endosteal reaction with a predilection for the ulna and fibula

A

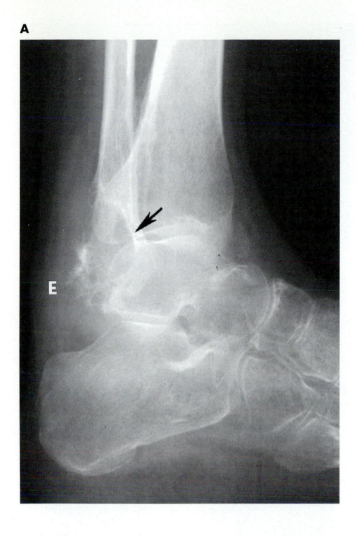

B

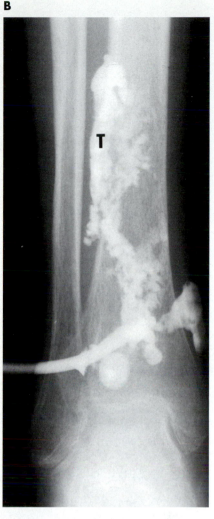

Fig. 2-17   Sinography in the staging of musculoskeletal infection. **A,** Lateral radiograph demonstrates expansile lytic destruction of the lateral malleolus (*arrow*) with adjacent soft-tissue edema (*E*), secondary to tuberculosis. **B,** Injection of a draining sinus in the area with contrast material reveals an extensive serpiginous tract (*T*) extending proximally into the interosseous region.

Box 2-8   Mycotic Osteomyelitis

- Hematogenous or secondary to adjacent soft-tissue involvement
- Cortical erosion or destruction
- Demarcation by sclerosis
- Minimal to absent periostitis
- Soft-tissue swelling, draining sinuses
- Sites of predilection
  Actinomycosis: mandible, chest wall
  Maduromycosis: feet
  Sporotrichosis: hands, feet
  Cryptococcosis: bony prominences
  Coccidioidomycosis: bony prominences, long-bone metaphyses

Box 2-9   Coccidioidomycosis

- 10% to 50% of patients with disseminated disease have osteomyelitis
- Any bone may be involved; predilection for bony prominences
- Long-bone lesions usually involve metaphyses
- Vertebral involvement common; relative sparing of discs
- Skull lesions circular or irregular; may involve both tables or meninges
- Arthritis occurs hematogenously or secondary to osteomyelitis

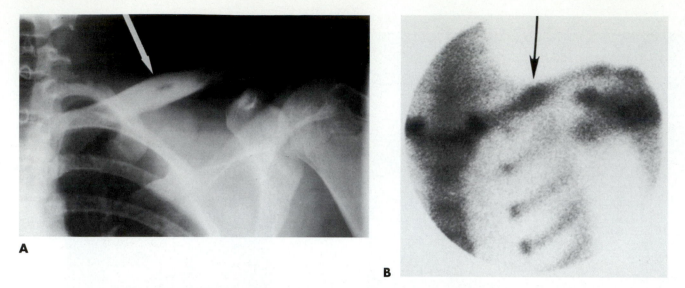

Fig. 2-18   Chronic recurrent multifocal osteomyelitis with clavicular involvement. **A,** Radiography demonstrates an expansile, predominantly osteosclerotic lesion (*arrow*). **B,** Skeletal scintigraphy reveals increased activity within the process (*arrow*). Differential diagnosis should include consideration of other causes of chronic osteomyelitis (Garré type), metastatic disease, Hodgkin disease, osteosarcoma, and fibrous dysplasia.

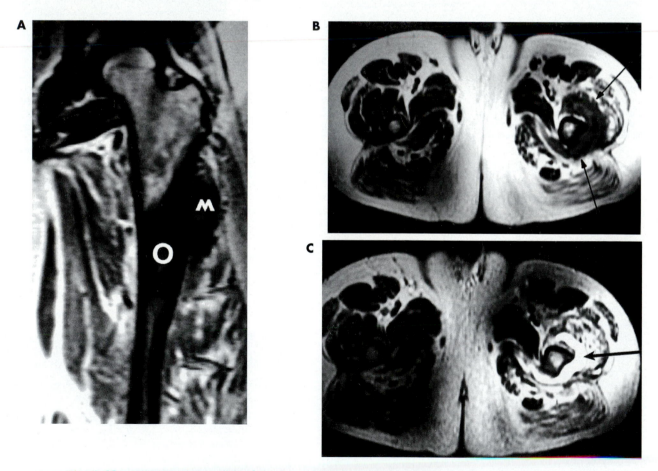

Fig. 2-19   Chronic osteomyelitis of the femoral diaphysis. **A,** Coronal T1-weighted MR image demonstrates an expansile area of low signal intensity (*O*) with an adjacent soft-tissue mass (*M*). **B,** Transaxial T1-weighted image reveals circumferential extension of the soft-tissue mass (*arrows*) around the affected bone. **C,** The mass (*arrow*) exhibits high signal intensity on a corresponding T2-weighted image. Differential diagnosis should include consideration of primary or metastatic osseous neoplasm with soft-tissue extension.

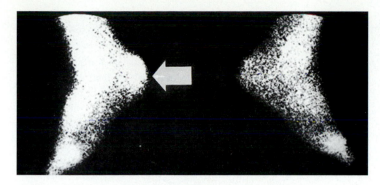

Fig. 2-20   On the blood pool phase of a radionuclide bone scan from a diabetic patient with a heel ulcer, increased activity (*arrow*) in the region could represent either cellulitis or osteomyelitis. Delayed static images would be necessary for reliable distinction.

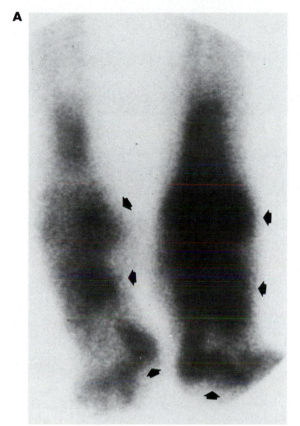

A

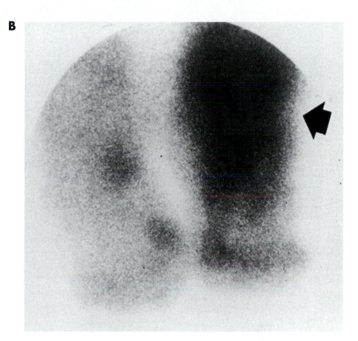

B

Fig. 2-21   Scintigraphy in the diabetic foot. **A,** Frontal gamma camera image from a radionuclide bone scan reveals multiple areas of abnormal increased activity bilaterally (*arrows*). The findings are nonspecific and could represent osteomyelitis, neuroarthropathy, postoperative reaction, or degenerative disease. **B,** Specificity is increased on a corresponding gallium citrate scan image, where intense uptake is present only in the left ankle (*arrow*). Differential diagnosis is limited to consideration of osteomyelitis and neuroarthropathy.

PEARL

Sporotrichosis is a chronic fungal disease acquired from plants through cutaneous wounds. It is an occupational hazard of florists and farmers. It characteristically causes septic arthritis with articular destruction. It is less likely to involve bones but may cause lytic osseous lesions. Bone involvement is uncommon in histoplasmosis. The incidence of tuberculous osteomyelitis following BCG vaccine (a vaccine for tuberculosis) is between 1 in 5,000 and 1 in 80,000. Brucellosis of the spine results in exuberant osteophyte formation.

9. Sclerotic osteomyelitis of a rib occurring in association with pleuritis and draining sinus tracts should suggest the diagnosis of actinomycosis.
10. Osseous destruction, sinus tracts, and widespread intraarticular osseous ankylosis in the foot should suggest the diagnosis of maduromycosis or mycetoma.
11. Osteomyelitis limited to the skull and facial bones in an immunocompromised patient with sinusitis should suggest the diagnosis of mucormycosis or aspergillosis.

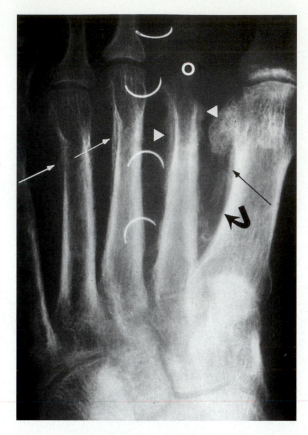

Fig. 2-22   Osteomyelitis in a diabetic foot. Frontal radiograph demonstrates poorly defined osteolysis (*o*) of the second metatarsal head and neck, in association with immature periostitis of its shaft (*arrowheads*). Additional findings include mature periostitis related to neuropathic disease (*straight arrows*) and small-vessel calcification (*curved arrow*) typical of diabetes and renal osteodystrophy.

SYPHILITIC OSTEOMYELITIS (Box 2-10)

1. Metaphyseal irregularity and a widened zone of provisional calcification are early radiographic manifestations of congenital syphilis.

Box 2-10   Radiographic Manifestations of Syphilis

- Congenital: metaphyseal irregularity, osteolysis with or without fracture, symmetric bilateral involvement of long bones, diaphyseal periostitis, lucent metaphyseal bands
- Juvenile: diaphyseal sclerosis, cortical thickening, bowing (saber shin tibia)
- Acquired: involvement of skull, spine, long bones; chronic osteomyelitis without sequestration, ill-defined lytic gummas, neuroarthropathy

PEARL

- Congenital syphilis
  - Hepatosplenomegaly is observed.
  - Celery stalking is most commonly seen in congenital rubella.
  - The Wimberger sign is bone destruction of the proximal medial tibial metaphysis adjacent to the physeal plate, and it occurs in over 50% of cases; metaphyseal fracture may result.
  - Periosteal reaction can occur in the metaphyseal area as well as in the diaphyses.
- With healing, thickening of the anterior tibial cortex is noted, leading to the saber shin deformity.
- The Wimberger line is seen in the epiphyses of patients with scurvy.

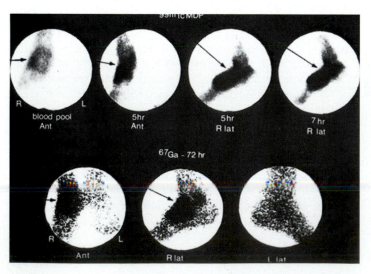

Fig. 2-23   Combined bone-gallium imaging in osteomyelitis of the diabetic foot. Sequential radionuclide bone scan images (*top*) reveal increased and persistent activity in the right midfoot and hindfoot regions (*arrows*). Gallium citrate scan images at 72 hours (*bottom*) demonstrate increased activity (*arrows*) commensurate with that of the bone scan. The findings are nonspecific and could also occur in neuroarthropathy.

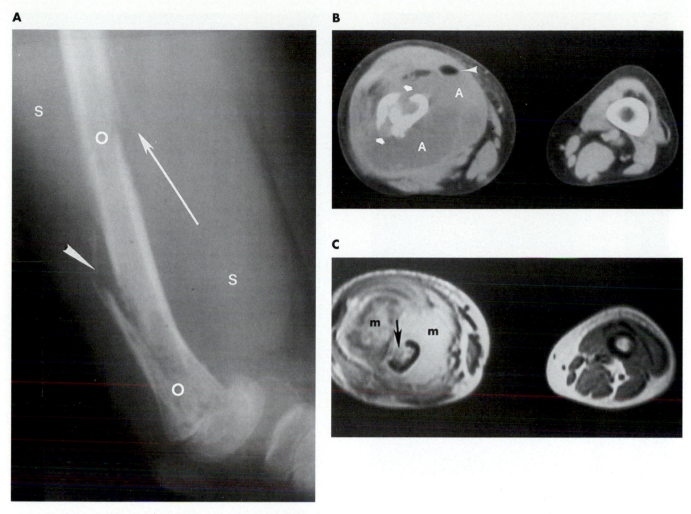

Fig. 2-24   Tuberculous abscess of the femur with pathologic fracture. **A,** Radiography demonstrates permeative osteolysis (*o*), periosteal reaction including Codman triangle (*arrow*), pathologic fracture (*arrowhead*), and soft-tissue swelling (*s*). **B,** CT reveals a large soft-tissue abscess (*A*) containing iatrogenic gas (*arrowhead*) as well as the fracture (*arrows*). **C,** The mass (*m*) is inhomogeneously bright on a transaxial T2-weighted MR image, which also depicts cortical destruction (*arrow*).

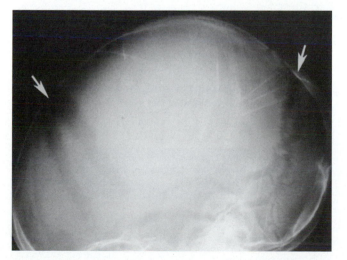

Fig. 2-25   Tuberculous meningitis. Lateral radiograph of the skull demonstrates generalized sutural widening (*arrows*). Differential diagnosis should include consideration of other causes of increased intracranial pressure in children.

2. Diaphyseal periostitis is a variable radiographic feature of congenital syphilis.

PEARL

Periosteal reaction may be seen in the newborn in congenital syphilis and as a physiologic normal variant. The periosteal reaction seen in leukemia and sickle cell disease does not occur in the newborn.

---

3. Slipped epiphysis may occasionally complicate physeal involvement by the disease.
4. The radiographic differential diagnosis of congenital syphilis should include consideration of neonatal viral infections such as rubella, small round cell neoplasms, battered child syndrome, and metabolic bone disease.

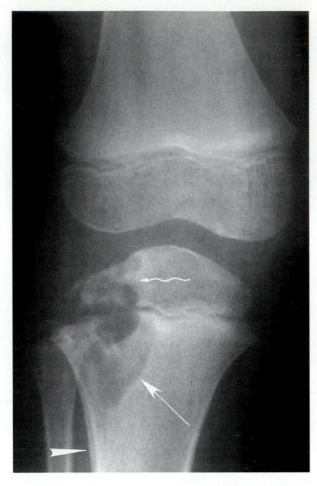

Fig. 2-26   Tuberculous osteomyelitis. The lesion is characteristically well defined with periosteal reaction (*arrowhead*) and manifests involvement of both metaphysis (*straight arrow*) and epiphysis (*wavy arrow*). Differential diagnosis should include other indolent infections, particularly fungal; physeal penetration is far less likely with neoplastic disease.

5. Acquired and reactivated congenital syphilis demonstrate similar radiographic abnormalities in the musculoskeletal system.
6. The radiographic manifestations of acquired syphilis include mixed osteolysis and osteosclerosis, periosteal and endosteal reaction, osseous enlargement, and bowing of long bones.
7. Anterior convex tibial bowing is characteristic of the saber shin deformity.
8. Acquired syphilitic osteomyelitis frequently involves the skull and flat bones.
9. Acquired syphilitic osteomyelitis is usually limited to the long-bone diaphyses; this feature may assist in distinguishing the process from Paget disease, which usually begins in the epiphysis or metaphysis and eventually involves the entire bone.
10. The radiographic differential diagnosis of acquired syphilitic osteomyelitis should include consideration of Paget disease and other treponemal infections.

11. Neuroarthropathy secondary to syphilis most characteristically affects the knee; a neuropathic shoulder should instead suggest the diagnosis of syringomyelia.

SEPTIC ARTHRITIS (Box 2-11)

Age-related Considerations

1. Pyogenic arthritis secondary to genitourinary infection occurs in adults following hematogenous spread, and antibiotics are the mainstay of treatment.
2. The knee and hip are the most commonly involved sites in the perinatal period.
3. As a result of epiphyseal and metaphyseal vascular connections in adults and infants, metaphyseal osteomyelitis commonly spreads to the joint space in these age groups.
4. Subluxation and dislocation caused by increased joint fluid and pressure are complications of pyogenic arthritis in infants.
5. Locations of septic arthritis among children, in order of decreasing frequency, include the knee, hip, ankle, elbow, and wrist.

Box 2-11   Radiographic Manifestations of Septic Arthritis

- Soft-tissue swelling
- Joint effusion
- Osteoporosis
- Joint space loss
- Marginal and central osseous erosion
- Bony ankylosis

A

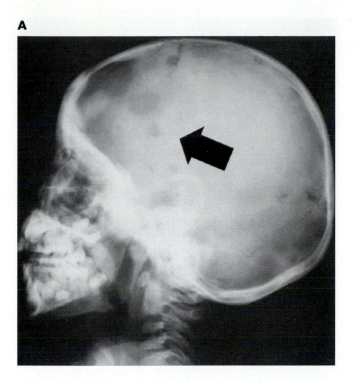

B

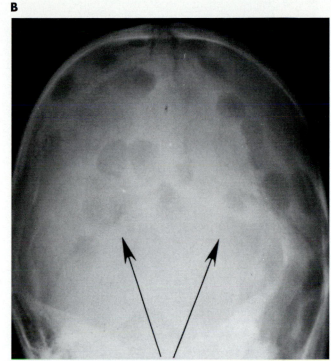

C

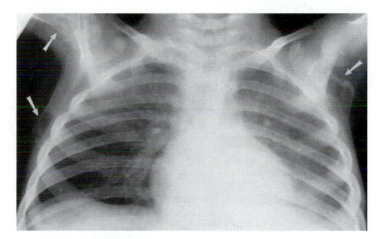

Fig. 2-27   Disseminated tuberculosis in a child. A and **B,** Lateral and Towne projections of the skull demonstrate multiple well-defined lytic lesions without sclerotic margins (*arrows*). **C** to **E,** Similar lesions are present in the scapulae, right humerus, and pelvic region (*arrows*). Differential diagnosis should include consideration of other granulomatous infections, metastatic disease, Langerhans cell histiocytosis, and angiomatosis of bone.

Continued

Radiographic Features of Bacterial Arthritis

1. Monoarticular involvement
2. Joint effusion (pyarthrosis) (Fig. 2-33)
3. Periarticular demineralization secondary to hyperemia induced by inflammation (Fig. 2-34)
4. Joint space narrowing secondary to cartilage destruction (Fig. 2-35)
5. Marginal, followed by central, osseous erosions (Fig. 2-36)

D

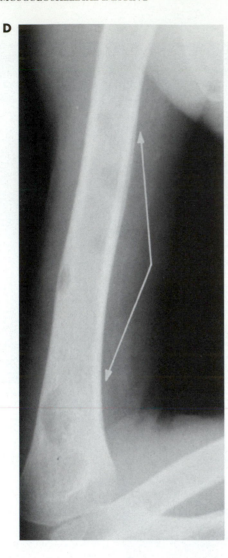

E

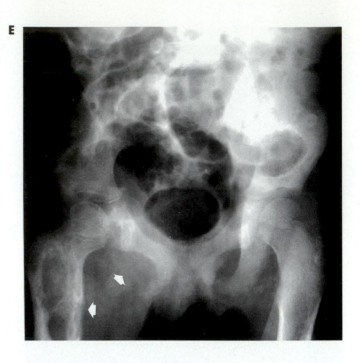

Fig. 2-27, cont'd.   For legend see p. 89.

PEARL

Joint effusions and fat pads:
- Unilateral hip widening in childhood occurs with septic arthritis and other causes of effusion.
- Iliopsoas and gluteus medius fat planes may be obscured with proximal femoral osteomyelitis.
- The earliest sign of knee effusion is thickening of the suprapatellar pouch between the prefemoral and suprapatellar fat pads.
- The ankle joint does not widen with effusion, but the fluid extends anteriorly and posteriorly with displacement of the periarticular fat.

6. Reactive sclerosis in periarticular host bone (Fig. 2-37)
7. Articular ankylosis
8. Complications including osteomyelitis and draining sinus tracts

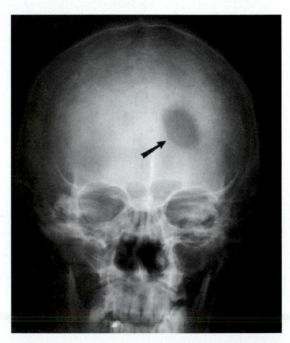

Fig. 2-28   Coccidioidomycosis. A well-defined lytic lesion (*arrow*) is present in the left frontal bone. Other possible causes of this appearance include other fungal infections, tuberculosis, metastatic disease, and eosinophilic granuloma.

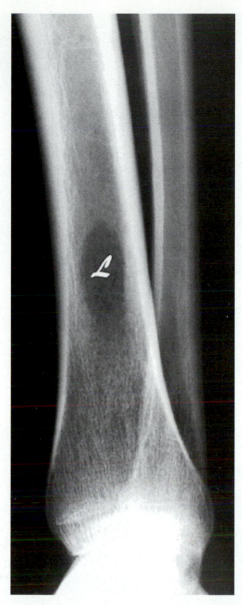

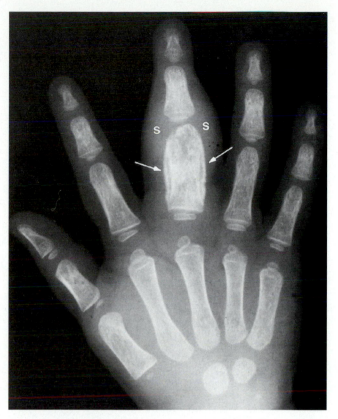

Fig. 2-30   Tuberculous dactylitis (spina ventosa) in a child. Radiographic findings include expansion and increased density of the bone, periosteal reaction (*arrows*), and soft-tissue swelling (*s*). Differential diagnosis should include consideration of sickle cell dactylitis and small round cell neoplasms of childhood.

Fig. 2-29   Coccidioidomycosis. A well-defined lytic lesion (*L*) without marginal sclerosis involves the distal tibial diaphysis. Differential diagnosis should include consideration of other hematogenous infections, metastatic disease, and eosinophilic granuloma.

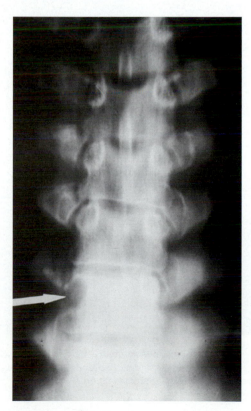

Fig. 2-31   Coccidioidomycosis. Frontal tomogram documents destruction of a midthoracic pedicle (*arrow*) on the right side. Differential diagnosis should include consideration of other granulomatous infections, metastatic disease, and aneurysmal bone cyst.

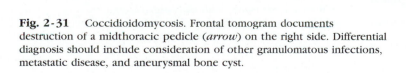

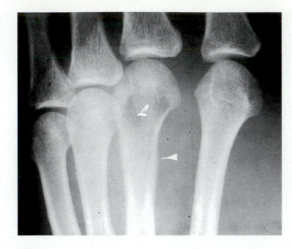

Fig. 2-32 Coccidioidomycosis. A lytic lesion (*L*) with periosteal reaction (*arrowhead*) involves the third metacarpal. Differential diagnosis should include consideration of other hematogenous infections (pyogenic, tuberculous, fungal), enchondroma with occult pathologic fracture, brown tumor, giant cell reparative granuloma, and metastatic disease.

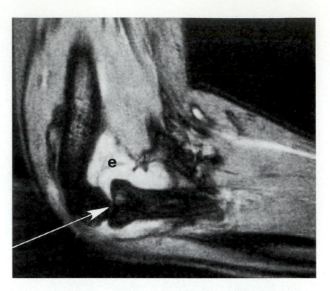

Fig. 2-33 Septic arthritis of the elbow. Sagittal gradient-echo MR image reveals a large joint effusion (*e*) and complete loss of articular cartilage (*arrow*) at the radiocapitellar articulation.

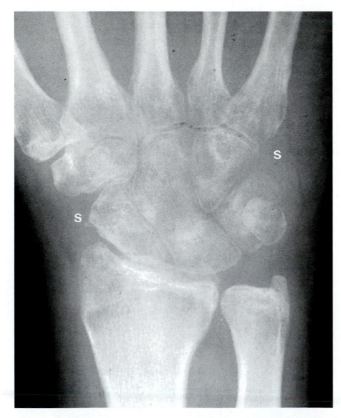

Fig. 2-34 Granulomatous synovitis affecting the wrist. Frontal radiograph demonstrates characteristic soft-tissue swelling (*S*) and prominent periarticular demineralization.

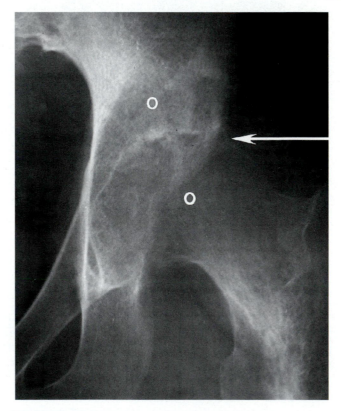

Fig. 2-35 Septic arthritis of the hip. Typical findings include joint space narrowing with early subchondral erosions (*arrow*) and periarticular osteopenia (*O*). Differential diagnosis should include consideration of rheumatoid arthritis and other inflammatory arthritides.

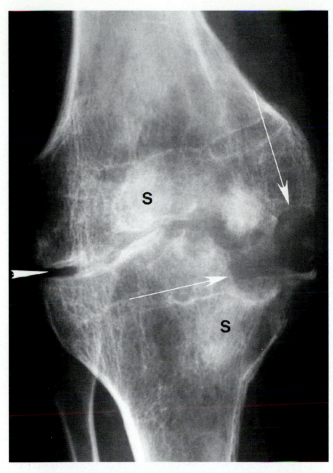

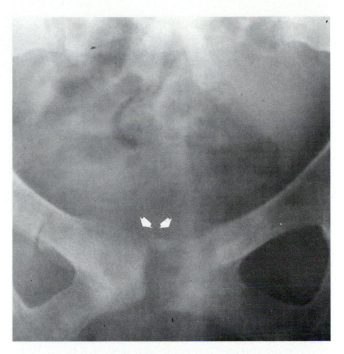

Fig. 2-37   Osteitis pubis secondary to infection. Poorly defined erosion of the opposing pubic bones (*arrows*) is evident, with mild reactive sclerosis. Differential diagnosis should include consideration of early ankylosing spondylitis, healing insufficiency fractures, prior transurethral prostate surgery, and trauma- or pregnancy-induced osteolysis.

Fig. 2-36   Septic arthritis of the knee. Characteristic findings include joint space narrowing (*arrowhead*), subchondral erosions with fragmentation (*arrows*), and reactive sclerosis (*S*). Differential diagnosis should include consideration of other inflammatory arthritides and neuroarthropathy.

PEARL

Septic arthritis:
- Radiographic findings are often absent during the first several days.
- The earliest signs of pyogenic arthritis are soft-tissue swelling and occasional joint space widening.
- Periarticular osteoporosis is often present.
- Subluxation and dislocation may occur.
- Subsequent joint space narrowing occurs within several days of clinical onset, and the joint cartilage is often completely destroyed.
- Subchondral bone erosion and destruction appear 10 to 14 days following initial clinical symptoms.
- Because pyogenic arthritis due to *Neisseria* and other organisms is often similar radiographically to staphylococcal or streptococcal infection, joint aspiration is often required for organism identification.

Radiographic Features of Tuberculous and Fungal Arthritis (Figs. 2-38, 2-39, 2-40; Table 2-3)

1. The Phemister triad is characteristic of tuberculous arthritis; this includes juxtaarticular osteoporosis, peripheral erosions, and gradual joint space narrowing due to slow progression.
2. Destruction of the articular margins is a frequent finding.
3. Single joint involvement is the most common pattern.
4. Fibrous ankylosis is much more common than bony ankylosis.
5. Destruction of cartilage progresses much more slowly than in pyogenic infection; hence, joint space narrowing occurs relatively late (Fig. 2-41).
6. Osseous erosions develop and progress more slowly.
7. Reactive host bone formation is minimal to absent (Fig. 2-42; Box 2-12).

Box 2-12   Radiographic Manifestations of Tuberculous Arthritis

- Joint effusion with occasional caseous calcification
- Periarticular osteoporosis
- Late destruction of cartilage and subchondral bone, beginning in non–weight-bearing areas
- Sharply circumscribed osteolysis with minimal reactive sclerosis

Septic Arthritis: General Concepts

1. Aspiration arthrography is the most definitive diagnostic imaging study, for both ordinary joints and arthroplasties.
2. The hip is a common site of septic arthritis in children, because of the intracapsular position of a frequent hematogenous focus of infection (the proximal femoral metaphysis) (Fig. 2-43).
3. In order of increasing frequency, the most common sites of involvement by tuberculous and fungal arthritis are the elbow, wrist, knee, and hip (Fig. 2-43).

Table 2-3   Comparison of tuberculous and pyogenic arthritis*

	Tuberculous arthritis	Pyogenic arthritis
Soft-tissue swelling	+	+
Osteoporosis	+	±
Joint space loss	Late	Early
Marginal erosions	+	+
Bony proliferation (sclerosis, periostitis)	±	+
Bony ankylosis	±	+
Slow progression	+	−

*+ = Common; ± = infrequent; − = rare or absent

4. False positive findings of bulging periarticular fat planes may be induced by flexion of the hip or rotation of the hip or pelvis.
5. Common sites of septic arthritis in intravenous drug abusers include the sacroiliac and sternoclavicular joints (Fig. 2-44).

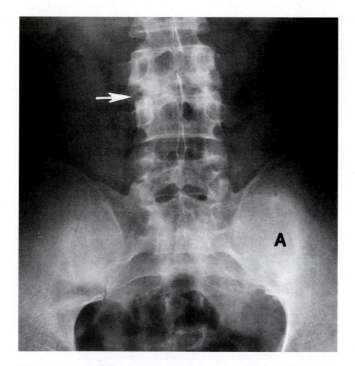

Fig. 2-38   Multifocal tuberculosis. Spinal infection is characterized by L3-4 disk space narrowing, vertebral endplate erosions, and reactive sclerosis (*arrow*). More advanced involvement of the left sacroiliac joint is characterized by osseous ankylosis (A). The latter finding is usually more typical of pyogenic than of tuberculous infection.

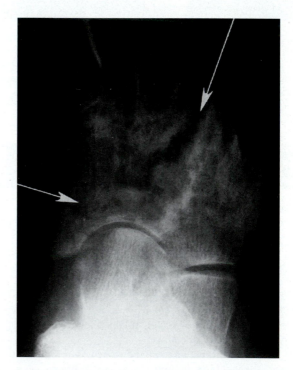

Fig. 2-39   Fungal arthritis secondary to coccidioidomycosis. Poorly defined subchondral erosions with reactive sclerosis (*arrows*) involve the articulations of the midfoot. Differential diagnosis should include consideration of other forms of septic arthritis, rheumatoid arthritis, and neuroarthropathy.

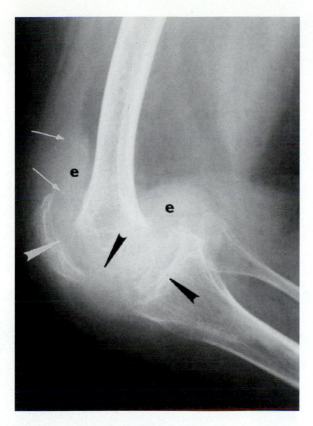

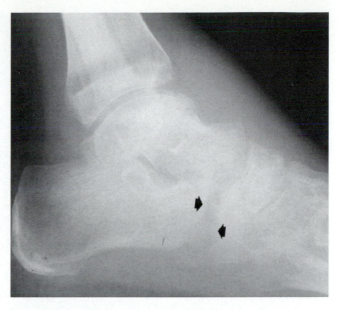

Fig. 2-42   Septic arthritis involving the calcaneocuboid joint. Characteristic features include a monoarticular process, subchondral bone erosion with apparent joint space widening (*arrows*), hyperemia-induced osteopenia, and diffuse soft-tissue swelling. The etiologic agent (tuberculosis) cannot be confidently ascertained without fluoroscopic joint aspiration.

Fig. 2-40   Tuberculous arthritis of the knee. A large joint effusion (*e*) containing caseous material (*arrows*) is evident, along with advanced subchondral erosions (*arrowheads*) and periarticular osteopenia.

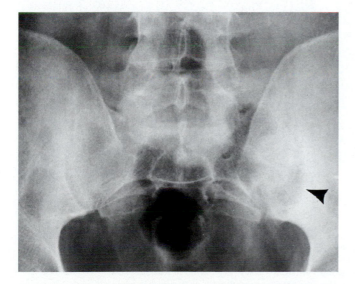

Fig. 2-41   Tuberculous sacroiliitis. Unilateral joint space widening and indistinctness with reactive sclerosis (*arrowhead*) are present. The radiographic pattern is indistinguishable from that of pyogenic or fungal infection and could also occur following repeated intraarticular steroid injections.

6. An increase in the femoral head–acetabular teardrop distance in children should suggest the presence of a hip joint effusion.
7. Radiographically, nonspecific lucency at a cement prosthesis or cement-bone interface is the most frequent abnormal finding in both infected and sterile loose arthroplasties.
8. Traction radiography of the hip joint in a child excludes the presence of an effusion if a vacuum phenomenon is observed.
9. Because widening and indistinctness are normal features of the sacroiliac joints in skeletally immature individuals, the early radiographic signs of septic arthritis are more difficult to recognize in this age group than in adults (Fig. 2-45).
10. Computed tomography is essential in suspected sacroiliac joint infection for the evaluation of possible soft-tissue extension of the process; whereas patients with infection limited to the joint should undergo aspiration arthrography with culture followed by appropriate antibiotic therapy, those with complicating pelvic abscess should instead undergo

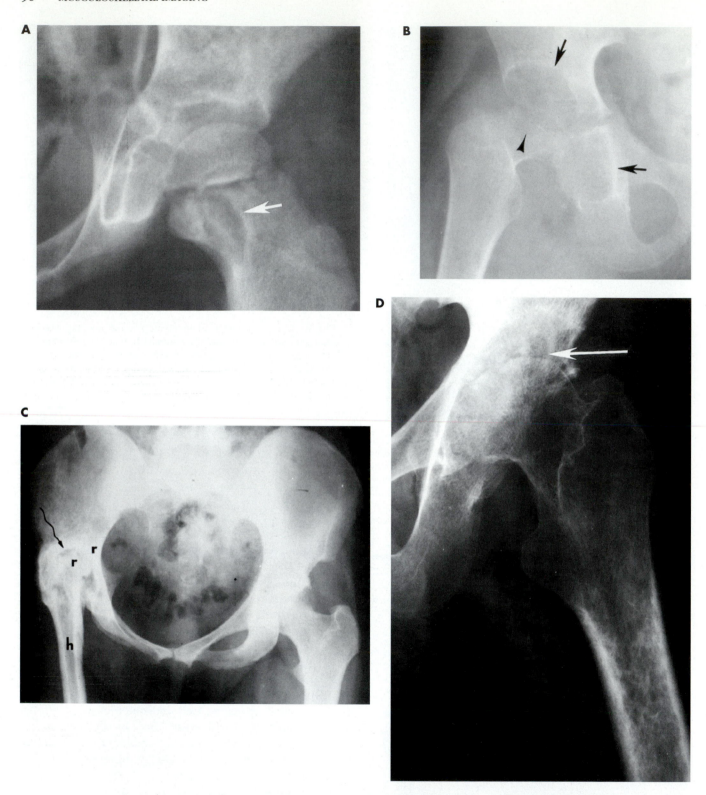

Fig. 2-43   Tuberculous involvement of the hip. **A,** A well-defined intracapsular osteolytic focus (*arrow*) involves the proximal femoral metaphysis and growth plate, indicating hematogenous osteomyelitis. **B,** In another child, destruction of the metaphysis, epiphysis, and growth plate (*arrowhead*) is associated with extensive acetabular erosion (*arrows*), indicating septic arthritis with osteomyelitis. **C,** An advanced childhood infection has resulted in femoral head and acetabular deformities (*arrow*), reactive sclerosis (*r*), and hypoplasia of the femur (*h*) secondary to longstanding disuse. **D,** Adult-onset hip disease is characterized by periarticular osteopenia, diffuse joint space narrowing (*arrow*), and early subchondral bone destruction. Other infectious agents could produce findings similar to these.

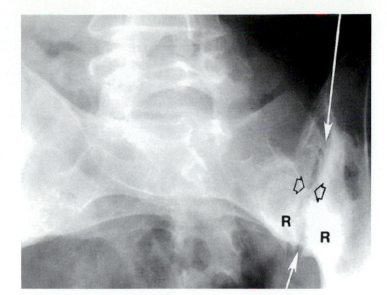

Fig. 2-44   Chronic sacroiliac joint infection. Abnormal findings include widening of the joint space (*solid arrows*), indistinctness of the subchondral cortex (*open arrows*), and reactive sclerosis (*R*).

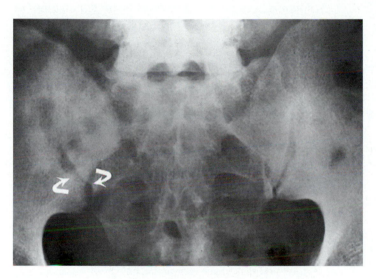

Fig. 2-45   Unilateral sacroiliitis, characterized by subchondral erosion with joint space widening (*arrows*), is virtually diagnostic of septic arthritis.

percutaneous drainage with culture followed by appropriate antibiotic therapy.

11. Immature periostitis with heterotopic ossification are suggestive of infection in the setting of total joint replacement, but these are relatively infrequent findings.

12. The differential diagnosis for septic arthritis of the hip in children should include consideration of juvenile chronic arthritis, transient synovitis of the hip, traumatic effusion, hemophilia, and early Legg-Calvé-Perthes disease (Fig. 2-46).

13. Computed tomography is superior to conventional tomography for characterizing the extent of infec-

tions involving the sternoclavicular joints because it can demonstrate the soft-tissue extent of the infection as well as osseous involvement.

14. Lyme arthritis is caused by a blood-borne spirochete that is transmitted by the bite of *Ixodes dammini* and related ixodid ticks.

15. Although the acute recurrent arthritis of Lyme disease manifests only soft-tissue swelling and effusions, a minority of patients develop a chronic arthritis characterized by periarticular demineralization, joint space narrowing, marginal osseous erosions, and osteophytes.

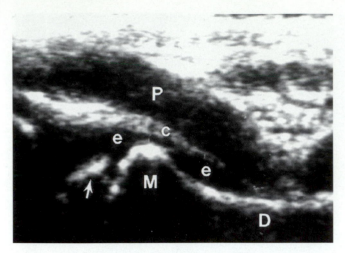

Fig. 2-46   Hip joint effusion (*e*) in transient synovitis of childhood documented by ultrasonography (*arrow = femoral head, M = femoral metaphysis, D = femoral diaphysis, P = iliopsoas muscle, c = joint capsule*). Differential diagnosis should include consideration of infection and juvenile chronic arthritis.

VIRAL AND PROTOZOAN INFECTIONS: MUSCULOSKELETAL MANIFESTATIONS (Box 2-13)

1. Variola (smallpox): nonsuppurative osteomyelitis, suppurative arthritis, and nonsuppurative arthritis
2. Rubella (postnatal): arthritis (wrists, hands, knees, and ankles)
3. Rubella (intrauterine), herpes simplex, *Toxoplasma gondii* infection (intrauterine): celery stalk appearance of metaphyses (Box 2-14)
4. Cytomegalovirus: celery stalk appearance of metaphyses with spontaneous pathologic fracture

5. Mumps: migratory polyarthritis predominating in large joints
6. Vaccinia: soft-tissue swelling, periostitis, hyperostosis, osteomyelitis, and metaphyseal irregularity
7. *T. gondii* infection (postnatal): periarticular swelling, tenosynovitis, and myositis

---

Box 2-13   Radiographic Manifestations of Poliomyelitis

- Osteoporosis
- Soft-tissue atrophy, ossification
- Rib erosions secondary to pressure from scapulae
- Intervertebral disk calcification
- Joint and disk space narrowing, ankylosis
- Gracile long and short tubular bones
- Coxa valga, pes cavus

PEARL

Transverse lucent metaphyseal zones can be seen in toxoplasmosis, rubella, cytomegalovirus, herpes, syphilis, other transplacental infections, scurvy, childhood leukemia, neuroblastoma, arthritis, and malnutrition. A useful mnemonic is the following:

L = leukemia
I = infection (toxoplasmosis, other infections, rubella, cytomegalovirus infection, and herpes simplex [TORCH]; syphilis)
N = neuroblastoma, nutritional deficiency
E = endocrine disorders (rickets, hypophosphatasia)
S = scurvy, Still disease (juvenile chronic arthritis)

Box 2-14   Radiographic Manifestations of Congenital Rubella

- Irregular, indistinct metaphyseal margins with coarsened trabeculae, longitudinal bands of sclerosis
- Normalization by 4 weeks of age with or without growth arrest or recovery lines
- Similar findings in congenital toxoplasmosis, cytomegalovirus, and herpes infections

HELMINTHIC INFECTION

Diseases

Helminths affecting the musculoskeletal system and their associated diseases are as follows:

1. *Loa loa:* loiasis
2. *Onchocerca volvulus:* river blindness
3. *Wuchereria bancrofti:* filariasis (Box 2-15)
4. *Dracunculus medinensis:* guinea worm disease (Box 2-16)
5. *Taenia solium:* cysticercosis (Box 2-17)
6. *Echinococcus granulosus:* hydatid disease (Box 2-18)
7. *Sarcocystis lindemanni:* sarcosporidiosis
8. *Armillifer armillatus:* porocephalosis
9. *Brugia malayi:* filariasis

The helminth *E. granulosus* does not produce calcification in musculoskeletal soft tissues but can cause osteomyelitis.

The helminths *L. loa, D. medinensis, T. solium,* and *S. lindemanni* commonly produce calcification in musculoskeletal soft tissues.

Box 2-15 Filariasis: Important Concepts

- Caused by *W. bancrofti, B. malayi*
- Chronic cases: periosteal reaction, irregular cortical thickening in limb bones
- Disuse osteoporosis in feet
- Soft-tissue calcifications (<4 mm)
- Clinical signs: nonpitting edema, skin thickening, soft-tissue hypertrophy, chyluria

Box 2-16 Dracunculiasis: Important Concepts

- Tropical Africa, Arabia, and India
- Calcified worms may fracture
- Soft-tissue calcification: convoluted, whorled, tangled "chain mail" calcification (dead female worms)
- Round or oval amorphous dystrophic calcification (inflammatory reaction around dead worms)
- Secondary infection of joints adjacent to inflammation around dead worms
- Large cysts or abscesses in groin, lower extremities, extradural sites
- *D. medinensis* (guinea worm), up to several feet long

Box 2-17 Cysticercosis: Important Concepts

- Caused by *Cysticercus cellulosae* (larval form of *T. solium*)
- Larval deposits in subcutaneous and muscular tissues, heart, brain, lung, liver, eye
- Long axis of calcifications (up to 25 mm in length) parallel to muscle bundles

Box 2-18 Echinococcosis: Important Concepts

- Caused by *E. granulosus*
- Bone lesions in 1% to 2% of cases
- Peak age for hydatid disease of bone: 30 to 60 years
- Osseous involvement usually hematogenous
- Most common sites: spine, pelvis, long bones, skull
- Medullary lesions without periosteal reaction
- Complications: pathologic fracture, secondary staphylococcal infection, rupture into spinal canal with neurologic compromise, transarticular extension with collapse and deformity, intrapelvic extension with organ compression, meningitis

Typical Locations of Soft-tissue Calcification

1. *L. loa:* widespread areas in subcutaneous tissue
2. *O. volvulus:* subcutaneous nodules in head, trunk, legs
3. *W. bancrofti:* subcutaneous tissue of scrotum, thighs, and legs
4. *D. medinensis:* extremities
5. *T. solium:* widespread areas in muscle
6. *S. lindemanni:* muscle and subcutaneous tissue of extremities
7. *A. armillatus:* thoracic and abdominal walls
8. *B. malayi:* subcutaneous tissue of scrotum, thighs, and legs

Typical Appearances of Soft-tissue Calcification

1. *L. loa:* varied in size, linear or beaded, extended or coiled
2. *O. volvulus:* small, linear or beaded, extended or coiled
3. *W. bancrofti:* small, straight or coiled
4. *D. medinensis:* long, extended or coiled

5. *T. solium:* numerous, varied in size, linear or oval, along muscle planes
6. *S. lindemanni:* numerous, varied in size and orientation, linear or oval
7. *A. armillatus:* multiple, crescentic or oval
8. *B. malayi:* small, straight or coiled

INFECTIONS OF THE SPINE

Pyogenic Infection

1. The most common organism encountered in pyogenic infections of the spine is *S. aureus.*
2. Infections of the genitourinary tract are a frequent source of hematogenous seeding to the spine.
3. The vertebral venous plexus of Batson is located within the spinal canal between the anterior column and the thecal sac and allows direct communication between the veins of the pelvis and thoracolumbar spine; it represents an important route for hematogenous spread of infection.
4. Pyogenic infections of the spine in adults begin in the endplate of the vertebral body, with subsequent spread to the intervertebral disk and the endplate of the next adjacent vertebral body.
5. Skeletal scintigraphy and magnetic resonance imaging are sensitive techniques for the early diagnosis of spinal infections (Fig. 2-47).
6. Conventional radiographic abnormalities generally begin 2 to 8 weeks after the onset of spinal infection (Fig. 2-48).
7. Radiographic features of pyogenic spinal infection include intervertebral disk space narrowing, irregular destruction of the adjacent vertebral endplates, and reactive sclerosis in the intact portions of the vertebral bodies (Figs. 2-49, 2-50).
8. Computed tomography or magnetic resonance imaging may be used to characterize the soft-tissue extent of spinal infections, which may include prevertebral masses and/or intraspinal mass (Fig. 2-51; Box 2-19; Tables 2-4, 2-5).
9. Percutaneous aspiration or biopsy of spinal infections to obtain material for culture may be performed under fluoroscopic or computed tomographic guidance.
10. The spine is a common site of pyogenic infection in patients with a history of intravenous drug abuse.
11. Gas-forming infections of the spine are rare and are best demonstrated by computed tomography.
12. The radiographic differential diagnosis of pyogenic spinal infection should include consideration of neuroarthropathy, spondyloarthropathy of hemodialysis, posttraumatic pseudoarthrosis in ankylosing

spondylitis, mechanical discovertebral erosion in rheumatoid arthritis, and degenerative discogenic or idiopathic segmental vertebral body sclerosis.

Box 2-19   Vertebral Osteomyelitis: CT Findings

- Paravertebral inflammatory masses
- Epidural soft-tissue extension with thecal sac deformity
- Vertebral endplate destruction
- Disk space narrowing
- Disk hypodensity

Tuberculous Spondylitis (Figs. 2-52, 2-53; Box 2-20)

1. Tuberculous spondylitis is also known as Pott disease.
2. Subligamentous spread with multilevel involvement, when present, is useful in distinguishing tuberculous from pyogenic spondylitis (Fig. 2-54).

Box 2-20   Characteristics of Tuberculous Spondylitis (Pott Disease)

- Begins in vertebral endplate
- Late involvement of disk
- Kyphosis (gibbus)
- Paravertebral abscess (with or without calcification)
- Subligamentous spread with multilevel involvement

PEARL

Destruction of an intervertebral disk space and two adjacent vertebral bodies is the most frequently encountered radiographic pattern of tuberculous spondylitis. Tuberculosis can also involve several contiguous levels and exhibit soft-tissue calcification, both of which are unusual in pyogenic infections.

---

3. Tuberculous spondylitis tends to involve the thoracolumbar junction (Fig. 2-55).
4. Slow progression with relative preservation of disk height and minimal to absent sclerotic response in host bone are typical characteristics (Fig. 2-56).
5. Angular kyphosis or gibbus deformity may be a late manifestation of the disease (Fig. 2-57).

Table 2-4   CT features of hematogenous infective spondylitis

	Site of predilection	Vertebral body	Disk	Paraspinal soft tissue	Epidural space
Pyogenic infections	Lower lumbar spine	Patchy destruction Reactive sclerosis Localization to endplate Involvement of two consecutive vertebrae Vertebral structure intact No gibbus	Collapse	Circumferential granulation tissue Patchy enhancement Localization to affected vertebrae	Infrequently involved Moderate encroachment caused by granulation tissue
Tuberculous infections	Thoracolumbar spine	Severe destruction Vertebral structure lost Minimal sclerosis Gibbus frequent Multiple sites frequent	Collapse Involvement of several levels	Large abscesses Extension paraspinally to distant location Rim enhancement	Frequently involved Severe encroachment caused by abscess and/or bone fragments
Brucella infections	Lower lumbar spine	Localized endplate destruction, simulating Schmorl nodes Involvement of two consecutive vertebrae Multiple sites infrequent Vertebral structure intact No gibbus	Collapse Diskal gas frequent	Moderate granulation tissue Patchy enhancement	Infrequently involved Moderate encroachment caused by granulation tissue

Table 2-5   MR imaging features of hematogenous infective spondylitis

	Vertebral body	Disk	Paraspinal soft tissue	Epidural space
Pyogenic infections	T1: decreased signal in two consecutive vertebrae T2: increased signal Vertebral structure intact Gd-DTPA: diffuse enhancement	T2: high signal, absent nuclear cleft Gd-DTPA: minimal enhancement	T2: moderately increased signal from granulation tissue Gd-DTPA: no definite encroachment	T1: obliteration of epidural space Gd-DTPA: meningeal enhancement
Tuberculous infections	T1: decreased signal in affected vertebra T2: diffuse increased signal Vertebral structure lost Gd-DTPA: rim enhancement demonstrating intraosseous abscesses	T2: high signal, absent nuclear cleft Gd-DTPA: minimal enhancement	T1: abscesses isointense with muscle T2: diffuse high signal from abscess collections Gd-DTPA: rim enhancement around abscesses	T1: obliteration by granulation tissue and bone fragments Gd-DTPA: meningeal enhancement frequent
Brucella infections	T1: decreased signal in two consecutive vertebrae T2: increased signal; Schmorl nodes can be seen Vertebral structure intact Gd-DTPA: diffuse enhancement	As in pyogenic	As in pyogenic	As in pyogenic

A

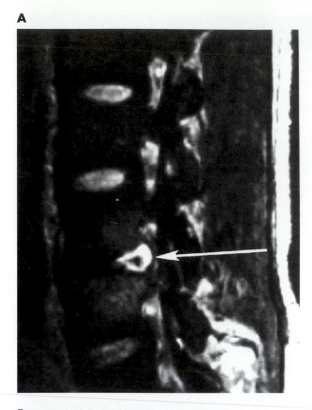

C

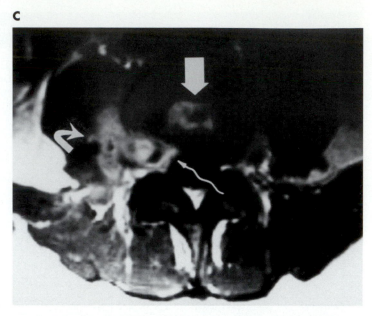

Fig. 2-47   Contrast material–enhanced MRI of early spinal infection. **A,** Sagittal inversion recovery image reveals high signal intensity in the posterior aspect of the L3-4 disk (*arrow*). **B,** Corresponding T1-weighted image following intravenous administration of gadolinium-DTPA demonstrates enhancement of the posterior disk margin (*arrowhead*), as well as the marrow of the adjacent vertebral bodies (*arrows*). **C,** Abnormal contrast material enhancement is also shown in the nucleus pulposus (*straight arrow*), annulus fibrosus (*wavy arrow*), and paravertebral soft tissues (*curved arrow*) on a transaxial T1-weighted image.

B

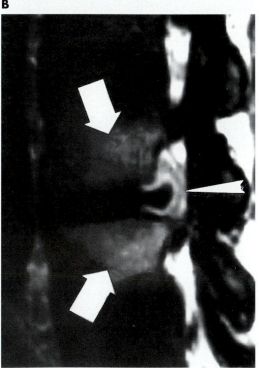

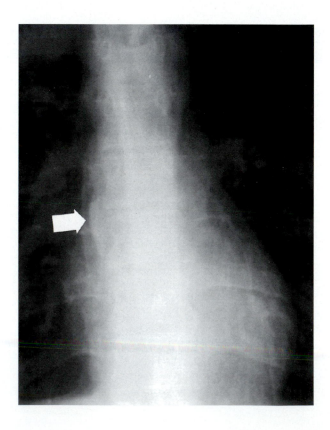

Fig. 2-48   Localized convexity of the paraspinal pleural reflection at the level of a disk space (*arrow*) can be an early radiographic sign of spinal infection.

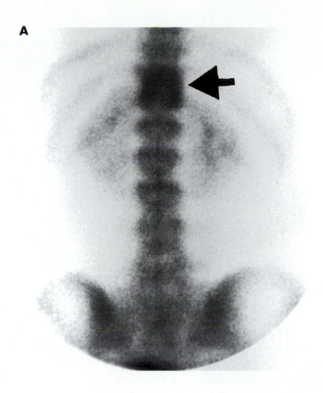

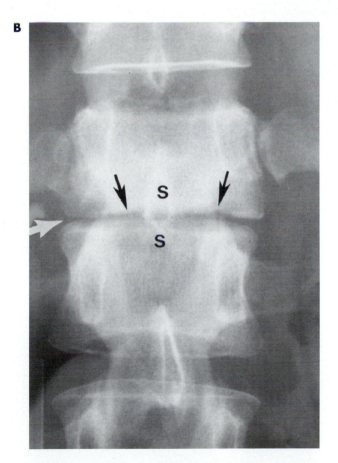

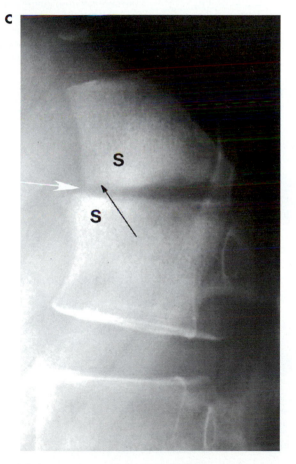

Fig. 2-49 Pyogenic infection of the spine. **A,** Radionuclide bone scan reveals increased activity (*arrow*) at the T12 to L1 levels. **B** and **C,** Frontal and lateral radiographs demonstrate typical disk space narrowing (*white arrows*), endplate erosions (*black arrows*), and sclerosis (*s*) of the adjacent bone.

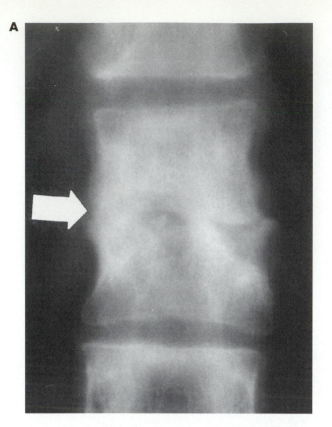

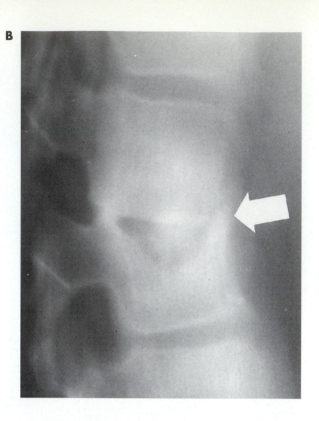

Fig. 2-50 Conventional tomography in spinal infection. Frontal (**A**) and lateral (**B**) images demonstrate intervertebral disk space narrowing, erosion and sclerosis of the vertebral endplates, and localized kyphosis at the level of involvement (*arrows*). This technique is superior to the use of CT reformations in depicting the relationship between diseased vertebrae because of its higher spatial resolution and freedom from motion artifact.

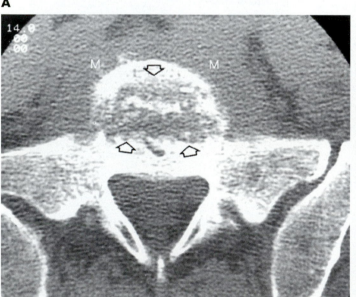

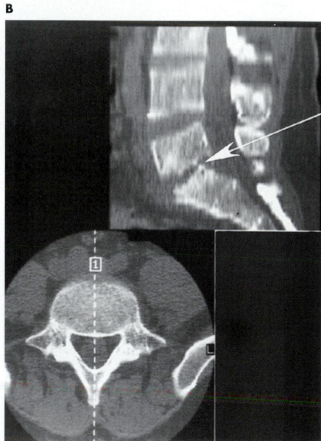

Fig. 2-51 Intervertebral disk space infection. **A**, CT demonstrates irregular destruction of the vertebral endplates (*arrows*) at L5-S1, in association with a large paraspinous soft-tissue mass (*M*). **B**, Sagittal reformation confirms these findings and reveals associated disk space narrowing (*arrow*).

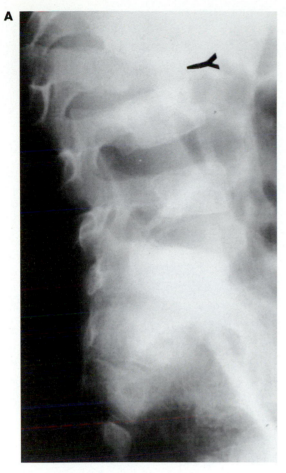

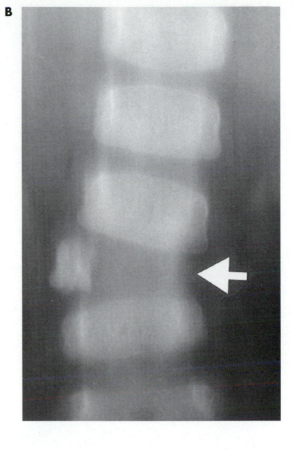

Fig. 2-52   Tuberculous spondylitis in a child. **A,** Lateral radiograph reveals anterior wedging of the L3 vertebral body with associated acute kyphosis (*arrow*). **B,** Frontal tomogram documents nearly complete destruction of the affected body (*arrow*), with angular scoliosis. Differential diagnosis should include consideration of fungal infection, metastatic disease, and eosinophilic granuloma.

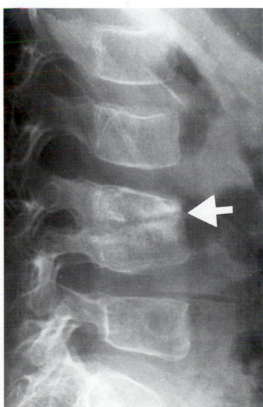

Fig. 2-53   Osseous ankylosis (*arrow*) following tuberculous infection of the spine in a child. Differential diagnostic consideration is virtually limited to congenital block vertebrae.

A
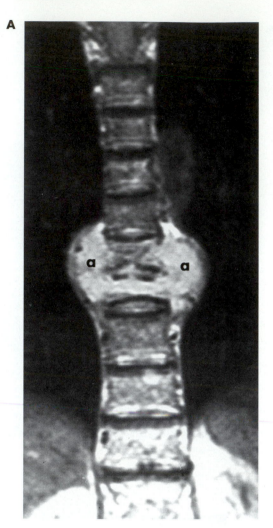

B
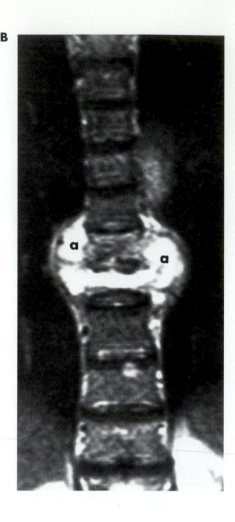

Fig. 2-54   Tuberculous spondylitis. **A** and **B,** Coronal proton density- and T2-weighted MR images reveal multilevel vertebral destruction due to subligamentous spread (unlikely with pyogenic infections), as well as prominent paraspinal abscesses (*a*).

6. Calcified psoas abscess is a variable feature that strongly suggests the diagnosis (Figs. 2-58, 2-59).
7. Pulmonary tuberculosis is usually present in the setting of tuberculous spondylitis.
8. Tuberculous arthritis of the hip may complicate the disease secondary to caudal extension along the iliopsoas muscle.
9. Fungal involvement of the spine is characterized by osteolytic lesions in the vertebral bodies and, less commonly, the posterior elements, with relative sparing of the intervertebral disks. The pattern tends to resemble metastatic disease as opposed to pyogenic or tuberculous infection (Figs. 2-60, 2-61).
10. Particularly early in the disease process, radiographic findings of tuberculosis and occasionally fungal infection may be indistinguishable from those of pyogenic spondylitis; hence, percutaneous biopsy with culture is mandatory (Fig. 2-62).

PEARL

Pyogenic and tuberculous infections of the spine commonly destroy two adjacent vertebral bodies and the disk space. Advanced infections may also involve an adjacent rib. Round cell neoplasms (such as Ewing sarcoma), osteoblastoma, and aneurysmal bone cyst may also involve two adjacent vertebrae.

A

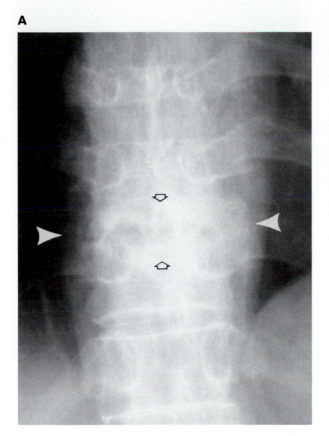

B

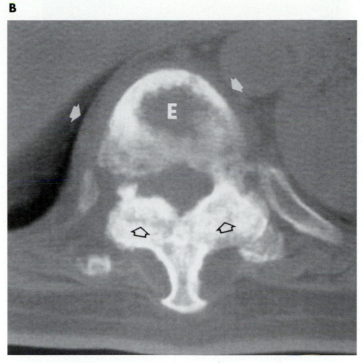

Fig. 2-55   Tuberculous infection in a patient with ankylosing spondylitis. **A,** Frontal radiograph demonstrates diffuse osteopenia secondary to corticosteroid therapy, along with paraspinal soft-tissue swelling (*arrowheads*) and a mixed pattern of osteolysis-osteosclerosis involving the opposing endplates at T9-10 (*arrows*). **B,** CT documents evolving ankylosis of the apophyseal joints (*open arrows*) typical of ankylosing spondylitis. Paraspinous soft-tissue swelling (*closed arrows*) and endplate lysis-sclerosis (*E*) are secondary to infection.

Childhood Diskitis

1. Childhood diskitis is often a self-limited disorder that occurs exclusively in skeletally immature patients.
2. The pathogenesis of the disease most likely involves direct hematogenous seeding of the intervertebral disk, because unlike the situation in adults, this structure possesses a blood supply during childhood.
3. Clinical features of childhood diskitis include back pain, low-grade fever, elevated erythrocyte sedimentation rate, and history of previous upper respiratory tract or other minor infection.
4. Routine cultures of blood or material obtained from percutaneous biopsy usually yield no growth, raising the possibility of a viral cause.
5. The most commonly isolated organism in childhood diskitis when cultures are positive is *S. aureus.*

6. Conventional radiographic findings in the disorder include intervertebral disk space narrowing, irregular destruction of adjacent vertebral endplates, and reactive sclerosis in intact portions of the vertebral bodies. These findings may also occur in early tuberculous and fungal infection, which must be excluded by clinical correlation.
7. Skeletal scintigraphy and magnetic resonance imaging are more sensitive than conventional radiography for establishing an early diagnosis (Fig. 2-63).
8. Cross-sectional imaging studies can confirm conventional radiographic findings and demonstrate associated paravertebral soft-tissue masses that tend to be smaller than those of adult disk space infections.
9. The treatment of choice for childhood diskitis is antibiotic therapy, although patients have also been known to recover with bed rest alone.
10. As compared to intervertebral disk space infections in adults, childhood diskitis is a less serious disorder with a much better long-term prognosis.

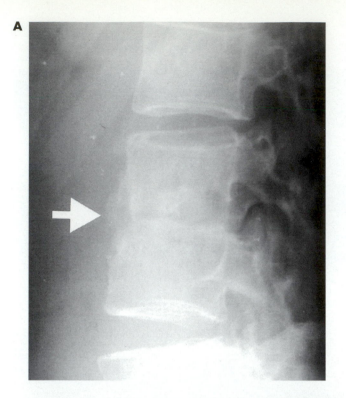

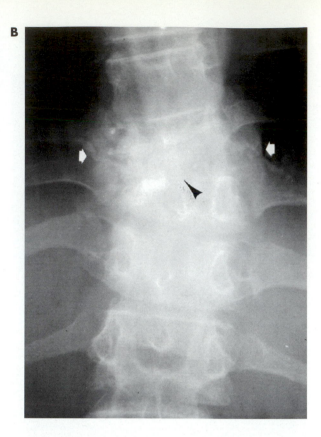

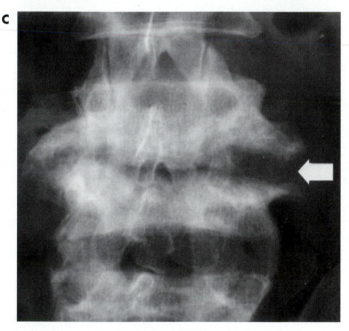

Fig. 2-56 Tuberculous spondylitis limited to a single level. **A,** Intervertebral disk space narrowing with erosion of the opposing endplates (*arrow*) and only sparse reactive sclerosis are typical. **B,** More specific clues for the diagnosis are flocculent calcifications within a paraspinal mass (*arrows*); intervertebral disk space narrowing (*arrowhead*) is associated with angular scoliosis. Although multilevel involvement and calcification are useful in distinguishing tuberculous infection from pyogenic infection, these findings are not always present; hence, percutaneous aspiration biopsy is frequently necessary. Neuroarthropathy and dialysis-associated spondyloarthropathy may simulate spinal infection. **C,** A more advanced case demonstrates severe endplate erosion with secondary degenerative changes including prominent osteophytosis and reactive sclerosis (*arrow*).

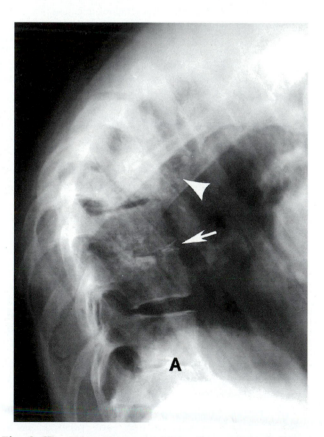

Fig. 2-57 Tuberculous spondylitis. Characteristic findings include multilevel involvement, disk space narrowing (*arrow*), angular kyphosis (*arrowhead*), flocculent calcification, and ankylosis (*A*).

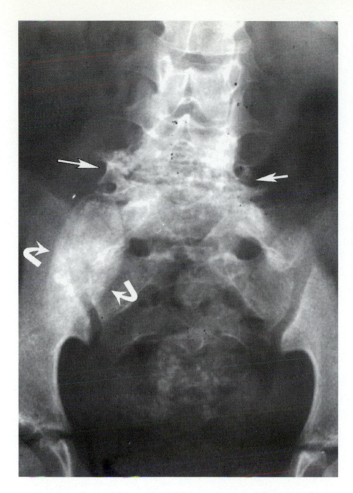

Fig. 2-58   Intervertebral disk space narrowing with associated flocculent calcification (*straight arrows*) and inferolateral extension (*curved arrows*) is pathognomonic of tuberculous spondylitis with psoas abscess.

A

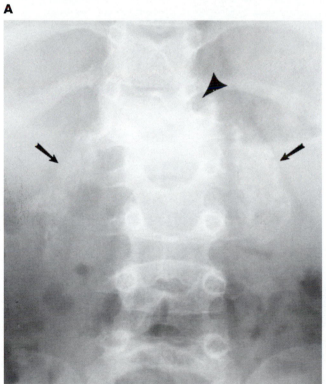

B

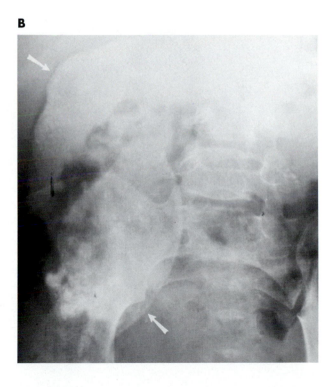

Fig. 2-59   Calcified iliopsoas abscesses in tuberculous spondylitis. **A,** Bilateral involvement (*arrows*) arises from an area of diskal destruction, endplate erosion, and spinal malalignment (*arrowhead*). **B,** A huge abscess (*arrows*) has tracked inferiorly along the course of the iliopsoas muscle. Contamination of the hip joint is a frequent complication of such extension.

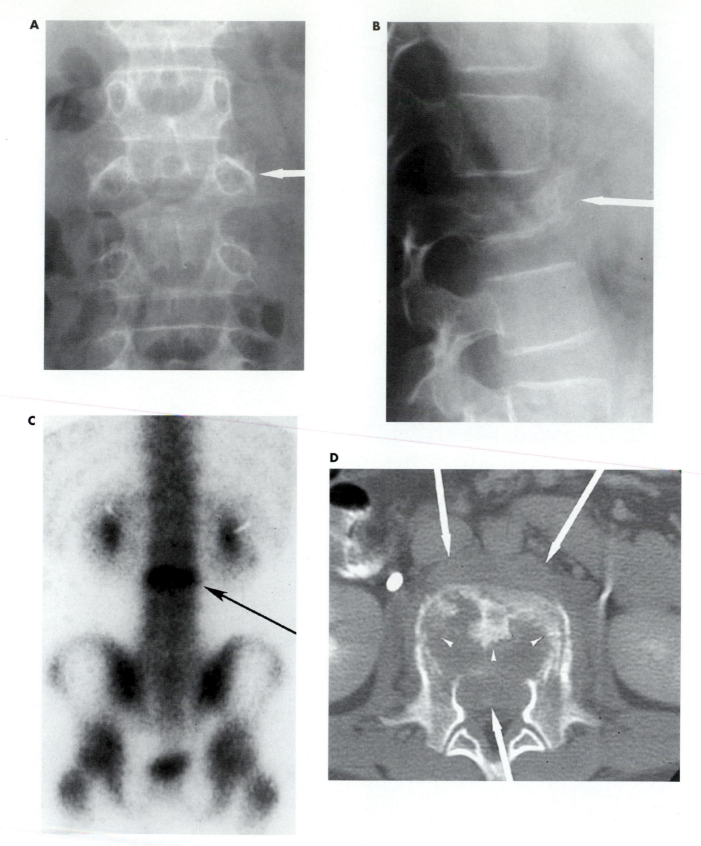

Fig. 2-60   Coccidioidomycosis of the spine. **A** and **B,** Frontal and lateral radiographs demonstrate pathologic fracture of the L3 vertebral body (*arrows*). **C,** The lesion exhibits increased activity (*arrow*) on a radionuclide bone scan. **D,** CT reveals paravertebral soft-tissue swelling (*arrows*) and osteolytic destruction of the vertebral body (*arrowheads*). Differential diagnosis should include consideration of metastatic disease, eosinophilic granuloma, and aneurysmal bone cyst.

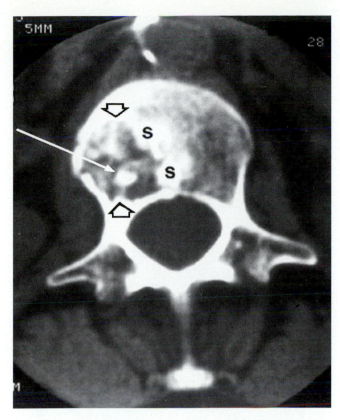

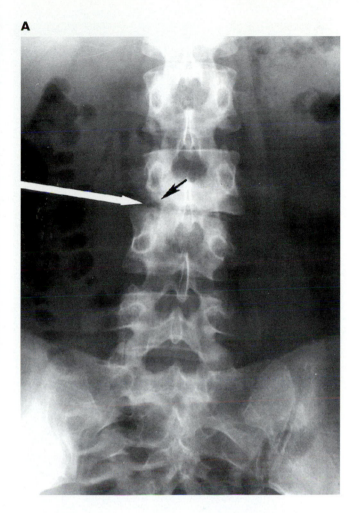

Fig. 2-61   Fungal osteomyelitis of the spine (coccidioidomyco-sis). CT demonstrates osteolysis (*black arrows*) with reactive sclerosis (*S*) and sequestration (*white arrow*). The adjacent disk spaces were unaffected; hence, the appearance resembles neo-plastic disease more strongly than pyogenic or tuberculous infec-tion.

Fig. 2-62   Tuberculous spondylitis. **A,** Frontal radiograph demonstrates severe intervertebral disk space narrowing (*white arrow*) and associated vertebral endplate erosions (*black arrow*). **B,** CT reveals osseous erosions (*arrows*), reactive sclerosis (*S*), and a large paraspinous soft-tissue mass (*M*). In this instance, the appearance is indistinguishable from that of pyogenic infection.

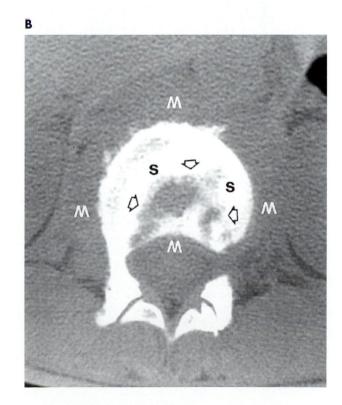

A

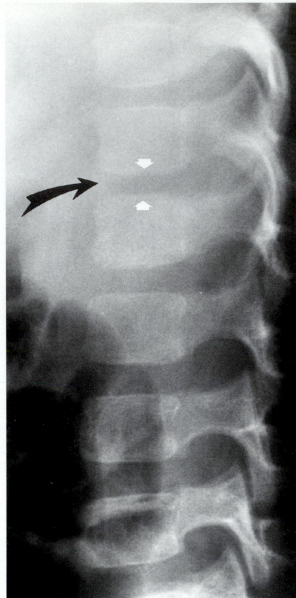

B

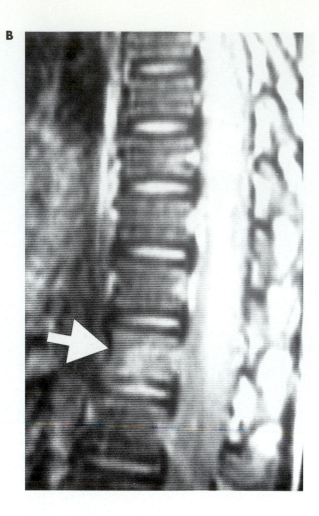

Fig. 2-63   Childhood diskitis. **A,** Lateral radiograph demonstrates mild disk space narrowing (*black arrow*) and indistinctness of the adjacent vertebral endplates (*white arrows*). **B,** T2-weighted MRI is more definitive, revealing high–signal intensity marrow (*arrow*) within one of the adjacent vertebral bodies.

SUGGESTED READINGS

1. Allwright SJ, Miller JH, and Gilsanz V: Subperiosteal abscesses in children: scintigraphic appearance, Radiology 179:725, 1991.
2. Al-Shahed MS et al: Imaging features of musculoskeletal brucellosis, Radiographics 14:333, 1994.
3. Dangman BC et al: Osteomyelitis in children: gadolinium-enhanced MR imaging, Radiology 182:743, 1992.
4. El-Desouki M: Skeletal brucellosis: assessment with bone scintigraphy, Radiology 181:415, 1991.
5. Erdman WA et al: Osteomyelitis: characteristics and pitfalls of diagnosis with MR imaging, Radiology 180:533, 1991.
6. Gold RH, Hawkins RA, and Katz RD: Bacterial osteomyelitis: findings on plain radiography, CT, MR, and scintigraphy, AJR 157:365, 1991.
7. Haygood TM and Williamson SI: Radiographic findings of extremity tuberculosis in childhood: back to the future? Radiographics 14:561, 1994.
8. Jacobson AF et al: Diagnosis of osteomyelitis in the presence of soft tissue infection and radiologic evidence of osseous abnormalities: value of leukocyte scintigraphy, AJR 157:807, 1991.
9. Klein, MA et al: MR imaging of septic sacroiliitis, J Comp Assist Tom 15:126, 1991.

10. Larcos G, Brown ML, and Sutton RT: Diagnosis of osteo-myelitis of the foot in diabetic patients: value of in-leukocyte scintigraphy, AJR 157:527, 1991.

11. LeBreton C et al: Infectious sacroiliitis: value of computed tomography (CT) and magnetic resonance imaging (MRI), Eur J Radiol 2:233, 1992.

12. Morrison WB et al: Diagnosis of osteomyelitis: utility of fat-suppressed contrast-enhanced MR imaging, Radiology 189:251, 1993.

13. Schauwecker DS: Scintigraphic diagnosis of osteomyelitis, AJR 158:9, 1992.

14. Shen W and Lee S: Chronic osteomyelitis with epidural abscess: CT and MR findings, J Comp Assist 15:839, 1991.

15. Sorsdahl OA et al: Quantitative bone-gallium scintigraphy in osteomyelitis, Skeletal Radiol 22:239, 1993.

16. Steinbach LS et al: Human immunodeficiency virus infec-tion: musculoskeletal manifestations, Radiology 186:833, 1993.

17. Yeo EE and Low JC: Scintigraphic changes in physis in osteomyelitis, Clin Nucl Med 16:686, 1991.

CHAPTER 3

Rheumatologic Disorders

114

IMAGING OF ARTHRITIS

Diagnostic imaging techniques are an important aspect of the appropriate examination and management of patients with rheumatologic disorders. Conventional radiography and tomography, computed tomography, arthrography, ultrasonography, scintigraphy, and magnetic resonance imaging are the available modalities. A thorough understanding of the specific roentgenographic signs and characteristic target areas of articular disease in both the axial and appendicular skeleton will permit a diagnosis to be established and allow therapy to be effectively monitored.

Radiographic examination is essential to proper analysis and treatment of patients with articular disorders. In some individuals whose clinical presentation is unclear, radiographic findings will allow a specific diagnosis to be made. In patients whose clinical diagnosis is readily made, radiography will document the extent and severity of articular involvement. In both situations, sequential radiographic examinations permit the efficacy of therapeutic regimens to be monitored.

Although conventional radiography remains the most important and most frequently employed technique in the evaluation of joints, an awareness of the specific indications for specialized imaging procedures should provide a more logical approach to the ordering of diagnostic tests.

Radiographic Signs of Articular Disease in the Appendicular Skeleton

Soft-tissue swelling Soft-tissue prominence may reflect an accumulation of intraarticular fluid, capsular distention, soft-tissue edema, or periarticular or intraarticular masses. In synovial inflammatory processes such as rheumatoid arthritis, psoriatic arthritis, Reiter syndrome, juvenile chronic arthritis, infection, and hemophilia, fusiform periarticular soft-tissue swelling is characteristic. In pigmented villonodular synovitis, gout, xanthomatosis, and amyloidosis, the soft-tissue swelling may appear more nodular, creating a lobulated radiopaque shadow around the involved joint. In gouty arthritis, the masses reflect the presence of eccentric tophi, whereas xanthomas commonly occur around tendinous structures, and prominence of the soft tissues around the shoulder (the "shoulder pad" sign) is a feature of amyloidosis.

Osteopenia Periarticular demineralization is a feature of certain articular disorders. In rheumatoid arthritis, synovitis with accompanying hyperemia leads to this phenomenon, especially in smaller joints. Perhaps due to the intermittent or episodic nature of synovial inflammation in the seronegative spondyloarthropathies, periarticular reduction in bone density is less common. Osteopenia rarely occurs in gout and osteoarthritis and is infrequent in pigmented villonodular synovitis and idiopathic synovial osteochondromatosis; it is a frequent finding in pyogenic arthritis, tuberculosis, and hemophilia.

Joint space narrowing Cartilaginous destruction is accompanied by loss of the interosseous space. In rheumatoid arthritis, early and widespread cartilaginous destruction is typical and leads to diffuse loss of joint space, particularly in the proximal interphalangeal and metacarpophalangeal joints of the hand, in the wrist, in the metatarsophalangeal joints of the foot, in the knee, and in the hip. Diffuse joint space loss is also common in ankylosing spondylitis, psoriatic arthritis, and Reiter syndrome. In gout, joint space diminution is less frequent and occurs in later stages of the disease.

Although joint space loss is also evident in degenerative joint disease, it is limited to stressed areas of the articulation. In the hip, joint space narrowing is usually maximal at the superolateral aspect of the articulation, whereas in the knee, the medial femorotibial space is predominantly affected. Uniform loss of articular space characterizes osteoarthritis of the interphalangeal and metacarpophalangeal joints of the hand and the first metatarsophalangeal joint of the foot.

Pyogenic arthritis is also associated with early and diffuse loss of joint space, whereas this phenomenon occurs late in the course of tuberculosis and fungal arthritis. Ischemic necrosis at such sites as the femoral and humeral heads is associated with subchondral osteolysis, osteosclerosis, and cyst formation without joint space narrowing until late in the process. This reflects the fact that articular cartilage derives its nutrition from the adjacent synovial fluid and is unaffected by disruption of blood supply to the subchondral bone. Disuse or immobilization, on the other hand, impairs normal nutrition of cartilage and results in diffuse loss of joint space. Chondrolysis also occurs in slipped capital femoral epiphysis or on an idiopathic basis. Joint space narrowing is not commonly observed in pigmented villonodular synovitis and idiopathic synovial osteochondromatosis.

Intraarticular bony ankylosis   Osseous fusion of joints occurs in rheumatoid arthritis, where it is usually limited to the carpal and tarsal areas, and in psoriatic arthritis and ankylosing spondylitis, where it may be more widespread. In psoriasis, the metacarpophalangeal, metatarsophalangeal, and interphalangeal joints of the hands and feet are affected; large joints such as the hip may fuse in ankylosing spondylitis. Intraarticular bony ankylosis also occurs in the interphalangeal joints of patients with inflammatory or erosive osteoarthritis. This phenomenon is rare in gout, neuroarthropathy, tuberculosis, pigmented villonodular synovitis, and idiopathic synovial osteochondromatosis, but may occur in juvenile chronic arthritis and subsequent to intraarticular infection.

Osseous erosion   Periarticular erosion may involve either marginal areas of the joint at sites where the synovium abuts bone that does not possess protective overlying articular cartilage ("bare areas"); at the central aspect of the articulation; or both. Marginal erosions are typical of rheumatoid arthritis and other processes characterized by synovial inflammation, such as psoriatic arthritis, Reiter syndrome, ankylosing spondylitis, and infection. Initial loss of definition of the marginal bone progresses to a ragged or irregular appearance related to destruction of portions of the subchondral bone by synovial inflammation. Transchondral extension of pannus can also create cystic lesions that usually communicate with the joint cavity, although they may appear enclosed on radiographs.

In erosive osteoarthritis, central erosions may be identified within the interphalangeal joints of the hand and, less commonly, the foot. Accompanying peripheral osteophytes at the capsular attachments of the joint create a "gull wing" or "seagull" configuration. The distinctive location of these osseous lesions is related to osseous collapse, a phenomenon also occasionally en-

countered in hyperparathyroidism, due to subchondral bone resorption. Both intraarticular and extraarticular osseous erosions occur in gout, frequently adjacent to tophaceous soft-tissue deposits of monosodium urate. Well-circumscribed, eccentric marginal erosions of variable size with sclerotic margins and elevated osseous spicules (the "overhanging edge" sign) are typical. Central osseous erosions also occur secondary to extension of cartilaginous deposits of monosodium urate into subchondral bone. Gouty erosions are frequently extraarticular and/or associated with joint space narrowing in advanced cases.

Intraarticular bony erosions may be evident in pigmented villonodular synovitis and idiopathic synovial osteochondromatosis, particularly in "tight" articulations such as the hip, elbow, and wrist. Single or multiple well-defined osseous defects occur in conjunction with soft-tissue swelling and, in the latter disease, intraarticular calcification and ossification.

Subchondral sclerosis, osteophytosis, and bony proliferation   Sclerosis or eburnation of subchondral bone is a characteristic feature of osteoarthritis, typically occurring in the stressed area of the articulation in association with joint space narrowing and cyst formation. Eburnation may be extensive in the hip and knee, but it is usually mild in the interphalangeal joints of the hand. Sclerosis is a prominent finding in neuroarthropathy, particularly secondary to syphilis and syringomyelia, and it also commonly occurs in calcium pyrophosphate dihydrate crystal deposition disease, osteonecrosis, and pyogenic arthritis.

Osteophyte formation is also a typical feature of osteoarthritis, occurring as a well-defined ridge of bone at the margin or nonstressed area of an articulation. Common sites include the medial aspect of the femoral head, the medial and lateral aspects of the distal femur and proximal tibia, and the posterior aspect of the patella. Osteoarthritis of the interphalangeal joints of the fingers is accompanied by osteophyte formation at the attachments of the joint capsule. In the hip, irritation by the synovial membrane can produce osteophytes at the head-neck junction that are frequently associated with proliferation of bone along the medial aspect of the femoral neck, a phenomenon known as buttressing.

Ill-defined, irregular osseous excrescences accompany bony erosions in the seronegative spondyloarthropathies, resulting in a "whiskered" appearance. This pattern often helps distinguish these disorders from rheumatoid arthritis, in which bone proliferation is rare. Similar ill-defined osseous excrescences also appear at sites of tendinous and ligamentous attachment to bone (entheses), including the femoral trochanters, humeral tuberosities, and calcanei (plantar aponeurosis and Achilles tendon insertions). Extensive erosions and bony reaction at the calcaneal entheses are particularly characteristic of Reiter syndrome. In diffuse idiopathic skeletal hyperostosis (ankylosing hyperostosis of Forrestier), well-defined bony outgrowths appear at entheses, including the olecranon process of the ulna, the anterior surface of the patella, and the plantar and posterior surfaces of the calcaneus.

Subchondral cyst formation   In rheumatoid arthritis and other processes characterized by synovial inflammation, transchondral extension of pannus creates subchondral radiolucent lesions of variable size, which may simulate neoplasms or lead to spontaneous fracture. Cyst formation also frequently occurs in stressed areas of articulations in osteoarthritis, subjacent to areas of cartilage loss. Multiple cysts involving both sides of the affected joint in association with bony eburnation are typical. Osteoclastic resorption of dead trabeculae occurs subsequent to necrosis of bone and marrow, resulting in cyst formation. Although also frequently associated with subchondral sclerosis, these cysts may occur in conjunction with a maintained joint space, unlike the situation in osteoarthritis.

Cyst formation is a recognized manifestation of pyrophosphate arthropathy, the structural joint disease that accompanies calcium pyrophosphate dihydrate (CPPD) crystal deposition disease. Although this arthropathy simulates osteoarthritis radiographically, with additional findings including eburnation, fragmentation of bone, and osteophytosis, the presence of chondrocalcinosis and the involvement of unusual articular sites such as the wrist and elbow permit an accurate diagnosis of CPPD crystal deposition disease in many instances. Multiple subchondral cysts are also a radiographic feature of hemophilia and pigmented villonodular synovitis.

Osseous fragmentation and collapse   Neuroarthropathy is associated with extensive fragmentation and collapse of apposing articular surfaces. Although the early radiographic appearance of this disorder can simulate that of osteoarthritis, with joint space narrowing and bony eburnation, rapid and progressive fragmentation subsequently occurs, resulting in a disorganized articulation. Fragments of bone and cartilage may be displaced into the articular cavity in both conditions, existing as "loose" bodies or becoming embedded in the synovial membrane at a distant site. Occasional migration into neighboring synovial cysts or soft-tissue planes occurs.

Pyrophosphate arthropathy can also be associated with rapidly progressive fragmentation of bone, but frequently manifests concomitant chondrocalcinosis. Fragmentation of subchondral bone is also a feature of acute osteochondral fractures, osteochondritis dissecans, and osteonecrosis, and it infrequently complicates repeated intraarticular steroid injections.

Intraarticular and periarticular calcification Calcification of hyaline cartilage or fibrocartilage or both (chondrocalcinosis) is a characteristic radiographic sign of idiopathic CPPD crystal deposition disease. Calcific deposits in fibrocartilage are amorphous and irregular and are most frequently observed in the menisci of the knee, triangular fibrocartilage of the wrist, and symphysis pubis. Hyaline cartilage calcification appears as thin curvilinear radiodensities that parallel the subchondral bone. Although any joint can be affected, the knee and wrist are favored sites. Synovial and capsular calcification may occur concomitantly. CPPD crystal deposition with chondrocalcinosis can be idiopathic or associated with primary hyperparathyroidism and hemochromatosis. Chondrocalcinosis is occasionally observed in gout, although the deposits are less extensive and usually periarticular.

Tendon calcification is a frequent radiographic sign of calcium hydroxyapatite crystal deposition disease and can also occur in CPPD crystal deposition disease. Calcific tendinitis related to the former is most common in the shoulder, although it may occur around other articulations such as the wrist, hip, and elbow; linear radiodensities are observed within the substance of the affected tendon. Calcium hydroxyapatite crystal deposition within joints has also been implicated in a destructive arthropathy reminiscent of that which occurs in CPPD crystal deposition disease.

Periarticular cloudlike deposits of calcification occur in renal osteodystrophy with secondary hyperparathyroidism; sarcoidosis; collagen vascular disorders such as scleroderma and dermatomyositis; hypervitaminosis D; and milk-alkali syndrome.

Tufted calcification, often associated with resorption of the distal phalanges, is a well-recognized manifestation of scleroderma, and linear calcification in subcutaneous and muscular tissue can occur in scleroderma, dermatomyositis, and systemic lupus erythematosus.

Radiographic Signs of Articular Disease in the Axial Skeleton

Intervertebral disk space narrowing   Degenerative disease of the nucleus pulposus (intervertebral osteochondrosis) leads to progressive narrowing of the disk space associated with collections of gas (vacuum phenomena) and sclerosis of the adjacent vertebral body margins. Osteomyelitis of the spine frequently begins within the vertebral endplate and spreads to involve the adjacent vertebra and intervening disk. Radiographically, as opposed to the situation in degenerative disease, the margins of the affected vertebral bodies are ill defined or destroyed and the intervertebral disk space is narrowed without a vacuum phenomenon unless a gas-forming organism is present.

Narrowing of the cervical intervertebral disks with irregularity of the vertebral body margins occurs in rheumatoid arthritis secondary to mechanical instability. Subluxation at one or more levels (including the atlantoaxial junction) and apophyseal joint erosions can occur, but osteophytes are absent. Rarely, similar changes can occur in the thoracic and lumbar spine. Alkaptonuria (ochronosis) is characterized by diffuse calcification of the intervertebral disks, with narrowing and frequently with multiple vacuum phenomena. Intervertebral disk space narrowing and irregularity of the adjacent vertebral body margins also occur in neuroarthropathy, intraosseous diskal herniation (cartilaginous or Schmorl node formation), trauma, and CPPD crystal deposition disease.

Osteophytes and other bony outgrowths   Degenerative disease of the annulus fibrosus (spondylosis deformans) results in widespread spinal osteophytosis. The outgrowths are typically horizontally oriented and extend in an anterolateral direction. When the disease is unassociated with intervertebral osteochondrosis, the neighboring intervertebral disks are usually normal in height, without vacuum phenomena.

Widespread vertebral excrescences having a predilection for the lower thoracic and upper lumbar spine are a characteristic feature of diffuse idiopathic skeletal hyperostosis (ankylosing hyperostosis of Forrestier). A flowing pattern of ossification along the anterolateral aspect of the spine, with a bumpy spinal contour and preservation of intervertebral disk height, is the typical radiographic pattern. Although the findings at a single level resemble those of spondylosis deformans, the preservation of disk space height and widespread distribution permit accurate diagnosis of diffuse idiopathic skeletal hyperostosis.

Ankylosing spondylitis is associated with thin vertical radiodense spicules of bone (syndesmophytes), which extend between adjacent vertebral bodies and occur initially at the thoracolumbar and lumbosacral junctions, later extending to involve the entire spine. Associated abnormalities include reactive sclerosis at the corners of the vertebral bodies (osteitis), straightening or squaring of the anterior vertebral margins, and ankylosis of the apophyseal and costovertebral joints. The well-known "bamboo" spine and "trolley track" sign are late manifestations resulting from widespread bone formation.

The spondylitis accompanying inflammatory bowel diseases such as ulcerative colitis and Crohn disease has identical radiographic manifestations. Paravertebral ossification in psoriatic arthritis and Reiter syndrome creates ill-defined radiodense shadows separated from the spine or well-defined excrescences that merge with it, particularly at the thoracolumbar junction. Their larger size and asymmetric distribution distinguish these outgrowths from the syndesmophytes of ankylosing spondylitis and inflammatory bowel disease. Fluorosis, acromegaly, and hypoparathyroidism also produce ligamentous ossification of the spine.

Intervertebral disk calcification   Extensive wafer-like calcification of multiple intervertebral disks is pathognomonic of alkaptonuria. Dystrophic globular diskal calcification at one or several levels can occur following trauma or infection. In children calcification of cervical intervertebral disks (diskitis) can be associated with significant self-limited clinical findings. Calcification of the outer fibers of the annulus fibrosus may occur in CPPD crystal deposition disease and simulate the appearance of syndesmophytes. Any disease that produces bony ankylosis between vertebral bodies or fusion in the corresponding apophyseal articulations, including ankylosing spondylitis, juvenile chronic arthritis, diffuse idiopathic skeletal hyperostosis, and long-standing poliomyelitis, can also be associated with intervertebral disk calcification.

Atlantoaxial subluxation   Synovial inflammation involving the transverse ligament of the atlas and the posterior surface of the odontoid process results in atlantoaxial subluxation, manifested as increased distance between the anterior arch of C1 and the odontoid process on lateral radiographs, a finding often accentuated by neck flexion. This phenomenon may occur with or without osseous erosion of the anterior or posterior aspect of the odontoid process in conditions such as rheumatoid arthritis, psoriatic arthritis, Reiter syndrome, and ankylosing spondylitis.

Paravertebral swelling   Fusiform soft-tissue masses around the lower thoracic and lumbar spine can occur in association with a pyogenic or tuberculous psoas abscess and may be calcified in the latter case. The differential diagnosis of paraspinal masses should also include consideration of posttraumatic hemorrhage and spinal neoplasia with soft-tissue extension.

ARTHRITIS IN THE AXIAL SKELETON

Degenerative Disorders

Osteoarthritis   Osteoarthritis is by far the most common type of arthritis seen in humans. By definition, osteoarthritis occurs in a synovial joint. In the spine, therefore, osteoarthritis occurs in the apophyseal (facet) joints, the uncovertebral joints (cervical spine), the costovertebral joints, and the sacroiliac joints. Osteoarthritis may be primary or secondary.

Degenerative nuclear disease (intervertebral osteochondrosis)   Another very common disorder is degeneration of the nucleus pulposus. With age, the nucleus tends to become more and more dehydrated and gradually begins to degenerate. As this happens, the intervertebral disk height begins to decrease. When this happens, the altered pattern of stresses may lead to marginal osteophytosis adjacent to the affected endplates. As the disk space decreases in height, increased stress is also placed on the facet joints, leading to the frequent association of osteoarthritis of the facet joints at the same level. In fact, an early sign of degenerative nuclear disease may be the presence of an unstable apophyseal joint.

Degenerative annular disease (spondylosis deformans) Yet another extremely common degenerative disorder involves the annulus fibrosus of the intervertebral disk. This leads to marginal osteophytosis at the vertebral endplates, especially in the thoracolumbar spine in many persons over 50 years of age. In the literature, this entity has been termed "spondylosis deformans" or "senile ankylosis." It is preferable to state that "marginal osteophytes are noted at multiple disk spaces in the spine" in dictations. The clinicians know what is being described, and neither they nor their patients are unduly concerned by the unfamiliar terminology used for this very common process.

Diffuse idiopathic skeletal hyperostosis Diffuse idiopathic skeletal hyperostosis (DISH) is an extremely common entity of unknown etiology that manifests itself by ossification of the anterior longitudinal ligament, which produces large flowing bony excrescences along the spine, especially the anterior aspect.

Inflammatory Spondyloarthropathies

Rheumatoid arthritis Rheumatoid arthritis is a disorder of unknown etiology characterized by synovial inflammation, pannus formation, and subsequent destruction of bone and cartilage.

Ankylosing spondylitis Ankylosing spondylitis is a chronic inflammatory disorder of unknown etiology that affects principally the axial skeleton. Alterations occur in synovial and cartilaginous articulations and at sites of tendon and ligament attachment to bone. Over 90% of spondylitic patients are HLA-B27 positive.

Psoriatic arthritis Psoriatic arthritis is a relatively uncommon arthropathy that occurs in about 2% to 6% of patients with cutaneous psoriasis. Approximately 25% to 60% of patients with psoriatic arthritis are HLA-B27 positive.

Reiter syndrome Reiter syndrome is a relatively uncommon arthropathy of uncertain etiology with the classic triad of urethritis, arthritis, and conjunctivitis. Of all of the rheumatic diseases, Reiter syndrome is most suspect for an infectious etiology. It appears likely that the disease can be transmitted in association with either epidemic dysentery or sexual intercourse. The syndrome frequently follows an infection of the bowel or lower genitourinary tract, and it seems likely that these

sites are the portals of entry for the causative agent. It has been suggested that the abnormalities of the vertebral column may be related to organisms extending directly to the sacroiliac joints and spine via the prostatic venous plexus or via the venous plexus of Batson. Implicated organisms include pleuropneumonia-like organisms (PPLO), the Bedsonia group of organisms, and viruses; although to date, no single agent has been definitely incriminated in this disease. Articular manifestations may involve cross-reactivity of infection-induced antibodies to synovium or intraarticular deposition of immune complexes. Approximately 75% to 95% of patients with Reiter syndrome are HLA-B27 positive.

Enteropathic arthropathy Enteropathic arthropathy is a condition occurring in about 1% to 25% of patients with ulcerative colitis or Crohn disease. Patients with Whipple disease may also be affected. The relationship between inflammatory intestinal diseases and arthritis is not fully understood. Infectious, immunologic, and genetic etiologies have been advanced. Approximately 90% of patients with ulcerative colitis or Crohn disease who develop spondylitis or sacroiliitis are HLA-B27 positive.

PEARL

Enteropathic arthropathies:
- Clubbing of the fingers is a recognized complication of ulcerative colitis and Crohn disease.
- Acute migratory episodic arthralgia and arthritis occur in 60% to 90% of patients with Whipple disease.
- Following intestinal bypass surgery for intractable obesity, symmetric polyarthritis can be observed, most frequently in the knees, ankles, fingers, and wrists.
- During recovery from Laënnec cirrhosis, inflammatory polyarthritis most commonly involves the shoulders, elbows, and knees.
- Erosive arthritis of the hands and wrists can occur in patients with primary biliary cirrhosis.
- Articular manifestations occur less commonly during the course of hepatitis A (infectious hepatitis) than that of hepatitis B (serum hepatitis).

Crystalline Arthritis

Gout Gout is the prototypic crystalline arthropathy, characterized by the deposition of monosodium urate crystals in the skin, subcutaneous tissues, and joints. This is more meaningfully classified as either idiopathic gout, encompassing the vast majority of individuals, or gout associated with known disorders caused by enzymatic defects.

Calcium pyrophosphate dihydrate crystal deposition (CPPD) disease *Calcium pyrophosphate* *dihydrate crystal deposition disease* is a general term for a disorder characterized by the deposition of calcium pyrophosphate dihydrate crystals in or around joints.

Pseudogout is a term applied to *one* of the clinical patterns that may be associated with CPPD crystal deposition disease. This pattern, characterized by intermittent acute attacks of arthritis, simulates the findings of gout.

Chondrocalcinosis is a term reserved for pathologically or radiologically evident cartilage calcification. In some cases, this calcification may not indicate deposits of CPPD crystals but rather accumulations of some other crystal.

Pyrophosphate arthropathy is a term used to describe a peculiar pattern of structural joint damage occurring in CPPD crystal deposition disease which simulates, in many respects, degenerative joint disease but is characterized by distinctive features.

Calcium hydroxyapatite crystal deposition disease

Calcium hydroxyapatite crystal deposition disease is characterized by recurrent painful periarticular calcium hydroxyapatite deposits in tendons and soft tissues.

RADIOGRAPHIC HALLMARKS

Osteophytes

In a diarthrodial joint, the presence of osteophytes is the most characteristic finding in osteoarthritis. Osteophytes can be seen in both primary and secondary osteoarthritis. Bony outgrowths can also be seen at various entheses, often due to altered or increased stress at the entheses (traction enthesophytes).

Syndesmophytes

Syndesmophytes are caused by inflammation and ossification of the outer fibers of the annulus fibrosus, known as Sharpey fibers. Syndesmophytes are generally seen only in the seronegative spondyloarthropathies.

Disk Space Narrowing

Disk space narrowing almost always signifies degenerative nuclear disease or infection. These processes can often be distinguished by evaluation of the adjacent endplates. In degenerative disk disease, the endplates are often dense, sclerotic, and associated with osteophytosis. In infection, the subchondral line of the endplate often becomes ill defined and discontinuous.

Bony Proliferation

Bony proliferation is a striking feature of the seronegative spondyloarthropathies, particularly psoriatic arthritis. This bony proliferation occurs in association with erosions and probably relates to an exaggerated healing response by the injured bone. This proliferation may take the form of irregular excrescences, subperiosteal deposition of bone, or intraarticular osseous fusion.

Erosions

In general, the presence of erosions indicates some type of inflammatory disease, whether the erosions are due to synovial hypertrophy, crystalline deposits, or infection.

Crystal Deposition

In general, crystal deposition is indicative of one of the crystalline arthropathies—either gout, CPPD, calcium hydroxyapatite, or mixed crystal deposition disease.

Sclerosis

Sclerosis is not an especially specific finding in spinal arthropathy.

Ankylosis

Ankylosis may occur as a result of many degenerative and inflammatory processes in their later stages.

Subluxation

Stability of the spine is maintained by the spinal ligaments, articular capsules, and disks. Any arthropathy that causes degeneration or destruction of these structures may lead to instability of the spine and subluxation in several locations.

PATTERN APPROACH

Osteoarthritis

Osteoarthritis of the spine resembles osteoarthritis elsewhere in the body. Any of the spinal synovial joints can be affected, including the facet and uncovertebral joints, the costovertebral joints, and the sacroiliac joints. Findings include osteophytosis, joint space narrowing, subchondral sclerosis, and subchondral cyst formation.

Degenerative Disk Disease

Degeneration of either the nucleus pulposus or the annulus fibrosus, or both, may be present. While these processes can often be distinguished from each other, overlap in findings will be seen in most patients who have both processes occurring in the same disk.

The most important distinguishing characteristic between these two processes is disk space narrowing. If present, this is strongly suggestive of degenerative nuclear disease. This is often accompanied by endplate sclerosis and mild to moderate osteophytosis. On the other hand, degenerative annular disease is often not associated with significant disk space narrowing, and the marginal osteophytes seen with this entity are often much larger than those seen with degenerative nuclear disease. Some patients with this appearance may actually have an early stage of diffuse idiopathic skeletal hyperostosis.

Diffuse Idiopathic Skeletal Hyperostosis

Although there has been speculation that DISH may be a disorder of vitamin A metabolism, DISH remains an idiopathic phenomenon. As such, it lacks not only a known specific cause but also a known specific disease marker (such as monosodium urate crystals in gout). Therefore a diagnosis of DISH is necessarily one made by exclusion. Because DISH is diagnosed on a morphologic basis alone, there will be of necessity some overlap between certain cases of DISH and other disorders with similar radiographic features, such as ankylosing spondylitis and spondylosis deformans. To help minimize this overlap, certain arbitrary morphologic criteria have been proposed:

- Sacroiliac and facet joints must be normal, unlike ankylosing spondylitis.
- Disk spaces must be of normal height, unlike degenerative nuclear disease.
- Ossification must occur along four contiguous vertebral bodies, unlike degenerative annular disease.

Knowledge of the presence of DISH may alter therapy for other disorders. For example, DISH patients are prone to heterotopic bone formation at surgical sites. Because of this, many orthopedic surgeons will prophylactically treat DISH patients with radiation or bisphosphonates before performing a total joint arthroplasty, in an attempt to prevent or diminish the development of heterotopic bone formation after surgery.

Ankylosing Spondylitis

Ankylosing spondylitis affects synovial and cartilaginous joints, as well as sites of tendon and ligament attachment to bone (entheses). An overwhelming predilection exists for involvement of the axial skeleton, especially the sacroiliac, apophyseal, discovertebral, and costovertebral articulations. Classically, changes are initially noted in the sacroiliac joints and next appear at the thoracolumbar and lumbosacral junctions. With disease chronicity, the remainder of the vertebrae may become involved. However, this characteristic pattern of spinal ascent is by no means invariable; it may occur slowly or rapidly, and it is less frequent in spondylitis accompanying psoriasis and Reiter disease.

Sacroiliitis is the hallmark of ankylosing spondylitis. It occurs early in the course of the disease. Although an asymmetric or unilateral distribution can be evident on initial radiographic examination, radiographic changes at later stages of the disease are almost invariably bilateral and symmetric in distribution. This symmetric pattern is an important diagnostic clue in this disease and may permit its differentiation from other disorders that affect the sacroiliac articulation, such as rheumatoid arthritis, psoriasis, Reiter syndrome, and infection. Abnormalities in the sacroiliac joint involve in both the synovial and ligamentous (superior) portions, and predominate on the iliac side for reasons that are obscure.

The characteristic radiographic features of ankylosing spondylitis include erosions, sclerosis, syndesmophytosis, and ankylosis.

The classic histologic descriptions of synovial joint alterations in ankylosing spondylitis stress that the synovitis is similar or identical to that in rheumatoid arthritis. However, in general, the inflammatory process in ankylosing spondylitis is more discrete and of lesser intensity than that in rheumatoid arthritis. In ankylosing spondylitis the peripheral joint findings typically include joint space narrowing, osseous erosions, cysts, and bony proliferation.

Rheumatoid Arthritis

The general distribution of spinal abnormalities in this disease is usually quite distinct from that of ankylosing spondylitis. For example, rheumatoid arthritis predominantly involves the cervical spine, with apophyseal joint erosion and malalignment, intervertebral disk space narrowing with eburnation but minimal to absent osteophytosis, and multiple subluxations, especially at the atlantoaxial junction. Abnormalities of the thoracolumbar spine and sacroiliac joints are infrequent and less prominent than those of ankylosing spondylitis.

Other helpful differential findings are the relative absence of osteoporosis and the presence of bony proliferation and intraarticular osseous ankylosis in the seronegative spondyloarthropathies. The following gives a comparison of sacroiliac joint abnormalities in rheumatoid arthritis and ankylosing spondylitis:

Finding	Rheumatoid arthritis	Ankylosing spondylitis
Distribution	Asymmetric or unilateral	Bilateral and symmetric
Erosions	Superficial	Deep
Sclerosis	Mild or absent	Moderate or severe
Bony ankylosis	Rare, segmental	Common, diffuse

CPPD Crystal Deposition Disease

The spine is frequently involved in CPPD crystal deposition disease. Intervertebral diskal calcifications are frequent in the outer annular fibers and may mimic early syndesmophytes of ankylosing spondylitis because of their vertical orientation and slender appearance. These annular calcifications may be associated with back pain. Disk space narrowing is common in CPPD crystal deposition disease and may be extensive, widespread, and associated with considerable vertebral sclerosis. However, the nucleus pulposus is not commonly calcified. Calcification of the ligamentum flavum may also be noted. Occasionally, destructive abnormalities of the cervical spine are present. Chondrocalcinosis is only infrequently seen in the sacroiliac joints but is very common in the fibrocartilaginous joint of the pubic symphysis.

Gout

Spinal manifestations of gout are extremely uncommon, and documented monosodium urate deposition in the spine is exceedingly rare. When seen, spinal gout may manifest itself as erosions of the synovial joints or endplates, and disk space narrowing may be present.

The incidence of gout in the sacroiliac joint has been reported at 7% to 17%, although many of the changes ascribed to gout in the earlier literature were probably manifestations of osteoarthritis simulating those of gout. Sacroiliac joint involvement is seen more frequently with early-onset disease, and large cystic areas of erosion in the ilium and sacrum are the most specific findings of gout at this site.

Calcium Hydroxyapatite Crystal Deposition Disease

Calcium hydroxyapatite crystal deposition disease is most commonly seen around the shoulder. However, it may also occur within the longus colli muscle, which is the principal flexor of the cervical spine. Tendinitis in this region may result in acute neck and occipital pain, rigidity, and dysphagia. Calcifications particularly tend to occur in the superolateral fibers of the longus colli.

The typical radiographic findings of this disorder consist of prevertebral soft-tissue swelling in the upper cervical region, and amorphous calcification, usually anterior to C-2 and just below the anterior arch of C-1. Resolution of this calcification and soft-tissue swelling is common, especially following treatment with antiinflammatory drugs, and it may disappear completely in 1 to 2 weeks.

Psoriatic Arthritis

About 30% to 50% of patients with psoriatic arthritis develop sacroiliac joint alterations that are manifest radiographically. Bilateral sacroiliac joint abnormalities are much more frequent than unilateral changes, and although asymmetric findings may be apparent, symmetric abnormalities predominate. Radiographic sacroiliac joint changes include erosions and sclerosis, predominantly on the iliac side, and widening of the articular space. Although significant joint space diminution and bony ankylosis can occur, these findings, particularly ankylosis, occur less frequently than in classic ankylosing spondylitis or the spondylitis associated with inflammatory bowel disease. Sacroiliitis may occur without spondylitis, just as spondylitis may appear without sacroiliitis.

As in Reiter syndrome, paravertebral ossification around the lower thoracic and upper lumbar segments can occur in psoriatic arthritis, and it may represent an early manifestation of the disease. Such ossification appears as a thick and fluffy or thin and curvilinear radiodense region on one side of the spine, paralleling the lateral surface of the vertebral bodies and the intervertebral disks. Occasionally, slender, centrally located, and symmetric spinal outgrowths in psoriasis are identical in appearance to the syndesmophytes of ankylosing spondylitis. However, the greater size, the unilateral or asymmetric distribution, and the location farther away from the vertebral column are features that distinguish paravertebral ossification from the typical syndesmophytosis of ankylosing spondylitis or the spondylitis of inflammatory bowel disease.

In addition to the pattern and distribution of bony outgrowths, there are other features of psoriatic spondylitis that differ from those in classic ankylosing spondylitis. Osteitis and squaring of the anterior surfaces of the vertebral bodies are relatively infrequent in psoriasis. Although apophyseal joint space narrowing, sclerosis, and bony ankylosis may be seen, the incidence of these findings is much less than that in ankylosing spondylitis.

Cervical spine abnormalities may be striking in psoriatic arthritis, including apophyseal joint space narrowing and sclerosis, osseous irregularity at the discovertebral margins, and extensive proliferation along the anterior surface of the spine. Atlantoaxial subluxation can also be evident (up to 45% of patients with psoriatic spondylitis).

Reiter Syndrome

Reiter syndrome is associated with an asymmetric arthritis of the lower extremities, sacroiliitis, and, less commonly, spondylitis. Although its general features resemble those of ankylosing spondylitis and psoriatic arthritis, Reiter syndrome possesses a sufficiently characteristic articular distribution to allow accurate diagnosis.

Ankylosing spondylitis has a similar axial skeletal distribution (although cervical changes are more frequent in ankylosing spondylitis), but significant peripheral articular changes are more frequent in Reiter syndrome.

Psoriatic arthritis may lead to considerable alterations in the articulations of both the appendicular and the axial skeleton. However, in psoriasis, widespread involvement of the upper extremities may be apparent, and distal interphalangeal joint abnormalities in both upper and lower extremities are common. The sacroiliac and spinal changes of Reiter syndrome are virtually identical to those of psoriasis, although the incidence and severity of these abnormalities and the tendency to involve the cervical spine are greater in psoriasis.

Enteropathic Arthropathy

The spondylitis and sacroiliitis of inflammatory bowel disease are identical to those of classic ankylosing spondylitis. The history of inflammatory bowel disease can sometimes help to distinguish these entities, although spondylitis in ulcerative colitis is poorly correlated with activity of the bowel disease. In ulcerative colitis, spinal abnormalities may become manifest before, at the same time as, or following the onset of intestinal changes. In fact, spondylitis most commonly precedes the onset of colitis and may progress relentlessly without relation to exacerbation, remission, or treatment of the bowel disease. In Crohn disease, the joint abnormalities tend to occur in synchrony with the bowel disease.

Peripheral joint abnormalities tend to occur much more frequently with enteropathic arthropathy than with ankylosing spondylitis. When they do occur, they are usually self-limited and rarely cause lasting deformity of the joint.

Target Areas of Specific Articular Disorders

The distribution of rheumatologic disease within the axial and appendicular skeleton is as important as the specific features of individual articular lesions in arriving at an accurate radiographic diagnosis.

Hand Rheumatoid arthritis predominantly involves the metacarpophalangeal and proximal interphalangeal joints and the interphalangeal joint of the thumb; erosions in the distal interphalangeal joints occur only infrequently. Osteoarthritis primarily affects the distal interphalangeal, proximal interphalangeal, and less commonly the metacarpophalangeal joints. Although isolated involvement of the metacarpophalangeal articulations is rare, isolated interphalangeal disease is frequent. Erosive osteoarthritis most frequently produces alterations in the distal interphalangeal and proximal interphalangeal joints.

Structural joint disease secondary to CPPD crystal deposition and hemochromatosis predominates at the metacarpophalangeal joints but may also be observed with lesser severity in the interphalangeal articulations. Gouty arthritis may affect any articulation of the hand, whereas psoriatic arthritis typically manifests itself as a destructive arthritis of the distal interphalangeal and proximal interphalangeal joints.

Wrist Rheumatoid arthritis initially involves the radiocarpal, midcarpal, inferior radioulnar, and pisiform-triquetral joint compartments, as well as the tendinous sheaths of the wrist. Eventually, pancompartmental alterations are characteristic.

Osteoarthritis, including the erosive variety, produces alterations that are confined to the first carpometacarpal and trapezioscaphoid compartments in the absence of prior accidental or occupational trauma. Involvement at other sites that appears morphologically degenerative should suggest crystal deposition disease or prior trauma. CPPD crystal deposition disease usually involves the radiocarpal compartment, as well as the midcarpal and first carpometacarpal joints. Gout can be associated with pancompartmental disease but tends to affect the common carpometacarpal compartment most severely. Juvenile chronic arthritis may be pancompartmental, but the radiocarpal compartment is often spared.

Foot Rheumatoid arthritis typically affects the metatarsophalangeal joints, the interphalangeal joint of the great toe, and the articulations of the midfoot and hindfoot, particularly the talonavicular and calcaneocuboid joints. The earliest alterations often occur in the fifth metatarsophalangeal articulation. Involvement of the interphalangeal joint of the hallux is most prominent on its medial aspect. Additional findings include posterior calcaneal erosions and well-defined plantar calcaneal spurs.

Gouty arthritis predominates in the hallux at both the metatarsophalangeal and interphalangeal joints, although any joint in the foot may be involved, including the remaining interphalangeal articulations. Psoriasis and Reiter syndrome can also affect any joint, although sites of predilection include the metatarsophalangeal joints and interphalangeal joint of the great toe. Proliferative erosions on the plantar and posterior calcaneal surfaces may be associated. Osteoarthritis frequently involves the first metatarsophalangeal joint. CPPD crystal deposition disease and the neuroarthropathy accompanying diabetes mellitus selectively involve the talonavicular area, and both may also cause alterations in the forefoot.

Knee Rheumatoid arthritis and the seronegative spondyloarthropathies produce symmetric loss of joint space in the medial and lateral femorotibial spaces, with or without alterations in the patellofemoral space. Osteoarthritis typically produces asymmetric joint space narrowing, with the medial femorotibial space predominantly affected, the lateral femorotibial space spared, and variable involvement of the patellofemoral compartment.

Joint space narrowing confined to the patellofemoral compartment is typical of CPPD crystal deposition disease and hyperparathyroidism, but it can also occur in osteoarthritis.

Hip In rheumatoid arthritis, symmetric loss of joint space occurs with axial migration of the femoral head and protrusion of the acetabulum into the pelvis. A similar pattern is found in ankylosing spondylitis and CPPD crystal deposition disease, usually accompanied by osteophytes.

In osteoarthritis, joint space loss most frequently occurs in the superior weight-bearing aspect of the joint, resulting in superior migration of the femoral head with respect to the acetabulum. With the appearance of large osteophytes along the medial aspect of the femoral head, the resulting radiographic appearance may simulate symmetric loss of joint space or an untreated prior slipped capital femoral epiphysis.

Sacroiliac joint Erosion, sclerosis, and widening or narrowing of the sacroiliac joints are nonspecific radiographic manifestations of a variety of articular disorders; hence, the distribution of abnormalities is of primary importance. Unilateral involvement should always suggest an infectious process, although this may be an early feature of any of the seronegative spondyloarthropathies.

Ankylosing spondylitis is characteristically bilateral and symmetric, whereas symmetric or asymmetric involvement of the sacroiliac joints may be observed in psoriatic arthritis and Reiter syndrome. Involvement of the sacroiliac joints by rheumatoid arthritis and gout is uncommon; in rheumatoid arthritis asymmetric involvement is typical, whereas in gout the abnormalities may be symmetric or asymmetric. Bilateral symmetric alterations are also characteristic of hyperparathyroidism (due to subchondral bone resorption) and the arthritis associated with inflammatory bowel disease.

Radiographic Diagnosis in Rheumatology

Accurate interpretation of imaging studies in arthritis depends on two important factors: the structure of individual articular lesions, and their distribution within the skeleton. Together with the clinical presentation, a thorough understanding of the fundamental principles of radiographic interpretation should enable an accurate diagnosis to be established in most patients with rheumatologic complaints.

Diagnostic Imaging Techniques

Conventional radiography Radiographic examination generally includes more than one view or projection of a symptomatic joint. The radiographic examination should be tailored to the nature of the clinical presentation, with emphasis on symptomatic articulations and those that are frequent target sites of the disease in question.

In certain situations, radiographs obtained during weight bearing, stress, or traction can supplement the basic arthritic series. Weight-bearing radiographs allow more precise documentation of the integrity of the articular cartilage; the loss of joint space at the knee may be more pronounced when the patient is standing rather than supine. Furthermore, the amount of medial or lateral displacement of the tibia on the femur and the degree of varus or valgus angulation are best assessed by weight-bearing radiography.

Standing radiographs may occasionally be useful in evaluating the hip and ankle. Radiographs exposed during the application of stress across an articulation provide information regarding the integrity of surrounding ligaments and soft-tissue structures. Evaluation of the collateral and cruciate ligaments of the knee and the ligaments of the ankle and first metacarpophalangeal joint constitute the most frequent indications for this technique. Diagnosis of acromioclavicular joint separation following trauma may require the stress produced by hanging 5- or 10-lb weights at the wrists with side-to-side comparison.

Diagnosis of spondylolysis and spondylolisthesis may be facilitated by lateral radiographs of the lumbar spine obtained after prolonged standing with or without the application of additional weight. The release of gas, principally nitrogen, into an articulation during traction (vacuum phenomenon) usually indicates the absence of an effusion and permits crude assessment of the cartilaginous surfaces.

Conventional and computed tomography In certain locations such as the sternoclavicular joints, plain radiography is often inadequate, and computed tomography may be required. In the sacroiliac joint, either technique may help to identify the early alterations of sacroiliitis. At any site, conventional or computed tomography can detect subchondral alterations such as acute osteochondral fractures, osteochondritis dissecans, cysts, articular neoplasms, and early osseous extension of joint infections.

Computed tomography, with its higher-density discrimination capability as compared with conventional tomography, is useful in the evaluation of a variety of articular disorders, including intervertebral disk disease, ischemic necrosis of the femoral head, and tarsal coalition. The recent addition of helical CT data acquisition as well as multiplanar reformation and three-dimensional image reconstruction has enhanced the diagnostic capabilities of this modality, particularly at such sites as the hip and spine.

Arthrography Intraarticular injection of contrast media, air, or both is useful in the evaluation of a variety of disorders. Knee arthrography is employed in the diagnosis of recurrent meniscal injuries, articular cartilage alterations, synovial neoplasia, and synovial (popliteal) cysts. In the hip and knee, arthrography is particularly useful for evaluating joint prostheses for loosening or infection. Glenohumeral joint arthrography is helpful in the evaluation of rotator cuff tears, adhesive capsulitis, previous dislocations, synovial neoplasia, infection, and bicipital tendon abnormalities. Arthrography of other joints including the ankle, elbow, and wrist is helpful in the diagnosis of ligamentous injuries and lesions involving articular cartilage.

In certain clinical situations, arthrography is combined with CT, MRI, anesthetic or steroid injection, or diagnostic aspiration. Computed arthrotomography has found particular utility in the diagnosis of injuries to the glenoid labrum following shoulder dislocation and in the assessment of intraarticular osteocartilaginous bodies in the knee and other large articulations. Magnetic resonance arthrography utilizing intraarticular gadolinium-DTPA requires institutional review board approval in the United States, but is a useful technique for the postoperative knee, shoulder instability, rotator cuff pathology, and staging of osteochondritis dissecans.

Ultrasonography Diagnostic ultrasound often assists in the diagnosis of synovial cysts, particularly those arising from the hip or knee joints. At the latter site, ultrasound is a reliable means to distinguish a large popliteal cyst from thrombophlebitis. Sonography has also been used to diagnose rotator cuff tear in the shoulder and tendon rupture at other sites, particularly the Achilles region.

Skeletal scintigraphy Although the evaluation of joints with bone-seeking radiopharmaceuticals (particularly technetium-99m phosphate compounds) is a nonspecific endeavor, this procedure is far more sensitive than radiography in determining the activity of disease. Increased activity around diseased articulations is related to hyperemia and inflammation and can often be demonstrated prior to the appearance of radiographic abnormalities. In some cases, the distribution of abnormal joints as noted on a bone scan can yield insight into the specific diagnosis (the "target area" approach).

Magnetic resonance imaging This relatively expensive technology has proven useful in the diagnosis of a variety of specific articular disorders. Magnetic resonance imaging is more sensitive than skeletal scintigraphy in the early diagnosis of ischemic necrosis involving the femoral head and other sites, and its ability to visualize articular cartilage without contrast material has been helpful in the diagnosis of chondromalacia. The technique is also the preferred method for evaluating internal derangements of the temporomandibular joint, and is occasionally indicated in the inflammatory, crystal-induced, and degenerative arthritides.

Demographics

In clinical practice, patients often have more than one arthropathy. This is most commonly seen in patients with secondary osteoarthritis superimposed upon some other arthropathy.

Virtually any arthropathy that causes cartilage loss can lead to secondary osteoarthritis, with all of the classic signs of osteoarthritis, including osteophytosis. In fact, in certain patients, the changes from the primary arthropathy may be significantly obscured by those from the secondary osteoarthritis. A clue that this is happening is that the most distinctive sign of osteoarthritis, osteophytosis, is fairly minimal compared with other findings such as joint space narrowing or subchondral sclerosis. In fact, this is a very common presentation of rheumatoid arthritis of the knee: marked joint space narrowing and subchondral sclerosis, but no evident erosions and only minimal osteophytosis. In primary osteoarthritis, on the other hand, marked joint space narrowing (particularly in the medial femorotibial joint compartment) is usually accompanied by moderate or marked osteophytosis.

Other combinations of arthropathies are possible, such as gout with CPPD crystal deposition disease (mixed crystal deposition disease), gout with rheumatoid arthritis, and rheumatoid arthritis with DISH (RADISH). Therefore, when apparently contradictory findings are noted, remember that the law of parsimony is often broken.

Following is a demographic summary of the most important articular disorders:

TYPICAL AGE AT PRESENTATION	
<20 years	
Juvenile chronic arthritis	
Septic arthritis	
≥20 years	
Ankylosing spondylitis	15-35 years
Reiter syndrome	15-35 years
Enteropathic arthropathies	Young adults
Rheumatoid arthritis	25-55 years
Psoriatic arthritis	25-55 years
Osteoarthritis	Older patients
DISH	>55 years
CPPD crystal deposition disease	>55 years
Gout	35-60 years
SEX	
Male Predominance	
Ankylosing spondylitis	4:1 to 10:1
Psoriatic arthritis	2:1 to 3:1, but controversial
Reiter syndrome	5:1 to 50:1
Gout	20:1
DISH	5:1
CPPD crystal deposition disease	2:1
Osteoarthritis	Onset < 45 years
Enteropathic arthropathy	
Ulcerative colitis	4:1
Crohn disease	1:1
Female Predominance	
Rheumatoid arthritis	2:1 to 3:1
Osteoarthritis	Onset >45 years

NORMAL ARTICULAR ANATOMY

1. The temporomandibular joint is a typical synovial joint, but has two noncommunicating synovial spaces separated by an intervening fibrocartilaginous meniscus.
2. A synchondrosis is a temporary cartilaginous joint, such as the growth plates and cranial sutures.
3. A syndesmosis is an interosseous ligament such as that between the tibia and fibula.
4. An amphiarthrosis is a slightly movable articulation, such as the sacroiliac joint.
5. A ginglymus or hinge joint is a uniaxial joint in which a broad transversely cylindric convexity on one bone fits into a corresponding concavity on the other bone, allowing motion in one plane only (as in the elbow).
6. Diarthrodial joints include arthrodial, ginglymus, saddle, and ellipsoid types.
7. Arthrodial or gliding joints include the vertebral apophyseal articulations.
8. Ginglymus or hinge joints include the elbow and knee.
9. A saddle joint has movement about two axes and is exemplified by the first carpometacarpal joint.
10. An ellipsoid or modified ball-and-socket joint is the radiocarpal joint.
11. The iliopsoas bursa may communicate with the hip joint.
12. The subdeltoid bursa communicates with the subacromial bursa.
13. The ulnar bursa is associated with the flexor tendons of the wrist.
14. The radial bursa is associated with the tendon for the flexor pollicis longus.

Fundamental Concepts

In general, three major categories of arthritis exist: (1) inflammatory, exemplified by rheumatoid arthritis; (2) degenerative, exemplified by osteoarthritis; and (3) crystal induced, exemplified by gout. The radiographic diagnosis of a specific arthritic disorder depends upon both the structure and the distribution of joint abnormalities.

Any arthritic disorder may manifest monoarticular involvement early in its course. Radiographic assessment must be correlated with clinical findings (including pain, stiffness, swelling, deformity, and altered range of motion), epidemiologic factors (such as patient age and sex), and laboratory test results (including erythrocyte sedimentation rate, rheumatoid factor, antinuclear antibody, and HLA typing results). The inflammatory arthritides are characterized by the formation of granulation tissue known as pannus, which produces enzymes that destroy cartilage and bone. Degenerative arthritides involve altered biomechanical forces combined with host reactive processes that result in productive changes, including osteophytes, subchondral sclerosis, and buttressing. Crystal-induced arthritis includes both intraarticular and periarticular deposition; the three most commonly encountered crystals are monosodium urate, calcium hydroxyapatite, and calcium pyrophosphate dihydrate.

Morphologic features of arthritis that may be evident on radiographs include soft-tissue swelling, increased or decreased periarticular bone density, joint space narrowing or widening, poorly defined or well-defined erosions, subchondral cysts, bone proliferation including whiskering and enthesopathy, ankylosis, and ligamentous abnormalities including subluxation, contracture, and rupture. The most widely accepted classification system for rheumatologic disorders is that developed by the American Rheumatism Association.

SPECIFIC ARTHRITIDES

Rheumatologic Disorders: Definitions

1. Rheumatoid arthritis: a common form of polyarticular arthritis of uncertain etiology that is characterized by synovial inflammation and articular destruction.

PEARL

Rheumatoid arthritis may be similar histologically to Reiter syndrome, ankylosing spondylitis, psoriatic arthritis, and septic arthritis.

2. Juvenile chronic arthritis: a group of related disorders of uncertain etiology that begin during childhood and include the following specific symptom complexes: pauciarticular disease, seronegative polyarticular disease, Still disease, seropositive polyarticular disease, juvenile-onset ankylosing spondylitis, juvenile-onset psoriatic arthritis, and arthropathy of juvenile-onset inflammatory bowel disease.
3. Ankylosing spondylitis: the most common seronegative spondyloarthropathy, which is of uncertain etiology and tends to primarily involve the axial skeleton and large proximal joints.
4. Psoriatic arthritis: an arthropathy that occurs in approximately 2% to 6% of patients with cutaneous psoriasis and that manifests five distinct symptom complexes: polyarthritis, arthritis mutilans, symmetric distribution type, oligoarthritis, and spondyloarthropathy.
5. Reiter syndrome: a syndrome including the triad of arthritis, urethritis or cervicitis, and conjunctivitis, which may be incomplete or may occur with keratoderma blennorrhagia or balanitis.
6. Systemic lupus erythematosus: an immunologic disorder involving antinuclear antibody production resulting in severe and widely varied tissue injury with frequent remissions and exacerbations; musculoskeletal involvement is typified by a nonerosive deforming polyarthritis.

PEARL

Seronegative spondyloarthropathies:
- Synovial inflammatory changes are less severe than in rheumatoid arthritis
- Onset of ankylosing spondylitis: ages 15 to 35 (average, 26 to 27)
- Peripheral articular manifestations in ankylosing spondylitis: initially 10% to 20%, eventually up to 50%
- Extraskeletal manifestations of ankylosing spondylitis: iritis, aortic insufficiency, pericarditis, upper lobe fibrosis and/or cavitation, pleuritis, inflammatory bowel disease, amyloidosis
- Incidence of arthritis in psoriasis: 2% to 6%
- Prevalence of psoriasis in polyarticular arthritis: 3% to 5%
- "Romanus lesion" of ankylosing spondylitis: osteitis at anterosuperior, anteroinferior corners of vertebral bodies
- Periarticular osteoporosis: rheumatoid > psoriatic arthritis
- Reiter syndrome: follows gastrointestinal or genitourinary tract infection; shows predilection for young men; urethritis, conjunctivitis, arthritis
- Radiographic findings in Reiter syndrome: 60% to 80% of patients
- Sacroiliac involvement in Reiter syndrome: bilateral, symmetric or asymmetric

7. Progressive systemic sclerosis (scleroderma): a connective tissue disorder of unknown etiology, characterized by small-vessel disease and fibrosis that affect several organ systems including the skin.
8. Polymyositis-dermatomyositis: a connective tissue disorder of uncertain etiology that produces inflammation and muscle degeneration, resulting in symptoms of proximal muscle weakness and arthralgias, sometimes with an associated diffuse skin rash.
9. Amyloidosis: an infiltrative disorder that may involve abnormal deposits in bone, synovium, and associated soft tissues; the condition may be primary or occur secondarily in multiple myeloma, rheumatoid arthritis, connective tissue disorders, spondyloarthropathies, familial Mediterranean fever, or chronic infection.
10. Rheumatic fever: a syndrome that follows a group A beta-hemolytic streptococcal infection (particularly pharyngitis) and includes fever, valvular heart disease, polyarthritis, nonerosive deforming arthropathy, and systemic symptoms.

PEARL

Acute rheumatic fever:
- Equal sex incidence
- Peak age range: 5 to 15 years
- Abrupt onset of migratory arthritis involving larger joints (knee, ankle, wrist)
- Marked periarticular swelling, occasional joint effusions
- Carditis, skin rash, subcutaneous nodules, fever, leukocytosis, mild anemia

PEARL

Jaccoud arthropathy is also known as chronic post–rheumatic-fever arthropathy. It is related to rheumatic heart disease. Diagnostic criteria include a fever, reversible ulnar deviation and flexion deformity in the hand, fibular deviation in the foot, tendon crepitus, and absence of synovitis. Rarely, periarticular erosion occurs. Rheumatoid nodules and periarticular calcifications do not occur. Radiographically, it most closely resembles systemic lupus erythematosus.

11. Osteoarthritis: degenerative joint disease induced by one or more of the following factors: abnormal mechanical forces across a joint, normal mechanical forces placed on abnormal articular cartilage, weakening or collapse of subchondral bone.

12. Neuroarthropathy: a severely destructive and rapidly progressive arthropathy resulting from an initial alteration in sympathetic control of osseous blood flow leading to hyperemia and bone resorption, followed by a neurotraumatic mechanism that involves blunted pain and proprioception, instability, recurrent injury, fragmentation, and articular disorganization; the most common causes include diabetes mellitus, tabes dorsalis, and syringomyelia (Box 3-1).

Box 3-1   Neuroarthropathy: Important Features

- Pathogenesis: altered articular sensory nerves and recurrent trauma
- Loss of joint sensation inducing ligamentous laxity leading to marginal fractures, bony erosion inducing fragmentation leading to disarticulation, atrophy
- Fractures with exuberant callus
- Diabetes: feet, hands
- Syphilis: lower extremity, especially knee
- Syringomyelia: upper extremity, especially shoulder

PEARL

Causes of neuroarthropathy include syphilis, diabetes, syringomyelia, peripheral nerve injury, spinal cord tumors, spina bifida, and spinal cord injury. A neuropathic joint or neurotrophic arthropathy may occur if there is a change in sensory innervation. This, coupled with repeated minor stresses and trauma, creates the environment for an unstable joint. There is deprivation of accurate weight distribution and loss of protection by an intact nervous system.

The mechanism for neuroarthropathy is denervation of the joint. Of all patients with tabes dorsalis, 5% to 10% will manifest this condition, secondary to dorsal column degeneration in the spinal cord. No organisms are found in the joints, but disorganized bone and cartilage are typical.

13. Gout: monosodium urate crystal–induced synovial inflammation that may involve occasional acute attacks or a chronic arthropathy with crystal deposition in articular cartilage, subchondral bone, synovial and capsular tissue, and periarticular soft tissues; the disorder may be idiopathic or secondary to defective enzymes or chronic diseases (Box 3-2).

Box 3-2   Gout: Important Features

- Male/female ratio 20:1, primary or secondary
- Onset at age 35 to 60 years
- Elevated serum uric acid
- Asymptomatic hyperuricemia leading to acute monoarticular pain resulting in polyarticular involvement with articular destruction accompanied by monosodium urate deposits
- Monosodium urate crystal deposition inducing pannus, leading to erosion of cartilage and cortical margins with eventual ankylosis
- First metatarsophalangeal joint initially affected
- Ankle, knee, wrist, spine, sacroiliac involvement possible
- Minimal periarticular osteoporosis
- Destruction of first metatarsophalangeal joint with valgus deformity
- Soft-tissue tophi with or without calcification (ear, olecranon bursa)
- Well-defined erosions with sclerotic margins and overhanging edges, often distant from joints
- Periarticular cysts
- Late joint space narrowing
- Chondrocalcinosis
- Articular effusions

PEARL

It is convenient to divide gout into the idiopathic variety versus that associated with other disorders. These conditions include type 1 glycogen storage disease, Lesch-Nyhan syndrome, neoplastic and myeloproliferative disorders (polycythemia vera, lymphoma, leukemia, sickle cell disease), endocrine disorders (hyperparathyroidism, hypoparathyroidism), obesity, lead poisoning, and drug reaction (to diuretics, pyrazinamide, salicylates). Diabetes mellitus is not formally associated with gout, but it may occur in diabetes-induced renal failure or secondary to pyelonephritis.

14. Calcium pyrophosphate dihydrate crystal deposition disease: relatively common arthropathy resulting from calcium pyrophosphate dihydrate crystal deposition that is usually intraarticular; one or more of the following specific manifestations may occur: structural joint damage resembling that of osteoarthritis (pyrophosphate arthropathy), clinical presentation resembling that of gout (pseudogout), and hyaline cartilage and/or fibrocartilage calcification (chondrocalcinosis).

PEARL

Calcium pyrophosphate dihydrate crystal deposition disease:
- The condition has been associated with diabetes mellitus, degenerative joint disease, gout, hyperparathyroidism, hemochromatosis, Wilson disease, neuroarthropathy, ochronosis, hypophosphatasia, and a number of other conditions.
- Chondrocalcinosis occurs most frequently in the knees, wrists, symphysis pubis, elbows, and hips.
- Pyrophosphate arthropathy differs from degenerative joint disease in its unusual articular distribution, unusual intraarticular distribution, prominent subchondral cyst formation, severe and progressive destructive osseous alterations, and variable osteophyte formation.
- Familial cases have a female predominance with a relatively early age of onset.
- Diskal calcification related to CPPD crystal deposition occurs initially in the annulus fibrosus.
- In pyrophosphate arthropathy, the knee, wrist, and metacarpophalangeal joints are the most commonly involved articulations.
- The pseudogout syndrome affects 10% to 20% of patients and most frequently involves the knee.

15. Hemochromatosis: an arthropathy resulting from the intraarticular deposition of iron and/or calcium pyrophosphate dihydrate crystals; the underlying disease may be primary, resulting from enhanced gastrointestinal iron absorption, or secondary to excessive iron intake.

PEARL

Hemochromatosis:
- The primary form is 10 to 20 times more frequent in men than in women.
- Chondrocalcinosis occurs in up to 30% of patients.
- Osteoporosis occurs in 25% to 60% of patients.
- Arthropathy resembles that of CPPD crystal deposition disease in its involvement of unusual sites, formation of large subchondral cysts, and uniform joint space loss.
- Arthropathy differs from that in CPPD crystal deposition disease in its widespread involvement of the wrist and metacarpophalangeal joints, slower rate of progression, and presence of beaklike excrescences on the metacarpal heads.
- Onset of the disease generally occurs between the ages of 40 and 60 years.
- CPPD crystal deposition is a documented feature of the condition.

16. Wilson disease: an autosomal recessive disease characterized by abnormal tissue accumulation of copper.
17. Calcium hydroxyapatite crystal deposition disease: a crystal-induced disorder characterized by periarticular and occasionally intraarticular deposition of calcium hydroxyapatite; repetitive minor trauma and tissue necrosis with associated inflammation result in dystrophic calcification that is generally monoarticular.

PEARL

Calcium hydroxyapatite crystal deposition disease:
- Periarticular calcific deposits are usually monoarticular in distribution but can be polyarticular.
- Multiple joints are involved successively in about two thirds, and simultaneously in one third, of patients.
- Prevertebral soft-tissue swelling is a radiographic feature in the cervical region (longus colli tendon).
- Calcific deposits in the supraspinatus portion of the rotator cuff are best seen on external rotation views.
- Intraarticular crystal deposition with destructive alterations is less common than in CPPD crystal deposition disease.
- The disease is particularly common in patients between the ages of 40 and 70 years.
- Once deposited, periarticular calcifications can enlarge, remain stable in size, or regress.

18. Alkaptonuria (ochronosis): a hereditary metabolic disorder involving a defect in the enzyme homogentisic acid oxidase with secondary accumulation of homogentisic acid in organs and connective tissues.
19. Villonodular synovitis: a proliferative synovial disorder of unknown etiology, which may appear as an extraarticular soft-tissue mass with or without osseous erosions or as a diffuse intraarticular hemorrhagic process.
20. Synovial osteochondromatosis: a synovial metaplastic process of unknown etiology, characterized by the formation of cartilaginous nodules that may ossify and become free intraarticular bodies or remain attached to the synovium.

PEARL

Synovial osteochondromatosis is pathologically similar to osteochondroma. It is usually monoarticular in young or middle-aged adults, and it is rare in children. Male patients are more commonly affected. Osteoporosis is usually absent. Histologically, nodules of proliferating cartilage and synovium that grow in size and may calcify are present. The nodules of cartilage receive nutrition via the synovial fluid that bathes them.

21. Hypertrophic osteoarthropathy: a primary or secondary disorder that manifests itself clinically with symptoms of digital clubbing and painful, swollen joints; the major radiographic manifestation is a symmetric periostitis, the thickness and extent of which depend upon the duration of the underlying disease process (Box 3-3; Table 3-1).

Box 3-3   Secondary Hypertrophic Osteoarthropathy (Marie-Bamberger Syndrome): Important Features

- Arthritis in 30% to 40% of cases
- Increased blood flow induces linear periostitis
- Early: tibiae, fibulae, radii, ulnae
- Advanced: femora, humeri, metacarpals, metatarsals, phalanges
- Scintigraphy more sensitive than radiography to involvement of scapulae, clavicles, ribs, spine, skull, and facial bones
- Occurs in 1% to 12% of patients with bronchogenic carcinoma
- Occurs in up to 50% of patients with benign pleural mesothelioma
- Unilateral only with infected aortic graft

Box 3-3—cont'd

- Treatment: thoracotomy, hilar neurectomy, vagotomy (chemical or surgical), ipsilateral pulmonary artery occlusion, radiation or chemotherapy, intercostal nerve section, hypophysectomy, or laparotomy
- Most common sites of arthritis: knees, ankles, wrists, elbows, and metacarpophalangeal joints

Table 3-1   Major causes of secondary hypertrophic osteoarthropathy

Pulmonary	Bronchogenic carcinoma
	Abscess
	Bronchiectasis
	Emphysema
	Hodgkin disease
	Metastasis
	Cystic fibrosis
Pleural-diaphragmatic	Mesothelioma (benign)
Cardiac	Cyanotic congenital heart disease
Abdominal	Portal-biliary cirrhosis
	Ulcerative colitis
	Crohn disease
	Dysentery
	Neoplasms
	Biliary atresia
Miscellaneous	Nasopharyngeal carcinoma
	Esophageal carcinoma

22. Ischemic necrosis: epiphyseal bone and marrow cell death usually occurring secondary to traumatic interruption of blood supply, vascular compression, or intraluminal obstruction induced by a variety of specific disease processes (Boxes 3-4, 3-5, 3-6, 3-7; Table 3-2).

PEARL

The common feature of all causes of ischemic necrosis and bone infarction is an increase in the intraosseous marrow pressure.

"Creeping substitution" is the process by which necrotic bone is resorbed by a migrating reactive fibrous interface and supported by new bone formation in the surrounding viable zone. Bone scan results can be normal in ischemic necrosis if the scan is performed between "cold" and "hot" phases of disease.

Ischemic necrosis or bone infarction almost invariably occurs within areas of predominantly fatty marrow. Rarely, areas of normal active hematopoiesis are affected, as in sickle cell anemia and following complete traumatic disruption of the arterial supply. Ischemic necrosis of cortical bone is rare and occurs with extensive interruption of the blood supply, as in osteomyelitis (sequestration).

Box 3-4   Phases of Ischemic Necrosis

1. Cellular death and initial host response
2. Cell modulation in ischemic zone, hyperemia
3. Emergence of reactive interface
4. Remodeling of reactive interface
5. Crescent sign and articular collapse

Box 3-5   Zones of Bone Infarcts

- Central area of cell death
- Zone of ischemic injury
- Peripheral area of active hyperemia
- Normal surrounding tissue

Box 3-6   Complications of Ischemic Necrosis and Bone Infarction

- Secondary degenerative joint disease
- Intraarticular osteochondral bodies
- Malignant degeneration (usually fibrous neoplasms)

Box 3-7   Major Causes of Osteonecrosis (Conditions that GIVE INFARCTS)

- Gaucher disease, gout
- Idiopathic (spontaneous osteonecrosis of the knee, Chandler disease of the hip, Muller-Weiss disease of the tarsal navicular in adults, Legg-Calvé-Perthes disease)
- Vasculitides (systemic lupus erythematosus, thromboangiitis obliterans [Buerger disease])
- Environmental causes (thermal injury)
- Infection (sequestration)
- Neoplasms
- Fat (corticosteroid induced)
- Alcoholism
- Radiation
- Caisson disease
- Trauma
- Sickle cell anemia

Table 3-2   Ischemic necrosis

Cell type	Time to anoxic death
Hematopoietic elements	6-12 hrs
Osteocytes, osteoclasts, osteoblasts	12-48 hrs
Marrow fat cells	48 hrs to 5 days

23. Diffuse idiopathic skeletal hyperostosis (Forrestier disease): a progressive disorder of unknown etiology, characterized by ligamentous ossification in the spine and bone proliferation at entheses in the peripheral skeleton (Boxes 3-8, 3-9).

Box 3-8   Diffuse Idiopathic Skeletal Hyperostosis (Ankylosing Hyperostosis of Forrestier and Rotes-Querol): Important Features

- Flowing calcification and ossification along the anterolateral aspect of at least four contiguous vertebral bodies with or without localized pointed excrescences at intervening body-disk junctions
- Relative preservation of disk height and absence of degenerative changes (vacuum phenomena, sclerosis)
- Usually affecting middle-aged or elderly men
- Spinal stiffness, mild back pain
- Cervical dysphagia, tendinitis, heterotopic ossification following total joint replacement
- Most common in thoracic spine
- Degenerative enthesopathy at peripheral sites
- Associated HLA-B27 positivity

Box 3-9   Ossification of the Posterior Longitudinal Ligament (OPLL) of the Spine: Important Features

- More frequent in men
- Occurs in fifth to seventh decades
- Appearance of cord signs when ligament thickness exceeds 60% of sagittal cervical canal diameter
- Cervical (particularly C3 to C5), thoracic or lumbar sites
- Most common in Japan, 1.0% to 1.7% incidence
- Ossified mass attached to or separated from posterior aspect of vertebral body and disk
- Frequently associated with DISH
- Thoracic involvement usually at T4 to T7
- Clinical manifestations: cord signs, radiculopathy, neck pain, or asymptomatic incidental finding

24. Transient regional osteoporosis: a self-limited disorder that may be migratory, which demonstrates periarticular demineralization radiographically and manifests itself clinically as an arthritis.
25. Multicentric reticulohistiocytosis or lipoid dermatoarthritis is an uncommon condition resembling gout radiographically with a predilection for the interphalangeal joints of the hands and the atlantoaxial articulation.

PEARL

Multicentric reticulohistiocytosis:

- Peak age of onset is 40 to 50 years, and women are more frequently affected than men.
- Polyarthritis is the initial manifestation of the disease in 60% to 70% of patients.
- The most characteristic sites of involvement are the interphalangeal joints of the hands.
- Symmetric distribution of articular alterations is typical.
- Atlantoaxial subluxation can be an early manifestation of the disease.
- Marginal erosions and uncalcified soft-tissue nodules occur, without significant periarticular osteoporosis.
- Associated cutaneous lesions consist of granulomatous infiltration by multinucleated histiocytic giant cells containing PAS-positive material.

Epidemiology

1. Synovial osteochondromatosis affects predominantly men in their third to fifth decades.
2. Osteoarthritis has equal sex incidence, often appearing earlier in men; incidence increases with age.
3. Wilson disease appears in adolescents and young adults, but arthropathy may begin later; it is more common in men.
4. Transient regional osteoporosis is a self-limited, sometimes migratory disorder with a predilection for men.
5. Progressive systemic sclerosis (scleroderma) affects women three times as often as men, usually during the third to fifth decades.
6. Rheumatoid arthritis appears in young to middle-aged patients, with a female predominance of 2:1 to 3:1.
7. Gout affects predominantly middle-aged to elderly white men, with moderate hereditary predisposition; patients with renal disease or myeloproliferative disorders may have an early age of onset.
8. Polymyositis and dermatomyositis are most common among women in their third through fifth decades, although one form of the disease also affects children.

PEARL

Dermatomyositis is associated with an elevated incidence of carcinomas in adults, which increases with age. Hypogammaglobulinemia or leukemia may occur in children with the disease. Juvenile dermatomyositis often resolves spontaneously, but has a 40% mortality rate if untreated. Twenty percent of cases of dermatomyositis affect children or adolescents. Female-to-male ratio is 3:2. Proximal muscle weakness, extensive soft-tissue calcification, vasculitis, and joint contractures occur.

9. Ankylosing spondylitis is a familial disorder usually appearing between the ages of 15 and 35 years; it affects men 4 to 10 times as frequently as women.
10. Hypertrophic osteoarthropathy in its primary form is familial and far more common in men; onset occurs during adolescence, usually with spontaneous arrest of the disease in young adulthood (Box 3-10).

Box 3-10 Primary Hypertrophic Osteoarthropathy (Familial Pachydermoperiostosis or Touraine-Solente-Golé Syndrome): Important Features

- 3% to 5% of all cases of hypertrophic osteoarthropathy
- 85% men, usually black
- Onset in adolescence
- Bone marrow failure and/or extramedullary hematopoiesis may occur
- Clubbing, enlargement of hands and feet, facial and scalp skin coarsening with oiliness, hyperhidrosis, bone and joint pain, stiffness, decreased range of motion, kyphosis, neurologic symptoms secondary to compression by osseous overgrowth
- Irregular, thick, asymmetric periostitis involving diaphyses, metaphyses, and epiphyses
- Articular inflammation less severe than in secondary hypertrophic osteoarthropathy
- Shaggy, ill-defined bony excrescences
- Calvarial and skull base thickening, prominent frontal and sphenoid sinuses, mandibular enlargement
- Acro-osteolysis, ligamentous calcification or ossification, articular ankylosis from osseous bridging

---

11. Psoriatic arthritis has approximately equal sex incidence, with onset during young adulthood.
12. Diffuse idiopathic skeletal hyperostosis (Forrestier disease) is a common incidental finding among middle-aged and elderly individuals, particularly men.
13. Amyloidosis occurs during the fourth through eighth decades, with a predilection for men.
14. Systemic lupus erythematosus usually appears in young adults under age 40, with a female/male ratio of 5:1 to 10:1; the condition exhibits a predilection for black individuals.
15. Reiter syndrome affects young adults with a strong predilection for men.
16. Alkaptonuria (ochronosis) has equal sex incidence, with aural pigmentation occurring in early adulthood, before the onset of arthropathy.

17. Still disease is an acute systemic disorder occurring in children under 5 years of age with equal sex incidence.
18. Pigmented villonodular synovitis is a monoarticular process appearing in middle-aged to elderly individuals.
19. Pauciarticular juvenile rheumatoid arthritis involves one to three large joints, and predominantly occurs among young girls.
20. Adult-onset Still disease is pauciarticular arthritis with systemic manifestations occurring in patients over 18 years of age.
21. Calcium hydroxyapatite crystal deposition disease has equal sex incidence, with a tendency to affect middle-aged to elderly patients.
22. Seropositive polyarticular juvenile rheumatoid arthritis (juvenile-onset adult-type rheumatoid arthritis) is a polyarticular disease resembling rheumatoid arthritis and generally occurring among teenage girls.
23. Hemochromatosis has its onset in middle age, with men affected much more commonly than women.
24. Juvenile-onset ankylosing spondylitis generally occurs among adolescent male individuals with back complaints.
25. Calcium pyrophosphate dihydrate crystal deposition disease affects middle-aged to elderly men and women, and may coexist with gout (mixed crystal deposition disease).
26. Seronegative polyarticular juvenile rheumatoid arthritis occurs at any age with female predominance and symmetric involvement of both large and small joints.

Clinical description

1. Polymyositis-dermatomyositis: skin rash, muscle weakness, arthralgias, Raynaud phenomenon, pulmonary fibrosis, pericarditis, abdominal pain, dysphagia
2. Rheumatoid arthritis: chronic or episodic morning stiffness, joint pain and swelling, muscle wasting, tendon contractures and rupture, subcutaneous or tendon sheath nodules, tenosynovitis, bursitis, inflammatory enthesopathy, pleural effusion, pulmonary nodules, diffuse interstitial pneumonitis
3. Synovial osteochondromatosis: pain and limited range of motion, or asymptomatic
4. Pseudogout syndrome: acute self-limited symptoms simulating gout, infection, or rheumatoid arthritis
5. Progressive systemic sclerosis (scleroderma): Raynaud phenomenon; cutaneous alterations on the hands, feet, or face (edema, thickening, fibrosis, inelasticity, shininess, atrophy); acral tapering of digits; distal joint pain and stiffness; proximal myopathy; dysphagia; esophageal atrophy, fibrosis, and dysmotility; colonic pseudosacculations; pulmonary and renal fibrosis; pericarditis; myocarditis

6. Reiter syndrome: polyarticular arthritis with heel pain, lower back pain, urethritis, conjunctivitis, pulmonary fibrosis, valvular heart disease, diarrhea
7. Pyrophosphate arthropathy: chronic progressive arthropathy with or without acute exacerbations, simulating osteoarthritis
8. Diffuse idiopathic skeletal hyperostosis (Forrestier disease): mild back pain, mildly decreased range of spinal mobility, dysphagia, tendinitis
9. Ischemic necrosis: joint pain with weight bearing, decreased range of motion
10. Primary hypertrophic osteoarthropathy (pachydermoperiostosis): painful, swollen joints; clubbed digits; skin thickening on the face and hands

---

PEARL

Acro-osteolysis in pachydermoperiostosis is associated with digital clubbing and occurs late; the osseous hallmark of the syndrome is periosteal new bone formation. Other causes of acro-osteolysis include vasculitis, collagen-vascular disease, psoriasis, epidermolysis bullosa, frostbite, burns, secondary hypertrophic osteoarthropathy, septic shock, multicentric reticulohistiocytosis, neuroarthropathy, leprosy, polyvinyl chloride exposure, trauma, and familial forms.

---

11. Still disease: high fever, hepatosplenomegaly, lymphadenopathy, polyarthritis
12. Pigmented villonodular synovitis: relatively painless joint effusion
13. Amyloidosis: joint pain, stiffness, swelling, and contractures; carpal tunnel syndrome; organomegaly; renal disease; cardiac failure; pulmonary septal infiltration; decreased gastrointestinal peristalsis with submucosal thickening
14. Rheumatic fever: self-limited, recurrent, migratory joint pain, inflammation, and swelling; fever; valvular heart disease
15. Pauciarticular juvenile chronic arthritis: mild involvement of one to three joints, with chronic iridocyclitis in 25% of patients
16. Ankylosing spondylitis: lower back pain, spinal stiffness, increased thoracic kyphosis, decreased lumbar lordosis, limited chest expansion, iritis, aortic insufficiency and other cardiac disease, upper lobe pulmonary interstitial disease with fibrosis
17. Jaccoud arthropathy: asymptomatic condition following multiple episodes of polyarthritis
18. Psoriatic arthritis: joint pain, swelling, and limited range of motion in the hands and feet; sausage digits; thickening, pitting, or discoloration of nails; lower back pain; characteristic skin disease

19. Osteoarthritis: joint pain, limited range of motion, crepitus, and subluxation (especially genu varus); Heberden and Bouchard nodes
20. Systemic lupus erythematosus: polyarthritis; myositis; characteristic skin rashes; malaise; weakness; neurologic dysfunction; pulmonary vasculitis, fibrosis, and effusions; pericarditis; cardiomyopathy; nephritis
21. Neuroarthropathy: joint swelling and instability, with pain in 30% of patients
22. Hemochromatosis: mild joint pain, swelling, and stiffness, sometimes with acute attacks; bronze skin; cirrhosis; diabetes; heart disease
23. Gout: red, swollen, extremely painful monoarticular or oligoarticular arthritis; tophaceous deposits in bursae (especially olecranon, prepatellar bursae), aural helix, or other soft tissues
24. Wilson disease: asymptomatic; or osteoarthritis-like joint involvement, basal ganglia degeneration, cirrhosis, corneal Kayser-Fleischer rings
25. Calcium hydroxyapatite crystal deposition disease: single painful joint
26. Alkaptonuria (ochronosis): mild joint pain and decreased range of motion, aural helix pigmentation

Laboratory findings

1. Alkaptonuria (ochronosis): urinary homogentisic acid
2. Calcium hydroxyapatite crystal deposition disease: hydroxyapatite crystals present on light microscopy (with Wright stain) or electron microscopy of aspirated material
3. Rheumatoid arthritis: erythrocyte sedimentation rate elevated and correlated with disease activity; rheumatoid factor test may be negative early but is eventually positive in up to 95% of patients

---

PEARL

Rheumatoid factor is an antibody against IgG. Most rheumatoid factor is IgM, but small amounts of IgG and IgA may be present. Latex fixation for rheumatoid factor is a highly sensitive test that may give false positive results, especially in the older population. A more specific test is agglutination of red blood cells of sheep that have been coated with rabbit IgG. Rheumatoid factor is not present in ankylosing spondylitis, nor is it associated with HLA-B27 positivity.

Two to six percent of the normal population has a positive rheumatoid factor (usually low titer); prevalence increases with age.

---

4. Wilson disease: evidence of renal tubular abnormalities

5. Still disease: anemia, polymorphonuclear leukocytosis
6. Adult-onset Still disease: rheumatoid factor negative; antinuclear antibody negative
7. Hemochromatosis: increased serum iron levels and iron-binding capacity
8. Pauciarticular juvenile chronic arthritis: rheumatoid factor negative; often antinuclear antibody positive
9. Calcium pyrophosphate dihydrate crystal deposition disease: calcium pyrophosphate crystals present on polarizing microscopy of synovial fluid

---

PEARL

Calcium pyrophosphate dihydrate crystals demonstrate weakly positive birefringence, and joint aspiration is diagnostic. Acute clinical episodes of joint pain caused by the disease are known as pseudogout and most commonly affect the knee. The wrist is frequently affected by chondrocalcinosis and pyrophosphate arthropathy.

---

10. Ankylosing spondylitis: erythrocyte sedimentation rate parallels disease activity; over 90% of patients are HLA-B27 positive; rheumatoid factor negative

---

PEARL

HLA-B27 positivity is associated with ankylosing spondylitis, psoriatic arthritis, Reiter syndrome, and reactive arthritis secondary to infection by *Yersinia,* gonococcus, or *Salmonella* organisms. Seven to ten percent of the white population has HLA-B27 positivity, and the incidence is lower in the black population. Psoriatic arthritis with spinal involvement has a higher percentage of HLA-B27 positivity than if only peripheral joints are involved. This is also true in the arthritis associated with inflammatory bowel disease.

---

11. Psoriatic arthritis: elevated erythrocyte sedimentation rate; 25% to 60% of patients are HLA-B27 positive; rheumatoid factor negative
12. Gout: hyperuricemia; monosodium urate crystals present on polarizing microscopy of synovial fluid
13. Systemic lupus erythematosus: antinuclear antibody positive; positive lupus erythematosus cell preparation; possible false positive rheumatoid factor

PEARL

The presence of extractable nuclear antigen is diagnostic of systemic lupus erythematosus. However, patients with dermatomyositis and scleroderma may also test positive for an antinuclear antibody.

14. Neuroarthropathy: hyperglycemia and glycosuria, or positive Venereal Disease Research Laboratories (VDRL) test results, may be present
15. Reiter syndrome: elevated erythrocyte sedimentation rate; 50% to 80% of patients are HLA-B27 positive; rheumatoid factor negative
16. Polymyositis or dermatomyositis: elevated muscle enzyme levels during active disease stages
17. Amyloidosis: biopsy results positive for abnormal protein characterized by fine fibrils on electron microscopy, dark brown staining with iodine, and birefringent staining with Congo red
18. Erosive (inflammatory) osteoarthritis: erythrocyte sedimentation rate normal; rheumatoid factor negative
19. Progressive systemic sclerosis (scleroderma): elevated erythrocyte sedimentation rate in 70% of patients; positive antinuclear antibody in up to 95% of patients; positive rheumatoid factor in up to 40% of patients
20. Rheumatic fever: throat culture positive for group A beta-hemolytic streptococcus; rheumatoid factor negative
21. Ischemic necrosis: anemia with sickle-shaped erythrocytes, or marrow infiltration by Gaucher cells, may be present
22. Hypertrophic osteoarthropathy secondary to cyanotic congenital heart disease: reduced serum hemoglobin levels exceed 5 g/100 ml
23. Pigmented villonodular synovitis: hemorrhagic joint aspirate

General radiographic abnormalities

1. Polymyositis or dermatomyositis: muscle edema, atrophy, and calcification; subcutaneous, periarticular, and sheetlike calcification along fascial or muscle planes; transient periarticular demineralization; flexion deformities (Fig. 3-1)
2. Psoriatic arthritis: periarticular or diffuse (sausage digit) soft-tissue swelling; generally normal periarticular bone density; joint space narrowing; marginal and subchondral (pencil-in-cup) erosions; bone proliferation or whiskering; phalangeal periosteal reaction; inflammatory enthesopathy; ankylosis of small joints (Fig. 3-2; Box 3-11)

Box 3-11 Psoriatic Arthritis: Important Features

- Nail abnormalities
- Asymmetric distribution
- Bilateral symmetric or asymmetric sacroiliitis
- Broad irregular asymmetric paravertebral ossification
- Atlantoaxial subluxation
- Hips: axial migration with bone proliferation at femoral head-neck junctions
- Minimal periarticular osteoporosis
- Carpal, metacarpophalangeal, proximal interphalangeal, distal interphalangeal joint involvement
- Arthritis mutilans with pencil-in-cup deformities
- Poorly defined bone proliferation around erosions (whiskering)
- Ankylosis
- Sausage digits
- Acro-osteolysis
- Bone sclerosis ("ivory phalanx")
- Articular effusions

3. Neuroarthropathy: large joint effusions; osseous debris and fragmentation; normal to increased periarticular bone density; joint space narrowing; coexistent erosive and productive alterations; ligamentous laxity with articular subluxation or dislocation; rapid disease progression (Figs. 3-3, 3-4)

PEARL

Early neuropathic joint disease must be differentiated from osteoarthritis, CPPD crystal deposition disease, osteonecrosis, and chronic infection. Later in the disease the joint becomes grossly disorganized with subluxation, fractures, and periarticular osseous fragments. Periarticular bone resorption tends to be sharply marginated, in contrast to the situation in infection. Destruction, deformity, debris, dislocation, and dense bones are characteristic of neuroarthropathy. Periarticular osteoporosis is not a typical feature.

4. Synovial osteochondromatosis: usually no effusion; multiple intraarticular bodies of similar size and variable mineralization, sometimes lamellar or trabecular; mechanical destruction of cartilage and bone with secondary osteoarthritis
5. Systemic lupus erythematosus: periarticular soft-tissue swelling; subcutaneous calcifications in fewer than 10% of cases, usually involving the lower extremities; occasional periarticular demineralization; ligamentous laxity in 10% of cases, including reducible ulnar deviation at metacarpophalangeal

A

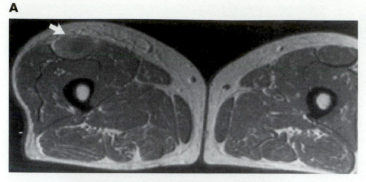

B

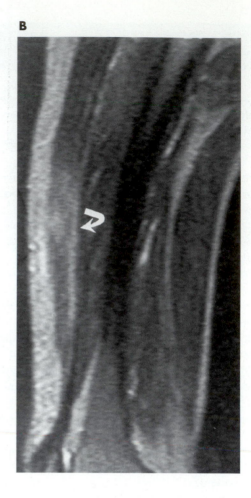

C

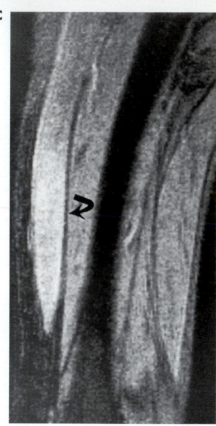

Fig. 3-1   Inflammatory myopathy of the rectus femoris. **A,** Transaxial T1-weighted MR image reveals enlargement and mildly decreased signal intensity of the affected muscle on the left (*arrow*). **B,** Sagittal T1-weighted image demonstrates similar findings (*arrow*). **C,** Corresponding gradient-echo image exhibits high signal intensity within the affected muscle (*arrow*).

joints, subluxation at first carpometacarpal joint, and flexion or extension deformities at interphalangeal joints (Box 3-12)

Box 3-12   Systemic Lupus Erythematosus: Important Features

- Subluxation (first carpometacarpal joint)
- Ulnar deviation (metacarpophalangeal joints)
- Female predominance
- Osteoporosis, predominately secondary to corticosteroid therapy
- Osteonecrosis, secondary to vasculitis and corticosteroid therapy

---

6. Rheumatoid arthritis: periarticular swelling secondary to effusions and synovitis; large synovial cysts, often dissecting; periarticular and generalized osteopenia; joint space narrowing; marginal and subchondral erosions; subchondral cysts, which may be large in the hips and knees; rare carpal and tarsal ankylosis; ligamentous laxity, tendon ruptures, and contractures leading to deformities and altered function; secondary osteoarthritic alterations (Fig. 3-5; Box 3-13)

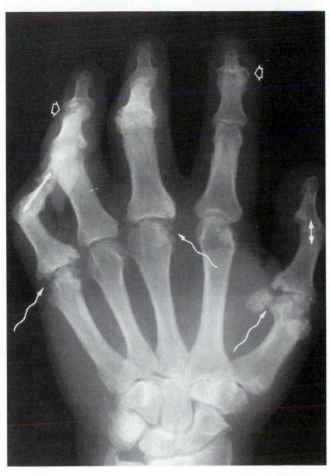

Fig. 3-2 Joint space narrowing, marginal and central osseous erosions with bone proliferation (*wavy arrows*), ankylosis (*double-headed arrow*), relative preservation of bone density, and a patchy distribution with distal interphalangeal joint involvement (*open arrows*) are typical of psoriatic arthritis.

Box 3-13   Rheumatoid Arthritis

- Onset occurs between 25 and 55 years of age, female/male ratio is approximately 3:1.
- Bilateral symmetry; centripetal involvement beginning in proximal interphalangeal joints
- Acute synovial inflammation leading to granulation tissue, osseous erosion, articular destruction, and eventual ankylosis
- Early: periarticular soft-tissue swelling, narrowed joint space, poorly defined erosions, pseudocysts, subluxations
- Late: flexion or extension contractures; ulnar deviation at metacarpophalangeal joints; cartilage and subchondral bone destruction; fourth and fifth metatarsophalangeal joint involvement, atlantoaxial and cervical apophyseal joint subluxation
- Distal interphalangeal joint involvement uncommon
- Proximal interphalangeal joint involvement usual

Box 3-13—cont'd

- Metacarpophalangeal joint involvement frequent
- Distal radioulnar joint often involved
- Erosions of ulnar and radial styloid processes, scaphoid, metacarpal heads, phalanges; pancompartmental involvement of large joints (knees, elbows, hips); fibrous or bony ankylosis
- Subcutaneous nodules (pressure areas)
- Articular effusions
- Osteoporosis (generalized and periarticular)
- Muscle atrophy
- Positive rheumatoid factor test

PEARL

Rheumatoid arthritis:
- Active disease is characterized by a highly vascularized synovium, periarticular osteoporosis, juxtaarticular bone erosions, and enlargement of the epiphyses in childhood.
- Pseudocystic form (associated with increased physical activity) simulates gout or juxtaarticular neoplasia.
- Advanced disease may produce fragments of cartilage and/or bone embedded in synovium.
- Bursal involvement occurs in over 5% of cases (popliteal region, wrist, foot, olecranon, subacromial-subdeltoid, retrocalcaneal bursae).
- Involvement of cartilaginous joints (symphysis pubis, manubriosternal, discovertebral), and entheses is less likely and less severe than in seronegative spondyloarthropathies.
- Complications include synovial cysts, tendon rupture, subcutaneous nodules (20% of patients), cutaneous joint fistulae, necrotizing arteritis, and fractures.
- Osteoporosis is caused by multiple factors, including hyperemia (periarticular), disuse, corticosteroid therapy, the disease itself, and reflex sympathetic dystrophy syndrome.

---

7. Jaccoud arthropathy: reversible deformity including ulnar deviation and flexion at metacarpophalangeal joints with fibular deviation and flexion at metatarsophalangeal joints; late mechanical cartilage destruction secondary to subluxation; late pressure erosions on metacarpal heads radially; occasional periarticular demineralization
8. Alkaptonuria (ochronosis): dystrophic calcification (calcium hydroxyapatite) of disks, cartilage, tendons, and ligaments; osteopenia; cartilage fragmentation; subchondral cysts and sclerosis; small osteophytes; ligamentous rupture (Box 3-14)

A

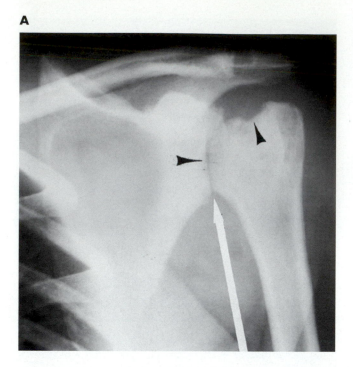

B

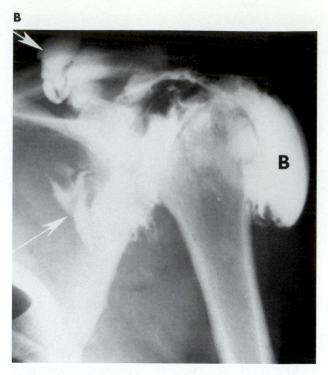

Fig. 3-3   Syringomyelia with neuroarthropathy (atrophic type). **A,** Radiography of the shoulder reveals joint space narrowing (*arrow*) with erosion of the humeral head and glenoid (*arrowheads*). **B,** Arthrography demonstrates opacification of the subacromial-subdeltoid bursa (*B*) indicative of rotator cuff disruption, as well as synovial cyst formation (*arrows*). Differential diagnosis should include consideration of rheumatoid and septic arthritis.

Box 3-14   Ochronosis (Alkaptonuria): Important Features

- Homogentisic acid (oxygenase) deficiency
- Homogentisic acid deposition in connective tissue and cartilage leading to degeneration and calcification
- Premature degenerative joint disease (knee, shoulder, hip, spine) with osteochondral bodies
- Intervertebral disk space narrowing
- Osteoporosis
- Calcification of intervertebral disks; meniscal, aural, and nasal cartilages; joint capsules; bursae
- Degenerative enthesopathy

Box 3-15   Ankylosing Spondylitis: Important Features

- Male predominance; HLA-B27 positivity
- Symmetric sacroiliitis: sclerosis, widening, fusion
- Spine: squaring of vertebral bodies, shiny corner sign, bridging syndesmophytes, bamboo spine, trolley track sign
- Peripheral inflammatory arthritis with whiskering, especially in hips

9. Ankylosing spondylitis: syndesmophyte formation in the annulus fibrosus of disks; disuse osteopenia subsequent to ankylosis, particularly in sacroiliac joints and spine (anterior and posterior columns); joint space narrowing; small erosions with bone proliferation; subchondral cysts; enthesopathy affecting pelvis, calcaneus, patella, and other sites; calcification of spinal longitudinal ligaments; rare instability and periostitis (Fig. 3-6; Box 3-15)

10. Gout: soft-tissue tophi, which may be calcified; chondrocalcinosis; relative preservation of joint space; intraarticular or paraarticular erosions with sclerotic margins and often overhanging edges; enlargement of phalangeal ends; secondary osteoarthritic alterations; enthesopathy; occasional ligament rupture (Fig. 3-7; Box 3-16)

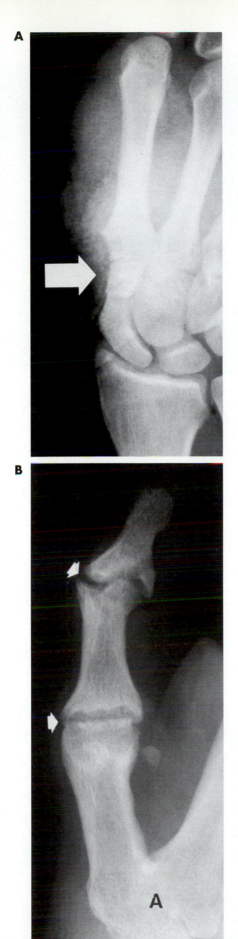

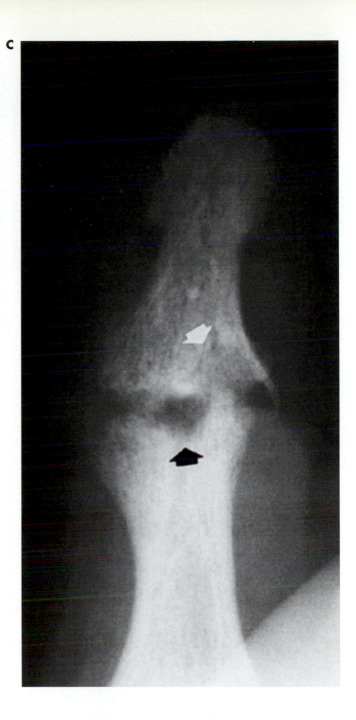

Fig. 3-4   Neuroarthropathy following thumb replantation. **A,** Amputation has occurred through portions of the second metacarpal base, trapezoid, and scaphoid (*arrow*). **B,** Following osseous ankylosis (*A*) at the replantation site, early flattening of the articular surfaces of the metacarpophalangeal and interphalangeal joints is evident (*arrows*). **C,** Rapidly progressive osteolysis at the interphalangeal joint (*arrows*) is evident on a follow-up radiograph. Other potential complications of replantation include osteonecrosis, nonunion, and secondary infection.

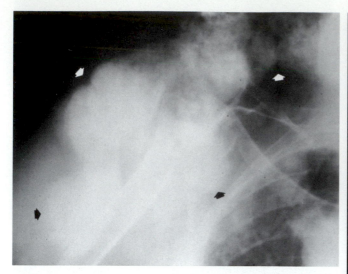

Fig. 3-5   Dissecting synovial cyst (*arrows*) of the glenohumeral joint in advanced rheumatoid arthritis (postarthrography).

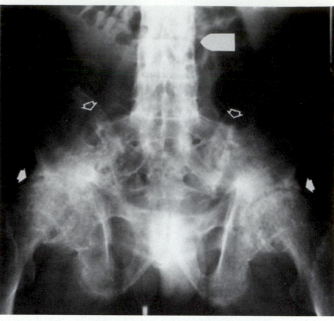

Fig. 3-6   Evolving fusion of the spine (*arrowhead*), sacroiliac joints (*open arrows*), and hips (*closed arrows*) in advanced ankylosing spondylitis.

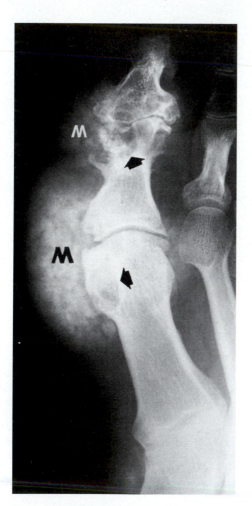

Fig. 3-7   Gout with renal involvement. Calcified tophaceous masses (*M*) with associated well-defined osseous erosions (*arrows*), some with overhanging edges, are characteristic.

Box 3-16   Gout: Important Features

- Articular, osseous, tendinous, and/or bursal monosodium urate deposits
- Soft-tissue tophi (with or without calcification)
- Subarticular cysts
- Well-circumscribed osseous erosions with overhanging edges and sclerotic marginsPredilection for hands and feet (particularly first metatarsophalangeal joints)
- Relative joint space preservation; minimal periarticular osteoporosis
- Late articular destruction
- Not necessarily periarticular
- Chondrocalcinosis
- Bursal swelling (especially olecranon)

PEARL

The elbows, wrists, and knees are affected frequently in gout. It is a crystal-induced synovial process producing subchondral bone erosions. Soft-tissue swelling of the digits (hands and feet) tends to be eccentric. The cartilage of an affected joint is often destroyed in late stages. Normal bone density in the periarticular region is characteristic. Chondrocalcinosis is frequent.

11. Diffuse idiopathic skeletal hyperostosis (Forrestier disease): calcification of anterior longitudinal ligament, paravertebral soft tissue, and occasionally posterior longitudinal ligament; large spinal osteophytes; enthesopathy, especially in pelvis, calcaneus (Achilles and plantar aponeurosis insertions), and patella (quadriceps insertion); calcification of pelvic ligaments (Fig. 3-8)

12. Reiter syndrome: periarticular or diffuse soft-tissue swelling (sausage digit); periarticular demineralization; joint space narrowing; erosions with mild bone proliferation; phalangeal periostitis; enthesopathy, especially in pelvis and calcaneus; ankylosis of sacroiliac joints and rarely, others (Fig. 3-9; Box 3-17)

Box 3-17 Reiter Syndrome: Important Features

- Urethritis, conjunctivitis, polyarthritis: venereal (men, ages 20 to 40) or postdysenteric (women, children)
- Acute stage: effusions in weight-bearing joints, soft-tissue swelling, poorly defined calcaneal enthesophytes
- Chronic stage: marginal erosions, bilateral symmetric or asymmetric sacroiliitis
- Paravertebral ossification unusual; interphalangeal joint of great toe often involved
- Poorly defined bone proliferation (whiskering)
- Lower extremity predominance
- Dactylitis (sausage digit)
- Asymmetric distribution
- Minimal periarticular osteoporosis
- Bone sclerosis

13. Hemochromatosis: mild soft-tissue swelling; chondrocalcinosis; osteopenia; focal cartilage loss; prominent beaklike osteophytes, particularly on metacarpal heads; prominent subchondral cysts

14. Osteoarthritis: small joint effusions; variable chondrocalcinosis; periarticular deposition of calcium hydroxyapatite; intraarticular bodies secondary to synovial metaplasia or fractured osteophytes; joint space

A

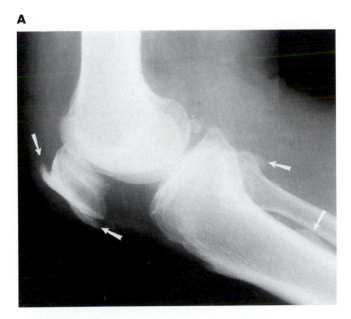

B

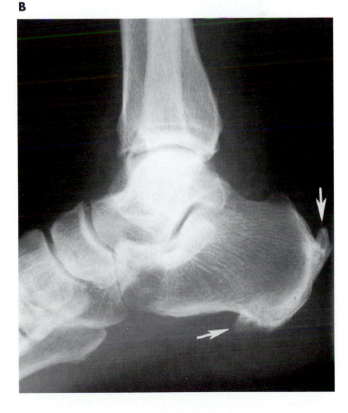

Fig. 3-8 Well-defined enthesopathy (*arrows*) involving the knee and ankle in diffuse idiopathic skeletal hyperostosis. Differential diagnosis should include consideration of acromegaly, hypoparathyroidism and its variants, fluorosis, and X-linked hypophosphatemic vitamin D–resistant osteomalacia. **A,** Knee. **B,** Ankle.

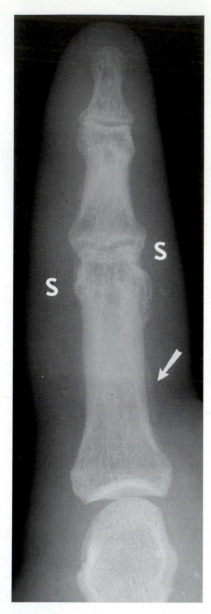

Fig. 3-9   Sausage digit in Reiter syndrome. Diffuse soft-tissue swelling (*s*) is associated with early bone proliferation (*arrow*). Differential diagnosis should include consideration of psoriatic arthritis and infection.

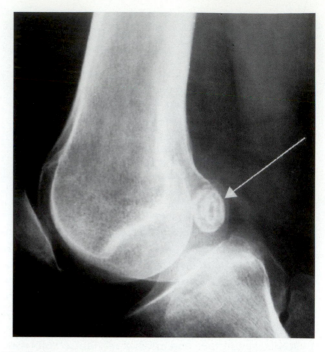

Fig. 3-10   Secondary osteochondral body formation (*arrow*) in osteoarthritis. Differential diagnosis of intraarticular osteochondral bodies should include consideration of pyrophosphate arthropathy, neuroarthropathy, osteonecrosis, osteochondritis dissecans, trauma, and inflammatory arthritides.

Box 3-18   Osteoarthritis: Important Features

- Female/male ratio 1:1; usual onset at 50 to 60 years of age; primary or secondary to prior articular insult
- Subchondral cysts, sclerosis
- Heberden nodes at distal interphalangeal joints, Bouchard nodes at proximal interphalangeal joints
- Superior joint space narrowing of hips
- Apophyseal joint space narrowing, sclerosis, subluxation
- Minimal periarticular osteoporosis
- First carpometacarpal joint involvement
- Order of decreasing frequency and severity: distal interphalangeal, proximal interphalangeal, metacarpophalangeal joint involvement
- Joint space narrowing
- Osteophytes
- Medial joint space narrowing of knees
- Articular effusions
- Gull-wing deformity of interphalangeal joints

---

narrowing, particularly in weight-bearing areas; intraarticular or marginal osteophytes; subchondral sclerosis; cortical buttressing, particularly in the femoral neck; subchondral cysts that may or may not communicate with the joint; enthesopathy, especially in the patella as well as pelvic and femoral apophyses; ligamentous contractures, laxity, and instability secondary to joint deformity (Fig. 3-10; Box 3-18)

15. Wilson disease: chondrocalcinosis; osteopenia; joint space narrowing; indistinct, irregular subchondral bone with small fragments or ossicles; subchondral sclerosis and cysts; periostitis; enthesopathy (Box 3-19)

Box 3-19 Wilson Disease (Hepatolenticular Degeneration): Important Features

- Congenital defect in copper metabolism causing tissue deposition
- Osteomalacia or rickets with osteopenia and pseudofractures secondary to renal tubular dysfunction
- Subarticular cysts, irregularity, and fragmentation
- Predilection for wrists, metacarpophalangeal joints, knees, and hips
- Premature degenerative joint disease
- Chondrocalcinosis

16. Amyloidosis: bulky soft-tissue nodules, especially around the wrists, elbows, and shoulders (shoulder pad sign); focal or diffuse osteopenia; joint space widening secondary to infiltration; well-defined erosions; subchondral cysts; joint contractures
17. Calcium hydroxyapatite crystal deposition disease: soft-tissue swelling; homogeneous, cloudlike periarticular calcification, involving tendons, ligaments, capsules, and bursae; occasional intraarticular in-

volvement, resulting in chondrocalcinosis, small subchondral cysts, and Milwaukee shoulder
18. Pigmented villonodular synovitis: hemorrhagic effusion; mild periarticular demineralization; joint space preservation until late in course; well-defined erosions; subchondral cysts (Fig. 3-11)

PEARL

Gout, psoriatic arthritis, Reiter syndrome, and pigmented villonodular synovitis exhibit minimal to absent periarticular osteoporosis in contrast to rheumatoid arthritis. Calcification in a synovial mass suggests idiopathic synovial osteochondromatosis or synovial sarcoma (30% of these lesions calcify) rather than pigmented villonodular synovitis. An increase in density of the synovial mass occurs in the disease, as well as in hemophilic arthropathy, because of hemosiderin deposition. Relative lack of periarticular osteoporosis and preservation of articular cartilage distinguishes the condition from rheumatoid and septic arthritis.

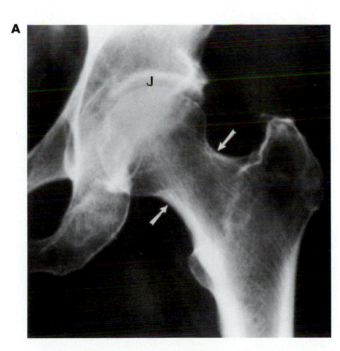

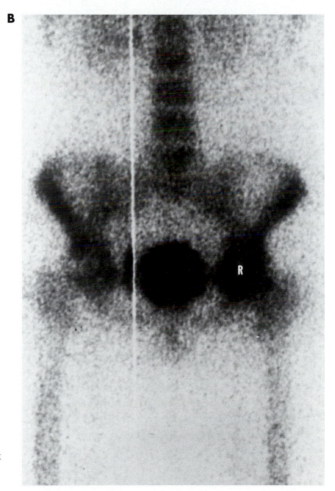

Fig. 3-11 Diffuse pigmented villonodular synovitis of the hip. **A,** Characteristic findings on a conventional frontal radiograph include preservation of the joint space (*J*) and constriction of the femoral neck (*arrows*). **B,** The process exhibits increased uptake of radiopharmaceutical (*R*) on a bone scan image.

19. Erosive (inflammatory) osteoarthritis: soft-tissue swelling, joint space narrowing; subchondral sclerosis; central erosions with marginal osteophytes (gull-wing appearance); occasional ankylosis (Box 3-20)

Box 3-20   Inflammatory (Erosive) Osteoarthritis: Important Features

- Minimal periarticular osteoporosis
- Soft-tissue swelling
- Order of decreasing frequency and severity: distal interphalangeal, proximal interphalangeal, metacarpophalangeal joint involvement
- First carpometacarpal joint involvement
- Joint space narrowing with osteophytes, sclerosis, central erosions, and articular destruction
- Gull-wing deformity of interphalangeal joints

PEARL

Erosive osteoarthritis, also known as inflammatory osteoarthritis, is most often seen in postmenopausal women. It is characterized by acute inflammatory episodes followed by eventual joint ankylosis. It predominantly involves the hands, especially the interphalangeal and first carpometacarpal joints. Radiographically, the joint space is narrowed, and subchondral sclerosis with central erosion occurs. Periarticular osteoporosis is inconstant. Subchondral cysts occur in late stages. Joint involvement is symmetric, which aids in differentiation from psoriatic arthritis. The disease is rarely reported in the hips and knees.

---

20. Progressive systemic sclerosis (scleroderma): acral tapering of digits; extensive subcutaneous, periarticular, and occasionally intraarticular calcification (calcinosis); periarticular demineralization; occasional joint space narrowing; erosions in up to 50% of patients; mild bone proliferation; ankylosis; flexion contractures, especially in hands, wrists, and elbows (Fig. 3-12)
21. Ischemic necrosis: joint space preservation until late in course; relative and reactive sclerosis; linear subchondral fracture (crescent sign); collapse and deformity of articular surface secondary to repair (creeping substitution); secondary osteoarthritic alterations (Fig. 3-13)

PEARL

Ischemic necrosis increases bone density because of resistance of dead bone to hyperemia-induced osteopenia of adjacent viable bone. Actual new bone formation (buttressing) also occurs in adjacent viable bone around the reactive zone of repair or creeping substitution. A later finding is compression of the dead bone, which causes increased density that is apparent radiographically.

---

22. Hypertrophic osteoarthropathy: digital clubbing; periarticular soft-tissue swelling; occasional effusions; symmetric periostitis, with thickness and extent depending upon duration of disease (Fig. 3-14; Table 3-3)
23. Calcium pyrophosphate dihydrate crystal deposition disease: periarticular soft-tissue swelling; chondrocalcinosis of hyaline cartilage or fibrocartilage, especially in the menisci of the knee, triangular fibrocartilage of the wrist, symphysis pubis, and acetabular labrum; occasional calcification of synovium, capsule, tendons, and ligaments; joint space narrowing; subchondral sclerosis; intraarticular osteochondral bodies; osseous excrescences, particularly at metacarpal heads; subchondral cysts, which may be large; ligamentous instability in wrist (Box 3-21)

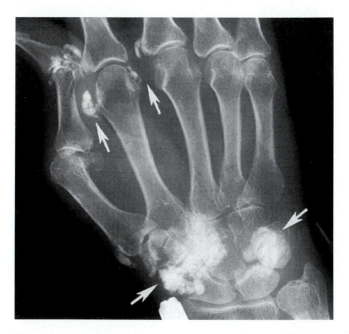

Fig. 3-12   Calcinosis circumscripta (*arrows*) and diffuse osteopenia in scleroderma. Differential diagnosis should include consideration of other collagen-vascular diseases, as well as the numerous causes of metastatic and dystrophic soft-tissue calcification.

Table 3-3 Some causes of diffuse periostitis

Disease	Location	Characteristics
Primary hypertrophic osteoarthropathy (pachydermoperiostosis)	Tibia, fibula, radius, ulna (less commonly, carpus, tarsus, metacarpals, metatarsals, phalanges, pelvis, ribs, clavicle)	Diaphyseal, metaphyseal, and epiphyseal involvement Shaggy, irregular excrescences Diaphyseal expansion Clubbing Ligamentous ossification Cranial and facial alterations
Secondary hypertrophic osteoarthropathy	Tibia, fibula, radius, ulna (less commonly, femur, humerus, metacarpals, metatarsals, phalanges)	Diaphyseal and metaphyseal involvement Monolayered or laminated, regular or irregular proliferation Clubbing Periarticular osteoporosis Soft-tissue swelling Underlying primary lesion
Thyroid acropachy	Metacarpals, metatarsals, phalanges, (less commonly, long tubular bones)	Diaphyseal involvement Radial side predilection Dense, solid, and spiculated periostitis Soft-tissue swelling, clubbing Thyroid gland abnormalities
Venous stasis	Tibia, fibula, femur, metatarsals, and phalanges of lower extremities	Diaphyseal and metaphyseal involvement Undulating osseous contour Cortical thickening Soft-tissue swelling Ulceration Phleboliths
Hypervitaminosis A	Ulna, metatarsals, clavicle, tibia, fibula	Diaphyseal involvement Undulating contour Epiphyseal deformities Soft-tissue nodules Intracranial hypertension
Infantile cortical hyperostosis (Caffey disease)	Mandible, clavicle, scapula, ribs, tubular bones	Periostitis and cortical hyperostosis Soft-tissue swelling may become severe and lead to deformities

---

Box 3-21 Calcium Pyrophosphate Dihydrate Crystal Deposition Disease: Important Features

- Male/female ratio 2:1
- Onset after 55 years of age
- Pseudogout syndrome (soft-tissue swelling, usually affecting the knee)
- Pyrophosphate arthropathy (radiocarpal, trapezio-scaphoid, and metacarpophalangeal joints; knee; spine)
- Calcification in symphysis pubis, annulus fibrosus of disks
- Calcification in hyaline cartilage, meniscal fibrocartilage, joint capsules, and at entheses
- Minimal periarticular involvement, as compared with calcium hydroxyapatite crystal deposition disease
- Articular effusions

PEARL

Periarticular calcification in CPPD crystal deposition disease can involve cartilage, synovium, capsule, tendons, bursae, ligaments, soft tissue, and vessels. Subarticular cysts are an important feature and are often large. Marginal erosions are not seen, which helps to differentiate the condition from psoriatic or rheumatoid arthritis.

24. Juvenile chronic arthritis: periarticular soft-tissue swelling; muscle wasting with severe disease; osteopenia secondary to hyperemia and disuse; radiolucent metaphyseal bands; joint space narrowing; erosions late in course; periostitis early in course; ankylosis, especially in wrist and interphalangeal joints; joint contractures (Figs. 3-15, 3-16; Box 3-22)

A

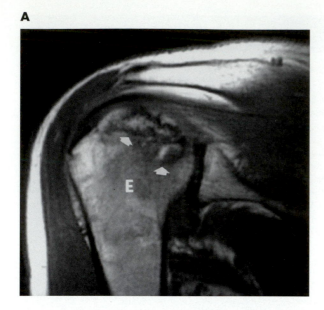

B

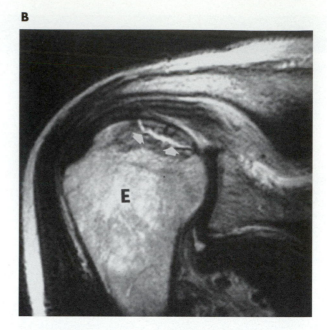

Fig. 3-13   Osteonecrosis of the humeral head. **A,** Coronal T1-weighted MR image reveals inhomogeneity of the subchondral bone (*arrows*) with diffuse low signal intensity in the metaphyseal marrow (*E*). **B,** Corresponding T2-weighted image documents subchondral fracture (crescent sign) of high signal intensity (*arrows*) along with brightening of the adjacent viable marrow secondary to edema (*E*). Differential diagnosis should include consideration of acute osteochondral fracture, osteochondritis dissecans, and early neuroarthropathy.

Box 3-22   Juvenile Chronic Arthritis: Important Features

- Classification:
  - Adult type: rheumatoid factor test positive
  - Seronegative group: Still disease, pauciarticular form with or without iridocyclitis
  - Polyarthritis with ankylosing spondylitis
  - Polyarthritis with psoriasis
- Female predominance except in polyarthritic forms
- Usually appears prior to age 7 years
- Growth arrest or recovery lines
- Late joint space narrowing
- Ankylosis in carpus, cervical spine, sacroiliac joints
- Early closure of physeal plates
- Epiphyseal overgrowth
- Osteoporosis (generalized and periarticular)
- Predilection for carpal, metacarpophalangeal, and proximal interphalangeal joints, as well as interphalangeal joints of thumbs
- Poorly defined erosions

25. Transient regional osteoporosis: self-limited periarticular demineralization, which may be migratory

PEARL

Periarticular osteopenia is a classic finding in rheumatoid and septic arthritis. Reiter syndrome causes soft-tissue swelling, periarticular osteopenia, joint space narrowing, and erosions, especially in the lower extremities. Osteoporosis is not a prominent feature of psoriatic arthritis but can be seen in the acute phase. Periarticular osteopenia is not typically seen in pigmented villonodular synovitis or gout.

26. Polyarthritis of rheumatic fever: self-limited periarticular soft-tissue swelling and demineralization, which is often migratory; associated valvular heart disease (Fig. 3-17)

See Table 3-4 for a summary of radiographic signs caused by arthritis of the appendicular skeleton in specific disorders.

Target sites of joint involvement

1. Diffuse idiopathic skeletal hyperostosis (Forrestier disease): thoracic spine, cervical spine, lumbar spine, superior (nonarticular) portions of sacroiliac joints

Table 3-4 Radiographic signs of arthritis of the appendicular skeleton in specific disorders*

Disorder	Soft-tissue swelling	Osteopenia	Joint space narrowing	Ankylosis	Erosions	Sclerosis	Osteophytosis or proliferation	Subchondral cysts	Bony fragmentation, collapse	Intraarticular calcification	Periarticular calcification
Ankylosing spondylitis	Fusiform	±	±	+	Marginal	±	Ill-defined proliferation	±	−	−	−
Calcium hydroxyapatite crystal deposition disease	+	±	±	−	−	±	±	+	+	±	+
CPPD crystal deposition disease	±	−	Diffuse	−	−	+	+	+	+	+	+
Gout	Lobulated	−	Late	−	Eccentric	+	Overhanging edges	±	−	±	−
Idiopathic synovial osteochondromatosis	±	±	±	−	±	−	−	+	−	+	−
Infection	+	+	+	±	+	±	−	±	±	±	±
Inflammatory osteoarthritis	+	−	Diffuse	+	Central	+	Capsular osteophytes	−	±	−	−
Neuroarthropathy	+	−	±	−	−	Extreme	+	+	+	Osseous debris	−
Osteoarthritis large joints	−	−	Asymmetric or segmental	−	−	+	Marginal osteophytes	+	±	−	−
small joints	+	−	Diffuse	−	−	+	Capsular osteophytes	+	−	−	−
Osteonecrosis	−	±	−	−	−	+	−	+	+	±	±
Pigmented villonodular synovitis	+	±	±	−	+	−	−	+	−	±	−
Psoriatic arthritis, Reiter syndrome	Fusiform	During acute episodes	Early, diffuse	+	Marginal	±	Ill-defined proliferation	±	±	−	−
Rheumatoid arthritis	Fusiform	+	Early, diffuse	Carpus; tarsus	Marginal	−	−	+	−	−	−

*+, Common; ±, uncommon; −, rare or absent.

147

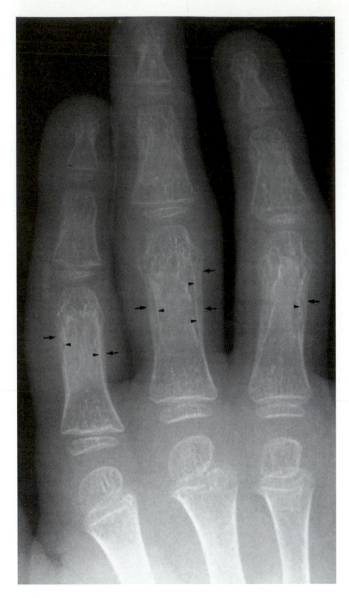

Fig. 3-14 Mature periostitis (*arrows*) in hypertrophic osteoarthropathy secondary to cyanotic congenital heart disease (*arrowheads = underlying cortices*). Differential diagnosis in this age group should include consideration of cystic fibrosis, biliary atresia, and pachydermoperiostosis.

2. Pyrophosphate arthropathy: knee (patellofemoral more common and severe than lateral more common and severe than medial), wrist (radiocarpal more common and severe than scapholunate), hand (second and third metacarpophalangeal), hip, shoulder, elbow joints

---

PEARL

The patellofemoral joint is classically involved in calcium pyrophosphate dihydrate crystal deposition disease of the knee. Hyaline cartilage and fibrocartilage are affected by chondrocalcinosis. The wrist is also a frequent site of this phenomenon. The end stage of pyrophosphate arthropathy in any articulation can resemble a Charcot joint. Chondrocalcinosis also occurs in Wilson disease, hemochromatosis, hyperparathyroidism, acromegaly, osteoarthritis, ochronosis, and gout, as well as idiopathically. Although the knee is most commonly involved, pseudogout can affect the metacarpophalangeal joints as well.

---

3. Psoriatic arthritis: hand (distal interphalangeal more common and severe than proximal interphalangeal, metacarpophalangeal), wrist, foot (interphalangeal, metatarsophalangeal), ankle, knee, hip, shoulder, sacroiliac joints, spine (thoracolumbar junction more common and severe than cervical)

PEARL

Involvement of the distal interphalangeal joints, less severe periarticular osteoporosis, asymmetric distribution, bone proliferation (whiskering), asymmetric paravertebral ossification, and sacroiliitis are useful in distinguishing psoriatic arthritis from rheumatoid arthritis. A fluffy periostitis can be seen in psoriasis, with the upper extremity being more often involved than the lower extremity.

---

4. Gout: foot (first metatarsophalangeal more common and severe than second through fifth metatarsophalangeal, interphalangeal, midfoot and hindfoot joints), ankle, knee, hand, wrist (interphalangeal, intercarpal more common and severe than metacarpophalangeal), elbow joints

5. Rheumatoid arthritis: hand (metacarpophalangeal, proximal interphalangeal more common and severe than distal interphalangeal), wrist, elbow, shoulder (acromioclavicular, glenohumeral), sternomanubrial, sternoclavicular, foot (metatarsophalangeal, interphalangeal, intertarsal), knee (medial, lateral, patellofemoral), hip, cervical spine (atlantoaxial, apophyseal, uncovertebral, discovertebral), ankle joints

6. Polyarticular juvenile chronic arthritis: hand (metacarpophalangeal, proximal interphalangeal), wrist (midcarpal more common and severe than radiocarpal), elbow, shoulder, foot (metatarsophalangeal, tarsal), ankle, knee, hip, cervical spine (apophyseal, atlantoaxial), temporomandibular joints

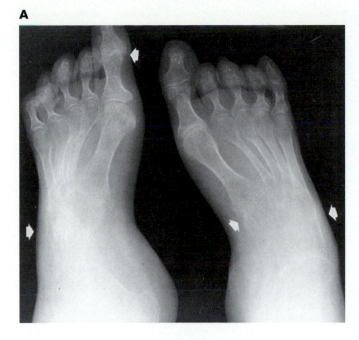

A

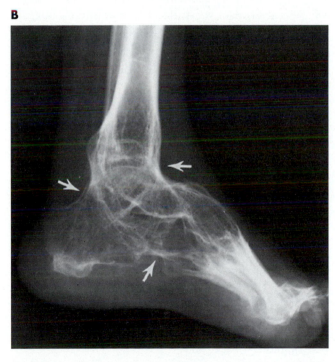

B

Fig. 3-15 Juvenile chronic arthritis. **A,** Symmetric findings include generalized osteopenia, diaphyseal overtubulation, widespread joint space narrowing, evolving ankylosis (*arrows*), and diffuse muscle wasting. **B,** Another patient exhibits widespread ankylosis (*arrows*) in the hindfoot and midfoot with cavus deformity, secondary to the disease itself as well as stabilizing surgery. Osteopenia and muscle atrophy are also evident. Differential diagnosis should include consideration of other causes of disuse beginning in childhood, as well as certain congenital foot deformities.

7. Pauciarticular juvenile chronic arthritis: knee, ankle, elbow joints
8. Synovial osteochondromatosis: knee, hip, elbow, shoulder joints

---

PEARL

Synovial osteochondromatosis is a monoarticular process. The most common joints affected are the knee, hip, and elbow. The process may also occur primarily in bursae. The osteochondral proliferation usually causes articular cartilage destruction resulting in secondary osteoarthritis. The degree of calcification of the bodies is proportional to the duration of the process.

---

9. Calcium hydroxyapatite crystal deposition disease: shoulder, elbow, wrist, hand (metacarpophalangeal, interphalangeal joints)
10. Reiter syndrome: foot (metatarsophalangeal, first interphalangeal more common and severe than tarsal), ankle, knee, hip, sacroiliac joints, spine (thoracolumbar)

PEARL

Reiter disease involves the lower extremities most commonly but can involve the hands. The sacroiliac and spinal abnormalities resemble those of psoriatic arthritis. Classic Reiter syndrome is the clinical triad of urethritis, conjunctivitis, and arthritis.

---

11. Posttraumatic ischemic necrosis: hip, wrist, ankle joints
12. Pigmented villonodular synovitis: knee, hip, elbow joints

---

PEARL

Pigmented villonodular synovitis is a monoarticular process, and the knee is most commonly involved. Periarticular erosions of bone are relatively common when the hip is affected, and these are characterized by well-defined sclerotic margins. Joint space narrowing is a relatively late finding. Periarticular osteoporosis is not a typical feature of the disease.

---

13. Osteoarthritis: hand (interphalangeal more common and severe than metacarpophalangeal), wrist (first carpometacarpal more common and severe than trapezioscaphoid), acromioclavicular, sacroiliac, hip, knee (medial more common and severe than

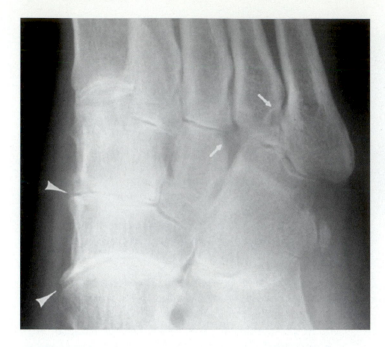

Fig. 3-16   Juvenile chronic arthritis in a skeletally mature patient. Abnormalities in the midfoot include generalized joint space narrowing, marginal osseous erosions (*arrows*), and secondary degenerative alterations (*arrowheads*). Only age and clinical history permit distinction from adult-onset rheumatoid arthritis.

patellofemoral more common and severe than lateral compartment), foot (first metatarsophalangeal, first tarsometatarsal), spine (lower cervical, lumbosacral more common and severe than elsewhere; discovertebral, apophyseal, uncovertebral joints)

14. Erosive (inflammatory) osteoarthritis: hand (interphalangeal more common and severe than first carpometacarpal joints)

15. Neuroarthropathy: knee, foot (talonavicular, tarsometatarsal, metatarsophalangeal), shoulder joints, spine

---

PEARL

Neuropathic skeletal disorders:
- Syringomyelia usually affects the upper extremities and spine.
- Syphilis usually involves the spine and large joints of the lower extremities.
- Diabetes and alcoholism usually involve the feet.
- Congenital insensitivity to pain usually involves the lower extremities.
- Multiple joints may be affected in any of these conditions.
- Superimposed infection occurs in diabetes, alcoholism, and syphilis.

---

16. Amyloidosis: wrist, elbow, shoulder more common and severe than knee, hip joints

17. Polyarthritis of rheumatic fever: knee, ankle joints

18. Wilson disease: wrist, hand (metacarpophalangeal), foot, hip, shoulder, elbow, knee joints

19. Alkaptonuria (ochronosis): spine more common and severe than sacroiliac, symphysis pubis, large peripheral joints

20. Jaccoud arthritis: hand (metacarpophalangeal), foot (metatarsophalangeal joints)

21. Ankylosing spondylitis: sacroiliac, spine (thoracolumbar more common and severe than cervical; discovertebral, apophyseal) more common and severe than hip more common and severe than glenohumeral more common and severe than symphysis pubis, sternomanubrial, costovertebral joints

22. Hemochromatosis: hand (second and third metacarpophalangeal), wrist (radiocarpal), knee (patellofemoral more common and severe than medial, lateral compartments), hip, shoulder joints

23. Progressive systemic sclerosis (scleroderma): hand (interphalangeal joints)

24. Systemic lupus erythematosus: hand, wrist, knee joints

25. Transient regional osteoporosis: large joints of lower extremity

26. Polymyositis or dermatomyositis: hand, wrist, knee joints (arthralgia without arthritis)

Spine   Rheumatologic disorders of the vertebral column and their most closely associated radiographic findings (Table 3-5):

Table 3-5   Radiographic signs of arthritis of the axial skeleton in specific disorders*

	Intervertebral disk space narrowing	Vacuum phenomena	Bony outgrowths	Apophyseal joint erosion	Apophyseal joint ankylosis	Intervertebral disk space calcification	Atlantoaxial subluxation
Alkaptonuria (ochronosis)	+	+	Syndesmophytes (rare)	–	+	+	–
Ankylosing spondylitis	±	–	Syndesmophytes	+	+	±	+
Diffuse idiopathic skeletal hyperostosis	–	–	Flowing anterolateral outgrowths	–	–	±	–
Infection	+	–	–	–	–	–	–
Intervertebral osteochondrosis	+	+	–	–	–	–	–
Juvenile rheumatoid arthritis	+	–	–	+	+	±	+
Psoriatic arthritis, Reiter syndrome	±	–	Paravertebral ossification	±	±	–	+
Rheumatoid arthritis	+	–	–	+	±	–	+
Spondylosis deformans	–	–	Osteophytes	–	–	–	–

*+, Common; ±, uncommon; –, rare or absent.

PEARL

Arthritic disorders of the spine:
- Juvenile chronic arthritis is associated with poorly developed vertebral bodies and ankylosis of the apophyseal joints.
- Ankylosing spondylitis causes osteitis with syndesmophytes.
- Rheumatoid arthritis predominantly involves the cervical spine causing atlantoaxial subluxation, odontoid process erosion, spondylolisthesis, apophyseal joint erosion, disk space narrowing with endplate irregularity and sclerosis, and spinous process erosion.
- Psoriatic spondylitis is associated with asymmetric paravertebral ossification.
- Diffuse idiopathic skeletal hyperostosis causes flowing anterolateral ossification.

1. Ankylosing spondylitis: reactive sclerosis at sites of osteitis (shiny corner sign)
2. Juvenile chronic arthritis: cervical vertebral body hypoplasia with ankylosis (Fig. 3-18)

PEARL

In juvenile chronic arthritis, 25% of patients have cervical fusion. Ankylosis of the apophyseal and uncovertebral joints of Luschka is common in this disease. It usually occurs in the upper cervical spine, and if it is present in the lower portion, the upper portion is almost always involved. There may be associated hypoplasia of the vertebral bodies and disks.

3. Ischemic necrosis: vacuum phenomenon within collapsed vertebral body
4. Diffuse idiopathic skeletal hyperostosis (Forrestier disease): flowing ossification of anterior longitudinal ligament spanning at least four contiguous vertebral bodies with relative preservation of disk height (Figs. 3-19, 3-20)
5. Alkaptonuria (ochronosis): osteoporosis with dense diskal calcification and degenerative signs
6. Ankylosing spondylitis: syndesmophytes (thin vertical ossifications) in the annulus fibrosus at discovertebral junctions (Fig. 3-21)

A

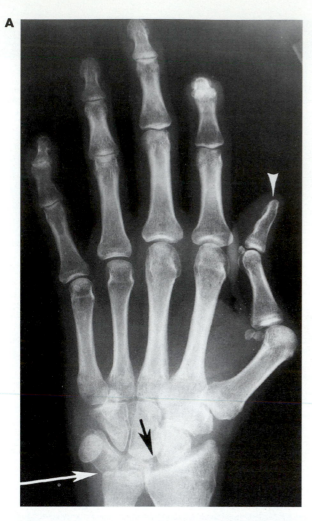

B

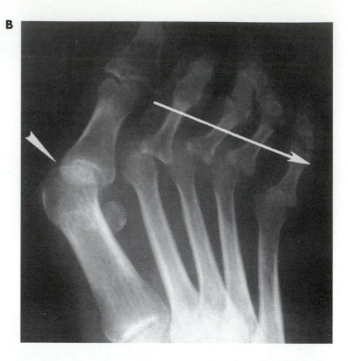

Fig. 3-17   Along with rheumatic fever, acquired immune deficiency syndrome (AIDS) and Lyme disease are other infectious disorders associated with arthritis. Arthropathy is associated with human immunodeficiency virus infection. **A,** Carpal subluxation (*white arrow*), scapholunate dissociation (*black arrow*), and boutonnière deformity of the thumb (*arrowhead*). **B,** Severe hallux valgus (*arrowhead*) is associated with lateral dislocation of the second through fifth metatarsophalangeal joints (*arrow*). Differential diagnosis should include consideration of inflammatory arthritides, collagen-vascular disease, and other causes of ligamentous laxity.

PEARL

The large asymmetric bony buttresses seen in Reiter syndrome and psoriatic arthritis have been termed *non-marginal syndesmophytes*. The syndesmophytes of ankylosing spondylitis and inflammatory bowel disease are thin, vertically oriented bony excrescences spanning the disk space and attaching at the margins of the vertebral endplates.

7. Intervertebral osteochondrosis: disk space narrowing, vacuum phenomenon, and reactive sclerosis of adjacent vertebral endplates (Fig. 3-22; Box 3-23; Table 3-6)

Box 3-23   CT of Degenerative Intervertebral Disk Disease

CT CRITERIA FOR DISK DISEASE

- Abnormal annular contour
- Abnormal soft-tissue structures in spinal canal
- Mass effect on fat, nerves, dura
- Variable features: disk calcification, extruded fragment
- Bulging annulus: generalized extension of disk margin beyond vertebral body margins
- Herniated nucleus pulposus: focal, usually posterolateral, protrusion of disk margin

CT FEATURES OF LATERAL DISK HERNIATION

- Focal protrusion of disk margin or soft-tissue mass within or lateral to intervertebral foramen
- Displacement of epidural fat within intervertebral foramen
- Absence of dural sac deformity
- More effectively visualized with CT or MRI than with myelography

Box 3-23—cont'd

CT OF EXTRUDED (FREE) DISK FRAGMENTS

- Epidural mass distinguishable from tumor or anomalous root sheath by density measurements and analysis of adjacent bone
- Normal posterior disk margin does not exclude extrusion

CT FEATURES OF THE POSTOPERATIVE SPINE

- Eccentric vacuum phenomenon
- Gas within spinal canal
- Dural sac retraction toward surgical site
- Intraspinal and extraspinal granulation or fibrous tissue accumulation
- Effacement of epidural fat
- Indistinct ipsilateral nerve root
- Decreased height of surgically violated disk space in absence of diskectomy with anterior fusion
- Loss of definition of intraspinal contents
- Intravenous contrast-enhanced CT of the postoperative lumbar spine may improve identification of recurrent disk herniation, scar, arachnoiditis, and diskitis

CONJOINED NERVE ROOTS

- Most commonly L5, S1
- May cause sciatica in absence of disk herniation if spinal stenosis present
- Intrathecal contrast administration or MRI is diagnostic if CT density measurements are equivocal

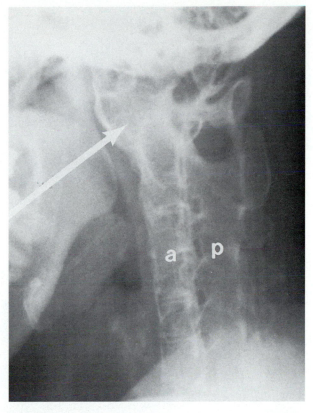

Fig. 3-18 Fusion of the anterior (*a*) and posterior (*p*) columns with vertebral hypoplasia and atlantoaxial subluxation (*arrow*) in juvenile-onset ankylosing spondylitis. Differential diagnosis might include consideration of Klippel-Feil anomaly or radiation therapy during childhood.

Table 3-6   Degenerative disorders of the vertebral column associated with aging

	Intervertebral osteochondrosis	Spondylosis deformans	Osteoarthritis
Site of predominant involvement	Nucleus pulposus	Annulus fibrosus	Apophyseal joints, costovertebral joints
Apophyseal and costovertebral joints	Normal	Normal	Joint space narrowing, sclerosis, osteophytosis
Intervertebral disk space	Moderate to severe narrowing, vacuum phenomena	Normal or slight narrowing	Normal
Vertebral body	Eburnation of superior and inferior endplates, cartilaginous nodes	Osteophytosis	Normal

8. Psoriatic arthritis or Reiter syndrome: prominent asymmetric paravertebral ossification at thoracolumbar junction (Fig. 3-23)
9. Ossification of the posterior longitudinal ligament: ossification of posterior longitudinal ligament, usually in the cervical region
10. Neuroarthropathy: disk space narrowing with bone destruction and fragmentation at first mobile level adjacent to surgical fusion following spinal cord trauma
11. Spondylosis deformans: traction osteophytes arising several millimeters from vertebral body endplates (Table 3-6)
12. Rheumatoid arthritis: atlantoaxial subluxation with cranial settling and odontoid erosion (Fig. 3-24)

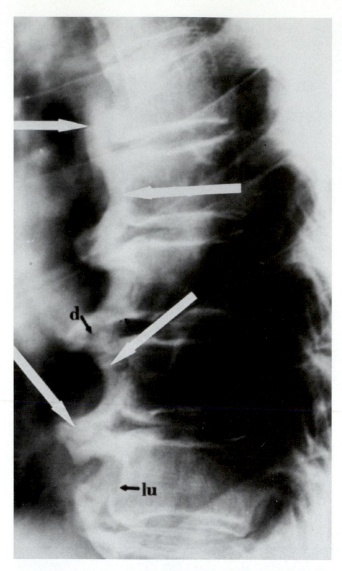

Fig. 3-19 Flowing ossification (*arrows*) along the anterior aspect of the spine with preservation of disk space height is typical of diffuse idiopathic skeletal hyperostosis. Areas of relative radiolucency within the process occur anterior to the vertebral bodies (*lu*) and at the level of the disks (*d*).

PEARL

Rheumatoid arthritis of the cervical spine most commonly involves the atlantoaxial joint, followed by the apophyseal joints. Gross subluxation may occur, with a stepladder appearance. Flexion-extension views showing subluxations may be useful in establishing the diagnosis. Cross-sectional imaging is frequently helpful in patients with quadriplegia or other neurologic deficits.

---

13. Ankylosing spondylitis: loss of normal anterior concavity of vertebral body (squaring)

14. Osteoarthritis of apophyseal joints: spondylolisthesis without spondylolysis (Table 3-6)
15. Uncovertebral osteoarthrosis: osteophytic encroachment upon anterior aspects of cervical neural foramina
16. Ankylosing spondylitis: fusion of apophyseal joints and anterior column (bamboo spine) (Fig. 3-25)
17. Rheumatoid arthritis: discovertebral lesions secondary to osteopenia and apophyseal joint instability
18. Spinal stenosis (acquired): disk protrusion with hypertrophy of apophyseal joints and ligamenta flavum
19. Ankylosing spondylitis: posttraumatic fractures leading to pseudoarthroses at cervicothoracic and thoracolumbar junctions (Fig. 3-20)
20. Rheumatoid arthritis: mechanical erosion of cervical spinous processes
21. Diffuse idiopathic skeletal hyperostosis (Forrestier disease): ossification of supraspinous ligament and other entheses (Fig. 3-26)
22. Calcium hydroxyapatite crystal deposition disease: homogeneous cloudlike calcification of longus colli tendon in cervical region
23. Ankylosing spondylitis: osseous fusion between adjacent spinous processes (trolley track sign) (Fig. 3-27)

See Table 3-7 for differential diagnoses associated with disk space narrowing.

Sacroiliac joints Rheumatologic disorders and their most closely associated radiographic findings in these articulations:

PEARL

Ankylosing spondylitis causes bilateral sacroiliac disease that is similar in presentation to inflammatory bowel disease such as Whipple disease. The iliac portion of the middle and lower thirds of the joint (synovial portion) is involved first. Osteoarthritis can result in unilateral or bilateral fusion (usually periarticular) of the sacroiliac joints. Bilateral symmetric subchondral bone resorption simulating inflammatory sacroiliitis occurs in hyperparathyroidism. Psoriatic arthritis, Reiter syndrome, ulcerative colitis, and rheumatoid arthritis can cause sacroiliac joint disease. Psoriatic arthritis and Reiter syndrome can cause unilateral or bilateral sacroiliac disease. Rheumatoid arthritis rarely involves the sacroiliac joint; when there is sacroiliac involvement, it is usually mild, asymmetric or unilateral, and similar to that in degenerative disease. Ulcerative colitis and other forms of inflammatory bowel disease characteristically involve the sacroiliac joints symmetrically, similar to ankylosing spondylitis.

---

1. Rheumatoid arthritis: infrequent, mild, unilateral or asymmetric erosive involvement

Table 3-7 Differential diagnosis of disk space narrowing

Disorder	Discovertebral margin	Sclerosis	Vacuum phenomena	Osteophytosis	Other findings
Infection	Poorly defined	Pyogenic more common and more extensive than in tuberculous	Rare	Absent	Vertebral lysis, soft-tissue mass
Intervertebral osteochondrosis	Well defined	Prominent	Present	Present	Cartilaginous nodes
Rheumatoid arthritis	Mechanical erosions	Variable	Absent	Minimal	Apophyseal joint abnormalities, subluxation
CPPD crystal deposition disease	Poorly or well defined	Prominent	Variable	Variable	Fragmentation, subluxation
Ochronosis	Poorly or well defined	Absent	Variable	Minimal	Osteopenia, disk calcification
Neuroarthropathy	Poorly or well defined	Prominent	Variable	Prominent	Fragmentation, subluxation, disorganization
Trauma	Well defined	Variable	Variable	Variable	Fracture, soft-tissue mass
Dialysis spondyloarthropathy	Poorly or well defined	Variable	Rare	Variable	Osteomalacia, secondary hyperparathyroidism

2. Diffuse idiopathic skeletal hyperostosis (Forrestier disease): bridging of superior (nonarticular) portions of joints by ligamentous calcification

3. Osteoarthritis: subchondral sclerosis and osteophytes anteroinferior and anterosuperior to synovial portions of joints

4. Reiter syndrome: bilateral symmetric or asymmetric (early) widening, sclerosis, erosions, and/or fusion in 10% to 40% of cases

5. Juvenile chronic arthritis: rare asymmetric erosive involvement

6. Ankylosing spondylitis: bilateral symmetric (may be asymmetric early) erosions and widening, progressing to sclerosis and fusion (site of initial disease involvement) (Fig. 3-28)

7. Psoriatic arthritis: bilateral symmetric or asymmetric erosions, widening, and sclerosis progressing to fusion (present in 30% to 50% of cases) (Figs. 3-29, 3-30)

8. Gout: scalloped erosions with adjacent reactive sclerosis, occurring occasionally

9. Arthritis of inflammatory bowel disease: involvement identical clinically and radiographically to that of ankylosing spondylitis in 20% to 30% of cases, which does not generally correlate with gastrointestinal disease activity

10. Septic arthritis: unilateral erosions, widening, and sclerosis (commonly associated with intravenous drug abuse)

11. Osteoarthritis: anterior osteophytic bridging of the joints, distinguished from osteoblastic metastases by computed tomography

12. Osteitis condensans ilii: triangular sclerosis on the iliac sides of the joints inferiorly, occurring most commonly in multiparous women

13. Alkaptonuria (ochronosis): involvement indistinguishable from that found in degenerative joint disease, usually following involvement of the spine

14. Familial Mediterranean fever: sacroiliitis occurs in 2% to 25% of patients despite an absence of HLA-B27 positivity

PEARL

Familial Mediterranean fever:
- It is inherited as an autosomal recessive trait with complete penetrance.
- Clinically, episodes of fever with abdominal, thoracic, or joint pain occur as a result of inflammation secondary to the disease.
- Amyloidosis is a recognized complication.
- Musculoskeletal manifestations occur in 60% to 70% of patients.
- Osteoporosis can develop rapidly and become severe. Epiphyseal overgrowth, joint effusion, joint space narrowing, juxtaarticular erosions, ankylosis, sclerosis, osteophytosis, and osteonecrosis can be observed in affected joints.

A   **B**

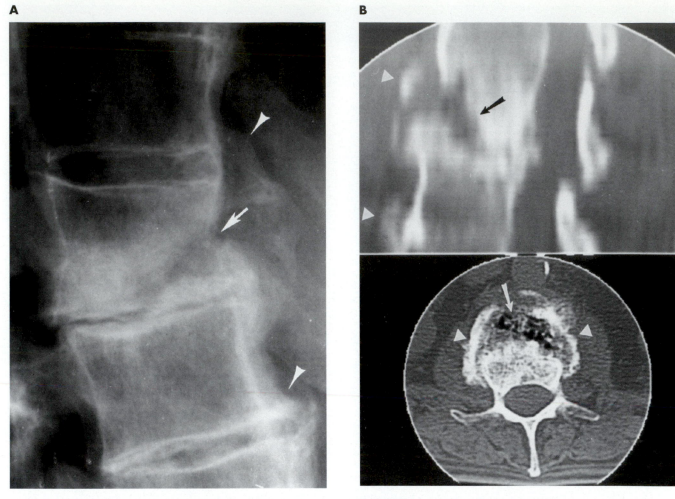

Fig. 3-20   Spinal fractures in diffuse idiopathic skeletal hyperostosis. **A,** Lateral radiograph demonstrates typical flowing ossification along the anterior aspect of the thoracolumbar spine (*arrowheads*) resulting in rigidity. The fracture (*arrow*) involves the anterior inferior corner of a vertebral body and exits posteriorly through the disk and posterior elements. **B,** CT with sagittal reformation in another case reveals ligamentous ossification (*arrowheads*) with a vacuum phenomenon within the fracture cleft (*arrows*). Fracture through a bamboo spine is more common in ankylosing spondylitis than in this disease.

Hip   Rheumatologic disorders and their most closely associated radiographic findings at this site (Fig. 3-31):

1. Pyrophosphate arthropathy: large subchondral cysts and labral chondrocalcinosis associated with typical alterations of degenerative joint disease
2. Ischemic necrosis: linear subchondral fracture of femoral head (crescent sign) (Fig. 3-32)
3. Transient osteoporosis of the hip: reversible demineralization on both sides of the joint
4. Osteoarthritis: superior (usually superolateral) joint space narrowing in 80% of cases (Fig. 3-33)
5. Ankylosing spondylitis: concentric joint space narrowing, small erosions, annular osteophytes, and

protrusio acetabuli; involvement often bilateral but may be asymmetric (Fig. 3-34)
6. Pigmented villonodular synovitis: well-defined erosions, subchondral cysts, and preservation of joint space until late in the disease process (Fig. 3-35)
7. Psoriatic arthritis or Reiter syndrome: combined erosive and productive pattern resembling that of ankylosing spondylitis
8. Ischemic necrosis: subchondral sclerosis with joint space preservation until collapse and secondary degenerative alterations occur (Figs. 3-36, 3-37, 3-38; Box 3-24)

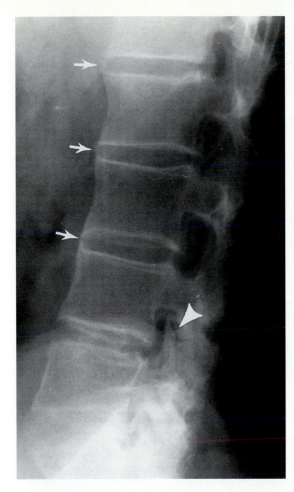

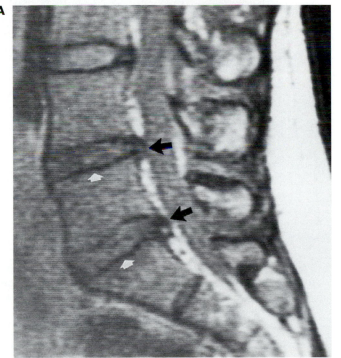

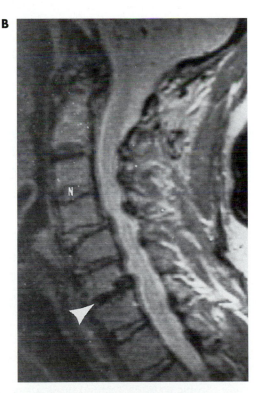

Fig. 3-21 Syndesmophyte formation (*arrows*) and evolving apophyseal joint fusion (*arrowhead*) in ankylosing spondylitis. Differential diagnosis should include consideration of long-term immobilization, as in poliomyelitis.

Fig. 3-22 MRI of degenerative disk disease. **A,** On a T1-weighted sagittal image, abnormal findings include posterior disk protrusion (*black arrows*) and decreased signal intensity in the nucleus pulposus (*white arrows*) at L4-5 and L5-S1. **B,** T2-weighted sagittal image of the cervical spine reveals multiple subluxations with posterior ligamentous hypertrophy. There is loss of normal nuclear signal (*N*) from the C6-7 disk (*arrowhead*), along with posterior protrusion.

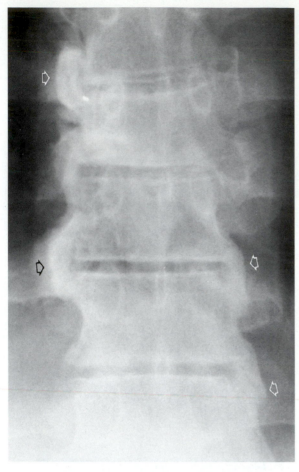

Fig. 3-23 Asymmetric paravertebral ossification (*arrows*) in psoriatic arthritis. Differential diagnosis should include consideration of Reiter syndrome and diffuse idiopathic skeletal hyperostosis.

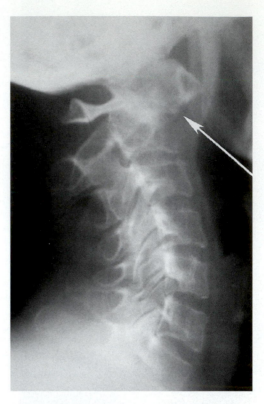

Fig. 3-24 Atlantoaxial subluxation (*arrow*) in rheumatoid arthritis. Differential diagnosis should include consideration of other inflammatory arthritides, trauma, septic arthritis, Griesl syndrome, Down syndrome, Morquio syndrome, and other conditions associated with ligamentous laxity.

Box 3-24 Spontaneous Osteonecrosis of the Femoral Head in Adults: Important Features

- Also known as Chandler disease or coronary disease of the hip
- Usually affects men
- Peak age range: fourth through seventh decades
- 35% to 75% of cases bilateral; presenting symptoms usually unilateral
- Patient has no known risk factors for osteonecrosis

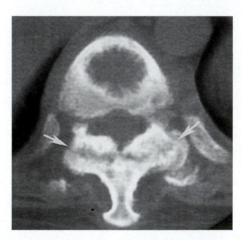

Fig. 3-25 Evolving apophyseal joint fusion (*arrows*) in ankylosing spondylitis. Differential diagnosis should include consideration of other causes of inflammatory arthritis, as well as long-term immobilization.

9. Osteoarthritis: medial or axial joint space loss in up to 20% of cases
10. Juvenile chronic arthritis: femoral head enlargement, short femoral neck, coxa valga, protrusio acetabuli
11. Rheumatoid arthritis: concentric joint space narrowing and protrusio acetabuli, often with secondary degenerative joint disease

12. Ankylosing spondylitis: bone proliferation (whiskering) at femoral head-neck junctions with axial joint space loss (approximately 50% of cases, more often than any other appendicular joint involved by the disease)
13. Osteoarthritis: femoral head remodeling with prominent medial osteophyte formation, sometimes simulating slipped capital femoral epiphysis

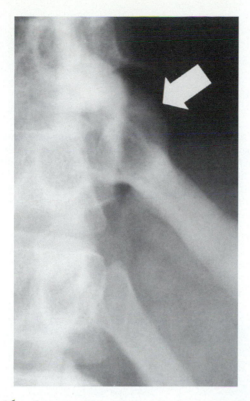

Fig. 3-26   Costovertebral enthesopathy (*arrow*) in diffuse idiopathic skeletal hyperostosis. Differential diagnosis of ligamentous calcification or ossification should include consideration of fluorosis, hypoparathyroidism and its variants, acromegaly, hypervitaminosis A, and X-linked hypophosphatemic vitamin D–resistant osteomalacia.

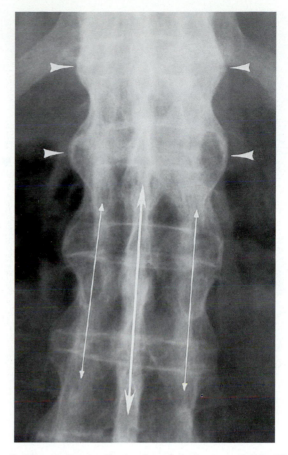

Fig. 3-27   Trolley track sign in ankylosing spondylitis. Three parallel bands of sclerosis represent fusion of the paired apophyseal joints (*short double-headed arrows*) and spinous processes (*long double-headed arrow*); bridging syndesmophytes (*arrowheads*) produce the bamboo appearance.

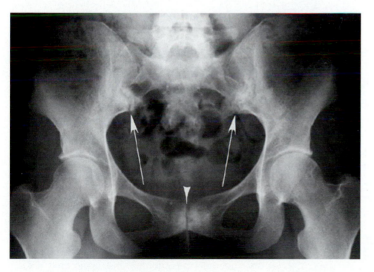

Fig. 3-28   Bilateral and symmetric sacroiliitis (*arrows*) with osteitis pubis (*arrowhead*) in early ankylosing spondylitis. Differential diagnosis should include consideration of other seronegative spondyloarthropathies and inflammatory bowel disease.

14. Synovial osteochondromatosis: multiple rounded radiopaque densities of similar size in a periarticular distribution, without joint space narrowing
15. Osteoarthritis: buttressing in the region of the calcar femorale occurring early in the disease process (Fig. 3-33)

16. Diffuse idiopathic skeletal hyperostosis (Forrestier disease): calcification of pelvic ligaments with well-defined enthesopathy involving femoral trochanters and pelvic apophyses

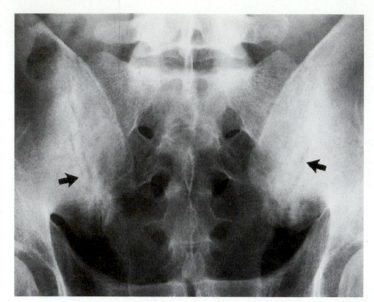

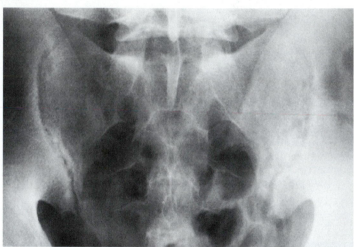

Fig. 3-29 Bilateral and symmetric sacroiliitis (*arrows*) in psoriatic arthritis. Differential diagnosis should include consideration of other seronegative spondyloarthropathies and inflammatory bowel disease.

Fig. 3-30 The sacroiliitis of relapsing polychondritis is indistinguishable from that seen in the seronegative spondyloarthropathies and is manifested as joint space narrowing and indistinct articular margins with variable sclerosis.

Knee Rheumatologic disorders and their most closely associated radiographic findings in this articulation (Figs. 3-39, 3-40):

1. Osteoarthritis: single-compartment disease is most commonly medial, with varus deformity and lateral tibial subluxation, although the patellofemoral compartment is also frequently affected (Figs. 3-41, 3-42; Box 3-25)

PEARL

In degenerative joint disease of the knee, weight-bearing alignment films of the lower extremity are preferred, in order to determine the relationship between the tibial spines and the mechanical axis (line joining centers of femoral head and talar dome).

Box 3-25   Chondromalacia of the Patella: MRI Findings

- Stage I: Focal areas of decreased signal intensity on T1-weighted images, usually small, and not extending to or deforming the outer surface.
- Stage II: Larger focal areas of decreased signal extending to the cartilage surface but with preservation of a sharp cartilage margin.
- Stage III: Focal signal abnormality extending to the cartilage surface with loss of its sharp margin. Focal or irregular cartilage narrowing may be present, without significant subchondral bone exposure. Decreased signal within the adjacent subchondral bone may be present.
- Stage IV: Focal decreased signal extending from subchondral bone to the cartilage surface over a significant area or cartilage thinning with exposure of subchondral bone. Decreased signal within subchondral bone is often present at this stage.

A

B

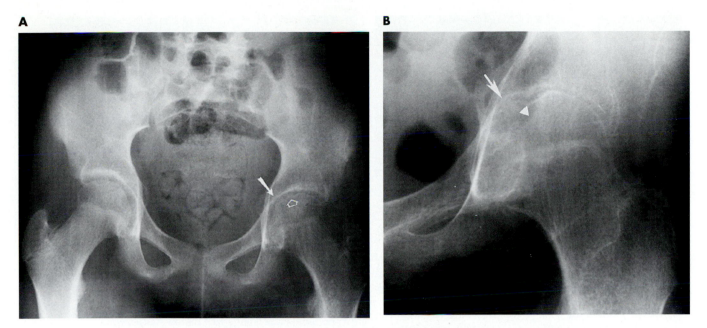

Fig. 3-31   Idiopathic chondrolysis of the hip. **A,** Frontal radiograph of the pelvis demonstrates characteristic periarticular demineralization, diffuse joint space narrowing (*closed arrow*), and irregularity of the femoral subchondral bone (*open arrow*) affecting the left hip. The right hip is within normal limits. **B,** Frontal radiograph from another patient depicts typical periarticular demineralization, severe joint space diminution (*arrow*), and subchondral erosion (*arrowhead*) affecting the weight-bearing aspect of the femoral head. Differential diagnosis includes septic arthritis, juvenile chronic arthritis, and transient synovitis of the hip.

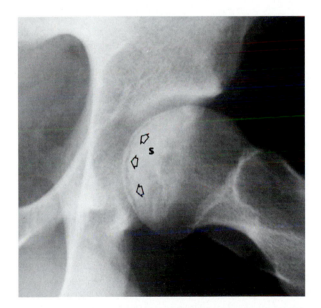

Fig. 3-32   Crescent sign in grade 3 osteonecrosis of the femoral head. Lateral radiograph demonstrates a subchondral radiolucency (*arrows*) traversing the sclerotic (*s*) femoral head that represents a fracture through necrotic bone and implies impending collapse of the articular surface (grade 4).

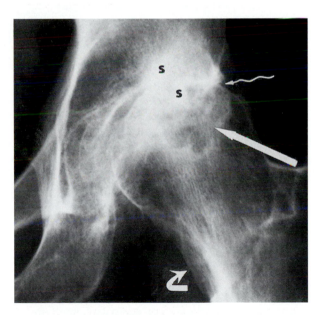

Fig. 3-33   Osteoarthritis of the hip. Typical findings include superior joint space narrowing (*wavy arrow*), subchondral sclerosis (*s*) and cyst formation (*straight arrow*), and buttressing of the femoral neck (*curved arrow*).

2. Gout: well-defined marginal erosions or erosions on the articular surface of the patella

3. Calcium pyrophosphate dihydrate crystal deposition disease: chondrocalcinosis involving menisci and hyaline cartilage (Fig. 3-43)

4. Rheumatoid arthritis: symmetric joint space narrowing, erosions, and subchondral cysts in a tricompartmental distribution

5. Ischemic necrosis: idiopathic variety most commonly involves the weight-bearing aspect of the medial femoral condyle (Box 3-26)

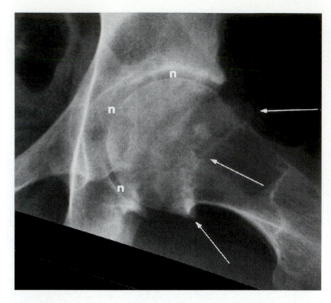

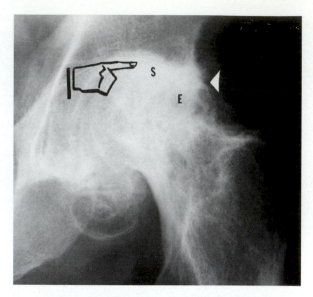

Fig. 3-34   Concentric joint space narrowing (*n*) with bone proliferation along the femoral head-neck junction (*arrows*) is characteristic of hip involvement by the seronegative spondyloarthropathies (ankylosing spondylitis in this instance).

Fig. 3-35   Advanced pigmented villonodular synovitis of the hip. Severe joint space narrowing (*pointing finger*) is associated with constriction of the femoral head and neck, well-defined erosive alterations (*E*), and degenerative changes including subchondral sclerosis (*S*) and osteophytes (*arrowhead*).

Box 3-26   Spontaneous Osteonecrosis of the Knee: Important Features

- Usually affects middle-aged to older women
- Unilateral involvement more common than bilateral disease
- Abrupt onset of pain, localized tenderness and stiffness, effusion, restricted range of motion
- Flattening of femoral condyle, radiolucent zone with adjacent sclerosis, intraarticular bodies, secondary osteoarthritis
- Medial femoral condyle, weight-bearing surface; occasionally affects lateral femoral condyle
- Associated medial meniscal tear; cause-effect relationship unclear

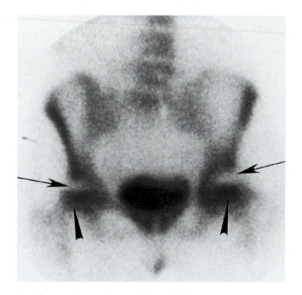

Fig. 3-36   Skeletal scintigraphy in bilateral osteonecrosis of the femoral heads. Photopenic areas (*arrows*) correspond to the ischemic weight-bearing aspects, whereas subjacent zones of increased activity (*arrowheads*) represent the reactive fibrous interface and buttressing of the reparative process. Positive radiographic findings would be anticipated at this stage of the disease.

6. Diffuse idiopathic skeletal hyperostosis (Forrestier disease): well-defined enthesopathy involving the quadriceps and patellar tendon attachments to the patella
7. Systemic lupus erythematosus: nonspecific periarticular swelling or nonerosive ligamentous deformity
8. Osteoarthritis: can be treated with high tibial valgus osteotomy, unicompartmental prosthesis, or total knee arthroplasty
9. Osteochondritis dissecans: may give rise to intraarticular osteochondral bodies and most commonly involves the lateral aspect of the medial femoral condyle
10. Rheumatoid arthritis: articular effusion with large dissecting popliteal cyst (Figs. 3-44, 3-45)
11. Pyrophosphate arthropathy: tricompartmental degenerative alterations that are most prominent in the patellofemoral compartment

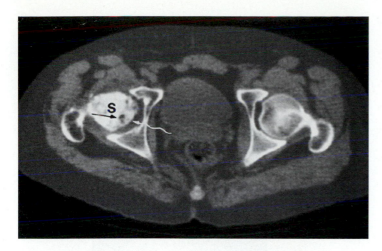

Fig. 3-37   Grade 4 osteonecrosis of the femoral head. CT demonstrates sclerosis (*s*), cyst formation (*straight arrow*), and collapse (*wavy arrow*) of the right femoral head. Contiguous thin sections with coronal and sagittal reformations are optimal for demonstrating subtle collapse, which precludes treatment by core decompression.

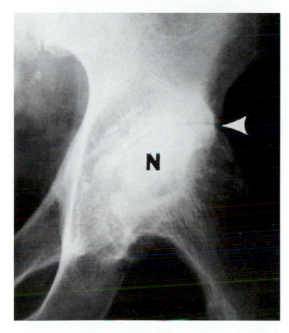

Fig. 3-38   Severe loss of articular cartilage (*arrowhead*) is evident in a patient with grade 5 ischemic necrosis (*N*) of the femoral head, which exhibits flattening.

12. Hemochromatosis: radiographic findings identical to those of calcium pyrophosphate dihydrate crystal deposition disease
13. Seronegative spondyloarthropathies: erosive alterations with whiskering
14. Rheumatoid arthritis: mechanical erosion of the anterior aspect of the distal femoral shaft secondary to patellar pressure
15. Juvenile chronic arthritis: joint effusion, uniform joint space narrowing, erosions, widened intercondylar notch, epiphyseal overgrowth, and squaring of the patella inferiorly (Fig. 3-46)

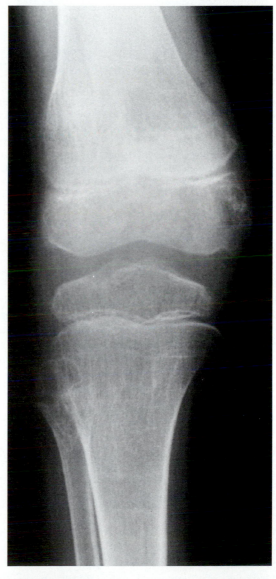

Fig. 3-39   Relapsing polychondritis affecting the knee. Frontal radiograph demonstrates characteristic striking periarticular demineralization, resembling the early appearance of juvenile chronic arthritis.

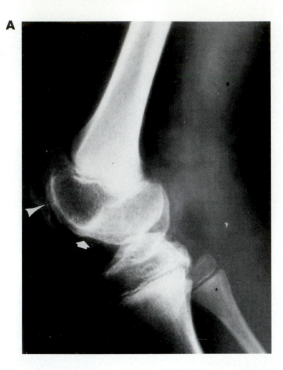

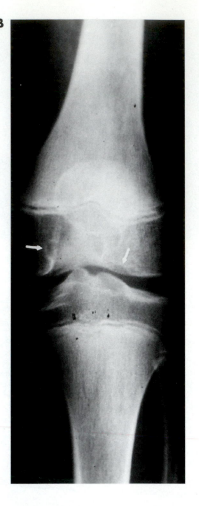

Fig. 3-40   Relapsing polychondritis affecting the knee joint. **A,** Lateral radiograph demonstrates marked narrowing of the patellofemoral joint space (*arrowhead*) along with irregularity of the femoral articular surface (*arrow*). **B,** Frontal radiograph reveals atypical well-defined erosions (*arrows*) involving the intercondylar notch region of the femur. Weight-bearing radiographs documented associated femorotibial joint space narrowing. Differential diagnosis includes juvenile chronic arthritis, hemophilia, and chronic indolent infection.

PEARL

Enlarged epiphyses are seen in juvenile chronic arthritis, hemophilia, septic arthritis, osteoid osteoma, and dysplasia epiphysealis hemimelica (Trevor disease). Juvenile chronic arthritis and hemophilia have similar radiographic appearances. Septic arthritis, especially tuberculous and fungal, causes enlarged epiphyses. Intracapsular osteoid osteoma enlarges the epiphysis secondary to hyperemia. Other diseases causing epiphyseal overgrowth include healed Legg-Calvé-Perthes disease, fibrous dysplasia, and Winchester syndrome. Hypothyroidism causes fragmentation and stippling of the epiphyses. The epiphyses are small with early fusion in thalassemia. Fairbank disease or multiple epiphyseal dysplasia has small fragmented epiphyses.

16. Pigmented villonodular synovitis: well-defined erosions, subchondral cysts, and preservation of joint space until late in the disease process (Fig. 3-47)
17. Osteoarthritis: marginal osteophytes in all three compartments of the knee and on the tibial spines
18. Neuroarthropathy: pattern of involvement is usually hypertrophic and secondary to tabes dorsalis

19. Rheumatoid arthritis: valgus malalignment deformity is most common
20. Synovial osteochondromatosis: multiple intraarticular osteochondral bodies of similar size, usually without effusion or significant joint space narrowing
21. Polyarthritis of rheumatic fever: migratory knee pain, swelling, and inflammation

PEARL

Behçet syndrome:
• Mean age of onset is 25 to 30 years, with a predilection for men.
• Articular abnormalities occur in over half of patients.
• Ophthalmic lesions (iritis, conjunctivitis, keratitis, optic neuritis) occur less commonly than ulcerations in the mouth or pharynx.
• Joint space narrowing and osseous erosions are rarely observed in affected articulations.
• Sacroiliitis is an inconstant feature of the disease.
• Monoarticular or oligoarticular involvement is the rule, and the knee is the principally affected site.

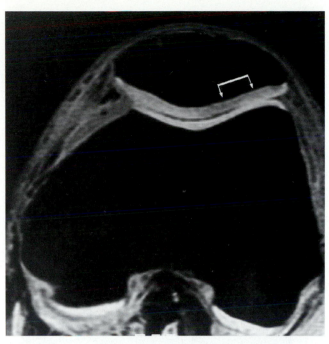

Fig. 3-41   Chondromalacia of the patella demonstrated by MRI. Transaxial gradient-echo image of the knee demonstrates areas of low signal intensity (*arrows*) within the normally bright hyaline cartilage of the lateral facet.

Ankle and foot   Rheumatologic disorders and their most closely associated radiographic findings in these areas (Fig. 3-48):

1. Jaccoud arthropathy: fibular deviation and flexion of the metatarsophalangeal joints
2. Osteoarthritis: joint space narrowing, subchondral cysts and sclerosis, and osteophyte formation affecting the first metatarsophalangeal (hallux rigidus) and first tarsometatarsal joints

> **PEARL**
>
> Hallux rigidus:
> - Obliterated articular space with subchondral sclerosis and a splayed appearance
> - Hallux valgus sometimes associated
> - A fairly common condition, frequently related to previous trauma or mechanical stress induced by certain types of footwear

3. Diffuse idiopathic skeletal hyperostosis (Forrestier disease): well-defined calcaneal enthesopathy at the Achilles and plantar aponeurosis insertions
4. Psoriatic arthritis or Reiter syndrome: erosions with whiskering involving the metatarsophalangeal and interphalangeal joints

> **PEARL**
>
> Psoriatic arthritis and Reiter syndrome:
> - Exuberant bone proliferation (malleoli, proximal corners of distal phalanges, proximal ends of proximal phalanges, cuneiforms, sesamoids of great toe)
> - Minimal osteoporosis
> - Fusion of interphalangeal joints
> - Ill-defined erosions around joints and at Achilles tendon insertion
> - Bone destruction (arthritis mutilans)

5. Rheumatoid arthritis: erosive alterations involving the metatarsophalangeal joints

> **PEARL**
>
> Rheumatoid arthritis:
> - Periarticular osteoporosis
> - Subluxations and erosions at metatarsophalangeal joints, some large and cystic (resembling gout or brown tumors of hyperparathyroidism)
> - Absence of whiskering

6. Hypertrophic osteoarthropathy: digital clubbing with metatarsal periostitis
7. Ischemic necrosis: relative sclerosis of the talar body following talar neck fracture; idiopathic form affects tarsal navicular in adults (Fig. 3-49)
8. Psoriatic arthritis: ivory phalangeal tufts
9. Rheumatoid arthritis, psoriatic arthritis, or Reiter syndrome: retrocalcaneal bursitis with obliteration of pre-Achilles fat (Fig. 3-50)
10. Neuroarthropathy: resorptive and proliferative alterations involving the talonavicular and metatarsophalangeal joints in a diabetic or (less commonly) alcoholic patient

> **PEARL**
>
> Charcot arthropathy in diabetes:
> - Fragmentation in midtarsal area
> - Destruction of distal ends of second and third metatarsals
> - Soft-tissue swelling
> - No osteoporosis
> - Calcified vessels
> - Soft-tissue edema
> - Fractures of multiple metatarsals
> - Possible associated infection

A

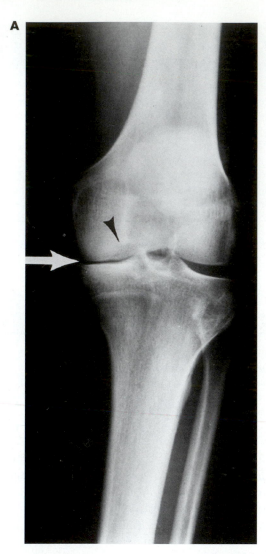

B

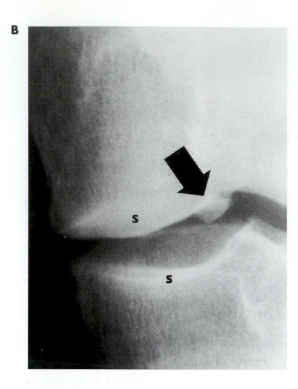

Fig. 3-42   **A** and **B,** Osteoarthritis of the knee is characterized by predominant narrowing of the medial femorotibial compartment (*white arrow*) with genu varus, marginal osteophytosis (*black arrow*), subchondral sclerosis (*s*), and cyst formation (*arrowhead*).

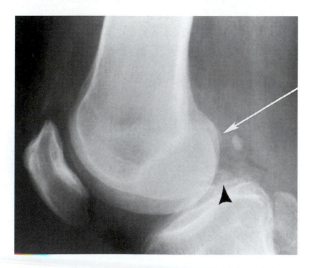

Fig. 3-43   Chondrocalcinosis involving hyaline cartilage (*arrow*) and fibrocartilage (*arrowhead*) in calcium pyrophosphate dihydrate crystal deposition disease. Differential diagnosis should include consideration of hyperparathyroidism, hemochromatosis, ochronosis, acromegaly, and degenerative disease.

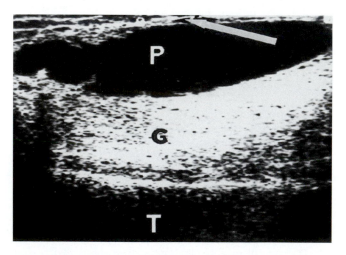

Fig. 3-44   Dissecting popliteal cyst (*P*) in rheumatoid arthritis characterized by ultrasonography. The abnormal anechoic fluid collection lies superficial to the gastrocnemius muscle (*G*) and tibia (*T*) (*arrow = subcutaneous tissue*).

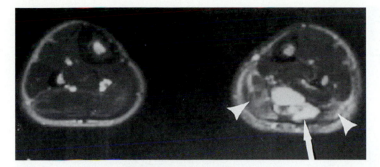

Fig. 3-45 Dissecting popliteal cyst in rheumatoid arthritis. Transaxial T2-weighted MR image demonstrates a lobulated mass of high signal intensity (*arrow*) with apparent leakage into the adjacent musculature (*arrowheads*).

A B

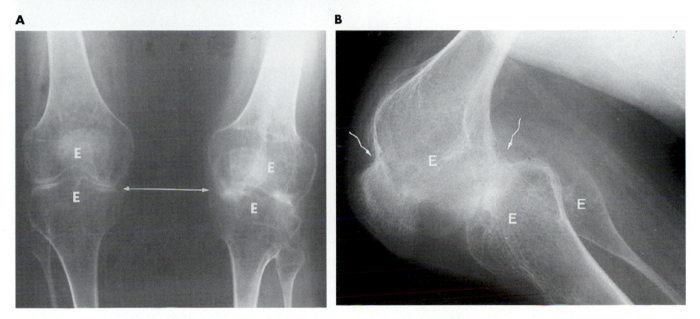

Fig. 3-46 Diffuse joint space narrowing (*arrows*), epiphyseal overgrowth (*E*), and generalized osteopenia in juvenile-onset rheumatoid arthritis. **A,** Frontal view. **B,** Lateral view. Differential diagnosis should include consideration of hemophilia and chronic joint infection, although the latter is usually monoarticular.

11. Juvenile chronic arthritis: joint space narrowing, demineralization, erosions, and epiphyseal overgrowth, with involvement of the tibiotalar, subtalar, tarsal, and metatarsophalangeal joints (Figs. 3-51, 3-52)
12. Rheumatoid arthritis: lateral deviation at the metatarsophalangeal joints, interphalangeal flexion deformities, and hyperextended metatarsophalangeal joints
13. Rheumatoid arthritis, psoriatic arthritis, or Reiter syndrome: inflammatory calcaneal enthesopathy at the Achilles and plantar aponeurosis insertions
14. Neuroarthropathy: Lisfranc tarsometatarsal fracture-dislocation
15. Rheumatoid arthritis, juvenile chronic arthritis, or Reiter syndrome: tarsal bone erosions with advanced disease
16. Ischemic necrosis: subchondral osteopenic band in the talar dome (Hawkins sign) implies an intact blood supply following trauma and excludes the presence of ischemia; unlike the crescent sign, it is a good rather than poor prognostic sign

17. Psoriatic arthritis or Reiter syndrome: sausage digit
18. Polyarthritis of rheumatic fever: migratory ankle pain, swelling, and inflammation
19. Gout: well-defined erosions with overhanging edges involving the metatarsophalangeal (especially the first), interphalangeal, midfoot, and hindfoot joints (Fig. 3-53)

PEARL

Gout:
- Soft-tissue masses, especially lateral to proximal end of fifth metatarsal
- Well-defined erosions with overhanging edges
- Subchondral cystic alterations
- Splayed appearance at first metatarsophalangeal joint
- Minimal osteoporosis
- Variable soft-tissue calcification

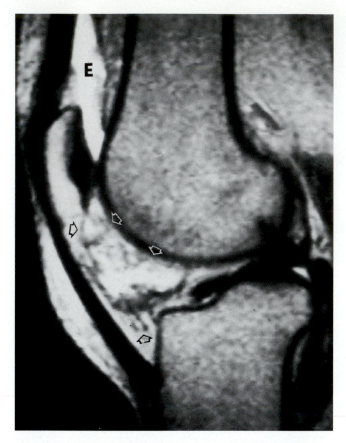

Fig. 3-47   Pigmented villonodular synovitis of the knee on a T2-weighted MR image. Irregularly thickened synovium (*arrows*) exhibits mixed areas of high and low signal intensity, with the latter representing hemosiderin deposition; a large serosanguinous effusion (*E*) is also present.

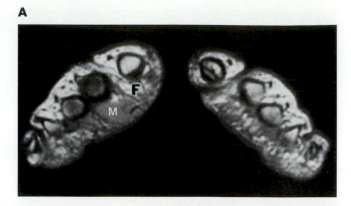

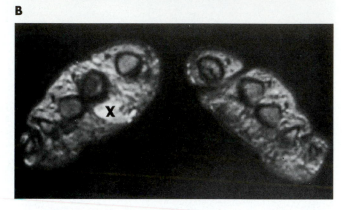

Fig. 3-48   Synovial inflammation affecting the plantar tendon sheaths of the foot. **A,** Axial T1-weighted magnetic resonance image demonstrates an irregular soft-tissue mass (*M*) of intermediate signal intensity within the plantar fat (*F*). **B,** On a corresponding T2-weighted image, the area exhibits extremely high signal intensity (*X*), compatible with inflammatory fluid.

20. Pseudogout syndrome: acute soft-tissue swelling around the first metatarsophalangeal joint
21. Gout: calcified soft-tissue masses in advanced cases with renal involvement, most with associated osseous erosion

Shoulder   Rheumatologic disorders and their most closely associated radiographic findings at this site (Fig. 3-54):

1. Amyloidosis: bulky soft-tissue nodules superimposed on atrophic musculature (shoulder pad sign)
2. Shoulder impingement syndrome: subacromial spur with sclerosis of the greater humeral tuberosity
3. Calcium hydroxyapatite crystal deposition disease: soft-tissue calcification involving the supraspinatus, infraspinatus, teres minor, subscapularis, and biceps tendons, as well as the subacromial and subdeltoid bursae
4. Synovial osteochondromatosis: multiple rounded intraarticular bodies of similar size and variable mineralization, often appearing lamellar or trabecular
5. Ischemic necrosis: subchondral sclerosis with linear fracture (crescent sign) (Fig. 3-55)
6. Rheumatoid arthritis: osteolysis of the distal clavicle, marginal erosions on the humeral head,

and clavicular erosion at the coracoclavicular ligament insertion are early findings
7. Juvenile chronic arthritis: enlargement of the glenoid and humeral head with late erosions
8. Ankylosing spondylitis: combined erosive and productive alterations, often bilateral and symmetric or asymmetric
9. Amyloidosis: joint space widening, well-defined erosions, subchondral cysts, and osteopenia
10. Calcium pyrophosphate dihydrate crystal deposition disease: chondrocalcinosis with changes of degenerative joint disease in a patient with no prior trauma
11. Rheumatoid arthritis: elevation of the humeral head with pressure erosion of the undersurface of the acromion and the medial aspect of the humeral surgical neck
12. Ankylosing spondylitis: common involvement (second most commonly involved appendicular articulation after the hip in this disease)
13. Osteoarthritis: marginal glenoid osteophytes with circumferential osteophytosis involving the humeral anatomic neck in a patient with prior trauma (Fig. 3-56)

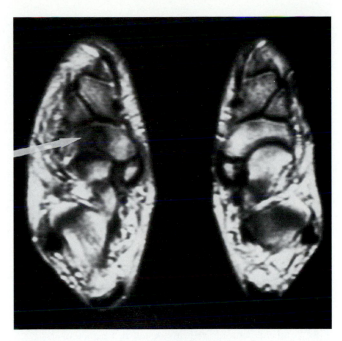

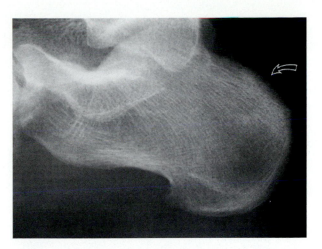

Fig. 3-50   Calcaneal erosion (*arrow*) secondary to pre-Achilles bursal inflammation in Reiter syndrome. Differential diagnosis should include consideration of other seronegative spondyloarthropathies, infection, hyperparathyroidism, and rheumatoid arthritis.

Fig. 3-49   Osteonecrosis of the tarsal navicular in an adult (Muller-Weiss syndrome). T1-weighted MR image in the plantar plane of the foot demonstrates low signal intensity in the lateral aspect of the bone (*arrow*). Osteonecrosis at this site may be isolated (secondary to trauma or abnormal stress across the midfoot) or associated with other areas of bone infarction in patients predisposed to osteonecrosis.

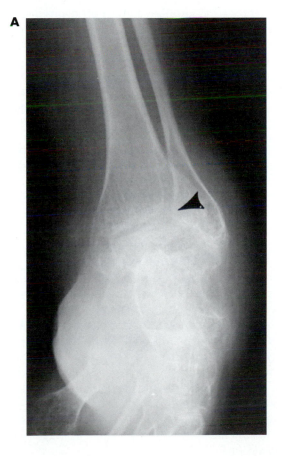

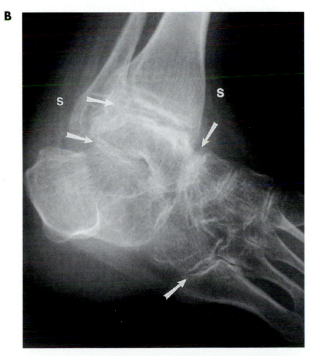

Fig. 3-51   Juvenile chronic arthritis. **A** and **B,** Typical abnormalities in the ankle and midfoot include generalized osteopenia, gracile diaphyses with relative epiphyseal overgrowth, widespread joint space narrowing with marginal and central osseous erosions (*arrows*), tibiotalar slant (*arrowhead*), soft-tissue swelling (*s*), and muscle atrophy. Differential diagnosis should include consideration of hemophilia, although the midfoot articulations are generally spared in this condition.

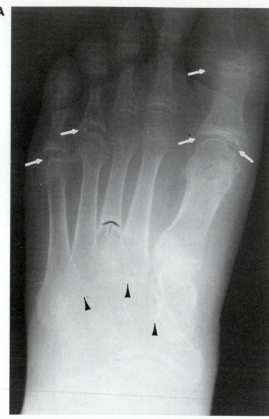

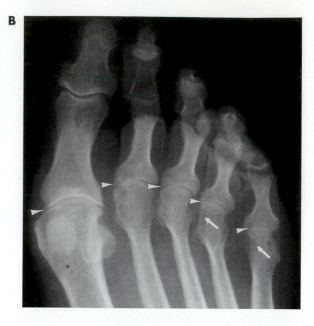

Fig. 3-52 Juvenile chronic arthritis. **A** and **B,** Characteristic findings in the forefoot of two different patients include generalized osteopenia, joint space narrowing (*arrowheads*), premature physeal closure with epiphyseal overgrowth, and marginal and central osseous erosions (*arrows*). Involvement of multiple small joints in a young patient is virtually diagnostic of this condition.

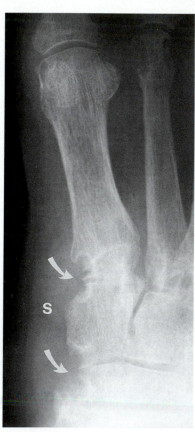

Fig. 3-53 Well-defined erosions, some with overhanging edges (*arrows*), and adjacent soft-tissue tophi (*S*) in the midfoot are characteristic of gout.

14. Neuroarthropathy: atrophic pattern of involvement, with extensive resorption of the humeral head and neck in a patient with syringomyelia (Figs. 3-57, 3-58)
15. Calcium hydroxyapatite crystal deposition disease: Milwaukee shoulder
16. Osteoarthritis: joint space narrowing, osteophytosis, and subchondral cysts or sclerosis affecting the acromioclavicular joint

Elbow Rheumatologic disorders and their most closely associated radiographic findings in this articulation (Fig. 3-59):

1. Rheumatoid arthritis: joint effusion with erosions of the distal humerus, radial head, and coronoid process
2. Juvenile chronic arthritis: uniform joint space narrowing, erosions, effusion, and enlargement of the radial head and trochlear notch
3. Amyloidosis: bulky soft-tissue nodules, joint space widening, well-defined erosions, subchondral cysts, contracture, and atrophy of bone and muscle
4. Gout: tophaceous deposits in the olecranon bursa; well-defined osseous erosions with overhanging edges (Fig. 3-60)
5. Calcium pyrophosphate dihydrate crystal deposition disease: chondrocalcinosis and degenerative joint disease in a patient with no prior trauma
6. Calcium hydroxyapatite crystal deposition disease: soft-tissue calcification at triceps and collateral ligament insertions

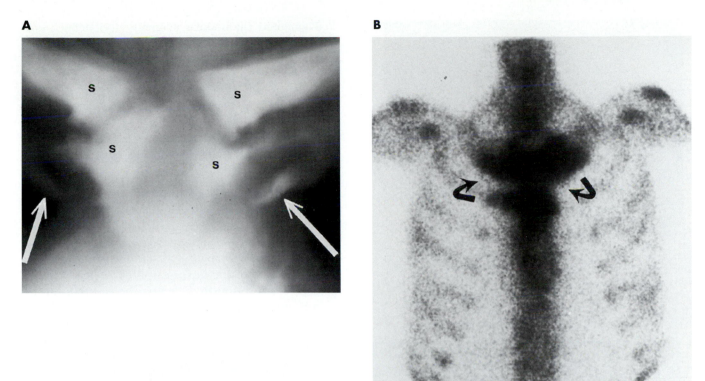

Fig. 3-54   Sternocostoclavicular hyperostosis. **A,** Frontal tomogram demonstrates sclerosis (*s*) of the manubrium and clavicular heads with irregular heterotopic ossification (*arrows*). **B,** Diffuse increased activity (*arrows*) is present in the region on a radionuclide bone scan.

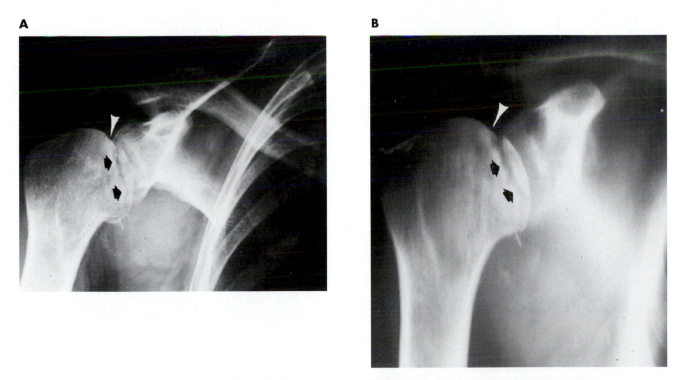

Fig. 3-55   Osteonecrosis of the humeral head. **A** and **B,** Conventional radiography and tomography demonstrate inhomogeneous density in the subchondral bone, with subchondral fracture (*arrows*) and early collapse (*arrowheads*). Differential diagnosis should include consideration of neuroarthropathy, osteochondritis dissecans, and osteochondral fracture.

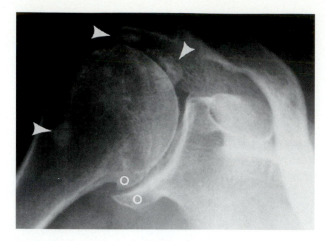

Fig. 3-56 Joint space narrowing, osteophytosis (*o*), and multiple osteochondral bodies (*arrowheads*) are typical of glenohumeral osteoarthritis.

7. Pigmented villonodular synovitis: well-defined erosions, subchondral cysts, hemorrhagic effusion, and joint space preservation until late in the disease process
8. Synovial osteochondromatosis: multiple rounded intraarticular bodies of similar size and variable mineralization
9. Progressive systemic sclerosis (scleroderma): flexion contracture with atrophy of bone, muscle, and cartilage

Wrist Rheumatologic disorders affecting this site and their most closely associated radiographic findings:
1. Ischemic necrosis: sclerosis and collapse of the proximal pole of the scaphoid following wrist fracture
2. Calcium hydroxyapatite crystal deposition disease: calcification of the flexor carpi ulnaris tendon adjacent to the pisiform, best seen on a carpal tunnel radiograph

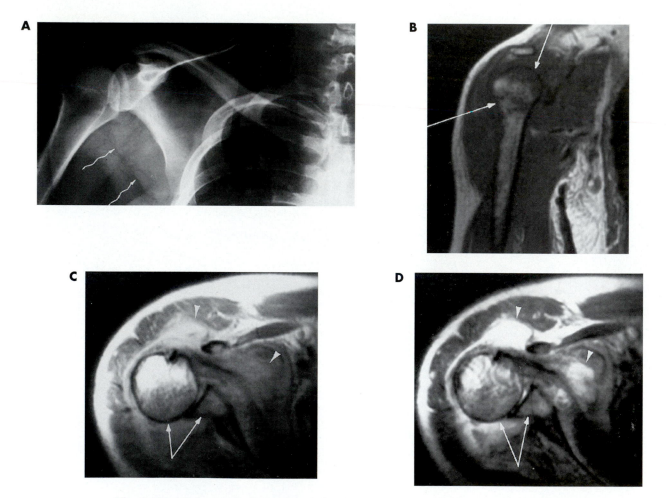

Fig. 3-57 Early neuroarthropathy of the shoulder in syringomyelia. **A,** Radiography demonstrates soft-tissue fullness in the axilla (*arrows*). **B,** Coronal T1-weighted MR image reveals patchy areas of low signal intensity (*arrows*) in the marrow of the proximal humerus. **C** and **D,** Transaxial proton density– and T2-weighted images confirm humeral and glenoid marrow abnormalities (*arrows*) and document anterior redundancy of the joint capsule (*arrowheads*) secondary to early instability. Differential diagnosis should include consideration of septic and rheumatoid arthritis, either of which can produce dissecting synovial cysts.

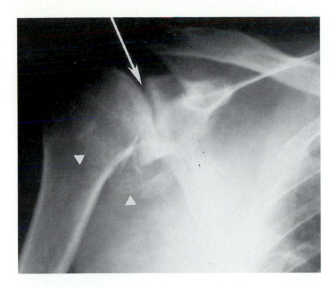

Fig. 3-58   Neuroarthropathy of the glenohumeral joint with early fragmentation. Joint space narrowing (*arrow*) is associated with multiple intraarticular osteochondral bodies (*arrowheads*). Differential diagnosis should include consideration of posttraumatic osteoarthritis, pyrophosphate arthropathy, osteonecrosis with collapse of the articular surface, osteochondritis dissecans, idiopathic synovial osteochondromatosis, and inflammatory arthritides.

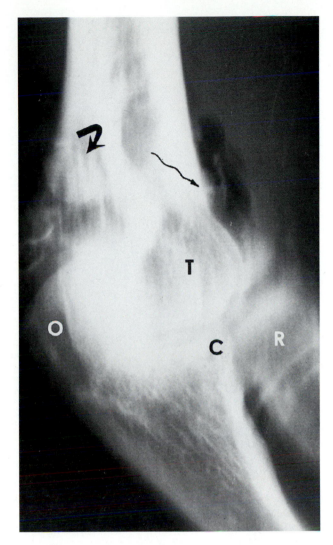

Fig. 3-59   Osteochondral bodies in the elbow tend to lodge within the olecranon (*curved arrow*) and coronoid (*wavy arrow*) recesses of the distal humerus, as demonstrated on a lateral arthrotomogram. CT arthrography is a preferred method in this situation because of its lower radiation exposure (*T = trochlea, C = coronoid process, O = olecranon process, R = radial head*).

3. Wilson disease: indistinct, irregular subchondral bone with fragmentation, subchondral cysts and sclerosis, joint space narrowing, chondrocalcinosis, and osteopenia

4. Hemochromatosis: chondrocalcinosis, osteopenia, joint space narrowing, and prominent subchondral cysts affecting the radiocarpal joint

5. Calcium pyrophosphate dihydrate crystal deposition disease: chondrocalcinosis in the triangular fibrocartilage and/or interosseous ligament calcification

6. Gout: well-defined erosions with overhanging edges affecting the intercarpal joints

7. Erosive (inflammatory) osteoarthritis: erosions and osteophyte formation affecting the first carpometacarpal and trapezioscaphoid joints

8. Osteoarthritis: sclerosis, osteophyte formation, and joint space narrowing in the first carpometacarpal articulation, often with radial subluxation of the thumb

9. Amyloidosis: bulky soft-tissue nodules, joint space widening, well-defined erosions, subchondral cysts, contracture, and atrophy of muscle and bone

10. Systemic lupus erythematosus: nonspecific swelling or nonerosive ligamentous deformity, particularly subluxation of the first carpometacarpal joint

11. Psoriatic arthritis: erosions and ill-defined bone proliferation affecting any compartment of the wrist, usually following involvement of the distal interphalangeal joints

12. Juvenile chronic arthritis: osteopenia, soft-tissue swelling, joint space narrowing, late erosive alterations, ankylosis, and contracture with muscle wasting, particularly involving the midcarpal compartment (Figs. 3-61, 3-62)

13. Adult-onset Still disease: pericapitate joint space narrowing and erosions in an adult

14. Rheumatoid arthritis: early erosions involving the distal radioulnar joint, ulnar and radial styloid processes, scaphoid wrist, triquetrum, and pisiform

15. Osteoarthritis: joint space narrowing, subchondral sclerosis, and osteophyte formation affecting the trapezioscaphoid joint as the second most common site of carpal involvement (Fig. 3-63)

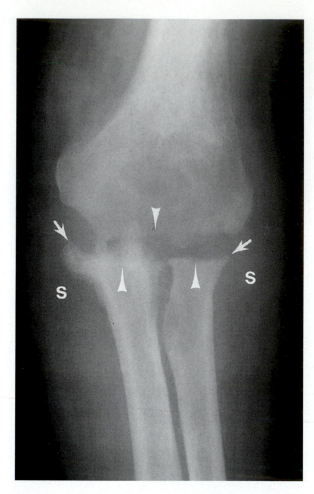

Fig. 3-60 Well-defined central erosions (*arrowheads*) with overhanging edges (*arrows*) and soft-tissue swelling (*s*) in gout. Differential diagnosis should include consideration of inflammatory arthritides and septic arthritis.

16. Pyrophosphate arthropathy: degenerative alterations in the radiocarpal compartment, with scapholunate dissociation and scaphoid erosion into the distal radial articular surface
17. Rheumatoid arthritis: joint space narrowing and erosions involving the intercarpal and intermetacarpal joints, occurring relatively late in the disease process
18. Gout: prominent calcified soft-tissue masses in a periarticular distribution
19. Pyrophosphate arthropathy: proximal migration of the capitate between the scaphoid and lunate, resulting in scapholunate advanced collapse (SLAC) wrist (Fig. 3-64)
20. Progressive systemic sclerosis (scleroderma): flexion contracture with atrophy of bone, muscle, and cartilage

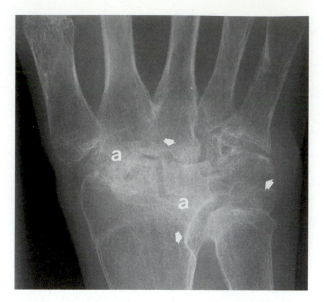

Fig. 3-61 Juvenile chronic arthritis of the wrist. Characteristic findings include generalized demineralization, joint space narrowing, erosive alterations (*arrows*), and evolving ankylosis (*a*).

21. Rheumatoid arthritis: carpal instability, including scapholunate dissociation, distal radioulnar dissociation, dorsiflexion and palmar flexion instability, radial deviation of the hand, and ulnar translocation (more than half of the lunate articulates with the ulna) (Fig. 3-65)

Hand Rheumatologic disorders and their most closely associated radiographic findings in these joints:

1. Rheumatoid arthritis: early joint space narrowing and marginal erosions affecting the metacarpophalangeal and proximal interphalangeal joints
2. Progressive systemic sclerosis (scleroderma): flexion contractures of the proximal and distal interphalangeal joints, with atrophy of bone, muscle, and cartilage
3. Jaccoud arthropathy: reversible ulnar deviation and flexion of the metacarpophalangeal joints
4. Osteoarthritis: joint space narrowing, subchondral sclerosis, and osteophyte formation affecting the interphalangeal joints
5. Calcium hydroxyapatite crystal deposition disease: periarticular calcification affecting the metacarpophalangeal and interphalangeal joints
6. Psoriatic arthritis: tuftal resorption and erosions with whiskering affecting the distal interphalangeal joints earlier and more severely than the proximal interphalangeal or metacarpophalangeal joints (Figs. 3-66, 3-67)
7. Gout: well-defined erosions with overhanging edges affecting the interphalangeal joints more frequently than the metacarpophalangeal joints

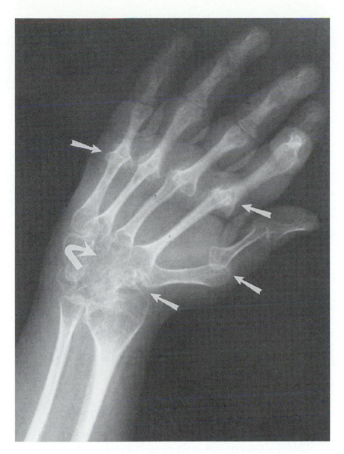

Fig. 3-62 Multiple joint subluxations with erosions (*straight arrows*), carpal ankylosis (*curved arrow*), gracile long and short tubular bones, and diffuse osteopenia are virtually diagnostic of juvenile-onset rheumatoid arthritis.

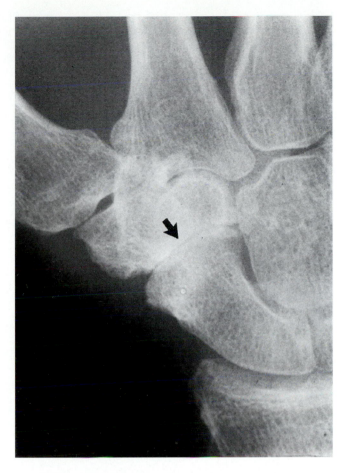

Fig. 3-63 Osteoarthritis involving the trapezioscaphoid articulation (*arrow*). This joint is also a target site in pyrophosphate arthropathy and rheumatoid arthritis.

8. Progressive systemic sclerosis (scleroderma): calcinosis circumscripta
9. Rheumatoid arthritis: occasional late erosions and joint space narrowing in the distal interphalangeal joints
10. Pyrophosphate arthropathy: joint space narrowing in the second and third metacarpophalangeal articulations
11. Wilson disease: indistinct, irregular subchondral bone with fragmentation, subchondral cysts and sclerosis, joint space narrowing, and osteopenia affecting the metacarpophalangeal joints
12. Systemic lupus erythematosus: reducible ulnar deviation at the metacarpophalangeal joints and flexion or extension deformities of the interphalangeal joints, with nonspecific polyarticular swelling (Fig. 3-68)
13. Progressive systemic sclerosis (scleroderma): severe bone resorption at the first carpometacarpal joint with radial subluxation of the first metacarpal
14. Hemochromatosis: joint space narrowing and subchondral cysts affecting the metacarpophalangeal joints, particularly the second and third, with

prominent hooklike osteophytes on the metacarpal heads (Fig. 3-69)
15. Gout: tophaceous soft-tissue deposits that may lie at some distance from the articulations (Fig. 3-70)
16. Rheumatoid arthritis: ulnar deviation and volar subluxation at the metacarpophalangeal joints, often with pressure erosions of the metacarpals and proximal phalanges
17. Osteoarthritis: joint space narrowing in the metacarpophalangeal articulations, with subchondral cysts in the metacarpal heads.
18. Erosive (inflammatory) osteoarthritis: joint space narrowing, subchondral sclerosis, central erosions, and marginal osteophytes (gull-wing appearance) affecting the interphalangeal joints
19. Rheumatoid arthritis: first metacarpophalangeal joint flexion with interphalangeal joint extension (hitchhiker thumb)
20. Juvenile chronic arthritis: soft-tissue swelling, osteoporosis, joint space narrowing, erosions, and ankylosis affecting the metacarpophalangeal and proximal interphalangeal joints (Fig. 3-71)

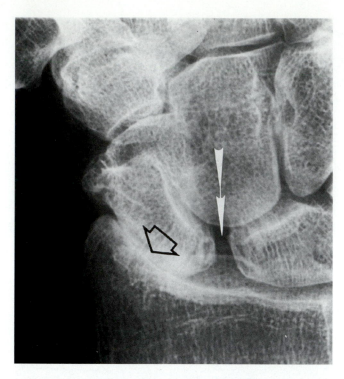

Fig. 3-64   Radiocarpal joint space loss (*open arrow*) and widening of the scapholunate space (*closed arrow*) in pyrophosphate arthropathy. Similar findings occur following trauma, as well as in inflammatory arthritides.

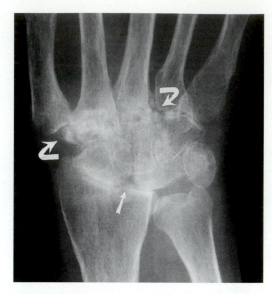

Fig. 3-65   Pancompartmental joint space narrowing, marginal and central osseous erosions (*curved arrows*), scapholunate dissociation (*straight arrow*) with carpal collapse, and diffuse osteopenia in rheumatoid arthritis. Differential diagnosis should include consideration of other inflammatory arthritides and septic arthritis.

21. Progressive systemic sclerosis (scleroderma): acroosteolysis, initially on the palmar aspect of the tufts (Fig. 3-72)
22. Rheumatoid arthritis: proximal interphalangeal joint hyperflexion with distal interphalangeal joint hyperextension (boutonnière deformity)
23. Osteoarthritis: Heberden and Bouchard nodes, secondary to prominent osteophytosis (Fig. 3-73)
24. Hypertrophic osteoarthropathy: digital clubbing with metacarpal periostitis
25. Rheumatoid arthritis: proximal interphalangeal joint hyperextension with distal interphalangeal joint hyperflexion (swan-neck deformity)
26. Osteoarthritis: occasional flexion, radial, or ulnar deformity at the interphalangeal joints

Important general concepts
1. Osteoarthritis or degenerative joint disease is the most common type of arthritis; in the temporomandibular joint, it occurs secondary to internal derangement (Figs. 3-74, 3-75, 3-76, 3-77).
2. Although cutaneous psoriasis usually precedes the arthropathy, among patients with both manifestations, approximately 20% have the presenting symptom of arthritis.
3. In Reiter syndrome, arthritis is usually the last clinical manifestation to appear.
4. The cervical spine is most commonly affected by the discovertebral lesions of rheumatoid arthritis.
5. Six to eight percent of the normal population is HLA-B27 positive.
6. Reactive tuftal sclerosis, known as the ivory phalanx, is most characteristic of psoriatic arthritis.
7. The four causes of osteopenia in rheumatoid arthritis are hyperemia secondary to articular inflammation, disuse, corticosteroid therapy, and the disease itself.
8. Joint effusion, orthopedic traction, acromegaly, and amyloidosis are four possible causes of joint space widening.
9. Marginal osteophyte formation is the cause of the clinically evident Heberden (distal interphalangeal joints) and Bouchard (proximal interphalangeal joints) nodes of osteoarthritis.
10. The subchondral cyst of Eggar occurs in the weight-bearing portion of the acetabulum and may be the first sign of osteoarthritis in the hip.
11. In inflammatory sacroiliitis, the iliac side of the joint is usually affected most prominently.
12. *Salmonella, Shigella,* and *Yersinia* organisms are most commonly implicated in gastrointestinal infection associated with self-limited polyarthritis (reactive arthritis), usually without radiographic findings but occasionally with sacroiliitis.
13. Ulcerative colitis, Crohn disease, and Whipple disease are three forms of inflammatory bowel disease that may be associated with sacroiliitis or peripheral arthropathy.

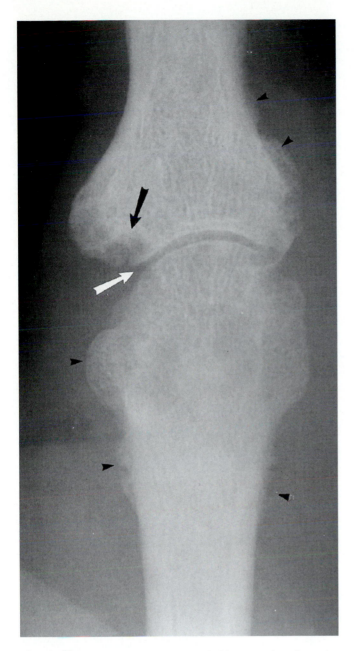

Fig. 3-66   Joint space narrowing (*white arrow*) and erosive arthritis (*black arrow*) associated with whiskering (*arrowheads*) are characteristic of seronegative spondyloarthropathies (psoriatic arthritis in this instance).

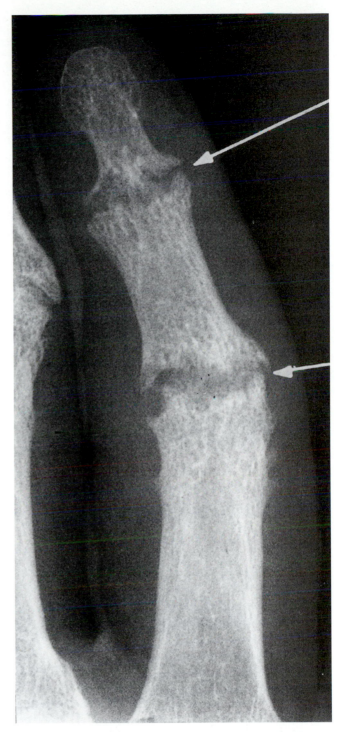

Fig. 3-67   Evolving interphalangeal joint ankylosis (*arrows*) with soft-tissue swelling in psoriatic arthritis. Differential diagnosis should include consideration of inflammatory or erosive osteoarthritis, juvenile chronic arthritis, other seronegative spondyloarthropathies, and infection (usually monoarticular).

14. Decreased activity in the acute phase and increased activity in the reparative phase are the characteristic features of ischemic necrosis on radionuclide bone scans.

15. Bronchogenic carcinoma, cyanotic congenital heart disease, inflammatory bowel disease, cystic fibrosis, chronic bronchitis or bronchiectasis, pneumoconiosis, hepatic or biliary cirrhosis, gastrointestinal carcinomas (esophagus, liver, colon), connective tissue disorders, and lymphoma are 10 of the numerous recognized causes of secondary hypertrophic osteoarthropathy.

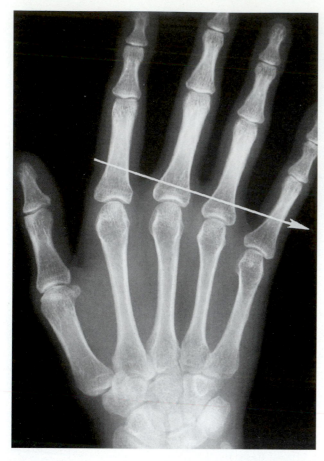

Fig. 3-68 Ulnar deviation at the metacarpophalangeal joints (*arrow*) is characteristic of the nonerosive deforming arthropathy of systemic lupus erythematosus. Differential diagnosis should include consideration of Jaccoud arthropathy following rheumatic fever and Ehlers-Danlos syndrome.

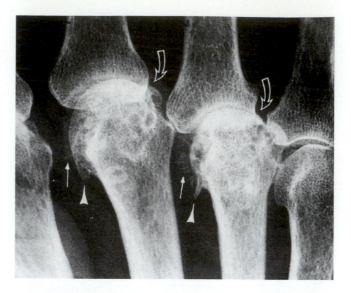

Fig. 3-69 Metacarpophalangeal joint space loss (*curved arrows*), chondrocalcinosis (*straight arrows*), and marginal excrescences on the metacarpal heads (*arrowheads*) in hemochromatosis. Differential diagnosis should include consideration of calcium pyrophosphate dihydrate crystal deposition disease and longstanding acromegaly.

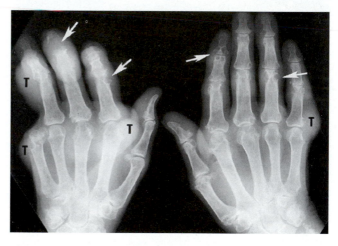

Fig. 3-70 Well-defined erosions (*arrows*), some with overhanging edges, soft-tissue tophi (*T*), asymmetric involvement, and relative preservation of joint spaces are characteristic of gout.

16. Clinically evident attacks of gout typically occur for several years before the appearance of radiographic abnormalities.

17. Nail thickening, pitting, or discoloration is common in psoriasis and strongly correlated with the severity of the arthropathy.

18. In osteoarthritis, the severity of radiographic change does not always correlate strongly with the experience of painful symptoms by the patient.

19. Giant cell tumor of tendon sheath is histologically indistinguishable from intraarticular diffuse pigmented villonodular synovitis.

20. Elderly patients may exhibit false positive test results for rheumatoid factor.

21. Approximately 40% of all neuropathic joints are both hypertrophic and atrophic, whereas 20% are predominantly hypertrophic and 40% are primarily atrophic.

22. Surgical arthrodesis is difficult in neuroarthropathy, and spontaneous ankylosis is rare.

23. In neuroarthropathy, joint effusions may decompress along fascial planes, carrying osseous debris away from the joint of origin (Fig. 3-78).

24. In Reiter syndrome, ankylosis occurs in the sacroiliac joints but is much less common at other sites than it is in psoriatic arthritis.

25. Joint space narrowing secondary to cartilage destruction frequently occurs later in the course of juvenile chronic arthritis than it does in adult-onset rheumatoid arthritis.

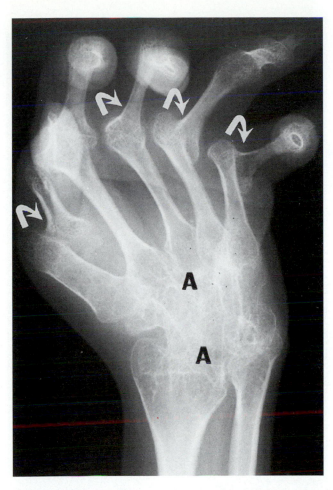

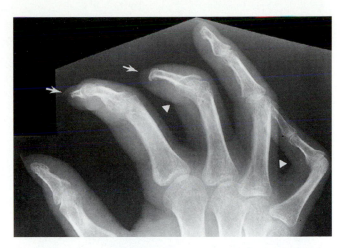

Fig. 3-72   Acro-osteolysis (*arrows*), flexion contractures (*arrowheads*), and diffuse osteopenia in scleroderma. This combination of findings also occurs in thermal injury and epidermolysis bullosa.

Fig. 3-71   Advanced juvenile chronic arthritis of the hand and wrist. The metacarpophalangeal joints manifest ulnar deviation with subluxation (*arrows*), whereas the carpus is ankylosed (*A*) in radial deviation. Severe osteopenia is evident in a generalized distribution.

26. Ankylosis is far more common in juvenile chronic arthritis than in adult-onset rheumatoid arthritis.
27. The clinically asymptomatic Jaccoud arthropathy occasionally follows multiple episodes of the much more common symptomatic polyarthritis of rheumatic fever.
28. Hip involvement is less common in Reiter syndrome than in the other seronegative spondyloarthropathies.
29. If large joints (hip, shoulder, knee, ankle) are involved in psoriatic arthritis, the small articulations of the hands and feet are almost always affected as well.
30. Erosions frequently occur at the sternomanubrial and sternoclavicular articulations in rheumatoid arthritis.
31. Rheumatoid arthritis tends to spare joints that are immobilized secondary to neurologic deficits.

32. Temporomandibular joint erosions, micrognathia, and a prominent antegonial notch are relatively common features of juvenile chronic arthritis.
33. Consideration of hyperparathyroidism should be included in the differential diagnosis of bilateral symmetric sacroiliitis.
34. Erosive or inflammatory osteoarthritis occurs predominantly in middle-aged women, who experience distinct episodes of articular inflammation, swelling, and erythema.
35. Spontaneous osteonecrosis of the femoral head in adults is more common in men than in women, although the opposite is true when the disease affects the knee.
36. Patients with rheumatoid arthritis who maintain normal levels of activity may manifest relatively normal bone density and large subchondral cysts.
37. Resorption of bone occurs in up to 80% of patients with scleroderma.
38. Thoracotomy may lead to clinically and radiographically evident remission in cases of secondary hypertrophic osteoarthropathy.
39. Because of the pulsating aorta on the left side of the spine, thoracic spinal ossification in diffuse idiopathic skeletal hyperostosis usually occurs anterolaterally on the right side.
40. Early sacroiliitis is difficult to recognize radiographically in children and adolescents because the joints are normally wide and manifest indistinct cortices.
41. Premature degenerative joint disease is a recognized complication of acromegaly and Paget disease.
42. The combination of connective tissue disease (frequently rheumatoid arthritis), xerostomia, and keratoconjunctivitis sicca is known as Sjögren syndrome.

A

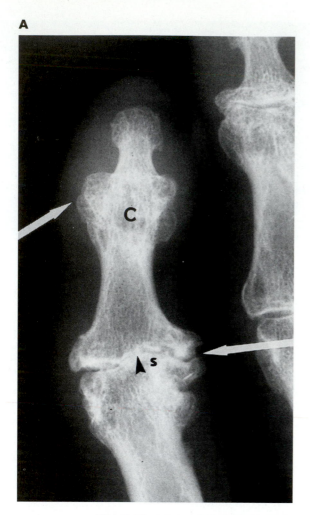

B

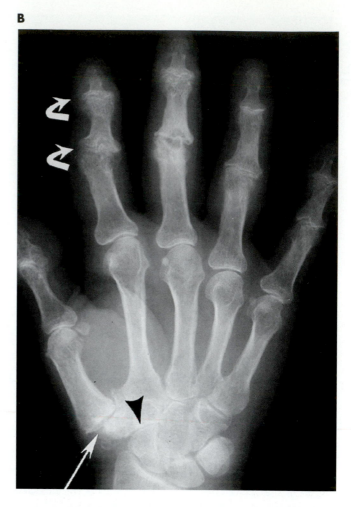

Fig. 3-73   **A,** Joint space narrowing (*arrowhead*), osteophytosis (*arrows*), and subchondral sclerosis (*s*) with cysts (*c*) in osteoarthritis. **B,** Characteristic target sites include the interphalangeal (*curved arrows*), first carpometacarpal (*straight arrow*), and trapezioscaphoid (*arrowhead*) articulations.

43. Ischemic necrosis affecting the humerus, talar dome, or bones of the hands or feet in the absence of trauma should suggest the diagnosis of systemic lupus erythematosus.

44. The combination of rheumatoid arthritis, splenomegaly, and leukopenia is known as Felty syndrome.

45. Resorption of the mandibular angle with facial skin disease occurs in progressive systemic sclerosis (scleroderma).

46. When dermatomyositis develops in older men, it may be associated with malignancy.

47. Calcific tendinitis is an example of the dystrophic category of soft-tissue calcification.

48. The most common cause of secondary hypertrophic osteoarthropathy is bronchogenic carcinoma.

49. Intercostal muscle atrophy resulting in resorption of the third through sixth ribs posteriorly is a recognized feature of progressive systemic sclerosis (scleroderma).

50. The condition with the highest frequency of secondary hypertrophic osteoarthropathy is benign pleural mesothelioma, which is unassociated with asbestos exposure.

51. The most sensitive imaging test for the early diagnosis of ischemic necrosis is magnetic resonance imaging.

52. Spontaneous osteonecrosis of the femoral head in adults is also known as Chandler disease or coronary disease of the hip.

53. The process of revascularization, repair, and remodeling that occurs in ischemic necrosis is known as creeping substitution.

54. Unilateral apophyseal joint erosion and collapse at the C1-2 level may lead to torticollis and ipsilateral facial pain in rheumatoid arthritis.

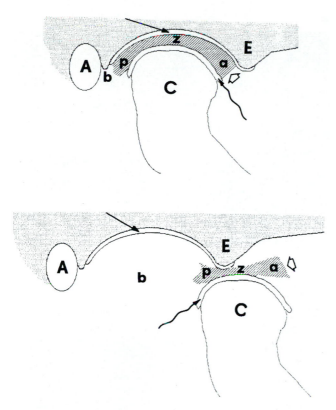

Fig. 3-74 Normal anatomy of the temporomandibular joint in the sagittal plane. *Top,* Closed-mouth position. *Bottom,* Open-mouth position. (*C = mandibular condyle; E = temporal eminence; A = external auditory canal; straight arrow = upper synovial space; wavy arrow = lower synovial space; open arrow = meniscus with anterior band [a], posterior band [p], and intermediate zone [z]; b = region of bilaminar zone.*)

55. Growth abnormalities (epiphyseal overgrowth, squaring of the patella as well as carpal and tarsal bones) and accelerated skeletal maturation with premature physeal closure, limb shortening, and limb length discrepancies are characteristic of juvenile chronic arthritis.
56. The differential diagnosis for knee involvement by juvenile chronic arthritis should include consideration of hemophilia and tuberculous arthritis.
57. Occult fractures with exuberant periostitis and callus formation frequently occur in patients with neuroarthropathy.
58. Xanthomatosis may simulate gout in that it produces soft-tissue masses on tendons that may be associated with osseous erosions.
59. Chronic venous stasis and thyroid acropachy may simulate hypertrophic osteoarthropathy in their radiographic features, in that they also cause periostitis.
60. In the spine, disk space infection and the spondyloarthropathy of dialysis therapy may simulate neuroarthropathy in their diagnostic imaging features.

61. The combination of rheumatoid arthritis, pulmonary nodules histologically similar to subcutaneous rheumatoid nodules, and pneumoconiosis in coal workers is known as Caplan syndrome.
62. The radiographic manifestations of ankylosing spondylitis in women tend to be neither as severe nor as classic in distribution as those in men with the disease.
63. The flares of peripheral arthropathy in inflammatory bowel disorders tend to correlate with disease activity, but radiographic abnormalities are milder than those of ankylosing spondylitis.
64. Ischemic necrosis occurs in up to one third of patients with systemic lupus erythematosus, and steroid therapy is more important in its pathogenesis than the collangen-vascular disease itself.
65. Erosions occur in some patients with scleroderma who demonstrate seronegativity for rheumatoid factor, and the proximal and distal interphalangeal joints are most commonly affected.
66. Myelofibrosis is a recognized risk factor for the development of ischemic necrosis.
67. The following rheumatologic disorders typically exhibit an asymmetric distribution of articular abnormalities:

   a. Psoriatic arthritis
   b. Reiter syndrome
   c. Neuroarthropathy
   d. Gout
   e. Pigmented villonodular synovitis—monoarticular
   f. Synovial osteochondromatosis—monoarticular

68. The following rheumatologic disorders involve the symphysis pubis:

   a. Ankylosing spondylitis
   b. Alkaptonuria (ochronosis)
   c. Osteitis pubis (Fig. 3-79)
   d. Septic arthritis
   e. Gout
   f. Calcium pyrophosphate dihydrate crystal deposition disease
   g. Diffuse idiopathic skeletal hyperostosis (Forrestier disease)
   h. Reiter syndrome
   i. Rheumatoid arthritis
   j. Hemochromatosis
   k. Degenerative joint disease

69. The following conditions are recognized causes of neuroarthropathy:

   a. Diabetes mellitus
   b. Tabes dorsalis
   c. Syringomyelia

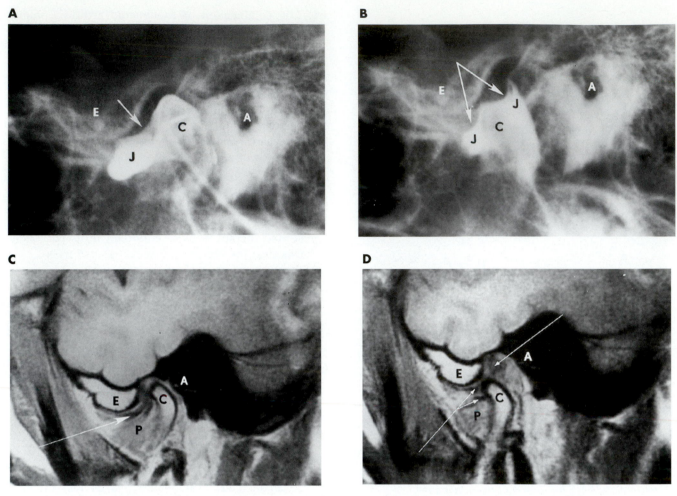

Fig. 3-75 Anterior meniscal displacement with recapture. **A,** Arthrographic image, closed-mouth position. **B,** Arthrographic image, open-mouth position. **C,** Sagittal T1-weighted MR image, closed-mouth position. **D,** Sagittal T1-weighted MR image, open-mouth position. Condylar excursion is extremely limited despite reduction of an anteriorly displaced meniscus with opening. (*Arrows = meniscus, C = mandibular condyle, E = temporal eminence, A = external auditory canal, J = lower synovial compartment, P = lateral pterygoid muscle.*)

d. Spinal cord injury
e. Meningomyelocele
f. Alcoholism
g. Congenital insensitivity to pain
h. Multiple sclerosis
i. Familial dysautonomia (Riley-Day syndrome)
j. Amyloidosis
k. Charcot-Marie-Tooth disease
l. Intraarticular corticosteroid injection
m. Leprosy
n. Traumatic denervation (Fig. 3-80)

70. The following conditions have been proven to occur in association with calcium pyrophosphate dihydrate crystal deposition disease:

a. Neuroarthropathy
b. Gout

c. Hemochromatosis
d. Osteoarthritis
e. Hyperparathyroidism
f. Amyloidosis
g. Long-term corticosteroid therapy

71. The following conditions are associated with deposition of calcium hydroxyapatite crystals:

a. Collagen-vascular disease
b. Renal osteodystrophy
c. Tumoral calcinosis
d. Hypoparathyroidism
e. Hypervitaminosis D
f. Milk-alkali syndrome
g. Dystrophic calcification secondary to trauma, neoplasm, or inflammation
h. Calcification in parasites

Osteochondroses

The osteochondroses are a heterogeneous group of disorders that share the common radiographic appearance of increased density, fragmentation, and/or flattening of an epiphysis or apophysis. Causes include osteonecrosis, normal variation, and trauma. With the exception of Freiberg disease, the conditions are more common in male individuals. In skeletally immature individuals, all of the symptomatic osteochondroses are accompanied by pain and decreased function of adjacent joints. Bilateral symmetric involvement is rare. Arthrography usually demonstrates that the articular cartilage is intact, although intraarticular osteochondral bodies may develop as a complication in some cases, particularly with involvement of the proximal femur and metatarsal heads.

Sites of involvement

1. Legg-Calvé-Perthes disease: femoral head epiphysis (Box 3-27)

Box 3-27 Legg-Calvé-Perthes Disease: Important Features

- Peak incidence between 4 and 8 years of age
- Male/female ratio approximately 5:1
- Onset before 4 years of age usually associated with good prognosis
- Rare in black children
- Bilateral involvement in 10% of cases
- Family history in 6% of cases
- When bilateral, both hips usually affected metachronously
- Osteochondritis dissecans complicates disease in 2% to 4% of cases, especially those with older age of onset

---

2. Blount disease: medial aspect of proximal tibial epiphysis (Box 3-28)

Box 3-28 Blount Disease (Tibia Vara): Important Features

- Evolves from physiologic bowing of infancy
- Bilateral in 60% of cases
- Arthrography, MRI findings: articular cartilage hypertrophy, hypermobility of medial meniscus, angulation of medial tibial articular surface
- No osteonecrosis
- Clinical signs: prominence of proximal medial tibia, fibular head; leg shortening; tibial torsion; pronated feet; obesity

---

Box 3-28—cont'd

- High incidence in Jamaicans, black African descendents
- Adolescent onset probably represents previous physeal injury or infection with osseous bridging

---

3. Freiberg disease: metatarsal head epiphysis (Box 3-29)

Box 3-29 Freiberg Infraction: Important Features

- Usually occurs in adolescents between ages 13 and 18 years
- Characterized histologically by osteonecrosis
- Most commonly involves second metatarsal head
- Predominates in women by a 3:1 to 4:1 ratio
- Complications: collapse, fragmentation, secondary osteoarthritis

---

4. Sever disease: calcaneal apophysis
5. Panner disease: epiphysis of humeral capitellum (Box 3-30)

Box 3-30 Panner Disease: Important Features

- Age range 4 to 16 years; peak incidence between 5 and 10 years
- Clinical signs: pain, stiffness, flexion contracture, tenderness, joint effusion
- Also termed Little Leaguer elbow
- Early radiographs may reveal subchondral radiolucent band
- Hyperemia may lead to accelerated maturation of capitellum, radial head

---

6. Köhler disease: tarsal navicular (Box 3-31)

Box 3-31 Köhler Disease: Important Features

- May be symptomatic or an incidental finding.
- Associated with Legg-Calvé-Perthes disease
- Male/female ratio 4:1 to 6:1
- Unilateral in 75% to 80% of cases
- Usually self-limited; may represent normal variation in tarsal ossification preceded by trauma in one third of cases

PEARL

Blount disease (tibia vara, osteochondrosis deformans tibiae): Characterized by disturbance of growth localized to medial portion of proximal tibia, involving physeal cartilage, metaphysis, and epiphysis. Consequence is abrupt angulation into varus deformity below knee associated with internal tibial torsion. Diaphysis is straight, and there is no bowing in the distal femur or tibia, but position of femoral condyles often causes slight compensation of varus resulting in bayonet deformity.

Two types described:

1. Infantile

   Deformity manifested in first 3 years of life. Usually bilateral. More common, more severe, more progressive than adolescent type. Characteristic history of normal physiologic bowing during first year, progressing to worsening rather than usual valgus. Earliest possible diagnosis: 18 months. Spontaneous recovery possible but rare after age 2. Internal tibial torsion progressive with bowing.

   Radiographic findings:

   a. Stages of Langenskiold:

       (1) Stage I—Age 2 to 3 years. Involves the medial one third of metaphyseal ossification zone. Protrusion of medial metaphysis causing distally directed beak. Medial epiphysis less developed than lateral. Islands of calcified tissue separated from metaphysis by radiolucent zones.

       (2) Stage II—Age 2½ to 4 years. Medial one third of metaphyseal ossification line slopes distally and medially into beak, proximal portion of which is less dense than remainder of metaphysis. Beak is seen as dense curved line on lateral view (also in stages III through VI). More severe wedging and maldevelopment of medial epiphysis than in stage I.

       (3) Stage III—Age 4 to 6 years. Deepening of cartilage-filled depression in metaphyseal beak causing "stair" in metaphysis. Medial epiphysis is wedge shaped with occasional small diffuse islands of calcification beneath its medial edge and an ill-defined border.

       (4) Stage IV—Age 9 to 11 years. Epiphysis occupies depression in medial metaphysis, often incompletely as an irregular separate ossification center. Border between growth plate and epiphysis forms stair parallel to that in stage III. Medial epiphyseal border irregular.

       (5) Stage V—Age 9 to 11 years. Stair of increased height. Deformed epiphysis and corresponding articular surface of intercondylar eminence, causing excessive laxity in knee joint. "Partially double epiphyseal plate" regularly seen: epiphysis divided medially by radiolucent band extending horizontally and medially from lateral epiphyseal plate to articular cartilage. Cartilage adjacent to epiphyseal island irregular.

       (6) Stage VI—Age 10 to 13 years. Ossification of both branches of duplicated medial portion of epiphyseal plate. Lateral physis continues growth.

   b. Varus of tibia secondary to abrupt angulation just below proximal epiphysis

   c. Cortical thickening of medial tibial shaft

   d. Secondary "tear" or stress fracture of medial beak occasionally seen

   e. MRI findings:

       (1) Early: large medial meniscus and hypertrophic epiphyseal cartilage

       (2) Progression: depression of epiphysis and metaphysis, beginning in medial epiphyseal cartilage and sparing anterolateral surface until later

       (3) Late: growth plate fuses with metaphysis, possibly secondary to retrograde ossification from metaphysis into unossified medial epiphysis

       (4) Indications:

           (a) Determination of degree of angulation necessary for osteotomy

           (b) Detection of posterior medial depression of tibial plateau, associated with laxity of medial collateral ligament, which is not correctable by osteotomy alone.

   Histologic findings: no evidence for infection or ischemic necrosis

Etiology:

- Genetic
- Cultural: infant carriage with knees in varus position
- Mechanical: early weight bearing on physiologically bowed legs in infants leads to slowing of tibial growth and consequent increase of differential length between fibula and tibia. Such differential growth can produce both varus and medial rotation, leading to established tibia vara or Blount disease.

Continued

PEARL—cont'd

Treatment:

- Early (18 months of age) mild deformity requires bowleg brace, only until legs straighten.
- Extreme angular deformity in child over 2 years of age requires osteotomy with overcorrection into physiologic valgus alignment.
- Stage VI deformity requires epiphysiodesis of lateral tibia, proximal fibula, and proximal ends of bones in opposite leg.
- Leg lengthening (Ilizarov method) may be required later.

2. Adolescent
   Deformity manifested between ages 8 and 15 years. Usually unilateral. Less severe, less frequent than infantile type. Prognosis worse with much higher incidence of recurrence than infantile type. Older patient, greater deformity translate to worse prognosis. Affected leg minimally shortened.

Radiographic findings:

a. Shape of epiphysis relatively normal
b. No stair in physeal plate
c. Narrowing of mid-portion of medial half of physis, with increased density of adjacent bone
d. MRI or tomography: osseous bridge between epiphysis and metaphysis
e. Late radiographic appearance of infantile type cannot be distinguished from that of the adolescent form

Etiology: secondary to trauma or other growth plate insult

Treatment:

- Valgus osteotomy suffices to correct both angulation and internal torsion.
- Permanent closure of medial tibial physeal plate may necessitate epiphysiodesis of lateral part of plate and head of fibula.
- Epiphysiodesis in contralateral limb or Ilizarov procedure may be indicated to equalize limb length.

---

7. Thiemann disease: phalangeal epiphysis of hand (Box 3-32)

Box 3-32 Thiemann Disease: Important Features

- Dominant trait, nearly complete penetrance
- Most commonly affects boys
- Painless swelling of proximal interphalangeal joints, digital shortening, deformity
- Peak incidence in second decade
- Simulates juvenile chronic arthritis, familial coalition of joints (Shrewsbury mark)

8. Kienböck disease: lunate bone (Box 3-33)

Box 3-33 Kienböck Disease: Important Features

- Predilection for dominant hand of manual laborers
- Peak incidence between ages 20 and 40 years
- Ulna-minus variant associated in up to 75% of cases
- Complications: disrupted carpal architecture, scapholunate dissociation, ulnar deviation of the triquetrum, secondary osteoarthritis
- Pathogenesis: osteonecrosis

9. Osgood-Schlatter disease: tibial tubercle apophysis (Box 3-34)

Box 3-34 Osgood-Schlatter Disease: Important Features

- Premature closure of the tibial tubercle physis with secondary genu recurvatum is a rare complication
- Athletic participation during the adolescent growth spurt predisposes to the condition
- Radiographic observation of a fragmented, irregular tibial tubercle ossification center is not diagnostic; soft-tissue swelling and pain must also be present
- Occurrence is bilateral in 75% of cases
- Nonunion of the tibial tubercle may rarely occur as a long-term complication
- Peak incidence in female individuals occurs at a younger age than in male individuals

10. Scheuermann disease: vertebral apophyses (Box 3-35)

Box 3-35 Scheuermann Disease: Important Features

- Caused by trauma superimposed on developmentally weak vertebral endplates
- Minimum criteria: involvement of three contiguous vertebrae with at least 5 degrees of anterior wedging
- Peak age range: 13 to 17 years

Continued

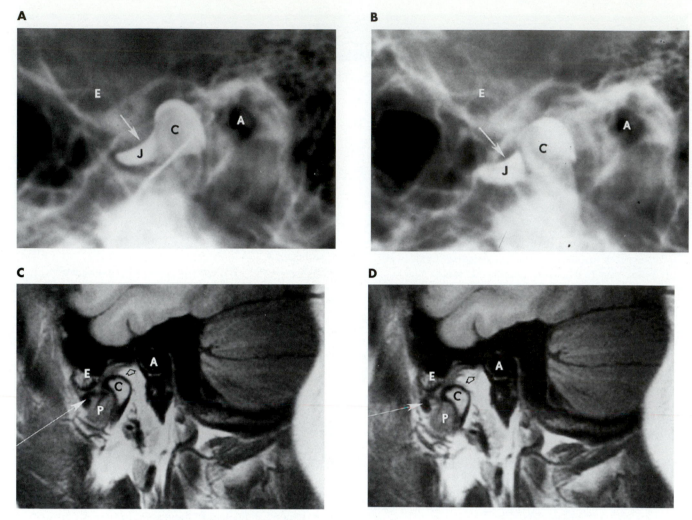

Fig. 3-76   Anterior meniscal displacement without recapture. **A,** Arthrographic image, closed-mouth position. **B,** Arthrographic image, open-mouth position. **C,** Sagittal T1-weighted MR image, semiopen position. **D,** Sagittal T1-weighted MR image, open-mouth position. Condylar excursion is extremely limited. (*Closed arrow = meniscus, C = mandibular condyle, E = temporal eminence, A = external auditory canal, J = lower synovial compartment, open arrow = bilaminar zone, P = lateral pterygoid muscle.*)

Box 3-35—cont'd

- Rare complications include cord compression and vertebral body synostosis
- Thoracic (75%) more common than thoracolumbar (20% to 25%) more common than lumbar (less than 5%) more common than cervical
- Irregular vertebral endplates, Schmorl nodes, disk space narrowing, limbus vertebrae
- Clinical signs: pain, tenderness, fatigue, poor posture, kyphosis, scoliosis

11. Sinding-Larsen-Johanssen disease: secondary patellar ossification center (Box 3-36)

Box 3-36   Sinding-Larsen-Johanssen Disease: Important Features

- May coexist with Osgood-Schlatter disease
- Peak age range: 10 to 14 years
- Posttraumatic or secondary to spastic paralysis
- Represents avulsion fracture of lower patellar pole, contusion, or tendinitis in the proximal patellar tendon
- No residual osseous or soft-tissue alterations

12. Van Neck disease: ischiopubic synchondrosis
13. Milch disease: ischial apophysis
14. Iselin disease: peroneal tubercle apophysis of fifth metatarsal

A

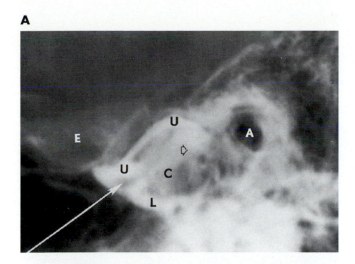

B

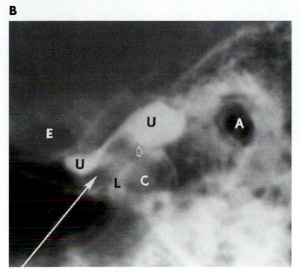

Fig. 3-77 Anterior meniscal displacement with perforation documented by arthrography. **A,** Closed-mouth position. **B,** Open-mouth position. No recapture is evident, and condylar excursion is extremely limited. (*Closed arrow = meniscus, C = mandibular condyle, E = temporal eminence, A = external auditory canal, L = lower synovial compartment, U = upper synovial compartment, open arrow = perforation through bilaminar zone.*)

A

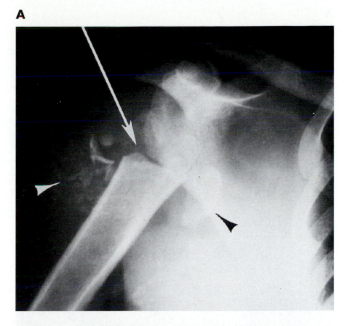

B

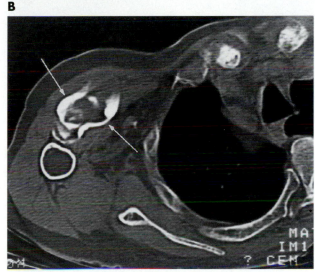

Fig. 3-78 Neuroarthropathy with dissecting synovial cyst. **A,** Radiography of the shoulder documents advanced osteolysis and subluxation with osseous fragmentation (*arrowheads*). **B,** Computed arthrotomography reveals inferior extension of contrast material and debris (*arrows*) into the anterior aspect of the upper arm. Longstanding septic or rheumatoid arthritis could yield similar findings, although fragmentation is usually less pronounced.

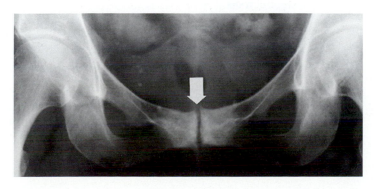

Fig. 3-79 Osteitis pubis (*arrow*) in a multiparous woman. Differential diagnosis of osteitis pubis should include consideration of inflammatory arthritides, infection, and, in men, prior prostate surgery.

A

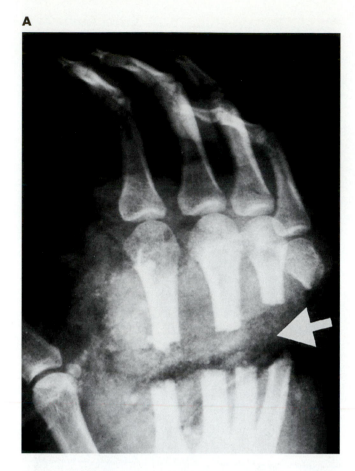

B

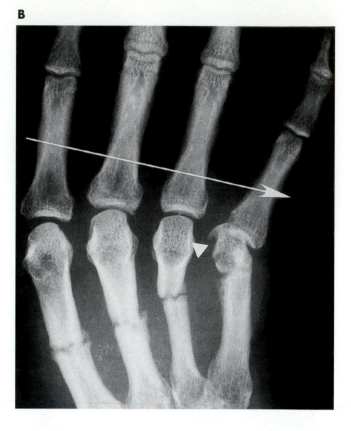

C

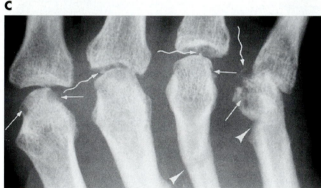

Fig. 3-80   Neuroarthropathy following digital replantation. **A,** Amputation has occurred through the second through fifth metacarpal diaphyses (*arrow*). **B,** Following replantation, ulnar deviation is evident at the metacarpophalangeal joints (*arrow*), particularly the fifth. Early erosion of the fifth metacarpal head (*arrowhead*) is present. **C,** Follow-up radiograph reveals progressive erosion (*straight arrows*), fragmentation (*wavy arrows*), and subluxation at the affected joints. The amputation sites have healed (*arrowheads*).

15. Diaz disease: talar dome
16. Buchman disease: iliac crest apophysis
17. Preiser disease: carpal navicular
18. Calvé disease: vertebral body

Classification by etiology

1. Trauma without osteonecrosis:

    a. Blount disease
    b. Osgood-Schlatter disease
    c. Scheuermann disease

    d. Sinding-Larsen-Johanssen disease
    e. Diaz disease (probably represents osteochondritis dissecans)

2. Osteonecrosis:

    a. Legg-Calvé-Perthes disease
    b. Freiberg disease
    c. Panner disease
    d. Köhler disease
    e. Thiemann disease
    f. Kienböck disease
    g. Preiser disease (follows traumatic scaphoid wrist fracture)

3. Normal variation:

    a. Sever disease

    b. Van Neck disease

    c. Milch disease

    d. Iselin disease

    e. Buchman disease

4. Other: Calvé disease (probably represents vertebra plana secondary to eosinophilic granuloma)

Important general concepts

1. The peak age range of incidence of Legg-Calvé-Perthes disease is 4 to 8 years, when the vascular supply to the proximal femoral epiphysis is most vulnerable.

2. The infantile type of Blount disease is more common than the adolescent form, and exhibits a predilection for black individuals.

3. Diaz disease probably represents posttraumatic osteochondritis dissecans rather than a true osteochondrosis.

4. Panner disease is observed most commonly in adolescent baseball pitchers.

5. Scheuermann disease involves a minimum of three contiguous vertebral bodies, each having at least 5 degrees of anterior wedging, and results in accentuated kyphosis.

6. Freiberg disease may be related to the trauma of weight bearing in high-heeled shoes during adolescence.

7. In Legg-Calvé-Perthes disease, prognosis is poorer for patients who are older at the time of diagnosis and for girls (as opposed to boys) all of whom are skeletally more mature at any age.

8. Calvé disease probably represents pathologic fracture secondary to eosinophilic granuloma.

9. In Osgood-Schlatter disease, soft-tissue swelling and pain must accompany radiographic evidence of fragmentation for the diagnosis to be made.

10. The adolescent type of Blount disease probably represents osseous bridging of the physis secondary to trauma or infection.

11. In Scheuermann disease, irregular vertebral endplates and multiple Schmorl nodes and/or limbus vertebrae are observed (Fig. 3-81).

12. The diagnosis of Köhler disease can be made only in the setting of pain and fragmentation of a previously normal bone.

13. Sinding-Larsen-Johanssen disease represents avulsion and fragmentation of the inferior aspect of a sesamoid (patella) (Fig. 3-82).

14. Blount disease evolves from normal physiologic bowing of the lower extremity with weight bearing, particularly in children who walk at an early age.

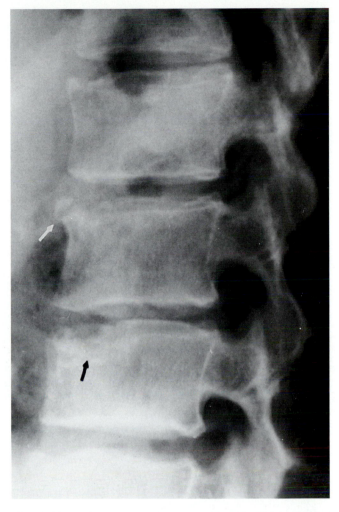

Fig. 3-81 Scheuermann disease involving the thoracolumbar spine. Characteristic findings include vertebral endplate irregularities, Schmorl nodes (*black arrow*), and limbus vertebrae (*white arrow*) involving multiple levels.

15. Only 10% of cases of Legg-Calvé-Perthes disease are bilateral, and these are characterized by metachronous involvement.

16. Over half of the infantile cases of Blount disease are bilateral.

17. Kienböck disease exhibits an association with ulna-minus variant that is probably not causative.

18. In Freiberg disease, the second ray is most commonly affected, followed in order by the third, fourth, and first (Fig. 3-83).

19. Consideration of hypothyroidism, epiphyseal dysplasias, and Legg-Calvé-Perthes disease should be included in the radiographic differential diagnosis of Meyer dysplasia of the femoral heads (Box 3-37).

A

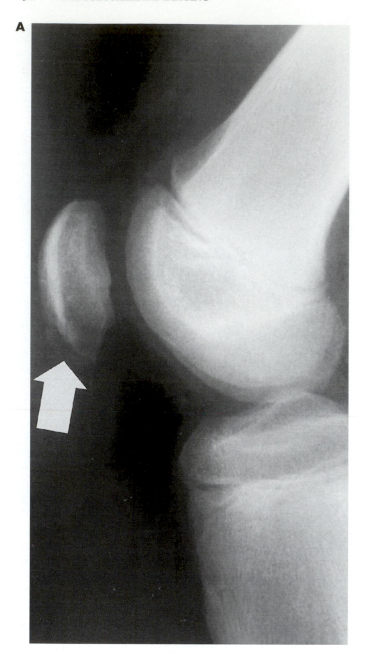

B

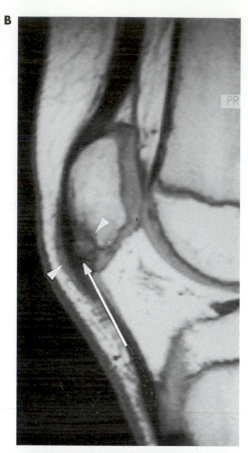

C

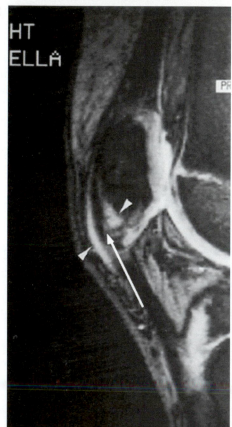

Fig. 3-82   Osteochondrosis of the inferior patellar pole (Sinding-Larsen-Johanssen disease). **A,** Lateral radiograph demonstrates fragmentation (*arrow*) of the inferior aspect of the patella. **B** and **C,** Sagittal T1-weighted and gradient-echo MR images confirm avulsion of the inferior patellar pole (*arrows*) and reveal edema both superficial and deep to the level of the displaced fragments (*arrowheads*).

20. Blount disease occurs unilaterally in the adolescent form.
21. Legg-Calvé-Perthes disease is extremely rare in black children.
22. The lower thoracic spine is affected in approximately 75% of cases of Scheuermann disease, and most patients are adolescents.
23. In Legg-Calvé-Perthes disease, as described by Gage, lateral extrusion of the epiphysis indicates a poor prognosis because of articular incongruity.
24. In Blount disease, genu varus results from involvement of both the epiphysis and metaphysis.
25. In Legg-Calvé-Perthes disease, metaphyseal irregularity and cysts indicate growth abnormality, which results in a short and broad metaphysis (Fig. 3-84).
26. Preiser disease may represent a fatigue fracture or healing fracture of the carpal navicular bone.
27. Scheuermann disease is most likely caused by chronic repetitive trauma superimposed upon congenitally weak bone.
28. Joint effusion is frequently the initial radiographic sign of Legg-Calvé-Perthes disease.
29. The osteochondrosis described by Iselin is distinguished from acute fracture by physeal orientation at the fifth metatarsal base.
30. The osteochondrosis described by Buchman (normal iliac crest apophysis) is useful in estimating the remaining growth potential of a skeletally immature individual, particularly in the setting of scoliosis.
31. Avulsion fractures frequently occur in adolescent athletes, particularly hurdlers, at the site of the osteochondrosis described by Milch (ischial apophysis).
32. The differential diagnosis of the osteochondrosis described by Van Neck at the ischiopubic synchondrosis should include consideration of stress fracture, posttraumatic osteolysis, and osteomyelitis.
33. The absence of the osteochondrosis described by Sever at the calcaneal apophysis indicates that the patient does not bear weight normally.

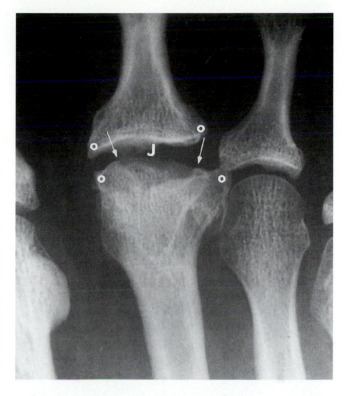

Fig. 3-83   Freiberg disease with secondary degenerative joint alterations. Flattening of the second metatarsal head (*arrows*) is associated with widening of the metatarsophalangeal joint (*J*) and osteophytosis (*o*). Differential diagnosis should include consideration of prior trauma, osteonecrosis (particularly in patients with systemic lupus erythematosus), and neuropathic disease (as in diabetes mellitus).

34. Approximately half of all patients with Legg-Calvé-Perthes disease manifest minimal, asymptomatic radiographic abnormalities on the contralateral side (Fig. 3-85).
35. The long-term prognosis for patients with Köhler disease is uniformly good.
36. The long-term prognosis for patients with Panner disease is usually good.
37. The patellar tendon is implicated in the pathogenesis of both Osgood-Schlatter disease and Sinding-Larsen-Johanssen disease.
38. Thiemann disease is a familial osteochondrosis, probably transmitted as a dominant trait with nearly complete penetrance.
39. Varus osteotomy of the proximal femur is frequently performed for treatment of advanced Legg-Calvé-Perthes disease, in order to improve congruence between the femoral head and acetabulum.
40. Valgus osteotomy of the tibia is frequently performed for treatment of advanced Blount disease, in order to improve alignment.

A

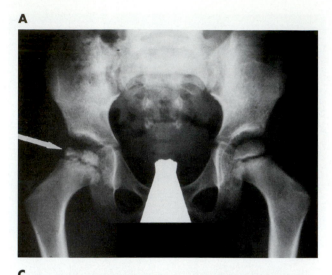

B

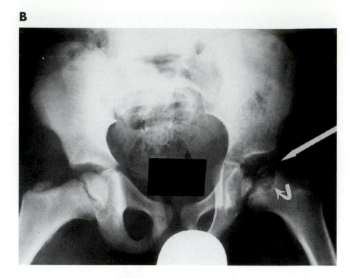

C

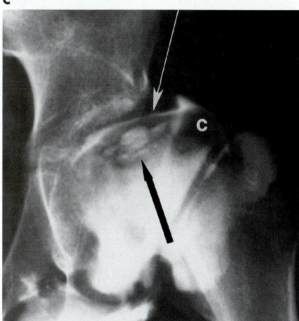

Fig. 3-84   Legg-Calvé-Perthes disease. **A,** Frontal radiograph demonstrates fragmentation and irregularity of the right femoral head (*arrow*). **B,** In another patient, more severe fragmentation (*straight arrow*) and metaphyseal cyst formation (*curved arrow*) (a poor prognostic sign) are evident. **C,** Arthrography in a third example documents fragmentation (*black arrow*), and lateral extrusion of the epiphyseal cartilage (*C*), also a poor prognostic sign. Differential diagnosis should include consideration of other causes of osteonecrosis in childhood, particularly corticosteroid therapy, sickle cell anemia, and Gaucher disease.

Fig. 3-85   MRI is a sensitive technique for use in the early diagnosis of Legg-Calvé-Perthes disease (*arrow = flattening and irregularity of the femoral head ossification center*), as well as childhood osteonecrosis induced by other disorders, such as developmental hip dysplasia, slipped capital femoral epiphysis, corticosteroid therapy, sickle cell anemia, and Gaucher disease.

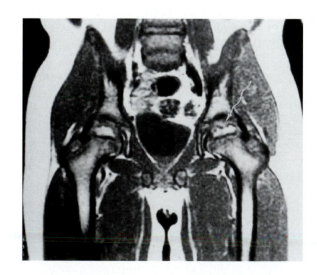

41. Sesamoiditis, a condition caused by trauma, osteonecrosis, or infection and resembling the osteochondroses radiographically, most commonly affects the bones within the flexor hallucis brevis tendons.
42. Panner disease is also commonly referred to as Little Leaguer elbow.
43. Magnetic resonance imaging is the procedure of choice for the early diagnosis of the osteochondroses that include osteonecrosis in their pathogenesis.
44. Earlier age of onset, bilateralism, and lack of progression are useful in distinguishing Meyer dysplasia of the femoral heads from Legg-Calvé-Perthes disease.
45. Consideration of systemic lupus erythematosus, diabetic neuroarthropathy, traumatic subchondral fracture, and normal variation should be included in the differential diagnosis of Freiberg disease.
46. Extrusion of disk material between the annular apophysis and body of a vertebra in Scheuermann disease results in a limbus vertebra.

See Table 3-8 for a summary of the common osteochondroses in terms of site, age, and probable mechanism.

SUGGESTED READINGS

1. Brower AC: Arthritis in black and white, Philadelphia, 1988, WB Saunders Co.
2. Cobby MJ and Martel W: Some commonly unrecognized manifestations of metabolic arthropathies, Clin Imaging 16:1, 1992.
3. Dussault RG, Samson L, and Fortin MT: Plain-film diagnosis of joint disease, Can Assoc Radiol J 42:87, 1991.
4. Freidman L et al: Limited, low-dose computed tomography protocol to examine the sacroiliac joints, Can Assoc Radiol J 44:267, 1993.
5. Ginalski JM et al: MR imaging of sternocostoclavicular arthro-osteitis with palmoplantar pustulosis, Eur J Radiol 14:221, 1992.
6. Hayes CW and Conway WF: Calcium hydroxyapatite deposition disease, Radiographics 10:1031, 1990.
7. Imaging of rheumatic diseases, Rheum Dis Clin North Am 17(3):457, 1991.
8. Kasperczyk A and Freyschmidt J: Pustulotic arthroosteitis: spectrum of bone lesions with palmoplantar pustulosis, Radiology 191:207, 1994.
9. Larheim TA, Smith HJ, and Aspestrand F: Temporomandibular joint abnormalities associated with rheumatic disease: comparison between MR imaging and arthrotomography, Radiology 183:221, 1992.

Table 3-8 Osteochondroses: Important features

Heterogeneous group of lesions sharing common radiographic features
Involvement of epiphyses or apophyses in the immature skeleton
Sclerosis
Fragmentation
Collapse
Reossification with partial or complete reconstitution of normal osseous contours

Disorder	Site	Age range (years)	Probable mechanism
Legg-Calvé-Perthes disease	Femoral head	4-8	Osteonecrosis, perhaps due to trauma
Freiberg infraction	Metatarsal head	13-18	Osteonecrosis due to trauma
Kienböck disease	Carpal lunate	20-40	Osteonecrosis due to trauma
Köhler disease	Tarsal navicular	3-7	Osteonecrosis or altered development
Panner disease	Capitellum of humerus	5-10	Osteonecrosis due to trauma
Thiemann disease	Phalanges of hand	11-19	Osteonecrosis, perhaps due to trauma
Osgood-Schlatter disease	Tibial tuberosity	11-15	Trauma
Blount disease	Proximal tibial epiphysis	1-3 (infantile) 8-15 (adolescent)	Trauma
Scheuermann disease	Discovertebral junction	13-17	Trauma
Sinding-Larsen-Johanssen disease	Patella	10-14	Trauma
Sever disease	Calcaneus	9-11	Normal variation in ossification
Van Neck disease	Ischiopubic synchondrosis	4-11	Normal variation in ossification

10. Laskin DM: Diagnosis of pathology of the temporomandibular joint: clinical and imaging perspectives, Radiol Clin North Am 31:135, 1993.

11. Murphey MD et al: Sacroiliitis: MR imaging findings, Radiology 180:239, 1991.

12. Oudjhane K et al: Computed tomography of the sacroiliac joints in children, Can Assoc Radiol J 44:313, 1993.

13. Recht MP and Resnick D: MR imaging of articular cartilage: current status and future directions, AJR 163:283, 1994.

14. Reiser MF and Naegele M: Inflammatory joint disease: static and dynamic gadolinium-enhanced MR imaging, JMRI 3:307, 1993.

15. Shellock FG et al: Patellofemoral joint: identification of abnormalities with active-movement, "unloaded" versus "loaded" kinematic MR imaging techniques, Radiology 188:575, 1993.

16. Vaatainen U, Pirinen A, and Makela A: Radiological evaluation of the acromioclavicular joint, Skeletal Radiol 20:115, 1991.

17. Whitten CG et al: Use of intravenous gadopentetate dimeglumine in magnetic resonance imaging of synovial lesions, Skeletal Radiol 21:215, 1992.

18. Winalski CS et al: Enhancement of joint fluid with intravenously administered gadopentetate dimeglumine: technique, rationale, and implications, Radiology 187:179, 1993.

CHAPTER 4

Neoplastic Disease

195

RADIOGRAPHIC APPROACH TO OSSEOUS AND ARTICULAR LESIONS

Distribution

The distribution of bone and joint lesions provides important clues to their specific etiology. Lesions may be monostotic or monoarticular (confined to one bone or joint), polyostotic or polyarticular (located in many bones or joints), or diffuse (involving virtually every bone or joint). Applying this distribution pattern to six basic pathologic categories—congenital, inflammatory, metabolic, neoplastic, traumatic, and vascular—allows one to narrow the diagnostic possibilities. For example, diffuse processes are either due to metabolic disease, developmental dysplasias, or diffuse neoplasms such as multiple myeloma or metastases. Further assistance may be obtained by defining target areas—the specific regions where particular diseases tend to occur. This is particularly useful in differentiating among certain types of primary bone tumors.

Behavior of Lesion

Bone lesions may be primarily osteoclastic (osteolytic, bone destroying), osteoblastic (bone forming, reactive, or reparative), or occasionally a mixture of the two. There are three forms of osteoclastic bone destruction—geographic, moth-eaten, and permeative or infiltrative. Geographic destruction consists of large areas of bone that have been destroyed and are easily visible with the unaided eye. Geographic lesions may either be aggressive or nonaggressive. The ultimate assessment of a geographic lesion will depend upon other predictable variables.

A moth-eaten appearance is one in which there are many discrete small holes throughout the bone as in a piece of clothing that has been ruined by moth larvae. A moth-eaten appearance suggests a more aggressive lesion. Certainly benign conditions such as osteomyelitis can induce a moth-eaten appearance. Bone affected by severe osteoporosis may also appear moth-eaten. Analysis of the texture of the bone will be more conclusive: in osteoporosis, there is simple loss of mineral, whereas in a moth-eaten destructive pattern, bone is actually destroyed. Very often, MRI will be required to determine if, in fact, there is marrow infiltration.

A permeative or infiltrative pattern is one in which there is fine bony destruction. Pathologically, this represents a very aggressive lesion diffusely infiltrating through the haversian system. In many instances, a magnifying lens is required to visualize the bone destruction. Permeative destruction indicates a very aggressive process such as a round cell tumor of bone (Ewing sarcoma, lymphoma, multiple myeloma). Occasionally, it may be seen with osteomyelitis.

Location in the Skeleton

Many diseases have a predilection for certain bones. For example, most of the malignant sarcomas favor the metaphyseal area of the distal femur or proximal tibia. Ossifying fibroma, chondromyxoid fibroma, and adamantinoma exhibit a strong predilection for the tibia.

Paget disease favors the spine, pelvis, skull, and tibia but tends to spare the fibula.

Locus within a Bone

The location of a lesion within a bone can provide an important clue as to its etiology. Many lesions have a predilection for the epiphysis, metaphysis, or diaphysis. In many instances, the actual location within the bone will be dependent upon the red marrow distribution within that bone. At birth, virtually all the marrow in the skeleton is red marrow. As the individual grows and approaches adulthood, hematopoietic marrow remains in the calvarium, sternum, vertebral bodies, pelvis, and the metaphyseal regions of long bones. It is not surprising that many neoplasms and infections are clustered in the red marrow areas. Although the majority of neoplasms occur in the metaphyseal region, certain lesions have a predilection for the epiphysis. These include chondroblastoma, giant cell tumor, clear cell chondrosarcoma, and intraosseous ganglion. Lesions in the diaphysis are often round cell neoplasms.

Age and Sex of Patient

The distribution of bone diseases is related to the patient's age and sex. Age is particularly useful in identifying many bone tumors. For example, a child with a permeative lesion of the diaphysis of a long bone is likely to have Ewing tumor. A similar-appearing lesion in a much older patient suggests malignant lymphoma of bone. It is a well-known fact that neuroblastoma is the most common type of bone tumor in patients under 1 year of age. In the first decade, Ewing tumor of long bones is the most common; between ages 10 and 30, osteosarcoma and Ewing tumor of flat bones predominate. Most of the malignant bone sarcomas occur between ages 30 and 40. Over age 40, the most common tumor encountered is metastatic carcinoma; multiple myeloma and chondrosarcoma also occur often in this age group.

Benign conditions also exhibit a definite age preference. Langerhans cell histiocytosis favors the younger age group, and occurs with fairly high frequency into young adulthood. Paget disease, on the other hand, occurs in the elderly.

There is also a definite difference in the occurrence of a certain lesion based on the patient's gender. Paget disease most commonly affects the elderly male patient. Many primary bone tumors as well as certain metastatic lesions occur with higher frequency in either men or women.

Margin of Lesion

In general, less aggressive lesions tend to produce a well-defined transition zone with the appearance of dense sclerosis clearly separating normal and abnormal bone. On the other hand, a broad or wide, poorly defined zone between normal and abnormal bone indicates a more aggressive process. This difference in appearance relates to the difference in growth rate between the two types of lesions. A slowly growing benign lesion such as chondromyxoid fibroma of bone or a focus of tuberculosis progresses at a rate slow enough to allow bone to react and attempt to wall off the lesion and reestablish architectural integrity. An aggressive lesion such as pyogenic osteomyelitis or a malignant tumor progresses at such a rapid rate that the bone is unable to respond adequately.

Shape of Lesion

The shape of the lesion assists in a fashion similar to that of the marginal characteristics. Any lesion that is longer than it is wide, that is oriented parallel to the diaphysis of the bone, is likely to be a nonaggressive benign process. In this situation, the lesion is growing with the bone and not faster than the bone. On the other hand, a lesion that is wider than the bone, that has broken out through the bony cortex, and that extends into the soft tissue is a highly aggressive lesion, usually one that is malignant.

Articular Involvement

If a lesion involves and crosses a joint space, it is most likely to be a benign inflammatory process. This is generally the case no matter how aggressive or malignant a process may appear. Common examples include an aggressive pyogenic osteomyelitis with septic arthritis and the aggressive form of tophaceous gout. Infectious processes and arthritides will extend across the joint space, but as a rule, tumors do not. Tumors that have a predilection for the ends of bones, such as chondroblastoma and giant cell tumor, extend to the joint space but will generally not cross it. Histologic examination of the end of the bone will usually exhibit a thin rim of cortical cells even though they may not be detected radiographically. Furthermore, even the most malignant of bone tumors generally respects the cartilage not only of the joint but also of the growth plate area. Exceptions to this rule are usually aggressive round cell neoplasms, such as myeloma and Ewing sarcoma.

Osseous Reaction

The bony reparative response to insult includes periosteal reaction, sclerosis, and buttressing. Periosteal reaction is of four varieties: (1) solid, (2) laminated or onion skin, (3) spiculated (sunburst or "hair on end"), and (4) the Codman triangle. Solid periosteal reaction is characterized by thick or wavy new bone deposition along the affected cortex. The reaction is well organized and not interrupted. As a rule, it indicates a benign process. It is most often seen in osteomyelitis and in fracture healing. A laminated or onion skin type of periosteal reaction indicates repetitive injury to bone. The character and morphology of the lamination must be used to determine whether one is dealing with an aggressive (usually malignant) lesion or a benign process. Multiple layers of well-organized solid periosteal reaction indicate a benign pathologic process with a repetitive pattern such as may be seen in the battered child. Poorly organized periosteal reaction with multiple thin layers containing many interruptions indicates a more aggressive process. This is the type of periosteal reaction seen with Ewing tumor and, in some instances, with osteogenic sarcoma.

A spiculated, sunburst, or hair-on-end appearance is almost always associated with a malignant bone lesion such as osteogenic sarcoma. In cases where this type of periosteal reaction occurs, there is an attempt by the bone to wall off the pathologic process. However, the aggressive nature of the lesion results in the destruction of the new periosteum as fingerlike extensions of the tumor erode through it. Thus, layer upon layer of spiculated projections of new bone is laid down. There are some instances in which the sunburst appearance may be seen with a benign process. In one such condition, subperiosteal hematoma, repetitive injury resulting in additional hemorrhage produces the pattern. The spicules of bone, however, are usually thicker than in a more malignant process. Finally, the Codman triangle is a triangular area of ossification at the junction of the periosteum and the cortex of the bone. This is a result of the pathologic process stripping periosteum away from the bone, with subsequent ossification at the juncture point. In the past, Codman triangles were thought to be pathognomonic of tumors; however, they occur in many benign conditions including the subperiosteal hemorrhage of scurvy and battered child syndrome.

Sclerosis is an attempt by the bone to wall off a diseased area. The area of sclerosis may be thin or thick. In either case, there is a sharply defined zone of transition between abnormal bone and normal bone.

Buttressing is an attempt by the bone to reestablish architectural integrity. The term is derived from the flying buttresses of Gothic architecture. Although it frequently occurs adjacent to a neoplasm, the most common example of this phenomenon occurs along the medial aspect of the femoral neck in degenerative arthritis of the hip.

Matrix Production

Matrix is a radiodense substance produced by certain bone tumors. It may be chondroid (cartilaginous), osteoid (bony), or mixed. Chondroid matrix appears as fine stippled calcifications or as multiple popcornlike calcifications. Very often it may be seen in bulky masses of tumor within the soft tissues. Osteoid matrix, on the other hand, is dense and usually of the same radiographic density as cortical bone. It occurs most commonly in the various subtypes of osteogenic sarcoma. Interestingly, it may also be seen in benign ossifying conditions such as myositis ossificans. Osteoid matrix may occur as a solid lump of new tissue or in a cloudlike amorphous pattern.

Soft-tissue Alterations

By analyzing the soft tissues on a radiograph, one may obtain important clues regarding an underlying disease process or a specific bone disease. For example, diffuse muscle wasting suggests a patient with paralysis, primary muscle disease, or severe cachexia due to disseminated neoplasm. The presence of soft-tissue swelling is indicative of trauma, a mass lesion, hemorrhage, inflammation, or edema. If a mass is present, it should be examined closely to determine if matrix is present. Calcifications may be present in osteoid or chondroid soft tissue or extended osseous neoplasms. The loss or displacement of fat lines that are normally seen in the soft tissues is another indication of adjacent abnormality. This is most often encountered in cases of trauma and infection.

History of Trauma

Trauma is the most common musculoskeletal disorder that is encountered by the radiologist. It is therefore very important to elicit a history of trauma whenever possible. A stress fracture may be misdiagnosed as a malignant bone tumor unless a specific history of pain with an unusual activity, worsening with that activity, and relief with rest is obtained. Occasionally, however, a history of trauma will be deliberately withheld, as in the case of a battered child or of a child who, prior to his or her injury, was doing something that was prohibited. Interestingly, many patients with primary bone tumors will provide a history of prior trauma. However, it is felt that a traumatic episode has simply called attention to an underlying lesion in these cases.

FUNDAMENTAL CONCEPTS

The purpose of learning the radiographic characteristics of bone tumors is to be a useful consultant in lesion identification and staging, decisions regarding operative resection, and appropriate management following therapy. In general, five categories of lesions can be identified:

1. Malignant neoplasms demanding preoperative clinical and diagnostic imaging evaluation followed by biopsy for histologic confirmation and definitive therapy
2. Lesions requiring biopsy for definitive determination of benign or malignant identity
3. Benign symptomatic neoplasms requiring elective surgical management and no further diagnostic tests
4. Lesions that are almost certainly benign and can be safely followed by imaging studies for confirmation of the suspected diagnosis
5. Benign lesions that can simply be left alone

The radiologist should understand the natural history of the lesion, as well as the therapeutic options, to provide a complete and accurate assessment of the extent of tumor involvement. In addition, the radiologist must be aware of which diagnostic modalities are most effective for tumor staging (Table 4-1), as well as for monitoring possible recurrence or complications.

Table 4-1   Enneking staging system for musculoskeletal neoplasms

Grade	Site
IA	Low grade, intracompartmental
IB	Low grade, extracompartmental
IIA	High grade, intracompartmental
IIB	High grade, extracompartmental
III	Distant metastases

Accurate assessment of several specific characteristics of a lesion usually affords an appropriate differential diagnosis.

Age of Patient

The following provides the most appropriate age category for each tumor listed:

1. Malignant fibrous histiocytoma: 30 to 60 years
2. Ewing sarcoma (flat bones): 10 to 30 years
3. Multiple myeloma: 50 to 80 years
4. Metastatic neuroblastoma: 1 year

5. Primary lymphoma of bone: 30 to 60 years
6. Aneurysmal bone cyst: 10 to 20 years
7. Fibrosarcoma: 30 to 60 years
8. Ewing sarcoma (tubular bones): 1 to 10 years
9. Giant cell tumor: skeletally mature to 50 years
10. Chondroblastoma: skeletally immature
11. Metastasis: 50 to 80 years
12. Osteosarcoma: 10 to 30 years
13. Chondrosarcoma: 30 to 60 years

Location of Lesion (Box 4-1)

1. Lesions are differentiated according to whether they occur (1) in flat or tubular bones and (2) in the axial or appendicular skeleton.
2. Eighty percent of metastatic lesions occur in the axial skeleton.
3. The following are examples of lesions most commonly found in various portions of a tubular bone:

   a. Epiphysis: chondroblastoma

PEARL

Epiphyseal lesions:
- Chondromyxoid fibroma originates in the metaphysis but can extend into the epiphysis.
- Aneurysmal bone cyst originates in the metaphysis but can extend to the epiphysis, particularly when associated with other tumors, such as giant cell tumor, chondroblastoma, and telangiectatic osteosarcoma.
- Giant cell tumor, chondroblastoma, and intraosseous ganglion are classically epiphyseal in location.
- Eosinophilic granuloma most commonly involves the diaphyseal and metaphyseal regions but can be epiphyseal, especially in children.

   b. Metaphysis: osteosarcoma
   c. Diaphysis: Ewing sarcoma

4. The following are examples of benign lesions that most commonly have their epicenter in the various regions of a tubular bone metaphysis:

   a. Central: simple bone cyst
   b. Eccentric: aneurysmal bone cyst
   c. Cortical: fibrous cortical defect

5. Lesions that are usually eccentric or cortical in position may appear centrally placed in tubular bones with small diameter, such as the ulna and fibula.
6. Three tumors that are found much more commonly in the tibia than elsewhere are ossifying fibroma, adamantinoma, and chondromyxoid fibroma (Table 4-2).

Box 4-1　Differential Diagnosis of Osteolytic Skull Lesions

1. Common
   a. Normal or iatrogenic
      (1) Venous lake
      (2) Emissary veins
      (3) Pacchionian granulations (usually near midline)
      (4) Parietal foramina
      (5) Glial rests (occipital bone)
      (6) Burr hole, craniotomy defect
   b. Epidermoid (ovoid and well circumscribed, with fine sclerotic border; many having lobulated periphery due to septations)
   c. Fibrous dysplasia (cystic variety; less common than sclerotic form)
   d. Fracture (especially depressed; also leptomeningeal cyst)
   e. Hemangioma (diploic, with well-defined nonsclerotic border; honeycomb or sunburst appearance)
   f. Langerhans cell histiocytosis (nonsclerotic border unless healing or treated; beveled edge or button sequestrum; most common in frontal bone)
   g. Meningocele, encephalocele (midline)
   h. Metastasis (poorly marginated)
   i. Myeloma (multiple punched out lesions of uniform size)
   j. Osteomyelitis
      (1) Bacterial osteomyelitis
      (2) Fungal osteomyelitis (actinomycosis, blastomycosis, coccidioidomycosis)
      (3) Tuberculosis, syphilis
   k. Paget disease (osteoporosis circumscripta)
2. Uncommon
   a. Benign bone tumors
      (1) Fibroma (slightly sclerotic borders)
      (2) Chondroma (paranasal and skull base, stippled calcification)
      (3) Giant cell tumor (skull base)
      (4) Meningioma (pressure erosion or osteosclerosis)
   b. Malignant bone tumors
      (1) Osteogenic sarcoma (lytic form more common in skull than in long bones)
      (2) Chondrosarcoma (skull base, stippled calcification)
      (3) Fibrosarcoma
      (4) Lymphoma
   c. Brown tumor (hyperparathyroidism)
   d. Neurofibromatosis (lambdoid or other sutural defect; sphenoid and orbital involvement)
   e. Calvarial "doughnut" lesion (parietal)
   f. Radiation osteonecrosis
   g. Scalp tumor or infection with underlying bone erosion

Table 4-2　Localization of bone tumors and tumorlike lesions

Tumor or lesion	First most common site	Second most common site
Malignant tumors		
Osteosarcoma	Femur	Tibia
Periosteal osteosarcoma	Femur	Humerus
Chondrosarcoma	Femur	Humerus and pelvis
Fibrosarcoma	Femur	Humerus
Ewing sarcoma	Pelvis and femur	Fibula
Lymphoma	Femur	Mandible
Angiosarcoma	Tibia	Femur
Chordoma	Sacrum	Clivus
Benign tumors		
Giant cell tumor	Femur	Tibia
Chondroblastoma	Humerus	Femur
Chondromyxoid fibroma	Tibia	Femur
Enchondroma	Phalanges of hand	Metacarpals
Solitary osteochondroma	Humerus	Femur and tibia
Osteoid osteoma	Femur	Tibia
Osteoblastoma	Tarsal bones	Vertebrae
Nonossifying fibroma	Tibia	Femur
Aneurysmal bone cyst	Femur	Vertebrae
Tumorlike lesions		
Solitary bone cyst	Humerus	Femur
Monostotic fibrous dysplasia	Ribs	Mandible
Solitary eosinophilic granuloma	Femur	Vertebrae and pelvis

Pattern of Bone Destruction

1. Permeative bone destruction is poorly demarcated and difficult to visualize; it implies the presence of a highly aggressive lesion (Fig. 4-1).
2. Moth-eaten bone destruction has a more well-defined margin; it implies a somewhat less aggressive lesion.
3. Geographic bone destruction has a well-defined margin; it implies the least aggressive type of lesion.

Margin of Lesion and Zone of Transition

1. A sclerotic margin and narrow zone of transition imply the presence of a nonaggressive lesion.
2. A nonsclerotic margin and wide zone of transition imply the presence of an aggressive lesion.

A

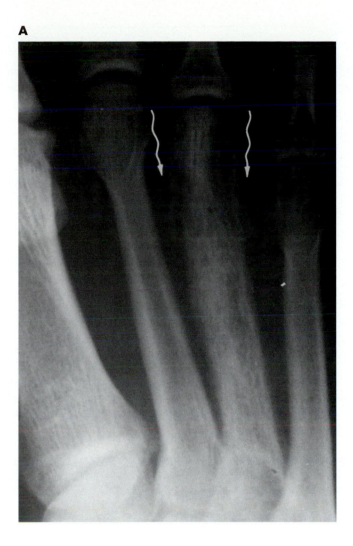

B

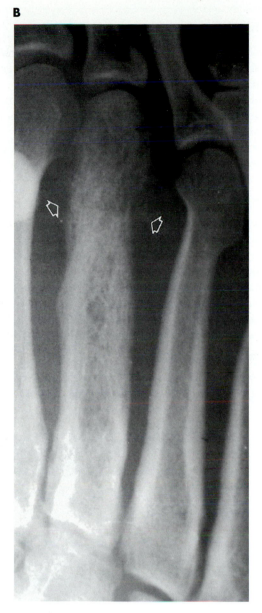

Fig. 4-1   Ewing sarcoma. Frontal (**A**) and oblique (**B**) radiographs demonstrate a permeative expansile process with poorly defined periostitis (*arrows*) involving the third metatarsal. Differential diagnosis should include consideration of osteomyelitis, fibrosarcoma, and evolving Paget disease.

3. Describing a lesion as aggressive or nonaggressive on the basis of the foregoing characteristics is useful, but these terms should not be equated with malignant or benign.
4. Plasmacytoma and metastasis from renal cell or thyroid carcinoma are examples of malignant conditions that may have a nonaggressive appearance.
5. Aneurysmal bone cyst, pyogenic infection, and eosinophilic granuloma are examples of benign conditions that may have an aggressive appearance.
6. Giant cell tumor and plasmacytoma are two lesions that may lack a sclerotic margin but have a narrow zone of transition.

Soft-tissue Involvement

1. Cortical breakthrough implies an aggressive lesion.
2. Neoplastic soft-tissue masses distort muscle and fat planes (unlike infection, which tends to obliterate them).
3. Benign soft-tissue neoplasms may appear infiltrative and locally aggressive (desmoid tumor).
4. Soft-tissue sarcomas often appear well defined and encapsulated; examples include liposarcoma, synovial sarcoma, and malignant fibrous histiocytoma.

Tumor Matrix

1. A matrix stippled with C and J shapes, which is denser than normal cancellous bone, implies a cartilaginous neoplasm.
2. Amorphous osteoid matrix, less dense than or as dense as normal bone, implies an aggressive bone-forming tumor.
3. Better-organized, dense, bony matrix implies a less aggressive bone-forming neoplasm.

Response of Host Bone

The following are possible responses of host bone to the presence of neoplastic disease:
1. Periosteal reaction
2. Cortical thickening
3. Cortical expansion (Fig. 4-2)
4. Cortical disruption
5. Sclerotic reaction

Size of Lesion

Size estimates (along with other basic characteristics of some specific bone lesions) are provided in Table 4-3.

Monostotic versus Polyostotic Involvement

1. Enchondromatosis (Ollier disease, Maffucci syndrome), multiple cartilaginous exostoses, polyostotic fibrous dysplasia, Langerhans cell histiocytosis, and polyostotic Paget disease are examples of benign conditions with polyostotic involvement.
2. Metastatic disease, multiple myeloma, and primary bone neoplasms with multifocality or metastasis (osteosarcoma, malignant giant cell tumor, Ewing sarcoma, and malignant fibrous histiocytoma) are examples of malignant conditions with polyostotic involvement.

Surgical Treatment Options

1. Curettage or intralesional excision: residual marginal tumor is present
2. Excisional biopsy or marginal excision: inadequate for malignant or recurrent benign lesions as dissection plane passes through reactive pseudocapsule
3. Wide excision: adequate for recurrent, aggressive benign lesions, low-grade sarcomas, and high-grade

Table 4-3   Some important characteristics of solitary bone lesions

Lesion	Age Range (years)	Age Peak (decades)	Size (cm) Range	Size (cm) Usual	Shape	Relation to axis of bone	Remarks
Osteosarcoma	7-70	2, 3, 7	3-10+	6-9	Round	75% central	70% in lower extremities, especially around knee; most common solitary malignant primary tumor of bone
Chondrosarcoma	10-80	2, 3, 5, 6	1-12+	5-10	Round to elongated	50% central; 50% eccentric or periosteal	30% occur in pelvis; 50% occur over age 40; represent 15-20% of all primary malignant bone tumors
Chondroblastoma	4-35	2	1-6	2-3	Round	67% eccentric; 33% central	25% occur after second decade; 10% occur before second decade; 50% around knee
Chondromyxoid fibroma	4-60	2, 3	1-9	3-6	Elongated	80% eccentric	80% in lower extremities; often mistaken for giant cell tumor or nonossifying fibroma
Giant cell tumor	4-80	3, 4	1-10+	3-8	Round	67% eccentric; 33% central	70% around knee and in distal radius; characteristically extends to subchondral bone
Fibrosarcoma	10-70	4, 5	3-10	6-10	Round to elongated but often indeterminate	67% central; 33% eccentric	Only bone tumor that frequently creates sequestra of dead bone

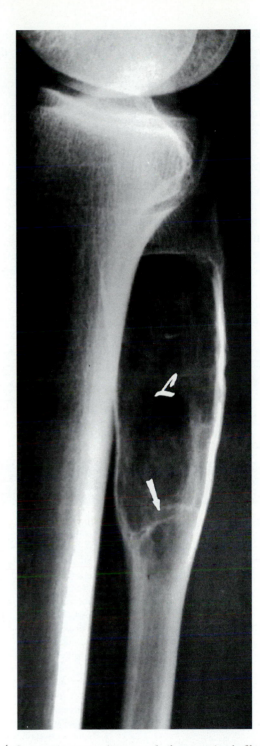

Fig. 4-2 Intraosseous lipoma of the proximal fibula. An expansile lytic lesion (*L*) with minimal septation (*arrow*) involves the metadiaphyseal region. Differential diagnosis should include consideration of aneurysmal bone cyst, chondromyxoid fibroma, fibrous dysplasia, plasmacytoma, metastatic disease (particularly secondary to renal and thyroid disease), and indolent infection.

sarcomas debulked by chemotherapy, because tumor is removed with a surrounding intact cuff of normal tissue

4. Radical resection: lesion is removed with all involved bones, muscles, and other tissues (in entirety) within affected compartment

5. En bloc limb salvage procedure: affords tumor control without sacrifice of limb

6. Amputation: involves tumor control with sacrifice of limb

Diagnostic Imaging Evaluation

1. Angiography is used if embolization of a highly vascular lesion is required before surgery.

2. Magnetic resonance imaging is the procedure of choice for defining extraosseous relationships because of superior contrast discrimination and multiplanar imaging capabilities.

3. Computed tomography is the procedure of choice for identifying and characterizing matrix calcification or ossification.

4. Magnetic resonance imaging is the procedure of choice for accurately defining the extent of a lesion in bone marrow.

5. Conventional radiography is the best method for providing bone detail, determining aggressiveness of a lesion, and establishing a differential diagnosis.

6. Conventional tomography is used only occasionally to provide more detailed depiction of bony architecture.

7. Technetium 99^m–methylenediphosphonate bone scanning is used to determine whether a lesion is monostotic or polyostotic.

8. Gallium-67 citrate scanning may help to distinguish a soft-tissue sarcoma from a benign soft-tissue lesion (sensitivity = 85%, specificity = 92%).

9. Computed tomography is the procedure of choice for determining the extent of cortical bone involvement.

10. Technetium 99^m–methylenediphosphonate bone scanning, computed tomography, and magnetic resonance imaging tend to overestimate the extent of a lesion.

11. In computed tomography and magnetic resonance imaging, intravenous contrast material may assist in more accurately defining tumor margins and identifying central necrosis (if present).

12. Percutaneous biopsy is most definitive in narrowing the differential diagnosis of a lesion.

13. Computed tomography is the procedure of choice for guiding percutaneous biopsy of a tumor, although many such procedures can be done safely under fluoroscopy if the lesion is radiographically evident.

14. Computed tomography is the best procedure for examining an entirely intraosseous lesion.
15. Magnetic resonance imaging is the best procedure for examining an entirely extraosseous lesion.
16. In an aggressive lesion, more information is generally gained from percutaneous biopsy of its periphery rather than its center, which may be necrotic.
17. The histologic diagnosis of metastatic disease or myeloma requires a relatively small biopsy sample.
18. The histologic diagnosis of primary benign and malignant bone tumors requires a relatively large biopsy sample.
19. In percutaneous biopsy of musculoskeletal lesions, multiple needle passes are recommended to increase diagnostic yield.
20. In skeletal scintigraphy using technetium-99^m methylenediphosphonate, the degree of abnormal uptake does not correlate with the histologic grade of the lesion.
21. Using quantitative computed tomography to define marrow extent of a lesion is most accurate in the diaphysis of a long bone.
22. T1-weighted or fat-suppressed T2-weighted magnetic resonance images enhance contrast between most soft-tissue lesions and normal fat.
23. T2-weighted magnetic resonance images enhance contrast between most soft-tissue lesions and normal muscle.
24. T1-weighted or inversion recovery magnetic resonance images are best for defining the marrow extent of a lesion.
25. Magnetic resonance imaging affords low specificity in the diagnosis of musculoskeletal neoplasms.
26. Musculoskeletal tumor staging usually requires either computed tomography or magnetic resonance imaging; both tests are seldom required.
27. An intraosseous lesion that extends into the soft tissues is best evaluated for staging and treatment planning by magnetic resonance imaging.

Potential Complications of Therapy

Radiation

1. Radiation-induced sarcoma (usually occurs 4 to 20 years after therapy, with histologies including osteosarcoma, fibrosarcoma, malignant fibrous histiocytoma, and chondrosarcoma)
2. Radiation-induced osteochondroma
3. Radiation osteonecrosis (usually occurs 7 to 10 years after therapy)
4. Pathologic fracture (usually occurs in the setting of radiation-induced sarcoma or radiation osteonecrosis, which may be radiographically indistinguishable)

5. Growth cessation and/or deformity (occurs secondary to epiphyseal or apophyseal irradiation during childhood)
6. Increased susceptibility to infection (also a complication of chemotherapy)
7. Tumor recurrence (also a complication of chemotherapy)

Limb salvage surgery with bone grafting

1. Delayed union (up to 2 years; less if vascularized fibular grafts are used)
2. Bone graft resorption
3. Failure of orthopedic hardware
4. Late fracture, often related to disuse osteopenia
5. Increased susceptibility to infection (hardware and devitalized graft serve as nidus in immunocompromised host)
6. Tumor recurrence

SPECIFIC MUSCULOSKELETAL NEOPLASMS

Benign Lesions

Table 4-4   Localized benign radiodense lesions

Lesion	Location	Appearance
Enostosis (bone island)	Medullary	Round or oblong, with thorny radiating spicules
Osteoma	Cortical protrusion	Homogeneous, smooth or lobular, extending from osseous surface
Osteochondroma	Cortical and medullary protrusion	Cortical bone and spongiosa are continuous with parent bone; calcified cap
Enchondroma	Medullary	Lucent, well circumscribed, endosteal scalloping, variable punctate calcification
Osteoid osteoma	Cortical, medullary, or subperiosteal	Cortical: lucent with or without calcification, surrounded by sclerosis; Medullary: lucent or calcified, with little sclerosis; Subperiosteal: scalloped excavation with or without calcification and sclerosis

Bone-forming tumors

Osteoma   This is a hamartomatous lesion characterized by abnormal bone proliferation on an osseous surface.

1. The lesion may be associated with Gardner syndrome (Box 4-2).
2. The calvarium and paranasal sinuses are favored sites.
3. The lesion is usually more than 2 cm in diameter.
4. The lesion may deform but does not involve soft tissue.
5. The entire lesion is sclerotic.
6. The conventional radiograph is usually diagnostic.
7. Multiple lesions may be observed.
8. No treatment is necessary.

Box 4-2   Gardner Syndrome: Important Features

- Multiple colonic polyps of varying size with tendency toward malignant degeneration
- Osteomas; most common in maxilla and mandible but occasionally occurring in long bones
- Diaphyses of long bones often exhibit irregularly thickened and undulating cortices
- Dental abnormalities (odontomas, dentigerous cysts, supernumerary and unerupted teeth)
- Soft-tissue swelling from underlying osteomas, sebaceous cysts, desmoids, fibromas, and lipomas

---

Enostosis or bone island   This is a hamartomatous proliferation of dense bone within the medullary space of a bone.

1. The lesion ranges in size from 2 mm to 2 cm in diameter.
2. The majority of lesions are scintigraphically negative.
3. Polyostotic involvement is seen in osteopoikilosis.
4. Areas of abundant trabecular bone are usually affected.
5. The lesion is very common and usually an incidental finding.

Osteoid osteoma   This is a lytic lesion composed of vascular osteoid tissue that may calcify (Box 4-3).

1. Most patients develop symptoms during the second or third decade of life.
2. The talus is among the most common sites of the subperiosteal variety.
3. The posterior elements are typically affected in the spine.

PEARL

Vertebral osteoid osteomas are usually located in the neural arch and cause painful scoliosis. Osteoid osteomas are angiographically hypervascular with blushing of the nidus.

---

4. Soft-tissue involvement occurs only in the rare subperiosteal variety.
5. The nidus is usually smaller than 2 cm in diameter.
6. The intracapsular variety occurs most commonly at the medial aspect of the femoral neck (Fig. 4-3).
7. The clinical presentation may include arthritis, muscle atrophy, painful synovitis or scoliosis, or localized aching that is worse at night and relieved by salicylates.
8. Sclerotic host reaction to the lesion may extend a great distance away from the nidus, except in small bones (Fig. 4-4).
9. Cortical involvement of a long bone is most typical.
10. In children, femoral lesions may cause irreversible growth deformities including valgus and overgrowth with limb length discrepancy.

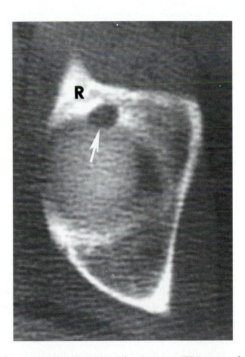

Fig. 4-3   Intracapsular osteoid osteoma. CT image of the hip reveals a radiolucent lesion (*arrow*) with associated sclerotic reaction (*R*) in the anterosuperior aspect of the acetabulum. Synovitis induced by such lesions is best demonstrated by MRI. Differential diagnosis should include consideration of indolent infection.

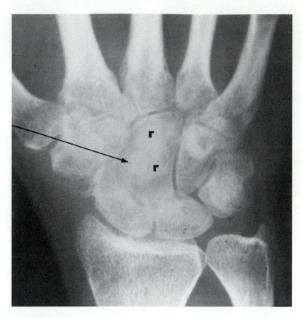

Fig. 4-4   Osteoid osteoma of the capitate bone. Frontal radiograph demonstrates a calcified nidus (*arrow*) with reactive sclerosis (*r*). Differential diagnosis should include consideration of multiple enostoses or a focal area of melorheostosis.

Box 4-3   Osteoid Osteoma: Important Features

- **Peak age:** second decade
- **Male/female ratio:** 3:1
- **Most common location:** proximal femur, in the region of the femoral neck
- **Major radiographic features:** a small, round lucency is identified and usually lies in the cortex; surrounding the lucency there is sclerosis and cortical reaction or thickening; at the center of the lucency or nidus, there may be central calcification; about 25% of tumors are not demonstrated on radiographs and therefore require isotope bone scan, tomography, computed tomography, or magnetic resonance imaging for identification; tumors in cancellous bone and subperiosteal or intracapsular lesions provoke little or no reactive sclerosis
- **Radiographic differential diagnosis:** Brodie abscess, stress fracture, osteoblastoma
- **Treatment:** resection of the nidus

PEARL

Osteoid osteoma may be associated with limb hypertrophy in a child, synovitis if near a joint, and radiographic features similar to those of Brodie abscess.

11. The lesion may have multiple nidi or be polyostotic (Fig. 4-5).
12. Localization of the nidus may be achieved by skeletal scintigraphy, conventional tomography, computed tomography, magnetic resonance imaging, or radiography of the surgically resected specimen (Figs. 4-6, 4-7).
13. Spontaneous healing has been observed in untreated lesions, suggesting an infectious etiology.
14. The treatment of choice is surgical resection of the nidus, although percutaneous removal under computed tomographic guidance has been recently described.
15. The lesion constitutes approximately 1.6% of excised primary bone neoplasms.

See Box 4-4 for a summary of important aspects of osteoid osteoma.

Box 4-4   Osteoid Osteoma: Summary

- Osteoblasts form vascularized osteoid that may calcify
- Long bones: lesions manifest reactive sclerosis and cortical thickening
- Small bone lesions of the hands and feet exhibit minimal to absent sclerosis
- Nidus localization can be accomplished by tomography, CT, MRI, scintigraphy
- Resection of nidus leads to resolution of reactive sclerosis
- Young male predilection
- Pain, accentuated at night, relieved with aspirin
- Femur, tibia most common sites
- Spinal lesions cause sclerosis and painful scoliosis
- Nidus consists of osteoid and fibrous tissue and may be radiodense or radiolucent
- Cortical lesions tend to occur in long bones
- Medullary lesions tend to occur in small bones
- Intracapsular lesions induce synovitis but little or no sclerosis
- Recurrence occurs following inadequate nidus resection
- Multiple nidi possible

Osteoblastoma   This is an expansile, generally nonaggressive lesion with lytic or sclerotic components that is histologically identical to osteoid osteoma (Box 4-5).

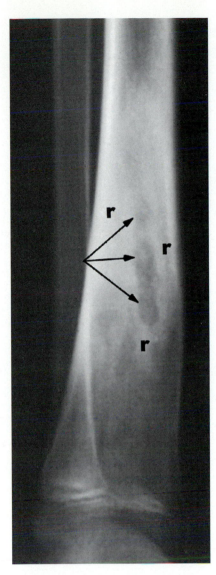

Fig. 4-5   Osteoid osteoma with multiple calcified nidi (*arrows*) and reactive sclerosis (*r*) in the distal tibia. Differential diagnosis should include consideration of chronic osteomyelitis, particularly the Brodie abscess.

PEARL

Osteoblastoma and osteoid osteoma have similar histologic features. Osteoblastoma is generally larger than 2 cm in diameter, whereas the nidus of an osteoid osteoma is less than 2 cm in size. Osteoblastomas may be purely lytic, mixed lytic-blastic, or predominantly blastic.

1. The lesion is rare, representing about 0.5% of all bone tumor biopsies.
2. Most lesions are more than 2 cm in diameter.

3. The tumor occurs chiefly during the second or third decade of life.
4. Over 40% of lesions occur in the posterior elements of the spine.
5. Soft-tissue involvement is rare.
6. Long-bone lesions are either metaphyseal or diaphyseal.
7. Clinically, dull aching pain is less severe than that produced by osteoid osteoma.
8. Aneurysmal bone cyst may be secondarily associated with the lesion, particularly in the spine.
9. Osteoblastoma tends to occur more centrally than primary aneurysmal bone cyst and giant cell tumor.
10. Malignant transformation and histologically aggressive lesions have been described.
11. Depending upon location, curettage with bone grafting or irradiation are therapeutic options.

Box 4-5   Osteoblastoma: Important Features

- **Peak age:** second decade of life (but may be seen at any age)
- **Male/female ratio:** 3:1
- **Most common location:** vertebra, involving the posterior elements
- **Major radiographic features:** radiographic features are sometimes similar to those of osteoid osteoma; appearance may be quite variable and nonspecific; 25% of cases exhibit features suggestive of a malignant neoplasm; in the vertebra, osteoblastoma usually results in expansion and is located in the posterior elements; 50% exhibit ossification radiographically; mandibular lesions (cementoblastoma) are radiodense, surrounded by a lucent halo, and located near a tooth root
- **Radiographic differential diagnosis:** osteoid osteoma, osteosarcoma, aneurysmal bone cyst
- **Treatment:** curettage and grafting or resection

Ossifying fibroma   This is an extremely uncommon benign lesion that is histologically similar to fibrous dysplasia or adamantinoma.

1. The lesion occurs almost exclusively in the tibia and facial bones.
2. Most patients seek treatment during the second or third decade.

PEARL

Ossifying fibroma typically involves the mandible, maxilla, and tibia in women during their second through fourth decades.

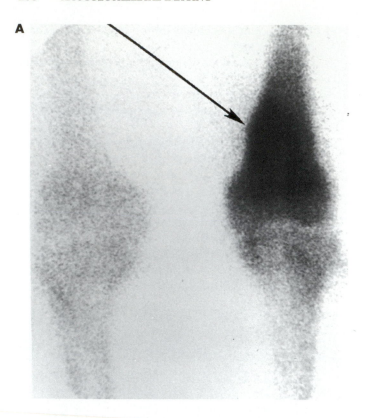

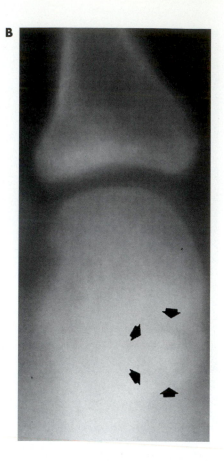

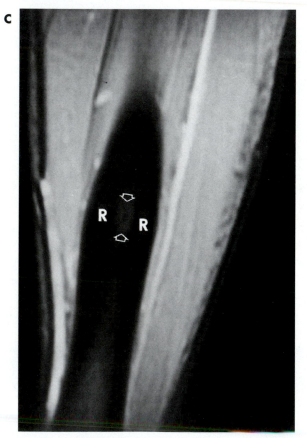

Fig. 4-6   The nidus of an osteoid osteoma can be localized by a variety of imaging methods. **A,** Skeletal scintigraphy of a distal femoral lesion reveals more intense uptake in the nidus (*arrow*) than in the adjacent reactive bone. **B,** Computed or conventional tomography optimally demonstrates calcified nidi (*arrows*), in this case localized within a metatarsal head. **C,** Coronal gradient-echo MR image documents a femoral diaphyseal nidus with central calcification (*arrows*) and reactive cortical thickening (*R*).

A

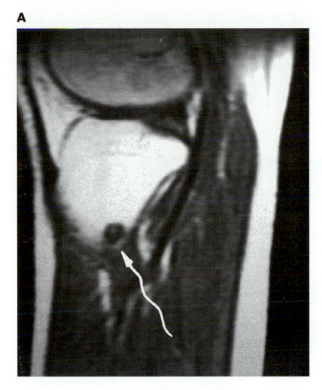

B

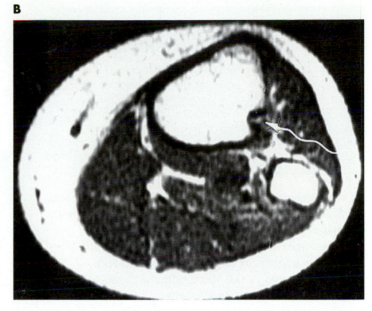

Fig. 4-7   Osteoid osteoma of the proximal tibial metaphysis. Sagittal (**A**) and transaxial (**B**) T1-weighted MR images demonstrate the subperiosteal nidus (*arrows*) of the lesion. Differential diagnosis should include consideration of cortical abscess, juxtacortical chondroma, fibrous cortical defect, and other benign periosteal lesions.

3. The lesion is expansile and cortically based.
4. The proximal third of the tibia is the most common site of involvement.
5. Anterior cortical bowing may be an associated feature of tibial lesions.
6. The lesion may become highly aggressive with recurrence.
7. Fibrous dysplasia and adamantinoma may manifest a radiographic appearance similar to that of ossifying fibroma.
8. Wide en bloc excision is indicated, because of the high recurrence rate with curettage or marginal excision.
9. Adamantinoma tends to occur more distally than ossifying fibroma in the tibia.
10. The lesion does not involve soft tissue.

PEARL

Pseudomalignant osseous tumor of soft tissue:
- It is a benign neoplastic process that produces bone.
- It usually appears during the second or third decade of life.
- Favored sites of involvement include the gluteal region and the extremities.
- Unlike fibrodysplasia ossificans progressiva, it is not associated with congenital anomalies of the thumb and great toe.
- Most lesions are under 6 cm in diameter at the time of presentation.

Cartilage-forming tumors

Enchondroma   This is a common asymptomatic medullary cartilaginous lesion (Box 4-6).
1. Complications of the lesion include pathologic fracture and malignant degeneration.

PEARL

Malignant degeneration of a solitary enchondroma is rare. Malignant degeneration is more commonly seen in Maffucci syndrome (up to 20% of cases) and Ollier disease (up to 5% of cases). Malignant degeneration in Maffucci syndrome includes hemangiosarcoma and lymphangiosarcoma. Malignant degeneration in Ollier disease includes chondroid chordoma. Chondrosarcoma is the most common malignancy in both conditions.

2. Long-bone lesions tend to be metaphyseal.
3. Multiple lesions occur in Ollier disease and Maffucci syndrome.

PEARL

Ollier disease is not hereditary, unlike hereditary multiple exostosis syndrome, which is an autosomal dominant trait. Both can undergo sarcomatous degeneration
Continued

PEARL—cont'd

and demonstrate shortened and/or deformed bones. They usually do not occur in the same patient; when they do, the condition is known as metachondromatosis. Ollier disease has a tendency to exhibit asymmetric or unilateral distribution, whereas hereditary multiple exostosis syndrome does not. The pseudo-Madelung deformity of the radius is seen with Ollier disease.

4. Confident distinction between enchondroma and bone infarct may at times be difficult because both manifest patchy increased medullary densities.
5. The tumor tends to be 3 to 4 cm in size, although smaller lesions occur in the hands and feet.
6. Enchondroma is the most common lytic lesion of small tubular bones in the hands and feet.
7. The lesion may either be purely lytic or manifest stippled, ringlike calcification.
8. The tumor is frequently an incidental finding during the third or fourth decade of life.
9. Early malignant degeneration is characterized histologically by either atypical enchondroma or low-grade chondrosarcoma.

PEARL

Enchondromas occasionally undergo sarcomatous malignant degeneration. Osteosarcoma, fibrosarcoma, and malignant fibrous histiocytoma are possible; however, chondrosarcoma is the most common histologic type.

10. Proximal lesions, including those occurring in the shoulder girdle and pelvis, have the highest incidence of malignant transformation.
11. Lesions of the hands or feet may exhibit aggressive histology but nearly always manifest benign behavior.
12. Local pain in the absence of fracture is more sensitive than imaging findings in predicting early malignant transformation.
13. Both enchondromas and chondrosarcomas exhibit highly variable uptake of bone-seeking radiopharmaceuticals.
14. In the setting of malignant degeneration, curettage with bone grafting is a viable option only in older patients unlikely to experience recurrence or metastasis during their remaining life; curative limb salvage is preferred in younger patients.
15. Maffucci syndrome includes enchondromatosis and soft-tissue hemangiomas, frequently with phleboliths.

16. Risk of malignant degeneration is higher in Maffucci syndrome than in Ollier disease.
17. Striated metaphyses, grotesque enchondromas, limb shortening, and growth deformities occur in both Ollier disease and Maffucci syndrome.
18. Neither Ollier disease nor Maffucci syndrome is hereditary or familial; the multiple cartilaginous exostosis syndrome is an autosomal dominant trait.

PEARL

Enchondromatosis is probably not hereditary, familial, or congenital. It represents failure of enchondral bone formation. Shortening of the bone and deformities resulting from eccentric growth and undertubulation are commonly seen. Ollier disease and Maffucci syndrome are the two forms of enchondromatosis. Both have high rates of malignant transformation (particularly Maffucci syndrome) and a tendency toward unilateral or asymmetric distribution.

Box 4-6 Enchondroma: Important Features

- **Peak age:** second decade (but found commonly in all age groups)
- **Male/female ratio:** 1:1
- **Most common location:** small bones of the hands (but commonly found in the femur and humerus as well)
- **Major radiographic features:** medullary location; nonaggressive features, especially sharp margination and expansion; punctate calcification; may be multiple
- **Radiographic differential diagnosis:** bone infarction, chondrosarcoma
- **Treatment:** curettage and bone grafting or observation

Osteochondroma   This is a common exophytic neoplasm resulting from displacement of growth plate cartilage to the metaphyseal region (Box 4-7).

1. Pedunculated and sessile morphologic subtypes exist.
2. Continuity exists between the periosteum, cortex, and marrow of the lesion and that of the host bone.
3. Sessile lesions may cause growth deformity, especially if they are multiple, whereas pedunculated lesions tend to induce mechanical problems.
4. The multiple cartilaginous exostosis syndrome is inherited as an autosomal dominant trait.

5. Lesion growth ceases with skeletal maturity; subsequent growth or pain suggests malignant degeneration to chondrosarcoma.

6. The multiple cartilaginous exostosis syndrome has an incidence of malignant transformation of up to 5%.

7. Painful bursitis may occur over a benign lesion in the setting of mechanical irritation.

8. Metaphyseal lesions usually point away from the adjacent joint.

9. Only about 5% of lesions occur in the axial skeleton.

10. Approximately one third of lesions occur around the knee.

11. Approximately 90% of lesions are solitary.

12. Malignant degeneration of an osteochondroma may be radiographically indistinguishable from benign osteochondroma.

13. Sessile lesions may resemble fibrous dysplasia or a metaphyseal dysplasia radiographically.

14. Fewer than 1% of solitary osteochondromas undergo malignant degeneration to chondrosarcoma.

15. Because osteochondroma exhibits mildly increased uptake and chondrosarcoma manifests variable uptake on skeletal scintigraphy, a single scan is not reliable in distinguishing between the two lesions, although serial scans may be useful.

16. If malignant degeneration is suspected, computed tomography or magnetic resonance imaging is indicated, as the thickness of the cartilaginous cap may assist in distinguishing benign from malignant lesions (greater than 1 cm implies the latter).

17. Local excision is the treatment of choice for benign osteochondromas with mechanical pain.

18. Malignant degeneration occurs earlier in life in the multiple cartilaginous exostosis syndrome than with solitary osteochondroma.

19. Primary soft-tissue osteochondromas are much less common than those arising from bone (Fig. 4-8).

Dysplasia Epiphysialis Hemimelica (Trevor-Fairbank Disease): Distinction from Multiple Cartilaginous Exostosis Syndrome

1. Dysplasia epiphysialis hemimelica (Trevor-Fairbank disease): Intraarticular osteochondromas arise from epiphyses.

PEARL

Dysplasia epiphysialis hemimelica (Trevor-Fairbank disease):

- During growth, irregular bone overgrowth on one side of one or more epiphyses and/or tarsal bones, sometimes with multiple ossification centers (most frequently involving distal femur and proximal tibia)
- Following growth, enlargement of one side of the end of one or more tubular and/or tarsal bones, with relatively normal trabecular pattern
- Unilateral distribution

A

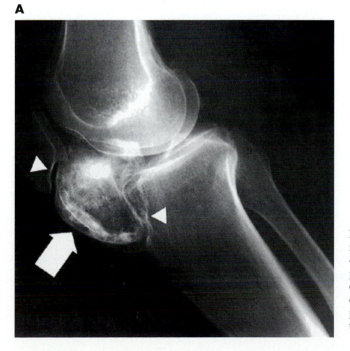

B

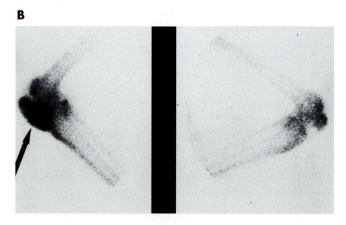

Fig. 4-8 Soft-tissue osteochondroma arising from the Hoffa fat-pad of the knee. **A,** Lateral radiograph reveals a characteristic ossific mass (*arrow*) with pressure erosion of the adjacent tibia and patella (*arrowheads*). **B,** The lesion exhibits intense activity (*arrow*) on a radionuclide bone scan. Opposite side shown for comparison. Differential diagnosis might include consideration of posttraumatic myositis ossificans, although the location is typical of this neoplasm.

2. Dysplasia epiphysialis hemimelica (Trevor-Fairbank disease): Knee and ankle are the most common sites of involvement.

3. Multiple cartilaginous exostosis syndrome: This is an autosomal dominant disorder.

4. Multiple cartilaginous exostosis syndrome, dysplasia epiphysialis hemimelica (Trevor-Fairbank disease): Multiple lesions with characteristic histologic features of osteochondroma occur.

5. Multiple cartilaginous exostosis syndrome: There is a relatively high incidence of sarcomatous degeneration, particularly in more proximal lesions.

6. Dysplasia epiphysialis hemimelica (Trevor-Fairbank disease): Joint deformity, pain, and limited range of motion are typical clinical features.

7. Multiple cartilaginous exostosis syndrome: Most of the lesions are broad based and sessile.

8. Multiple cartilaginous exostosis syndrome, dysplasia epiphysialis hemimelica (Trevor-Fairbank disease): Local resection is the treatment of choice for symptomatic lesions.

9. Multiple cartilaginous exostosis syndrome: Lumps around joints, short limbs, and deformities are typical clinical manifestations.

10. Dysplasia epiphysialis hemimelica (Trevor-Fairbank disease): Lesion distribution shows a predilection for one side of the body.

Box 4-7   Osteochondroma: Important Features

- **Peak age:** second decade
- **Male/female ratio:** 1.7:1
- **Most common location:** distal femur
- **Major radiographic features:** osseous growth projecting from bone surface; continuity of the cortical and cancellous bone between the underlying parent bone and the lesion; flaring of the metaphysis of the affected bone
- **Radiographic differential diagnosis:** soft-tissue osteochondroma, parosteal osteosarcoma, myositis ossificans
- **Treatment:** surgical resection if the lesion is symptomatic; observation if asymptomatic

---

Chondroblastoma   This is an uncommon lesion found almost exclusively in the epiphysis of a skeletally immature patient (Box 4-8).

PEARL

Chondroblastoma originates in and is usually confined to the unfused epiphysis. Radiographically, this tumor appears as a lytic lesion with sclerotic margins. More than 50% are calcified. Favorite sites include the proximal humerus, proximal tibia, distal femur, and proximal femur. Histologic features include the presence of multiple chondroblasts and giant cells. In contrast, giant cell tumors usually occur in skeletally mature patients, and osteoblastomas frequent the posterior elements of the spine or metadiaphyses of long bones.

---

1. The lesion may extend into the metaphysis, particularly after physeal closure.
2. Metaphyseal periosteal reaction occurs occasionally.
3. Calcified matrix occurs in approximately 50% of lesions.

PEARL

Chondroblastoma may exhibit focal areas of calcification and also has giant cells within it. Amyloidosis of bone also may demonstrate giant cells and microscopic calcification. Chondromyxoid fibroma demonstrates calcifications only in advanced lesions, and these are usually not visible radiographically. Giant cell tumor of bone and pigmented villonodular synovitis are not associated with calcification.

---

4. The lesion occurs most commonly in the proximal humerus.
5. Soft-tissue involvement does not usually occur.
6. The lesion generally ranges in size from 1.5 to 4 cm.
7. Common sites of involvement outside the humerus include the proximal femur, distal femur, and proximal tibia.

PEARL

Chondroblastoma is an epiphyseal lesion affecting male patients more commonly than females in a ratio of approximately 2:1. The tumor most commonly occurs in the second and third decades of life. The long bones are most frequently involved, particularly the proximal humerus, femur, or proximal tibia.

8. Aneurysmal bone cyst may accompany the lesion.
9. In the differential diagnosis, major consideration should be given to giant cell tumor and articular processes resulting in large subchondral cysts.
10. Malignant chondroblastomas with pulmonary metastases occur rarely.
11. Chondroblastoma is a monostotic lesion.
12. Clinically, joint pain may simulate that of an arthritic process.
13. Curettage with bone grafting is the treatment of choice.
14. Recurrence is more likely in the setting of a coexistent aneurysmal bone cyst.
15. Recurrent lesions tend to be more aggressive and may require wider excision.

3. Soft-tissue involvement does not occur.
4. The lesion is often greater than 5 cm in diameter at presentation.
5. Common sites of involvement outside the tibia include the proximal and distal femur, flat bones, and tarsal bones (Fig. 4-9).
6. Radiographically evident calcification is rare, occurring in only 2% of cases.
7. Endosteal sclerosis is common, whereas periosteal reaction is rare.
8. Malignant transformation is extremely rare.
9. Major differential diagnostic considerations include aneurysmal bone cyst, giant cell tumor, nonossifying fibroma, and enchondroma.

Box 4-8 Chondroblastoma: Important Features

- **Peak age:** second decade (with approximately 50% of tumors in this age range)
- **Male/female ratio:** 2:1
- **Most common location:** proximal humerus, followed by the distal femur and proximal tibia
- **Major radiographic features:** located in the epiphysis or an apophysis; small size and sharp margins; sclerotic rim and expansion of the affected bone are commonly seen; matrix calcification is present in 50% of cases; lesion may cross an open physis
- **Radiographic differential diagnosis:** giant cell tumor, ischemic necrosis, clear cell chondrosarcoma, intraosseous ganglion
- **Treatment:** curettage and bone grafting

Chondromyxoid fibroma This is a very uncommon lesion containing fibrous and myxoid tissue, as well as cartilage (Box 4-9).
1. Approximately one third of lesions occur in the proximal tibial metaphysis.
2. Most patients with this lesion are diagnosed during the second or third decade of life.

PEARL

Chondromyxoid fibroma most frequently involves the proximal tibia, proximal and distal ends of the femur, and small bones of the foot. Male patients in their second and third decades are typically affected. Chondromyxoid fibroma is usually a diametaphyseal lesion.

PEARL

Chondromyxoid fibromas are usually eccentrically situated diametaphyseal lesions that are lucent and elongated in shape. Cortical expansion, exuberant endosteal sclerosis, and coarse trabeculation are commonly noted. Enchondromas are expansile and cause endosteal scalloping, but do not usually result in significant sclerosis.

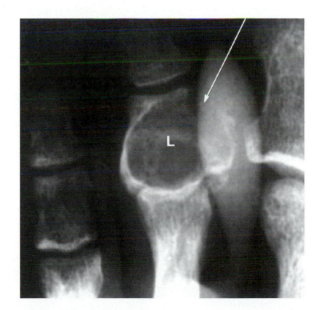

Fig. 4-9 Chondromyxoid fibroma involving the middle phalanx of the second toe. Frontal radiograph reveals a lytic, expansile lesion (*L*) with cortical disruption (*arrow*). Differential diagnosis should include consideration of enchondroma, aneurysmal bone cyst, giant cell tumor, giant cell reparative granuloma, fibrous dysplasia, indolent infection, sarcoidosis, and tuberous sclerosis.

10. The lesion is geographic and lobulated, resulting in a pseudoseptate or bubbly appearance.
11. Younger patients tend to have more aggressive lesions with a higher recurrence rate.
12. Curettage and bone grafting is associated with a 25% to 30% rate of local recurrence.
13. Wide excision is generally reserved for aggressive or recurrent lesions.

Box 4-9 Chondromyxoid Fibroma: Important Features

- **Peak age:** second or third decade
- **Male/female ratio:** 1.6:1
- **Most common location:** Around the knee, with the proximal tibia accounting for approximately one third of cases
- **Major radiographic features:** tibia most common site; eccentric diametaphyseal location; sharp, sclerotic, and scalloped margins; matrix calcification rare
- **Radiographic differential diagnosis:** aneurysmal bone cyst, giant cell tumor, nonossifying fibroma, and enchondroma
- **Treatment:** curettage and bone grafting

Juxtacortical chondroma This is a rare lesion arising from the periosteal surface of the bone (Fig. 4-10; Box 4-10).

1. Cortical scalloping and periosteal reaction may result in an aggressive appearance.
2. Soft-tissue mass is a typical feature of the lesion.
3. The lesion occurs in both children and adults, but most affected patients are under age 30.
4. Most lesions are less than 4 cm in size.
5. The lesion has a predilection for the femur or humerus.
6. Calcification occurs in approximately 50% of lesions.
7. The histologic features of juxtacortical chondroma frequently appear to be more aggressive than those of enchondroma.
8. Major differential diagnostic considerations include periosteal chondrosarcoma, periosteal osteosarcoma, and soft-tissue tumor with secondary osseous involvement.
9. Malignant degeneration may occur, with recurrence following therapy.
10. Computed tomography or magnetic resonance imaging may be useful in characterizing long-bone lesions.
11. En bloc excision is the preferred therapy.

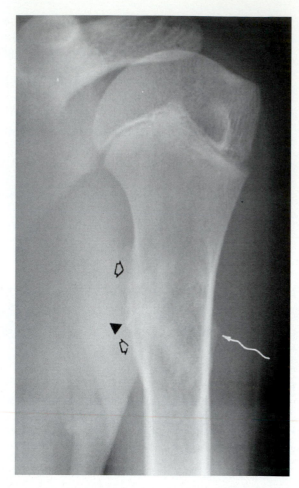

Fig. 4-10 Juxtacortical chondroma of the proximal humerus. Frontal radiograph demonstrates a cortically based, slightly expansile lesion (*open arrows*) with sparse central calcification (*arrowhead*) and circumferential periosteal reaction (*wavy arrow*). Differential diagnosis should include consideration of low-grade periosteal chondrosarcoma, periosteal osteosarcoma, and periosteal hemangioma.

Box 4-10 Juxtacortical (Periosteal) Chondroma: Important Features

- **Peak age:** second or third decade
- **Male/female ratio:** 2:1
- **Most common location:** femur or humerus
- **Major radiographic features:** small surface mass eroding the underlying cortical surface; metaphyseal or diaphyseal in location; marginal spicules or buttresses are present; approximately 50% are calcified
- **Radiographic differential diagnosis:** periosteal chondrosarcoma, periosteal osteosarcoma, soft-tissue neoplasm secondarily involving the cortical bone
- **Treatment:** conservative resection of the lesion or observation

Vascular tumors

Vascular tumors, although not uncommon in soft tissues, are extremely rare in bone, constituting less than 1% in most major series of bone tumors. Of significance are (1) their propensity to bleed profusely at biopsy and (2) the extreme aggressiveness of the malignant tumors of vascular origin.

Benign tumors are more common and include hemangioma and lymphangioma of bone, glomus tumor, and massive aneurysmal bone cyst. Malignant tumors include hemangiopericytoma and hemangioendothelioma.

Hemangioma   This is a hamartomatous lesion containing numerous vascular channels (Box 4-11).

1. Approximately 75% of lesions occur in the vertebral bodies.
2. Vertebral lesions exhibit a predilection for the thoracic spine.
3. Skull lesions typically affect the outer table and spare the inner table.
4. Radiographically, a coarsened trabecular pattern (vertically oriented in the spine and radially oriented in the skull) is characteristic (Fig. 4-11).
5. Soft-tissue extension of vertebral lesions may produce neurologic symptoms.
6. Most patients are diagnosed during the fourth and fifth decades of life.
7. The facial bones are the third most common site of involvement by hemangioma.
8. Most lesions are greater than 2 cm in size (Fig. 4-12).
9. Vertebral lesions may simulate Paget disease, although osseous enlargement is absent.
10. Computed tomography or magnetic resonance imaging may assist in defining epidural extent of symptomatic vertebral lesions.

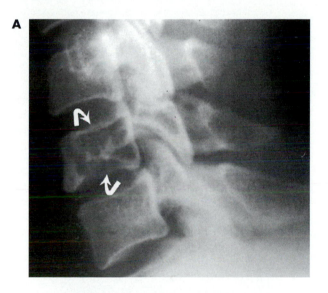

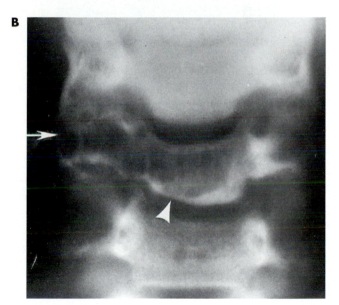

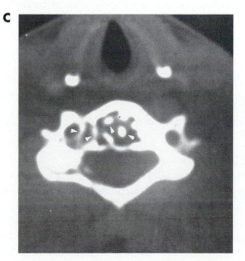

Fig. 4-11   Vertebral hemangioma. **A,** Lateral radiograph reveals a septate lytic lesion (*arrows*) involving the C5 vertebral body. **B,** Frontal tomogram demonstrates the typical corduroy appearance of the lesion, which involves both the vertebral body (*arrowhead*) and right pillar (*arrow*). **C,** Prominent vertical septations (*arrowheads*) are characteristic CT findings. Differential diagnosis might include consideration of Paget disease, although osseous enlargement and cortical thickening are absent in this patient.

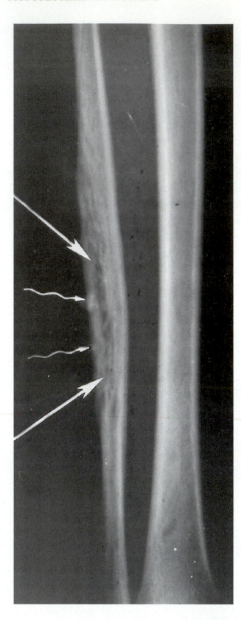

Fig. 4-12   Periosteal hemangioma. Frontal radiograph of the lower extremity reveals a cortically based lytic lesion (*straight arrows*) in the fibular diaphysis, which contains radiodense phleboliths (*wavy arrows*). Differential diagnosis should include consideration of juxtacortical chondroma and juxtacortical chondrosarcoma.

Box 4-11   Osseous Hemangioma: Important Features

- The lesion constitutes less than 1% of all osseous neoplasms.
- Sex distribution is approximately equal.
- All ages are affected, but the tumor is most common in adulthood (fourth and fifth decades).
- Over two thirds of cases occur in the vertebrae, skull, and facial bones (including the mandible).
- Hemangioma is often an asymptomatic incidental finding. Symptoms (when present) may include spasm, radiculopathy, or paraparesis. Local swelling may occur, particularly in association with skull lesions.
- Radiographic findings:
  - In the skull, ribs, mandible, or flat bones of the pelvis, a well-circumscribed lytic zone with central sclerotic bone radiating to the periphery in a sunburst pattern is typical.
  - In the vertebrae, rarefaction of the body with accentuation of the vertical striations is observed; the posterior elements are less frequently affected, but when they are involved, spinal cord compression is not common.
  - In long bones (very rare), the lesion favors the metaphysis and produces a lytic "soap bubble" appearance; the epiphyses and areas near nutrient arteries are usually spared.
  - Soft-tissue hemangiomas may occur close to bone causing local erosion, or, less commonly, a sclerotic reaction; the major radiographic difference between these and primary osseous lesions is the presence of phleboliths, which are usually present in soft-tissue hemangiomas, but not in hemangiomas of bone.
- Pathologic findings:
  - Grossly, the lesions usually have a bluish color, and sunburst lesions manifest associated bony trabeculae.
  - Microscopically, the lesion may be either a capillary or cavernous hemangioma, with the latter being more common.
- Treatment:
  - When indicated (usually when there is pain or deformity in a skull lesion, spinal canal encroachment in a vertebral lesion, or pathologic fracture in a long bone), surgical intervention may be necessary; hemorrhage is a serious problem and must be anticipated.
  - Radiation is an alternative form of therapy that may be used for inaccessible lesions or as an adjunct to surgery.

11. Treatment of choice for symptomatic lesions includes resection and/or radiation therapy.
12. The lesion does not significantly weaken the vertebral body; in cases of trauma, a vertebra above or below a hemangioma frequently fractures.

Soft-tissue hemangioma
1. Radiographically, a soft-tissue mass and calcified phleboliths are typical; synovial involvement occasionally occurs (Fig. 4-13).
2. Soft-tissue hemangiomas may invade and erode adjacent bone.

3. The lesions are associated with osseous enchondromatosis in the Maffucci syndrome.
4. Contrast material–enhanced computed tomography or magnetic resonance imaging demonstrates numerous tortuous dilated vessels.
5. The Klippel-Trenaunay-Weber syndrome includes hypertrophy of soft tissues and bone (usually monomelic) and a variety of vascular malformations including soft-tissue hemangiomas.

PEARL

Limb lengthening and hemangiomas are seen in Klippel-Trenaunay-Weber disease. Neurofibromatosis and macrodystrophia lipomatosa can also cause osseous and soft-tissue overgrowth.

6. An association between soft-tissue hemangioma and hemihypertrophy has been described.

Cystic angiomatosis (Box 4-12) This is a rare condition characterized by multifocal lymphangiomatosis or hemangiomatosis, frequently with severe visceral involvement.

1. Most patients are initially diagnosed during the first through third decades of life.
2. Soft-tissue involvement is characterized by masses with calcified phleboliths.
3. The femur, pelvis, ribs, humerus, skull, and vertebrae are the most commonly affected bones.
4. Large, expansile, aggressive-appearing lesions are typically observed.
5. Differential diagnostic considerations include polyostotic fibrous dysplasia, the Maffucci syndrome, Langerhans cell histiocytosis, and metastatic disease.
6. The 5-year survival rate is 50% in the presence of visceral lesions.
7. Recurrence is common following curettage or radiation therapy, the accepted forms of management.
8. Chylous effusions and death are common in the setting of visceral involvement.

Box 4-12 Cystic Angiomatosis: Important Features

- The condition is extremely rare, constituting fewer than 0.1% of bone tumors.
- Sex distribution is approximately equal.
- The condition may occur from the age of 3 months to the sixth decade; most affected patients are under age 20.

Box 4-12—cont'd

- The lesions may affect the entire skeleton, but the long bones, skull, and flat bones are most commonly involved.
- Clinical manifestations:
  - Symptoms include pain, recurrent pathologic fractures, and palpable masses.
  - The condition may cause bony overgrowth of involved long bones, and intramedullary lesions may cause cortical expansion, thickening, and obliteration of the medullary canal.
  - When extraosseous involvement is present, anemia, thrombocytopenia, and respiratory insufficiency may be present.
- Associated findings:
  - In 50% of cases, there are soft-tissue and visceral angiomas, most often found in the liver, spleen, lungs, lymph nodes, kidneys, skin, and brain.
  - Recurrent hemorrhage from these extraskeletal sites can lead to anemia and thrombocytopenia.
  - Chylous effusions can also occur, leading to respiratory insufficiency.
- Radiographic findings:
  - In younger patients, the lesions are lytic, whereas sclerotic lesions predominate in older patients.
  - Lesions typically begin at the site of entry of the nutrient arteries into bone.
  - Soft-tissue phleboliths may be associated with the bony lesions.
- The histopathologic structure is similar to that of hemangioma, but this syndrome can and often does involve a proliferation of lymphatic channels; most commonly, both lymphangiomas and hemangiomas are intermingled.
- Treatment options are curettage, cauterization, and irradiation.
- Prognosis is excellent except when there is visceral involvement, in which case the 5-year survival rate is about 50%, with 10-year figures down to 30%; death occurs from massive hemorrhage or cardiorespiratory failure.

Lymphangioma This is a hamartomatous lesion composed of numerous dilated lymphatic channels.
1. Polyostotic involvement occurs in lymphangiomatosis, along with chylous pleural effusions and lymphedema.
2. The majority of patients are adults.
3. Soft-tissue involvement by osseous lesions does not usually occur.
4. Major differential diagnostic considerations include fibrous dysplasia, solitary bone cyst, and desmoplastic fibroma.
5. Most lesions exceed 2 cm in size.

6. Curettage is the treatment of choice for monostotic lesions.
7. Lymphangiomas can occur in central, flat, or tubular bones.

PEARL

Lymphangioma is another rare entity that, like heman-gioma, can occur as a solitary lesion or with multiple-site involvement. Solitary lesions usually affect younger individuals and have the presenting symptom of patho-logic fracture. Pathologic examination reveals benign channels containing lymph. Treatment, if indicated, con-sists of curettage and bone grafting.

The multicentric form, cystic lymphangiomatosis, is usually classified with cystic hemangiomatosis. It is distinguished from the latter by its propensity to exhibit abnormal lymphangiographic images with extensive cystic spaces and scattered intraosseous collections of contrast material. The disease is progressive in some cases with multiple pathologic fractures, yet stable and asymptomatic in others. No satisfactory therapy exists (irradiation is relatively ineffective), and death may occur with generalized involvement.

Glomus tumor This is a rare benign vascular tumor that has a predilection for the distal phalanges. The lesion is lytic and well circumscribed radiographically. The tumor is painful and is usually found in adults. The treatment of choice is curettage or marginal excision. A lesion that does not commonly occur at the site of predilection for glomus tumor is enchondroma.

PEARL

Glomus tumor (angioblastoma) is classically a lytic le-sion in the terminal phalanx occurring during the fourth or fifth decade. It represents an osteolytic hamartoma-tous degeneration of the normal glomus body and usually involves the terminal phalanges of the hand, although it can occur anywhere. The glomus body is an arteriovenous shunt with small perivascular myocytes that are extensively innervated. These shunts are under autonomic nervous control, and they function to regu-late blood pressure and body temperature. Because of their rich nerve supply, glomus tumors are exquisitely sensitive to direct pressure. The lesion frequently occurs subungually, causing pain with pressure on the nail. The condition is slightly more common in men. Physical examination often reveals a small bluish punctum, which is very tender and cold sensitive. Surgical re-moval is curative, although recurrence is possible.

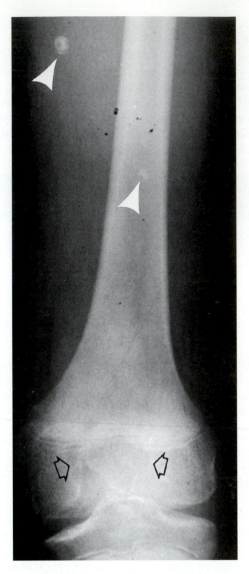

Fig. 4-13 Synovial hemangioma of the knee. Osteopenia, epiphyseal overgrowth, and widening of the intercondylar notch (*arrows*) in the setting of phleboliths (*arrowheads*) are character-istic. Differential diagnosis of the osseous abnormalities should include consideration of juvenile chronic arthritis, hemophilia, and granulomatous infection such as tuberculosis.

Massive osteolysis This is a form of angiomatosis with regional dissolution of bone (Box 4-13).

1. The condition is also known as Gorham disease.
2. The condition occurs most commonly in children and young adults.
3. Bone destruction is much more extensive than in simple angiomatosis.
4. Trauma usually precedes the onset of the condi-tion.

5. Rapid destruction of bone occurs, to the extent that the condition is also termed *vanishing bone disease*.
6. Bone dissolution spreads contiguously across articulations.
7. The shoulder and hip regions are the most common sites of involvement.
8. Radiation therapy may stabilize the process in some instances, but other cases continue to progress relentlessly.

Box 4-13 Massive Osteolysis: Important Features

- The condition is extremely rare.
- Sex distribution is equal.
- All ages are affected, but the condition most often occurs in patients under age 40.
- The condition is frequently associated with previous trauma to the affected area; pain may be present, but often the onset is insidious, with atrophy and limited range of motion.
- Radiographic findings include lysis of affected bones without periosteal reaction, new bone formation, or expansion; the process crosses joints to affect multiple bones.
- Pathologically there is extensive involvement of the affected bone with vascular tissue, which may be lymphangiomatous or hemangiomatous.
- Radiation therapy may be attempted but is usually unsuccessful.
- The course is self-limited in some cases, but severe thoracic involvement may lead to death from respiratory complications.

Vascular neoplasm with secondary osteomalacia

Soft-tissue or osseous vascular lesions may induce hypophosphatemic rickets or osteomalacia. Affected patients have progressive bone pain and tenderness, and usually no other cause for the metabolic bone disease is apparent. Treatment with vitamin D and phosphates is effective, but unless the vascular tumor is found and removed, relapse occurs upon cessation of therapy. The mechanism for this phenomenon is poorly understood, but it may relate to secretion of a humoral substance by the tumor.

Fibrous tumors

Fibrous cortical defect and nonossifying fibroma

These are cortically based metaphyseal lesions with identical histologic features characterized by benign fibrous tissue (Boxes 4-14, 4-15).

PEARL

Fibrous cortical defects are eccentric cortical defects in the metaphyses of long bones near the physes, occurring most commonly on the posteromedial aspect of the distal femur. Rounded in shape when small, they rapidly become ovoid with long axes parallel to the direction of bone growth. Cortical thinning and mild expansion may occur. The margin is usually sclerotic. Larger lesions are usually multilocular, and termed *nonossifying fibromas*. Regression occurs with progressively increasing sclerosis, and lesions migrate away from the physis as the bone grows.

Similarities and differences between these two lesions are summarized as follows:

1. Fibrous cortical defect: The lesion occurs in 30% to 40% of children over 2 years of age.
2. Fibrous cortical defect, nonossifying fibroma: The lesion is lytic with a well-defined sclerotic margin.
3. Nonossifying fibroma: The lesion is moderately expansile with pseudotrabeculation.
4. Fibrous cortical defect, nonossifying fibroma: Sclerosis replaces lysis during the spontaneous healing phase.
5. Fibrous cortical defect, nonossifying fibroma: 95% of lesions occur in patients under 20 years of age.
6. Fibrous cortical defect: Lesions are less than 2 cm in size.
7. Nonossifying fibroma: Lesions are greater than 2 cm in size.
8. Fibrous cortical defect, nonossifying fibroma: 80% of lesions occur in the long bones of the lower extremity.
9. Fibrous cortical defect, nonossifying fibroma: The lesion manifests a central location in thin bones such as the ulna and fibula.
10. Fibrous cortical defect, nonossifying fibroma: Lesions are polyostotic in 25% of affected patients.
11. Nonossifying fibroma: The lesion begins in the cortex but may enlarge to involve the medullary cavity of the bone.
12. Fibrous cortical defect, nonossifying fibroma: Periosteal reaction does not usually occur.
13. Fibrous cortical defect: The lesion is asymptomatic and discovered incidentally.
14. Fibrous cortical defect, nonossifying fibroma: The lesion may be associated with neurofibromatosis, particularly when multiple.
15. Nonossifying fibroma: Differential diagnosis should include consideration of aneurysmal bone cyst, chondromyxoid fibroma, and brown tumor of hyperparathyroidism.

16. Fibrous cortical defect: The lesion never requires treatment.
17. Nonossifying fibroma: In the setting of pain or pathologic fracture, curettage with bone graft packing is indicated.

Box 4-14   Fibrous Cortical Defect: Important Features

- Pathologically identical to nonossifying fibroma; smaller in size
- Begins adjacent to physis in childhood or adolescence, and subsequently migrates away from the epiphysis with growth
- Asymptomatic, incidental finding
- Involutes with sclerosis in young adulthood

Box 4-15   Nonossifying Fibroma: Important Features

- Affects long bones of extremities, especially the lower
- Eccentric, with cortical thinning and expansion
- Well-defined sclerotic border
- Asymptomatic unless pathologically fractured
- Lesion becomes sclerotic with age
- Larger than fibrous cortical defect with identical histologic features

---

Fibrous dysplasia   This is a common hamartomatous lesion composed of a fibrous stroma with islands of woven bone and osteoid (Box 4-16).

PEARL

Fibrous dysplasia results from replacement of portions of the medullary cavity by fibro-osseous tissue.

McCune-Albright syndrome is a clinical triad consisting of (1) unilateral or asymmetric polyostotic fibrous dysplasia, (2) café-au-lait spots that tend to be on the same side of the body as the bony abnormalities, and (3) endocrine dysfunction often resulting in precocious puberty among female patients.

Fibrous dysplasia is the most common cause of an asymptomatic lesion in a rib, which may be either lytic or sclerotic.

Involvement of the skull may produce lytic, sclerotic, or mixed lesions. Involvement of the sphenoid wing may be quite sclerotic and may simulate Paget disease. The age and sex of the patient are helpful differentiating clinical features. Sclerosis of the base of the skull and sphenoid bone may produce cranial and optic nerve compression. Facial involvement manifests itself clinically as leontiasis ossea, or the lion facies.

---

1. The lesion may be detected at any age, although most patients are diagnosed in the second or third decade of life.
2. Most lesions are greater than 5 cm in size.
3. Approximately 70% of cases are monostotic.
4. Lesions involving the skull base are expansile and sclerotic (Box 4-17).
5. Approximately 90% of polyostotic lesions are unilateral in distribution.

PEARL

Polyostotic fibrous dysplasia is usually unilateral or asymmetric in distribution and may be associated with precocious puberty. There is equivalent frequency among male and female patients. However, in the McCune-Albright syndrome (which includes precocious puberty and café-au-lait spots), girls are more often affected than boys.

---

6. Rib and long-bone lesions tend to be mildly expansile (Fig. 4-14).
7. Vertebral involvement is uncommon (Fig. 4-15).
8. Pelvic lesions tend to be large and bubbly in appearance.

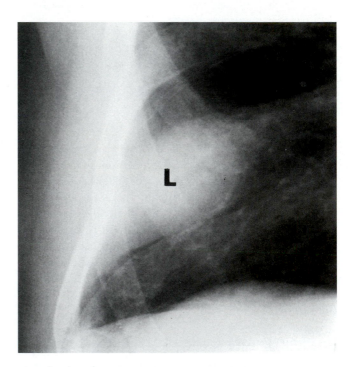

Fig. 4-14   Monostotic fibrous dysplasia of a rib, the most common benign asymptomatic neoplasm at this site. Frontal radiograph demonstrates an expansile radiodense lesion (*L*). Differential diagnosis should include consideration of Paget disease, chronic osteomyelitis, and metastatic disease (particularly from prostate carcinoma).

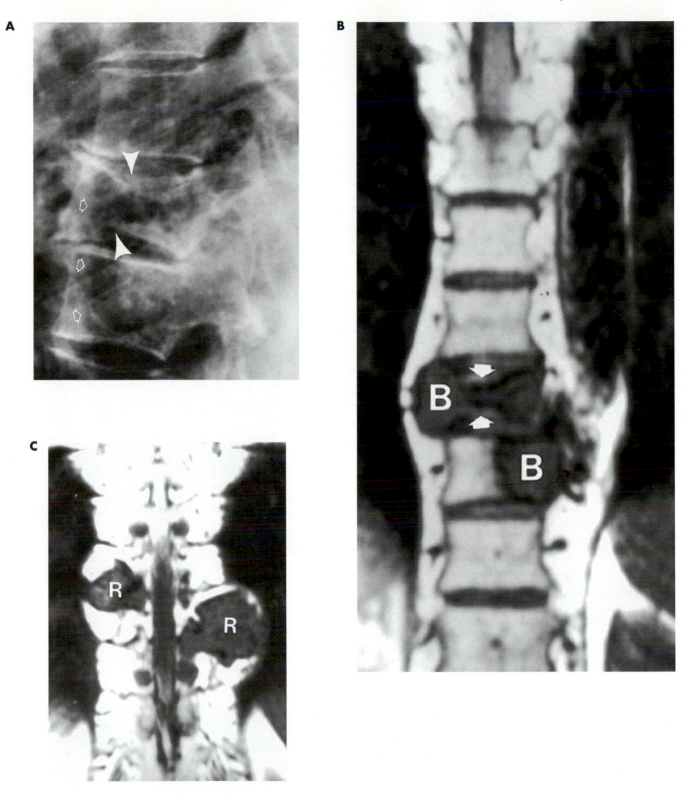

Fig. 4-15 Polyostotic fibrous dysplasia. **A,** Lateral radiograph demonstrates osteolytic involvement of two adjacent vertebral bodies (*arrows*), along with irregular endplate deformities (*arrowheads*). **B** and **C,** Coronal T1-weighted MR images reveal expansile low–signal intensity areas within the vertebral bodies (*B*) and proximal ribs (*R*) (*arrows = pathologic fracture*). Differential diagnosis should include consideration of metastatic disease and indolent infection.

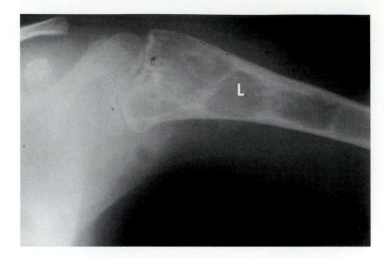

Fig. 4-16   Fibrous dysplasia of the proximal humerus in a child. A well-defined septate lytic lesion (*L*) with areas of ground-glass texture involves the metadiaphysis. Differential diagnosis should include consideration of eosinophilic granuloma, solitary bone cyst, and nonossifying fibroma.

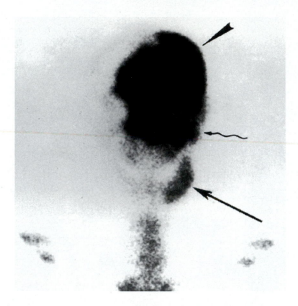

Fig. 4-17   Calvarial and facial involvement in fibrous dysplasia (leontiasis ossea). Skeletal scintigraphy demonstrates asymmetric abnormal increased activity in the skull (*arrowhead*), maxilla (*wavy arrow*), and mandible (*straight arrow*). Differential diagnosis should include consideration of Paget disease.

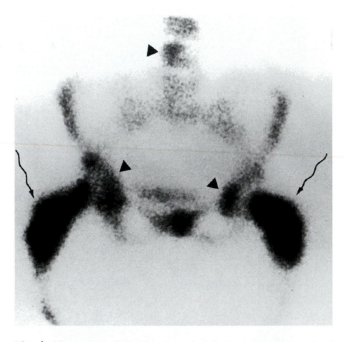

Fig. 4-18   Bilateral shepherd crook deformities of the proximal femurs (*arrows*) in polyostotic fibrous dysplasia. Multiple additional areas of involvement (*arrowheads*) are demonstrated on the skeletal scintigram.

9. The polyostotic form is usually diagnosed before the age of 10 years (Box 4-18).
10. Fibrous dysplasia constitutes the most common benign lesion of a rib.
11. A lytic, ground-glass, or sclerotic texture may be observed (Fig. 4-16).
12. Polyostotic lesions tend to be more aggressive than monostotic lesions.
13. The McCune-Albright syndrome, including polyostotic fibrous dysplasia, precocious puberty, and café-au-lait spots, constitutes approximately 3% of all cases of fibrous dysplasia (Box 4-19).

PEARL

McCune-Albright syndrome is a nonhereditary disorder consisting of endocrine dysfunction (precocious puberty, hyperthyroidism) and unilateral or asymmetric polyostotic fibrous dysplasia, which affects female patients more frequently than male individuals. The patients also have café-au-lait spots with irregular margins resembling the coast of Maine, frequently overlying affected bone. Unilateral bone involvement with ipsilateral skin pigmentation is the most typical pattern.

14. Periosteal reaction does not usually occur in the absence of pathologic fracture.
15. Only 5% of lesions continue to enlarge after skeletal maturation.
16. Pathologic fracture occurs in about 40% of cases.
17. Skeletal scintigraphy is a sensitive technique for establishing the extent of polyostotic involvement (Fig. 4-17).
18. Osteofibrous dysplasia is a variant of fibrous dysplasia occurring in children under 5 years of age that results in bowing, fracture, and pseudoarthrosis of the tibia.

PEARL

Osteofibrous dysplasia is typically manifested as a lytic, expansile cortical lesion of the lower tibial diaphysis (with or without similar changes in the adjacent fibula) with bowing deformity in an infant or child.

19. Malignant degeneration of fibrous dysplasia to fibrosarcoma or malignant fibrous histiocytoma occurs only rarely.
20. Cherubism is a variant of fibrous dysplasia occurring in children that is characterized by expansile lesions of the maxilla and mandible (Box 4-20).
21. Complications of fibrous dysplasia include bowing of long bones, shepherd crook (varus) deformity of the proximal femur, and leg length discrepancy (Fig. 4-18).

22. Skull base lesions may simulate Paget disease or sclerotic response to meningioma.
23. Variable areas of high signal intensity are observed on T2-weighted magnetic resonance images of fibrous dysplasia, despite its fibro-osseous histology. Rapid growth, high vascularity, pathologic fracture, and malignant degeneration are potential causes of this phenomenon (Fig. 4-19).
24. The differential diagnosis of rib lesions should include consideration of metastatic disease, Ewing sarcoma, and eosinophilic granuloma.
25. The lesion is histologically similar to ossifying fibroma.

PEARL

Histologically, fibrous dysplasia is fibro-osseous, whereas adamantinoma is epithelial and angioblastic. Radiographically, fibrous dysplasia has many appearances including purely lytic, ground glass, mixed lytic and sclerotic, and purely sclerotic (particularly at the skull base). It has several forms, including monostotic, polyostotic, and a craniofacial variant that involves the mandible and maxilla (cherubism). Rarely, fibrous dysplasia will undergo malignant degeneration to fibrosarcoma or malignant fibrous histiocytoma. Absence of osteoblasts distinguishes fibrous dysplasia from ossifying fibroma, which is otherwise histologically identical.

26. Polyostotic involvement may simulate metastatic disease, Langerhans cell histiocytosis, or enchondromatosis.

A

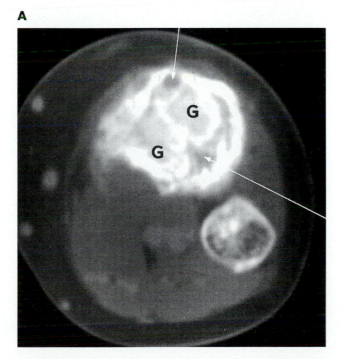

B

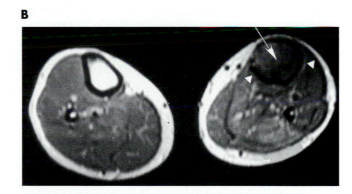

Fig. 4-19 Fibrous dysplasia of the tibia. **A,** CT demonstrates osseous expansion with mixed areas of osteolysis (*arrows*) and ground-glass texture (*G*). **B,** Transaxial T1-weighted MR image confirms osseous enlargement (*arrowheads*) and reveals replacement of normal marrow by low signal intensity (*arrow*). Differential diagnosis should include consideration of ossifying fibroma, osteofibrous dysplasia, adamantinoma, Paget disease, tertiary syphilis, and other forms of chronic osteomyelitis.

Box 4-16   Fibrous Dysplasia: Important Features

- Polyostotic: McCune-Albright syndrome (pigmentary disorders, unilateral or asymmetric predominance, precocious puberty in girls)
- Monostotic: proximal femur, tibia, ribs, and facial bones are the most common sites; diametaphyseal localization in tubular bones
- Variable radiographic appearance: trabecular obliteration; endosteal scalloping; sclerosis; loculation; expansion; deformity; homogeneous ground-glass appearance
- Shepherd crook: coxa vara deformity of proximal femur
- Leontiasis ossea: sclerotic involvement of skull base, sphenoid wings, and facial bones
- Calvarium: lytic, outer more involved than inner table, sclerotic margin, diploic widening
- Skull involvement: 50% of patients with polyostotic disease; 10% to 25% of patients with monostotic disease
- Pathologic fracture may occur
- Malignant degeneration to fibrosarcoma, malignant fibrous histiocytoma (less than 1% of cases), and osteosarcoma (postirradiation of facial, femoral lesions)
- Most common benign lesion of ribs
- Elevated serum alkaline phosphatase level in 30% of cases, usually polyostotic
- Only about 25% of polyostotic cases involve more than half of the skeleton

Box 4-17   Increased Density of Skull Base: Differential Diagnosis

- Fibrous dysplasia
- Paget disease
- Craniometaphyseal dysplasia or Pyle disease
- Engelmann disease (progressive diaphyseal dysplasia)
- Fluorosis
- Renal osteodystrophy
- Hypervitaminosis D, treated rickets
- Idiopathic hypercalcemia (Williams syndrome)
- Melorheostosis
- Meningioma
- Osteopetrosis
- Pyknodysostosis
- Ribbing disease (hereditary multiple diaphyseal sclerosis)
- Von Buchem disease (hyperostosis corticalis generalisata)

Box 4-18   Multiple Osteolytic Lesions: Differential Diagnosis

- Polyostotic fibrous dysplasia
- Enchondromatosis
- Langerhans cell histiocytosis
- Multiple myeloma, metastatic disease
- Brown tumors (hyperparathyroidism)
- Infection
- Cystic angiomatosis

Box 4-19   McCune-Albright Syndrome: Important Radiographic Features

- Polyostotic involvement of multiple bones with tendency toward unilateral predominance
- Medullary portions of long bones have homogeneous, ground-glass appearance that blends with cortical bone without sharp margins
- Medullary trabeculae are frequently obliterated; the endosteal cortex is thinned and scalloped
- Areas of endosteal sclerosis may produce a multilocular appearance
- Diaphyses are widened or expanded, with large lesions covered by a thin shell of bone
- Bone deformity is common, including coxa vara (shepherd crook)
- Diaphyses and metaphyses are most commonly involved; epiphyseal involvement may occur in children
- Cystic lesions in the outer table of the calvarium are surrounded by a sclerotic margin
- Sclerosis involves the skull base, sphenoid ridges, and facial bones (leontiasis ossea)

Box 4-20   Cherubism: Important Features

- Autosomal dominant trait, with complete penetrance in male individuals and 50% to 70% penetrance in female individuals
- Four phases: presentation (first few years of life), progression (early childhood), quiescence (puberty, adolescence), regression (adulthood)
- Mandible most frequently affected; maxilla never involved alone
- Bilateral symmetric involvement

Continued

Box 4-20—cont'd

- Multilocular cystlike lucencies with thin peripheral rims of cortical bone
- Occasional resorption of tooth roots
- Second and third molars often absent; other teeth displaced, transposed, impacted
- Histologic characteristics: cellular fibroblastic tissue resembling that seen in fibrous dysplasia, prominent vascularity, multinucleated giant cells

Desmoplastic fibroma This is a rare osseous lesion that is histologically identical to the soft-tissue fibromatoses.

PEARL

Desmoplastic fibroma typically involves the iliac bones, mandible, humerus, femur, tibia, and scapula in the second and third decades of life as an aggressive osteolytic process.

1. About 50% of lesions appear during the second decade of life.
2. Soft-tissue extension generally does not occur.
3. The lesion is often difficult to distinguish radiographically and histologically from a well-differentiated fibrosarcoma.
4. Although size is variable, most lesions exceed 5 cm in diameter.
5. The long bones are most commonly affected, particularly the distal femur and proximal tibia.
6. Most lesions are central and metaphyseal in location.
7. The pelvis and mandible are relatively common sites of involvement.
8. Major differential diagnostic considerations include giant cell tumor, aneurysmal bone cyst, and fibrosarcoma.
9. Extensive cryosurgery and curettage is the treatment of choice (Fig. 4-20).
10. Local recurrence is very common.
11. Radiographic appearance does not reliably predict tumor behavior.
12. Computed tomography or magnetic resonance imaging is indicated as part of the workup of the lesion.
13. Desmoplastic fibroma is a monostotic lesion.
14. Local recurrence occurs slowly, usually 1 to 2 years following initial therapy.
15. Wide excision is the treatment of choice for recurrent lesions.

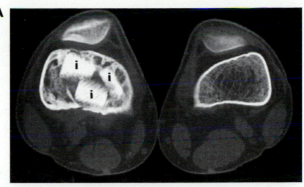

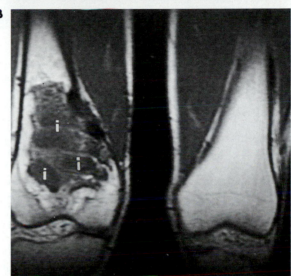

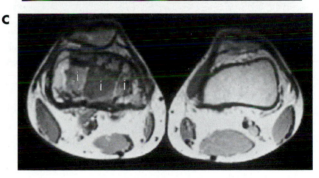

Fig. 4-20 Monitoring of bone graft incorporation with cross-sectional imaging. Computed tomographic image (**A**) depicts implanted bone graft material (*i*) within a large distal femoral metaphyseal defect resulting from prior curettage of a desmoplastic fibroma. Intermediate signal intensity within the implanted graft material (*i*) on coronal (**B**) and axial (**C**) T1-weighted MR images is indicative of early bone marrow ingrowth.

Cortical or periosteal desmoid (avulsive cortical irregularity) The cortical or periosteal desmoid is not a true desmoid tumor, but is characterized by proliferation of fibroblastic tissue. The lesion is posttraumatic and occurs at the insertion of the adductor magnus muscle. The radiographic features include exuberant periostitis, soft-tissue mass, and cortical erosion, which may simulate the features of a malignant neoplasm. The lesion may be misdiagnosed both histologically and radiographically as osteosarcoma, particularly the periosteal variety. The peak age range of detection is 15 to 20 years. The lesion is typically found on the posteromedial cortex of the distal femur adjacent to the medial femoral condyle, although similar posttraumatic abnormalities can be found at entheses in other portions of the skeleton.

PEARL

Cortical desmoid represents a benign posttraumatic tumorlike alteration of the periosteum characterized by fibroblastic proliferation analogous to that found in desmoplastic fibroma. It is usually seen in patients between the ages of 15 and 20 years. Almost all cases are localized to the posterior medial aspect of the distal femur at the adductor magnus insertion. Desmoid tumor of soft tissue is a true neoplasm characterized by benign abdominal and extraabdominal fibrous tissue proliferation. Desmoid tumors of soft tissue are associated with Gardner syndrome. Desmoplastic fibroma is an expansile lytic lesion of bone.

Soft-tissue fibromatoses This is a heterogeneous group of locally infiltrative lesions manifesting similar histologic characteristics. Important features of specific types of fibromatosis are as follows:

1. Desmoid tumor, infantile dermal fibromatosis, juvenile aponeurotic fibroma: Recurrence after excision is common.
2. Juvenile aponeurotic fibroma: This is a slowly infiltrative lesion arising in the aponeurotic tissue of the palms, wrists, and soles (Fig. 4-21).
3. Desmoid tumor: This lesion is also known as desmoid fibromatosis, aggressive fibromatosis, or desmoid-type grade I fibrosarcoma.
4. Congenital generalized fibromatosis, congenital multiple fibromatosis: The condition develops in utero.
5. Desmoid tumor: The lesion is painless and infiltrative, arising in abdominal or extraabdominal muscle.
6. Infantile dermal fibromatosis: The lesion infiltrates the extensor surfaces of digits.
7. Infantile dermal fibromatosis: The lesion always appears at 1 to 2 years of age.
8. Juvenile aponeurotic fibroma: Calcification may occur, especially in the interosseous membrane of the distal forearm.
9. Desmoid tumor, infantile dermal fibromatosis, juvenile aponeurotic fibroma: Bone erosion is rare.
10. Desmoid tumor: The lesion rarely metastasizes despite being histologically similar to low-grade fibrosarcoma.
11. Congenital generalized fibromatosis: The condition is uniformly fatal within several months.
12. Infantile dermal fibromatosis: The lesion manifests itself as multiple firm nodules attached to the skin, fascia, tendons, and periosteum.

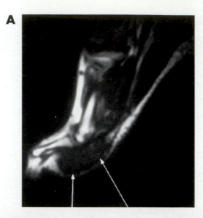

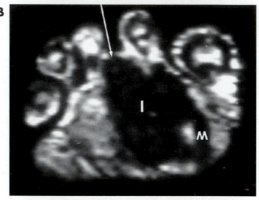

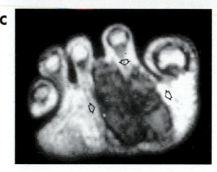

Fig. 4-21 Aggressive fibromatosis arising from the plantar aponeurosis (juvenile aponeurotic fibroma). **A,** Sagittal T1-weighted MR image demonstrates a well-circumscribed soft-tissue mass of low signal intensity (*arrows*) beneath the metatarsophalangeal joints. **B,** Transaxial T1-weighted image documents involvement of the medial (*M*) and intermediate (*I*) plantar compartments with extension to the dorsum of the foot (*arrow*). **C,** Corresponding T2-weighted image reveals inhomogeneous intermediate signal intensity within the lesion (*arrows*). Differential diagnosis should include consideration of malignant fibrous histiocytoma, fibrosarcoma, desmoid tumor, and other soft-tissue neoplasms of primarily fibrous composition.

13. Desmoid tumor, infantile dermal fibromatosis, juvenile aponeurotic fibroma, congenital generalized fibromatosis, congenital multiple fibromatosis: The lesion often infiltrates through soft-tissue compartmental barriers.

14. Desmoid tumor: Prominent frondlike osseous excrescences may emanate from a stimulated periosteum and radiate into the soft-tissue mass.

15. Juvenile aponeurotic fibroma: The lesion manifests itself as a painless soft-tissue mass that is generally under 4 cm in length.

16. Desmoid tumor: The lesion may be indolent for long periods of time.

17. Congenital generalized fibromatosis: Disseminated fibromatosis involves much of the viscera and musculature.

18. Desmoid tumor, infantile dermal fibromatosis, juvenile aponeurotic fibroma, congenital generalized fibromatosis, congenital multiple fibromatosis: The lesion often appears in children.

19. Desmoid tumor: Aggressive infiltration of adjacent muscles, tendons, nerves, and blood vessels occurs.

20. Congenital multiple fibromatosis: Disseminated fibromatosis involves much of the musculature but spares the viscera.

21. Desmoid tumor, infantile dermal fibromatosis, juvenile aponeurotic fibroma, congenital generalized fibromatosis, congenital multiple fibromatosis: No pseudocapsule is evident on computed tomographic or magnetic resonance imaging studies.

22. Congenital generalized fibromatosis, congenital multiple fibromatosis: Lytic metaphyseal osseous lesions occur.

PEARL

Fibrous tumors:
- Ossifying fibroma causes convex anterior tibial bowing.
- Desmoplastic fibroma never metastasizes but is locally aggressive.
- Juvenile aponeurotic fibroma typically occurs in the hand and wrist, or foot.
- Infantile myofibromatosis manifests itself as lytic lesions in the metaphyses of long bones; the condition occurs as a lethal generalized form and as an indolent diffuse form.
  - Type I (congenital generalized fibromatosis) involves bone and organs and manifests itself at birth.
  - Type II (congenital multiple fibromatosis) involves muscle and bone with metaphyseal lytic lesions and tends to be bilaterally symmetric; viscera and spared.

Fatty tumors

Xanthomatoses   Xanthomatoses is a heterogeneous group of conditions histologically characterized by foam cell deposition and frequently associated with altered lipid or cholesterol metabolism.

1. Tendinous xanthomas commonly occur around the fingers, heel, elbow, and knee and can erode subjacent bone.

2. Calcification occurs in 20% to 25% of tendinous xanthomas.

3. Intraosseous xanthomas produce nonspecific well-defined radiolucent lesions.

4. Fibrous xanthomas or xanthofibromas represent small thickenings of the corium covered by an intact epidermis.

5. Some xanthomas are associated with metabolic and endocrine disorders, such as diabetes mellitus and hypercholesterolemia.

6. Malignant xanthomatous lesions include xanthosarcoma and fibrous xanthosarcoma.

PEARL

Xanthoma of the tendon sheath is usually associated with multicentric reticulohistiocytosis and familial hyperlipidemia Type II. Diseases associated with tendon sheath xanthomas include hyperlipidemias, hypercholesterolemia, lipase deficiency, pseudoxanthoma elasticum, malignant disease of the reticuloendothelial system (multiple myeloma, lymphoma), and multicentric reticulohistiocytosis or lipoid dermatoarthritis. In the last condition, multiple cutaneous xanthomas are associated with secondary arthritis and hyperlipidemia.

Intraosseous lipoma (Figs. 4-22, 4-23) Intraosseous lipoma is a rare fat-containing lesion of bone that is usually asymptomatic. Computed tomography or magnetic resonance imaging is usually reliable in establishing the diagnosis (Figs. 4-24, 4-25, 4-26). The intramedullary form of the lesion is predominantly lytic and may have a thin well-defined sclerotic margin, a nonaggressive appearance, and a central calcified nidus. The lesion occurs in the metaphyses of long bones including the fibula. Particular sites of predilection include the proximal femur and anterior third of the calcaneus. The differential diagnosis of such lesions, which occur between major trabecular groups, should include consideration of fibrous dysplasia and solitary bone cyst. The epicenter of an intraosseous lipoma is occasionally parosteal, in which case the radiographic findings may include periosteal hyperostosis, a fat-density soft-tissue mass adjacent to the cortex, and

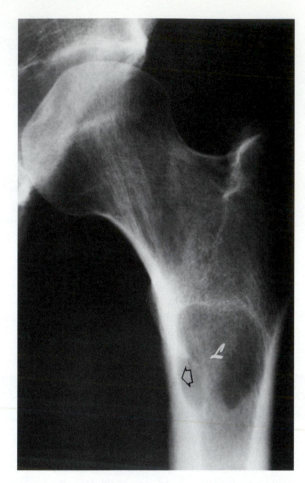

Fig. 4-22   Intraosseous lipoma of the proximal femur. A well-defined lytic lesion (*L*) with endosteal scalloping (*arrow*) is present in the subtrochanteric region. Differential diagnosis should include consideration of fibrous dysplasia, solitary bone cyst, eosinophilic granuloma, metastatic disease, and indolent infection.

A

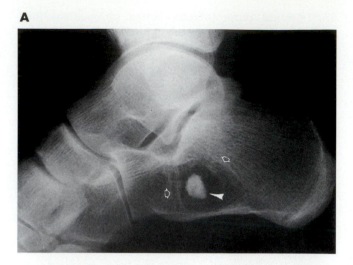

B

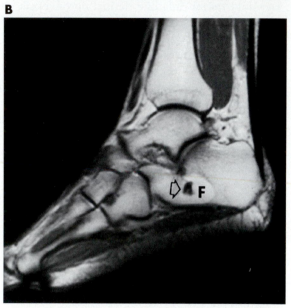

Fig. 4-23   Intraosseous lipoma of the calcaneus. **A,** Lateral radiograph demonstrates a typical well-marginated lytic lesion (*arrows*) with a radiodense center (*arrowhead*) in the anterior third of the bone. **B,** Sagittal T1-weighted MR image confirms the fatty (*F*) and calcific (*arrow*) composition of the lesion. Simple bone cyst and fibrous dysplasia also occur in this location but lack the radiodense center of many lipomas.

radiating spiculated periostitis. Pathologic fracture is not a common complication of the lesion.

PEARL

Intraosseous lipoma is characterized by a well-defined radiolucent lesion with or without septation, calcification, or ossification. It may affect any bone at any age in either sex, and is rare in the spine. It is considered a true neoplasm as opposed to a degenerative process secondary to trauma or infection.

Soft-tissue lipoma   Soft-tissue lipoma is a common lesion composed of fatty tissue. Approximately 80% of lesions occur in the subcutaneous tissue, with the remainder lying either within or between muscles. Rarely, the lesion may contain metaplastic cartilage with calcification. Lipomas typically manifest themselves as asymptomatic, soft, movable, and compressible masses. Multiple lesions occur in 5% of cases.

Conventional radiography or computed tomography reveals a well-encapsulated mass with density diagnostic of its fatty histologic composition. With magnetic resonance imaging, the lesion exhibits signal intensity higher than that of fluid on T1-weighted images (Figs. 4-27, 4-28).

Lipomatosis   Lipomatosis is either a localized or generalized disorder. Macrodystrophia lipomatosa is a localized condition characterized by overgrowth of both

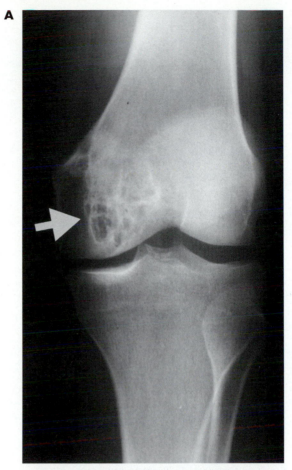

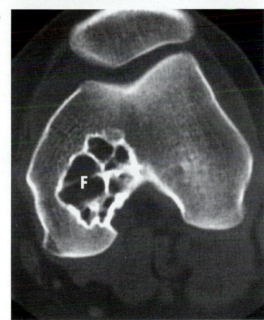

Fig. 4-24   Intraosseous lipoma of the distal femur. **A,** Frontal radiograph reveals a well-defined lytic and septated lesion (*arrow*) in the epiphysis. Differential diagnosis for lytic epiphyseal lesions should include consideration of chondroblastoma, giant cell tumor, intraosseous ganglion, clear cell chondrosarcoma, infection, and (rarely) eosinophilic granuloma. **B,** CT eliminates these possibilities by documenting low-density fat (*F*) within the lesion.

bone and soft tissues; it usually involves a hand or foot. Surgical management, the treatment of choice, is difficult. Regional gigantism can also be associated with neurofibromatosis, Wilms tumor, Klippel-Trenaunay-Weber syndrome, and Beckwith-Wiedemann syndrome. Multiple lipomas distributed either randomly or symmetrically over the body are a congenital form of lipomatosis.

PEARL

Macrodystrophia lipomatosa is a rare congenital malformation characterized by increased size of all histologic elements of a digit or extremity, particularly fat.

Lipoblastomatosis   Lipoblastomatosis is an embryonal adipose neoplasm that occurs in children. It is benign and simulates liposarcoma histologically. Recurrence is common following surgical excision. The peak age of occurrence for this tumor is younger than that for liposarcoma.

Giant cell lesions

Giant cell tumor of bone (osteoclastoma)   This is an uncommon lesion characterized histologically by multinucleated giant cells on a fibroid stromal background (Box 4-21).

PEARL

Giant cells (osteoclasts) can be seen in giant cell tumor, brown tumor of hyperparathyroidism, aneurysmal bone cyst, chondroblastoma, fibrous dysplasia, giant cell reparative granuloma, ossifying fibroma, osteosarcoma, chondromyxoid fibroma, and osteoblastoma.

---

1. The lesion constitutes approximately 5% of all primary osseous neoplasms.
2. The lesion nearly always occurs after physeal closure, with approximately 85% of cases appearing after age 20.
3. Cortical disruption with soft-tissue extension occurs in approximately 25% of patients (Fig. 4-29).
4. In tubular bones, the lesion usually begins eccentrically in the metaphysis and may later extend into the epiphysis or apophysis (Fig. 4-30).

A

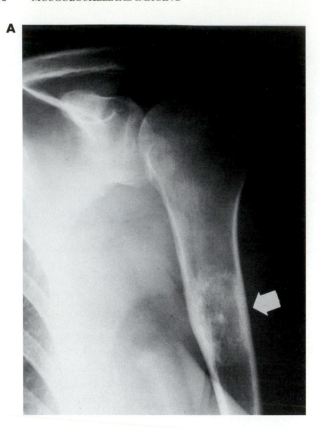

B

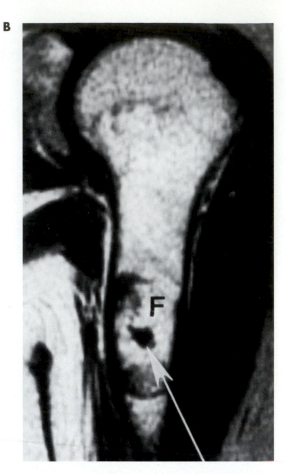

Fig. 4-25   Intraosseous lipoma. **A,** Radiography demonstrates a mixed-density lesion (*arrow*) in the humeral diaphysis. **B,** T1-weighted coronal MR image confirms the diagnosis by demonstrating fat (*F*) and a calcified center (*arrow*) within the lesion.

PEARL

An apophysis is a secondary ossification center that occurs at the growth site of a nonarticular bony prominence for tendon or ligament insertion. Examples are the apophyses of the calcaneus, tibial tubercle, and greater femoral trochanter.

5. Approximately 65% of lesions affect the distal femur, proximal tibia, distal radius, or distal ulna.
6. Histologically, the lesion resembles the brown tumor of hyperparathyroidism.
7. In the spine, the lesion often involves the sacrum or a vertebral body, but is rare in the posterior elements.
8. Approximately 15% of lesions exhibit malignant behavior, with metastases to the lungs.
9. Benign and malignant giant cell tumors cannot be reliably distinguished either radiographically or histologically (Fig. 4-31).

10. Multicentric involvement occurs, particularly in the skull and facial bones affected by Paget disease.

PEARL

Multicentric giant cell tumors are rare and difficult to distinguish from metastatic disease secondary to a solitary lesion. If associated with Paget disease, giant cell tumors are found in the skull, mandible, maxilla, innominate bone, and sometimes tibia. About 15% of all giant cell tumors are malignant.

11. Prognosis following surgical resection of pulmonary metastases is frequently good, with a mortality rate of approximately 25%.
12. Approximately 50% of lesions recur following curettage and bone grafting.
13. Only 10% to 15% of lesions recur following wide en bloc resection.

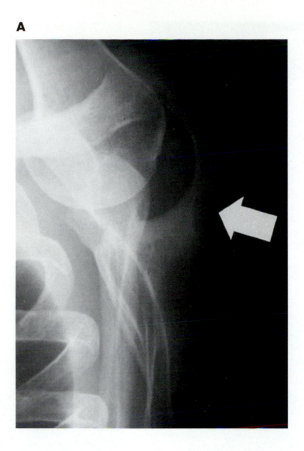

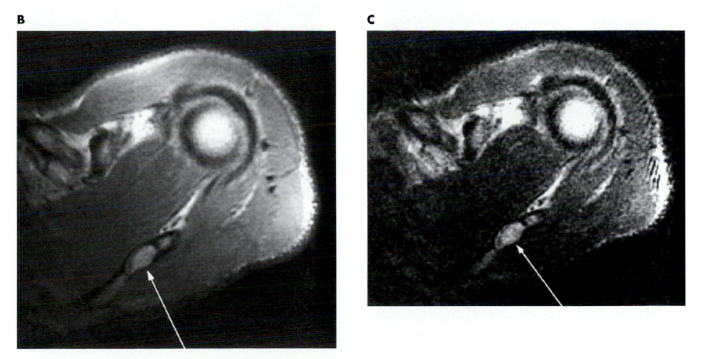

Fig. 4-26   Intraosseous lipoma. **A,** Tangential radiograph of the scapula demonstrates a well-defined lytic lesion (*arrow*) in the acromion. **B** and **C,** Transaxial proton density– and T2-weighted MR images confirm the diagnosis by documenting signal behavior identical to that of normal marrow fat (*arrows*).

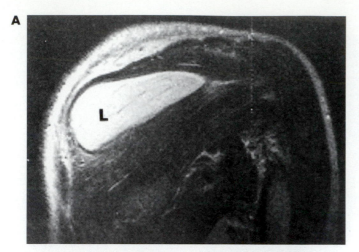

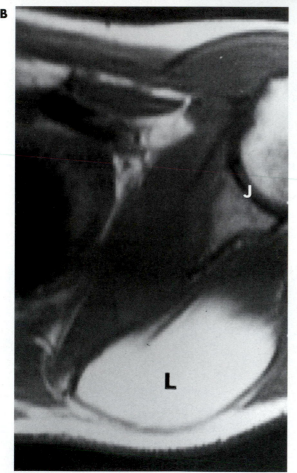

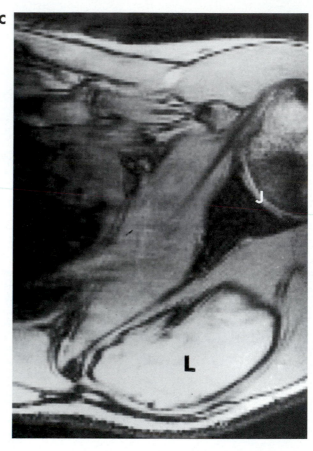

Fig. 4-27   Soft-tissue lipoma (*L*) posteromedial to the glenohumeral joint (*J*) demonstrated by MRI. **A,** Sagittal T1-weighted image. **B,** Transaxial T1-weighted image. **C,** Transaxial gradient-echo image. The lesion exhibits homogeneous high signal intensity identical to that of subcutaneous fat and occupies a common location.

14. Radiation therapy for giant cell tumor yields an unacceptably low cure rate of approximately 50%.
15. Radiation-induced sarcomas develop in 10% to 15% of irradiated giant cell tumors, resulting in extremely high mortality rates.
16. Other therapeutic options include curettage with cryosurgery, wide metaphyseal resection with subarticular curettage, joint resection with fusion, and total joint replacement.

17. Lesions in small tubular bones tend to be central rather than eccentric in location (Fig. 4-32).
18. Flat bones tend to be involved in epiphyseal-equivalent areas such as the acetabulum.
19. Periosteal reaction does not usually occur in the absence of pathologic fracture.
20. Chondroblastoma usually occurs in skeletally immature patients, in contrast to giant cell tumor.

A

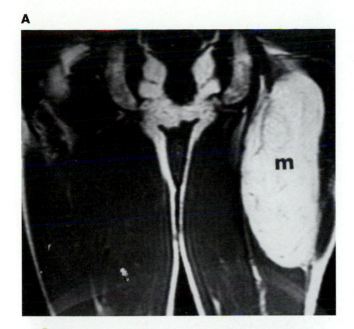

B

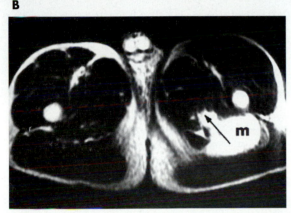

A

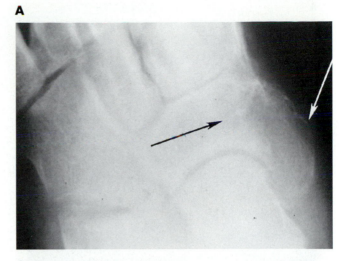

B

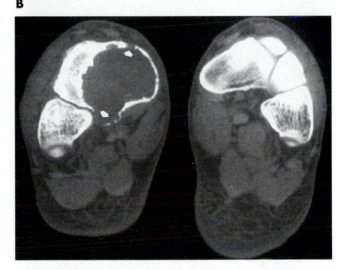

Fig. 4-28   Soft-tissue lipoma of the thigh. **A,** Coronal T1-weighted MR image documents a large relatively homogeneous fat-containing mass (*m*) in the posterior compartment. **B,** Trans-axial T2-weighted image demonstrates predominant displacement of the adductor magnus muscle (*arrow*) by the mass (*m*). Differential diagnosis should include consideration of well-differentiated liposarcoma.

Fig. 4-29   Giant cell tumor. **A,** Oblique radiograph reveals a bubbly, expansile lytic lesion (*arrows*) arising from the navicular tuberosity. **B,** CT documents the extent of the process and demonstrates cortical disruption in several areas (*arrows*). Differential diagnosis should include consideration of aneurysmal bone cyst, indolent infection, metastatic disease (renal and thyroid primaries in particular), and plasmacytoma.

21. Giant cell tumors are often large (up to 5 cm) at the time of diagnosis.
22. The differential diagnosis for flat bone or spinal lesions should include consideration of metastatic disease, plasmacytoma, osteoblastoma, aneurysmal bone cyst, and chordoma (spinal only).
23. Cross-sectional imaging studies may demonstrate fluid-fluid levels within giant cell tumors because of internal hemorrhage or associated aneurysmal bone cyst elements (Fig. 4-33).
24. Cross-sectional imaging studies may be indicated for preoperative planning (Fig. 4-34).
25. The lesion is also known as osteoclastoma.
26. Pathologic fracture is a recognized complication of the tumor.

PEARL

Giant cell tumor:
- Chondroblastoma, giant cell tumor, and Paget disease all contain large numbers of giant cells.
- Giant cell tumors are usually hypervascular.
- Giant cell tumors can be seen in Paget disease, but are not common.
- Eighty-five percent of chondroblastomas occur between the ages of 5 and 25 years, whereas giant cell tumors are predominately seen in 20- to 40-year-old patients (15% occur in those under age 20).

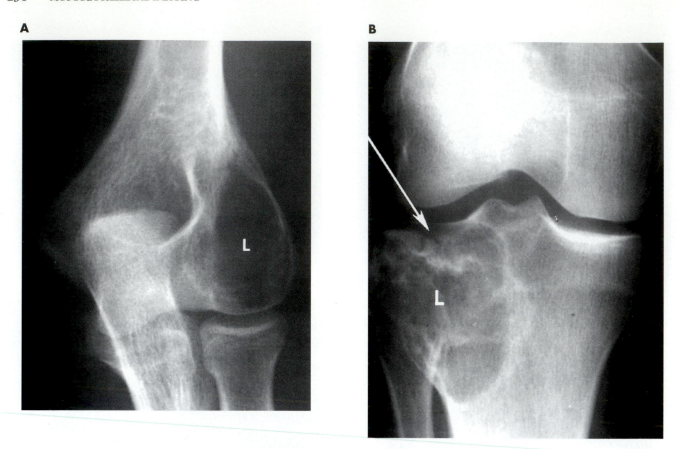

Fig. 4-30   Giant cell tumor is characteristically an osteolytic expansile lesion centered in the metaphysis with epiphyseal extension in a skeletally mature young adult. **A,** Distal humeral lesion (*L*) with capitellar extension. **B,** Proximal tibial lesion (*L*) with pathologic fracture (*arrow*).

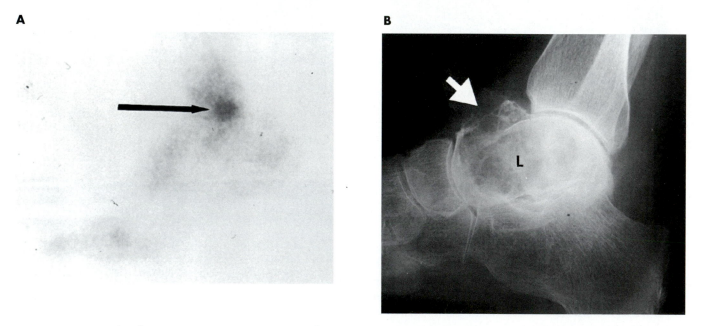

Fig. 4-31   Malignant giant cell tumor of the talus. **A,** The lesion exhibits abnormal increased activity (*arrow*) on a radionuclide bone scan. **B,** Lateral radiograph demonstrates an expansile septate lesion (*L*) involving most of the bone, with cortical disruption along the dorsal aspect of the neck and head (*arrow*). Differential diagnosis should include consideration of aneurysmal bone cyst, metastatic disease (particularly from renal and thyroid primaries), plasmacytoma, and indolent infection.

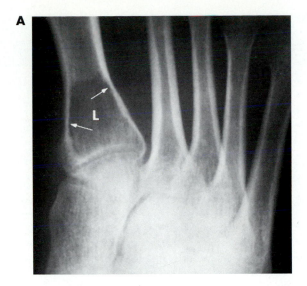

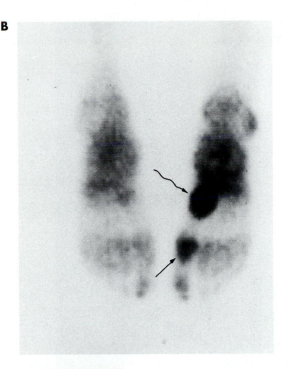

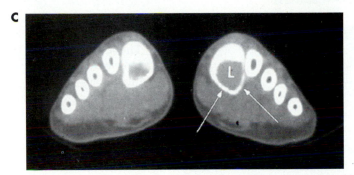

Fig. 4-32   Giant cell tumor. **A,** A well-defined lytic lesion (*L*) with endosteal scalloping (*arrows*) is present in the first metatarsal base. The lesion is centered in the metaphysis but extends into the epiphysis. **B,** Positive uptake is present both in the lesion (*wavy arrow*) and at a site of hallux rigidus (*straight arrow*) on a radionuclide bone scan. **C,** CT optimally demonstrates cortical thinning (*arrows*) and expansion (*L = lesion*). **D** and **E,** The lesion (*arrow in* **D,** *L in* **E**) exhibits low signal intensity on T1-weighted MR images in the plantar and sagittal planes. **F,** Sagittal T2-weighted image reveals high signal intensity within the lesion (*L*). Differential diagnosis should include consideration of enchondroma, giant cell reparative granuloma, indolent infection, fibrous dysplasia, and brown tumor of hyperparathyroidism.

Continued

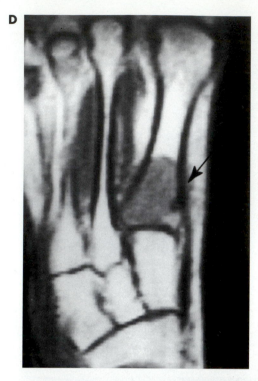

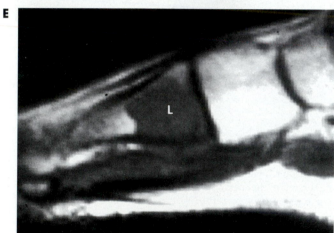

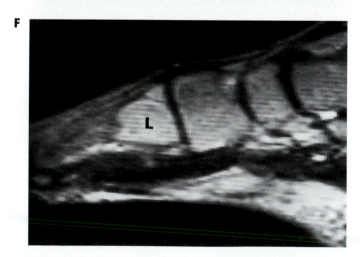

Fig. 4-32, cont'd. For legend see p. 235.

Giant cell tumor of tendon sheath   This is a proliferative disorder of synovium, the etiology of which is uncertain (neoplastic versus reactive).

1. The lesion is usually painless.
2. Most patients are between 30 and 50 years of age at initial diagnosis.
3. The lesion grows slowly.
4. The finger is the most common site of involvement (Fig. 4-35).
5. The localized soft-tissue mass is not typically centered around a joint.
6. Pressure erosion of bone occurs in only approximately 10% of cases.
7. Approximately 30% of lesions will recur following surgical resection.
8. The lesion is histologically identical to intraarticular pigmented villonodular synovitis.

Bone cysts

Simple bone cyst   This is a common fluid-filled bone lesion of childhood (Box 4-22).

1. The lesion is also known as solitary or unicameral bone cyst.
2. Soft-tissue extension does not occur.
3. Approximately half of all lesions occur in the proximal humerus.
4. The lesion begins in the metaphysis adjacent to the open physis, but later migrates to a diaphyseal location.
5. The lesion is usually central as opposed to eccentric in location.
6. Approximately 20% of lesions occur in the proximal femur.

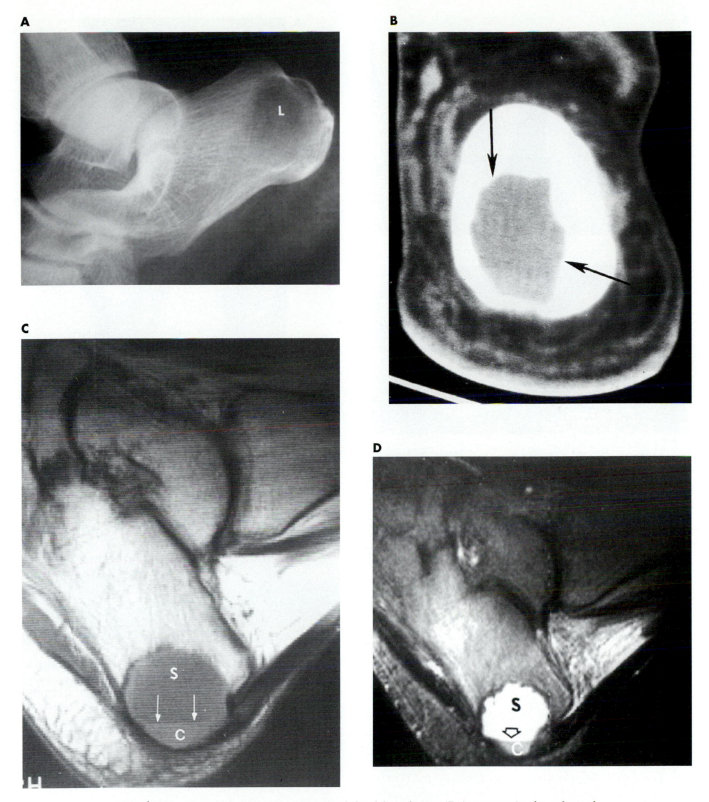

Fig. 4-33   Giant cell tumor. **A,** A poorly defined lytic lesion (*L*) is present in the calcaneal tuberosity. **B,** CT demonstrates a homogeneous tissue-density process (*arrows*) without radiodense matrix. **C,** Sagittal T1-weighted MR image reveals a fluid-fluid level (*arrows*) within the lesion, indicating a hematocrit effect (*S = serum, C = cells*). **D,** The finding (*arrow*) is confirmed on a sagittal T2-weighted image, where the serum (*S*) manifests high signal intensity (*C = cells*). Fluid-fluid levels indicate a blood-filled lesion, the differential diagnosis of which should include consideration of aneurysmal bone cyst, hemorrhagic metastasis, and other vascular tumors.

A

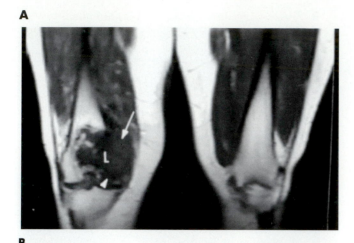

B

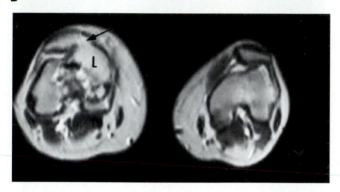

A

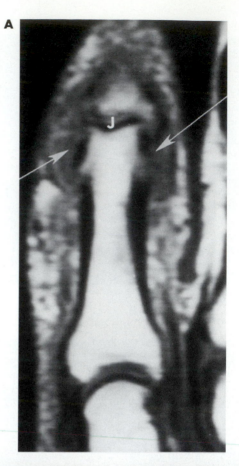

B

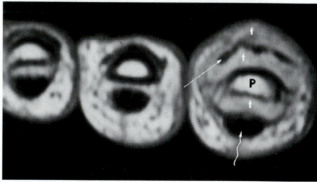

Fig. 4-34   Giant cell tumor of the distal femur. **A,** Coronal T1-weighted MR image demonstrates a metaphyseal lesion with epiphyseal and soft-tissue extension, manifesting low signal intensity. **B,** Transaxial T2-weighted image reveals predominantly high signal intensity within the lesion (*L*), which has extended into the patellofemoral joint (*arrow*). Differential diagnosis should include consideration of metastatic disease, plasmacytoma, clear cell chondrosarcoma, and indolent infection.

Fig. 4-35   Giant cell tumor of tendon sheath. **A,** Coronal T1-weighted MR image reveals a low–signal intensity mass (*arrows*) surrounding the proximal interphalangeal joint (*J*) of the affected finger. **B,** Transaxial proton density–weighted image demonstrates tumor infiltration (*t*) circumferentially around the bony phalanx (*P*) (*straight arrow = extensor tendon, wavy arrow = flexor tendon*). Differential diagnosis should include consideration of digital fibromatosis and synovial sarcoma.

7. Many lesions exceed 5 cm in size.
8. The lesion is asymptomatic in the absence of pathologic fracture.
9. Recurrence is twice as likely among patients under 10 years of age as compared with older individuals.
10. The lesion rarely occurs in adults.
11. Lesions occurring in older individuals tend to be found in unusual sites, such as the calcaneus or iliac wing.
12. Periosteal reaction does not occur in the absence of pathologic fracture.
13. The fallen fragment sign of pathologic fracture is important in that it indicates the fluid-filled nature of the lesion.
14. The radiographic differential diagnosis of the lesion should include consideration of aneurysmal bone cyst, eosinophilic granuloma, and fibrous dysplasia.
15. The documented recurrence rate following curettage with bone grafting is 35% to 50%.
16. Following multiple pathologic fractures, a lesion will occasionally heal spontaneously.
17. Intralesional injection of steroids is associated with a lower recurrence rate than curettage and bone grafting.

18. Cryosurgery is associated with a lower recurrence rate than curettage and bone grafting.
19. Accelerated or arrested growth may complicate surgery performed on a lesion that abuts the physeal plate.
20. Diaphyseal lesions tend to be less active and less prone to recurrence than those that fail to migrate away from the physis.
21. Both simple and aneurysmal bone cysts may appear multilocular; location (central versus eccentric) and degree of cortical expansion are more reliable distinguishing characteristics.

Box 4-22 Unicameral Bone Cyst: Important Features

- Peak age of incidence is 5 to 15 years
- Sex incidence is equivalent
- Contains straw-colored fluid with a thin lining membrane
- Occurs in a growing bone
- Diametaphyseal thinning of cortex is observed
- The lesion is moderately expansile
- Periosteal reaction occurs if fractured
- A well-defined margin is characteristic
- Humerus, femur, tibia, and pelvis are frequent sites
- "Fallen fragment" sign is diagnostic
- Curettage is frequently indicated

Aneurysmal bone cyst This is a vascular lesion composed of cystic blood-filled cavities.

PEARL

Aneurysmal bone cyst: Differential diagnosis
- Aneurysmal bone cyst is typically a metaphyseal lesion of the first and second decade. It is frequently a blown-out lesion with a soap bubble appearance. The cavity is filled with blood, and there are giant cells around the vascular channels.
- Fibrous dysplasia is often a metaphyseal lesion of the first and second decades. It can have a ground-glass appearance or manifest a lytic or blastic appearance. The monostotic form is most likely to be confused with aneurysmal bone cyst.
- Enchondroma is frequently a tumor that develops in the medullary cavity of the long or short tubular bones. It is seen in the second to fourth decades with equivalent sex incidence.
- Nonossifying fibroma may occasionally appear expansile and resemble an aneurysmal bone cyst.

---

1. The condition is associated with a preexisting osseous lesion in 30% to 50% of cases.

2. The lesion has been observed concomitantly with nonossifying fibroma, fibrous dysplasia, osteoblastoma, chondroblastoma, and giant cell tumor.

PEARL

Aneurysmal bone cysts arise de novo as a distinct non-neoplastic primary bone lesion (most commonly) or within a preexisting lesion such as giant cell tumor (about 50% of cases), osteoblastoma, chondroblastoma, nonossifying fibroma, fibrous dysplasia, or, occasionally, telangiectatic osteosarcoma. The lesion expands and causes extreme thinning of the cortex. It does not usually cross the physis unless it arose in a preexisting lesion.

---

3. One theory concerning the pathogenesis of aneurysmal bone cyst is that trauma induces an intraosseous vascular anomaly.
4. Approximately 70% of affected patients are between the ages of 5 and 20 years at the time of diagnosis.
5. The lesion is typically lytic, expansile, and eccentric in location (Fig. 4-36).
6. Rapid enlargement, aggressive recurrence, or pathologic fracture may elicit periosteal reaction.
7. Although the appearance is usually nonaggressive, rapidly enlarging or recurrent lesions may appear aggressive (Fig. 4-37).
8. Common sites of involvement include the pelvis, the posterior elements of the spine, and long-bone metaphyses (Fig. 4-38).

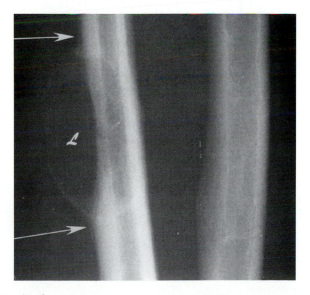

Fig. 4-36 Aneurysmal bone cyst. A markedly expansile lesion (*L*) with periosteal reaction (*arrows = Codman triangles*) arises from the distal ulna. Differential diagnosis should include consideration of metastatic disease (particularly from renal and thyroid primaries), plasmacytoma, and indolent infection.

A

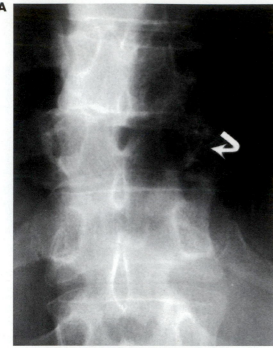

B

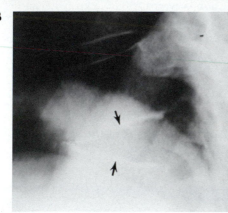

Fig. 4-37   Aneurysmal bone cyst. **A** and **B,** Frontal and lateral radiographs reveal expansile destruction of a lower thoracic vertebra (*curved arrow*), associated with collapse and severe kyphosis (*straight arrows*). Differential diagnosis should include consideration of metastatic disease, fungal infection, plasmacytoma, eosinophilic granuloma, and chordoma.

PEARL

Aneurysmal bone cyst preferentially affects the posterior elements when the spine is involved. These lesions can be seen in the first through third decades of life. Sex incidence is equal. Aneurysmal bone cyst exhibits a predilection for the spine, innominate bone, and metaphyses of long tubular bones, although it may arise elsewhere. The condition is considered a nonneoplastic process.

---

9. The associated underlying lesion occasionally is a telangiectatic osteosarcoma or other sarcoma.
10. Fluid-fluid levels may be observed within the lesion by the use of computed tomography or magnetic resonance imaging (Fig. 4-39).
11. Osteoblastoma is the major differential diagnostic consideration for spinal lesions.
12. Pelvic and long-bone lesions may resemble fibrous dysplasia or nonossifying fibroma radiographically.

PEARL

Lesions frequently confused radiographically with aneurysmal bone cyst include giant cell tumor, hemophilic pseudotumor, enchondroma, fibrous dysplasia, nonossifying fibroma, metastases (renal cell or thyroid), certain sarcomas, and echinococcal disease of bone.

---

13. Approximately 50% of lesions will recur following curettage.
14. Radiation therapy is indicated for surgically inaccessible lesions.
15. Cryosurgery is often used as an adjunct to curettage for treatment.
16. Rapid enlargement of an aneurysmal bone cyst may be mistaken for a malignant tumor radiographically, but does not signify malignant transformation.

Miscellaneous bone cysts (Box 4-32)

Epidermoid cyst, intraosseous ganglion (Fig. 4-40), and the latent mandibular bone cyst of Stafne are additional cystlike osseous lesions. Epidermoid cysts occur among patients in their second to fourth decades and are usually found in the calvarium, phalanges, or toes. These lesions are usually associated with trauma.

Box 4-23   Mandibular Developmental Defect (Latent Bone Cyst): Important Features

- Prevalence: 1/1000
- Male predilection
- Located at posterior aspect of mandible near angle, below mandibular canal
- May contain normal salivary gland tissue (in close proximity to submandibular gland) demonstrable by sialography or CT-sialography
- Appears as an elliptical, ovoid, or rounded lucency with a well-defined sclerotic border
- Asymptomatic normal variant
- Larger lesions are observed in older patients

A

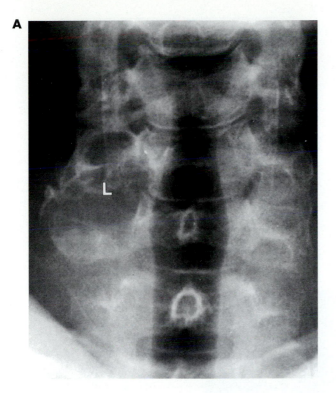

B

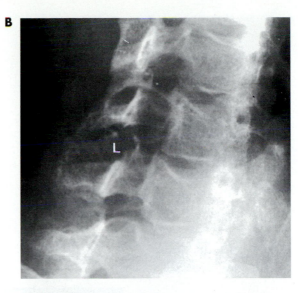

C

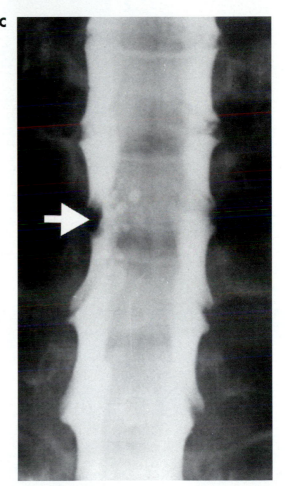

Fig. 4-38   Aneurysmal bone cyst arising from the posterior elements of the cervical spine. **A** and **B,** Frontal and oblique radiographs demonstrate expansile lytic areas involving the pillars and facet joints at C5-6. **C,** Myelography documents encroachment on the adjacent C6 nerve root (*arrow*). Differential diagnosis should include consideration of osteoblastoma, metastatic disease (particularly from renal and thyroid primaries), plasmacytoma, indolent infection, and tophaceous gout.

Myxomatoses   This is heterogeneous group of osseous and soft-tissue lesions possessing similar histologic characteristics.

1. Arthrography or tenography can outline communication between a soft-tissue ganglion and the underlying articular or tendinous structure (Fig. 4-41).

2. Myxomas of bone occur almost exclusively in the jaws.
3. Soft-tissue myxomas tend to invade striated muscle, have a tendency to recur, and have a rare malignant counterpart known as myxosarcoma.
4. Ganglia may arise either by synovial herniation or by tissue degeneration (Fig. 4-42).

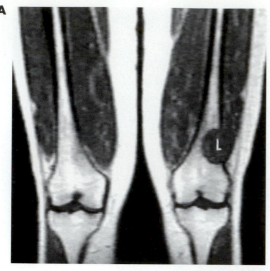

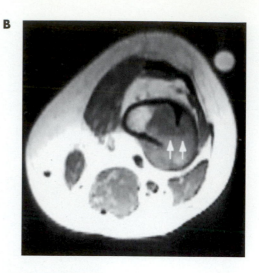

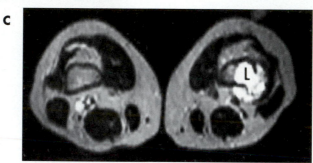

Fig. 4-39   Aneurysmal bone cyst. **A,** Coronal T1-weighted MR image reveals a well-defined lesion (*L*) of low signal intensity in the distal femoral metaphysis. **B,** Transaxial proton density–weighted image demonstrates the expansile nature of the process, which contains a fluid-fluid level (*arrows*). **C,** The lesion (*L*) exhibits high signal intensity on a corresponding T2-weighted image. Differential diagnosis should include consideration of hemorrhagic metastasis.

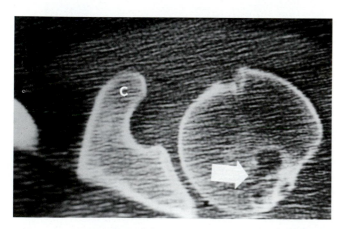

Fig. 4-40   Intraosseous ganglion of the proximal humerus. CT image reveals a lobulated, well-defined epiphyseal lesion of low density (*arrow*) in a skeletally mature patient without osteoarthritis. Other epiphyseal lesions include subchondral cyst or geode, chondroblastoma, giant cell tumor, and clear cell chondrosarcoma (*C = coracoid process*).

5. Imaging features of ganglia may include a soft-tissue mass, surface bony resorption, and periosteal new bone formation (Figs. 4-43, 4-44).

6. Soft-tissue myxomas are frequently solitary and rarely calcify.

7. Ganglia can arise directly from muscles, menisci, or articulations (Fig. 4-44).

8. Noncommunicating ganglia are more viscous and have a higher protein content than those which retain communication with an adjacent joint or tendon sheath (also known as synovial cysts).

9. The differential diagnosis of soft-tissue ganglion includes benign soft-tissue neoplasms such as neuromas (Fig. 4-45).

Langerhans cell disease (histiocytosis X)   Letterer-Siwe disease is the acute fulminant form of Langerhans cell histiocytosis, constituting approximately 10% of cases. The disease involves the skeleton, liver, spleen,

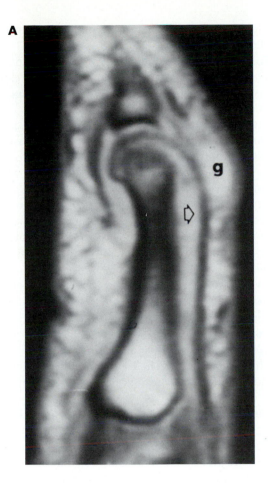

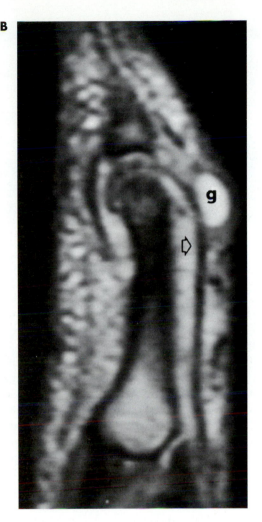

Fig. 4-41 Soft-tissue ganglion arising from digital extensor tendon sheath (*arrows*). Sagittal proton density–weighted (**A**) and T2-weighted (**B**) MR images demonstrate a well-circumscribed oval soft-tissue mass (*g*) that exhibits high signal intensity on both sequences.

lymph nodes, and skin. Most patients are under the age of 2 years. Radiographically, bone involvement may simulate Hand-Schüller-Christian disease, acute leukemia, eosinophilic granuloma, or neuroblastoma. The treatment of choice is chemotherapy. Most cases are fatal, although a minority of patients convert to the Hand-Schüller-Christian form and survive.

Box 4-24 Langerhans Cell Disease (Histiocytosis X): Important Features

- Lytic lesions originating in the medullary cavity of the skull, femur, or, most commonly, the spine; usually sharply defined (more rapidly growing lesions may have hazy borders)

Box 4-24—cont'd

- No periosteal reaction in flat bones; long bones may have extension into cortex with periosteal new bone formation (especially in children); usually no associated soft-tissue mass
- Predilection for anterior half of skull base; erosion of one or both tables; multiple large coalescent lesions produce "geographic skull"; rarely, small bony sequestrum remains in lytic area (button sequestration)
- Vertebral involvement may result in collapse (vertebra plana)
- Occasional diffuse marrow involvement causing generalized osteoporosis without focal lytic lesions

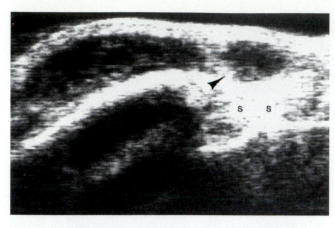

Fig. 4-42 Sonography of a plantar ganglion. Sagittal image documents a hypoechoic oval subcutaneous mass (*arrowhead*) with enhanced through-transmission of sound (*s*).

Hand-Schüller-Christian disease   Hand-Schüller-Christian disease is the chronic disseminated form of Langerhans cell histiocytosis, constituting fewer than 15% to 40% of cases. The disease affects the reticuloendothelial, genitourinary, and musculoskeletal systems as well as other visceral sites. The condition usually develops by the age of 5 to 10 years. Radiographically, osseous lesions are identical to those of eosinophilic granuloma. Although the prognosis is variable, morbidity rate is generally high and mortality rate is approximately 10%.

Eosinophilic granuloma
1. The lesion constitutes 60% to 80% of all cases of Langerhans cell histiocytosis.
2. Peak age of incidence is between 5 and 10 years.
3. Soft-tissue masses are common, particularly with skull lesions.
4. Most lesions range in size from 1 to 5 cm (Fig. 4-46).
5. The skull is involved in approximately half of all cases.
6. The ratio of axial skeleton to long-bone lesions is approximately 3:2.
7. The femur is the most commonly affected long bone, although any site can be involved (Fig. 4-47).
8. Vertebra plana is a pathologic fracture in a lesion involving a vertebral body.
9. The beveled edge sign is nonuniform involvement of the inner and outer tables by skull lesions.
10. The button sequestrum sign is a fragment of normal bone isolated centrally within a lytic lesion, usually in the calvarium.
11. Sclerosis is a frequent radiographic feature of healing lesions.
12. Periosteal reaction is relatively common.
13. Only 10% of patients develop polyostotic disease, usually within 6 months of the appearance of the initial lesion.
14. The lesion may progress more rapidly than osteomyelitis.
15. Eosinophilic granuloma is a painful lesion and occasionally may be associated with fever and elevated erythrocyte sedimentation rate, thus simulating infection.
16. The radiographic differential diagnosis should include consideration of metastatic disease, osteomyelitis, lymphoma, and Ewing sarcoma.
17. Eosinophilic granuloma may be multifocal without extraskeletal involvement.
18. Because a significant number of lesions fail to accumulate bone-seeking radiopharmaceuticals, radiographic skeletal survey is more reliable than scintigraphy for assessing polyostotic involvement (Fig. 4-48).
19. Eosinophilic granuloma is a self-limited condition that has been shown to undergo spontaneous healing in the absence of specific therapy.
20. Curettage and wide excision are associated with similar rates of healing and recurrence.
21. The healing rate of the lesion is similar for low-dose radiation therapy and intralesional injection of corticosteroids.
22. Partial reconstitution of height is the natural history of vertebra plana lesions.
23. The floating tooth sign has been described in eosinophilic granuloma, neuroblastoma, and a variety of other lytic lesions affecting the jaw bones.
24. The lesion may be geographic, permeative, or moth-eaten radiographically.
25. The lesion occurs with significant frequency among patients in the first through third decades of life.

Malignant Lesions

Bone-forming tumors

Classical osteosarcoma   This is an extremely aggressive neoplasm composed of malignant osteoid tissue. Histologically, approximately half of lesions are predominantly osteoid (osteoblastic type), whereas 25% are chiefly cartilaginous and 25% contain a preponderance of spindle cells (Box 4-25).
1. The lesion constitutes approximately 75% of all osteosarcomas.
2. Peak age of incidence is between 10 and 25 years.
3. Approximately 90% of long-bone lesions are metaphyseal, with the remaining 10% being diaphyseal.
4. In order of decreasing incidence, the three most common sites of involvement are the distal femur, proximal tibia, and proximal humerus (Fig. 4-49).
5. Radiographically, tumor bone matrix is evident in approximately 90% of cases.

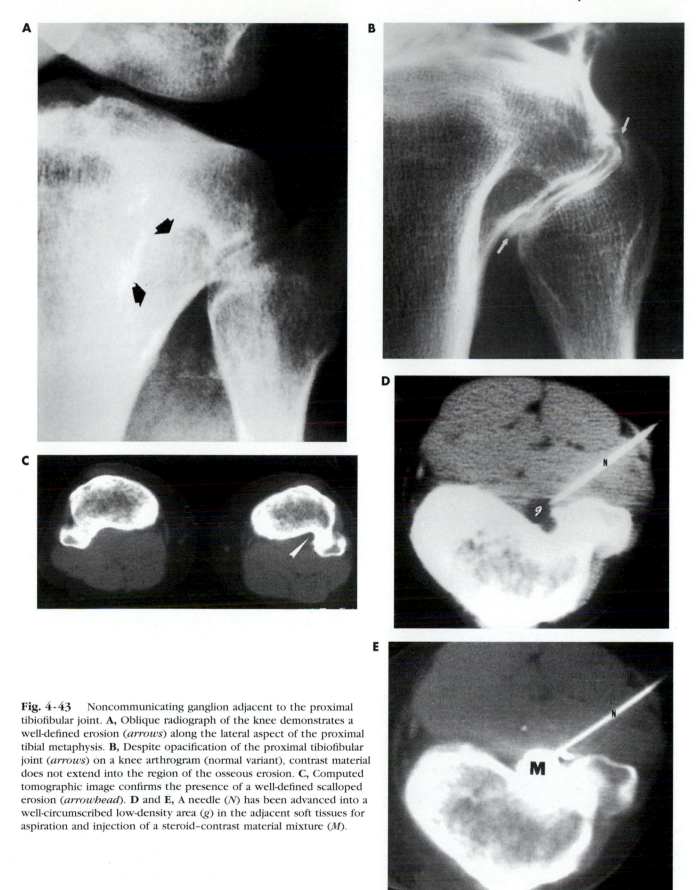

Fig. 4-43   Noncommunicating ganglion adjacent to the proximal tibiofibular joint. **A,** Oblique radiograph of the knee demonstrates a well-defined erosion (*arrows*) along the lateral aspect of the proximal tibial metaphysis. **B,** Despite opacification of the proximal tibiofibular joint (*arrows*) on a knee arthrogram (normal variant), contrast material does not extend into the region of the osseous erosion. **C,** Computed tomographic image confirms the presence of a well-defined scalloped erosion (*arrowhead*). **D** and **E,** A needle (*N*) has been advanced into a well-circumscribed low-density area (*g*) in the adjacent soft tissues for aspiration and injection of a steroid–contrast material mixture (*M*).

A

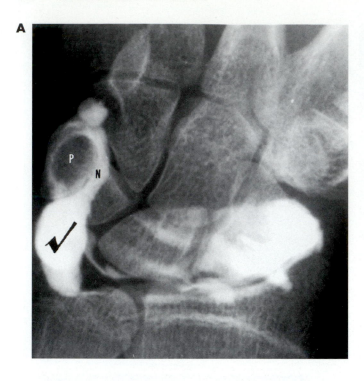

B

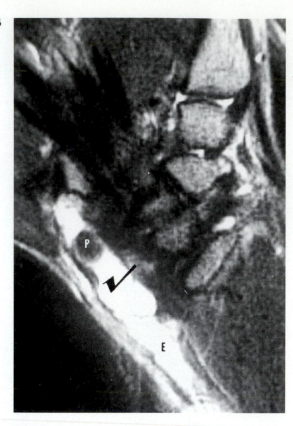

Fig. 4-44   Communicating synovial cyst following carpal ligament surgery. **A,** Frontal radiograph following radiocarpal arthrography demonstrates a prominent contrast material–filled structure (*check*) arising from the pisiform-triquetral joint via a narrow neck (*N*). **B,** The finding (*check*) is confirmed on a coronal T2-weighted MR image; in addition, some extravasation (*E*) from the cyst has occurred following exercise (*P* = *pisiform*).

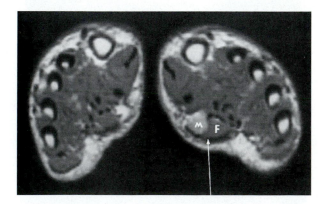

Fig. 4-45   Plantar neuroma. Transaxial proton density–weighted MR image reveals a well-circumscribed soft-tissue mass (*M*) deep to the plantar aponeurosis adjacent to the flexor digitorum brevis muscle (*F*). Such lesions exhibit high signal intensity on T2-weighted images. Differential diagnosis should include consideration of tendon sheath ganglion and other soft-tissue neoplasms.

6. Ewing sarcoma with extensive reactive bone formation may simulate classical osteosarcoma radiographically; conversely, a diaphyseal osteosarcoma without calcified matrix in its soft-tissue mass may resemble Ewing sarcoma.

7. A Codman triangle or sunburst periostitis may be observed on conventional radiographs.

8. Skip lesions of marrow involvement within the bone of origin occur in up to 10% of cases.

9. The current survival rate is approximately 50%.

10. The avulsive cortical irregularity or cortical desmoid is readily distinguished from classical osteosarcoma on the basis of its location on the posterior aspect of the medial femoral condyle.

11. Biopsy material obtained during the active reparative phase of an avulsive cortical irregularity or cortical desmoid is difficult to distinguish histologically from that of osteosarcoma; hence, differentiation is best accomplished radiographically.

12. The initial 4 to 8 weeks in the evolution of posttraumatic myositis ossificans are characterized by amor-

A

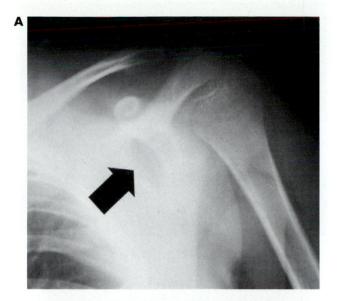

B

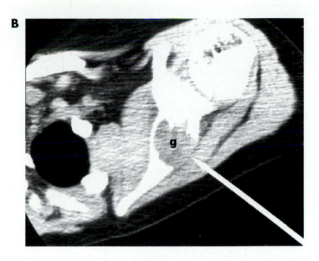

C

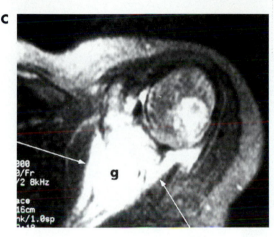

Fig. 4-46   Eosinophilic granuloma of the scapula. **A,** Radiography reveals a well-defined lytic lesion (*arrow*) inferior to the coracoid process. **B,** CT demonstrates posterior cortical disruption (*arrow*) caused by the process (*g*). **C,** Transaxial T2-weighted MR image documents high signal intensity within the lesion (*g*), which exhibits extensive adjacent edema (*arrows*). Differential diagnosis should include consideration of hematogenous osteomyelitis, Ewing sarcoma, and metastatic disease.

phous soft-tissue calcification and sometimes periosteal reaction, which may resemble osteosarcoma; clinical history and follow-up radiographs are thus important in distinction.

13. The initial 4 weeks in the histologic evolution of posttraumatic myositis ossificans are characterized by a pseudosarcomatous appearance centrally that may simulate malignancy.

14. Aggressive osteoblastoma is a rare lesion that may simulate osteosarcoma both radiographically and histologically.

15. Approximately 15% of patients with classical osteosarcoma manifest pulmonary or osseous metastases at presentation.

16. Following therapy, approximately 80% of relapses occur in the lungs, whereas the remainder occur in bone.

17. Relapses usually occur within 2 years of the initial diagnosis.

18. Osseous relapse has a worse prognosis than does pulmonary relapse.

19. The 5-year survival rate for pulmonary relapse is approximately 15% following chemotherapy with surgical resection of metastatic foci.

20. Magnetic resonance imaging is superior to computed tomography for staging of the primary lesion, particularly in determining the extent of bone marrow and soft-tissue involvement (Fig. 4-50).

21. Amputation and wide en bloc excision with limb salvage are therapeutic options in cases with resectable or absent metastases.

22. Adjuvant multidrug chemotherapy enhances short-term survival rate but may not influence overall cure rate.

23. Induction chemotherapy may facilitate limb salvage surgery by decreasing tumor bulk and controlling micrometastases until a suitable homograft or prosthesis becomes available.

24. The degree of tumor necrosis in the resection specimen indicates the effectiveness of induction chemotherapy.

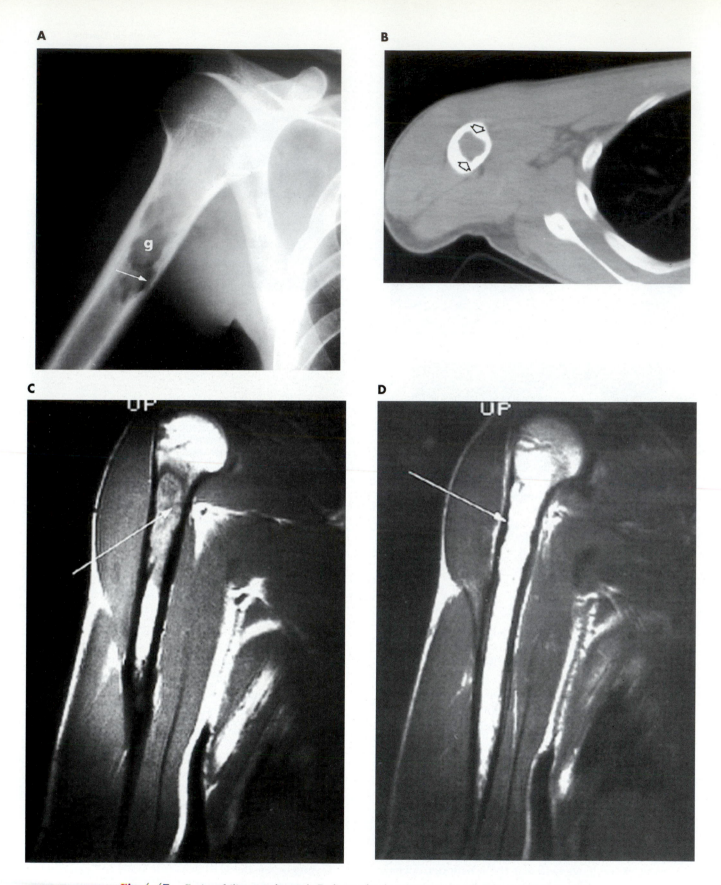

Fig. 4-47 Eosinophilic granuloma. **A,** Radiography demonstrates a well-defined lytic lesion (*g*) in the proximal humeral diaphysis with endosteal scalloping (*arrow*). **B,** CT documents endosteal erosion (*arrows*) to better advantage. **C,** Coronal T1-weighted MR image reveals predominantly low signal intensity (*arrow*) within the lesion. **D,** The lesion exhibits high signal intensity on a corresponding T2-weighted image. Differential diagnosis should include consideration of fibrous dysplasia, plasmacytoma, metastatic disease, and indolent infection.

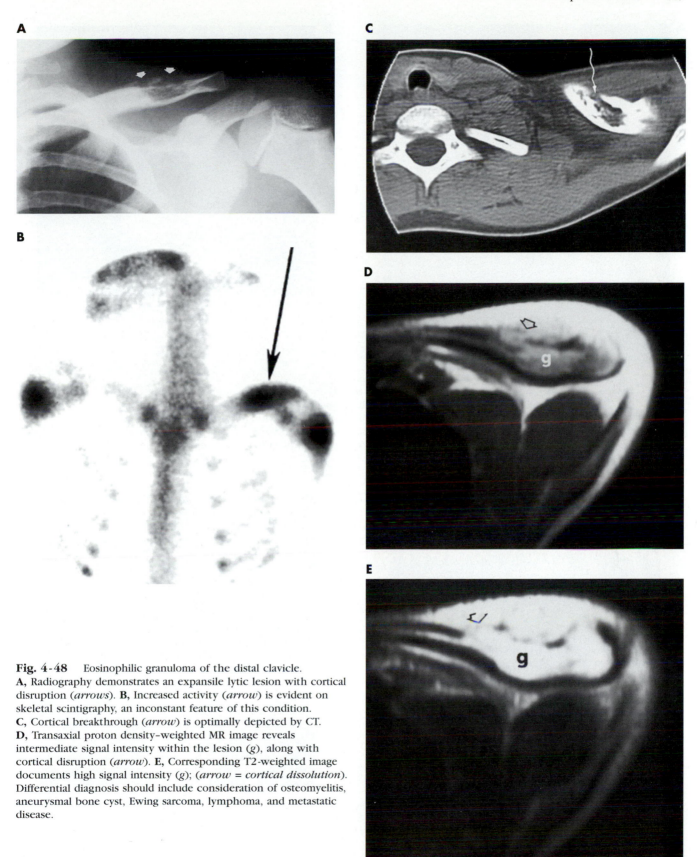

Fig. 4-48   Eosinophilic granuloma of the distal clavicle.
A, Radiography demonstrates an expansile lytic lesion with cortical
disruption (*arrows*). **B,** Increased activity (*arrow*) is evident on
skeletal scintigraphy, an inconstant feature of this condition.
C, Cortical breakthrough (*arrow*) is optimally depicted by CT.
D, Transaxial proton density–weighted MR image reveals
intermediate signal intensity within the lesion (*g*), along with
cortical disruption (*arrow*). **E,** Corresponding T2-weighted image
documents high signal intensity (*g*); (*arrow = cortical dissolution*).
Differential diagnosis should include consideration of osteomyelitis,
aneurysmal bone cyst, Ewing sarcoma, lymphoma, and metastatic
disease.

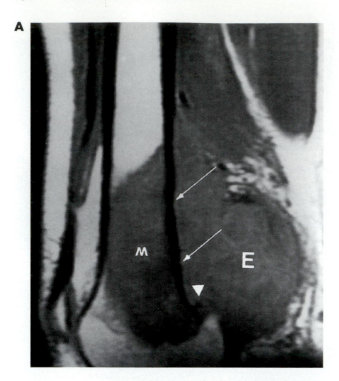

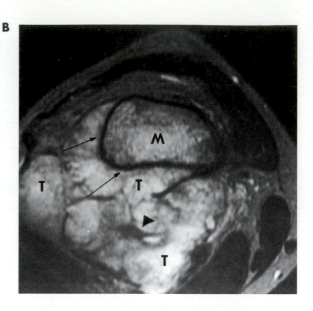

Fig. 4-49   Distal femoral osteosarcoma. **A,** Sagittal T1-weighted MR image demonstrates low–signal intensity marrow involvement (*M*) with prominent posterior extension (*E*). **B,** Transaxial T2-weighted image reveals predominantly high signal intensity within both the marrow (*M*) and soft-tissue (*T*) components of the lesion. Areas of intact cortex (*arrows*) are evident on both images, along with tumor bone formation (*arrowheads*) suggesting the correct diagnosis.

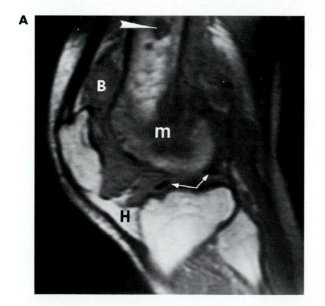

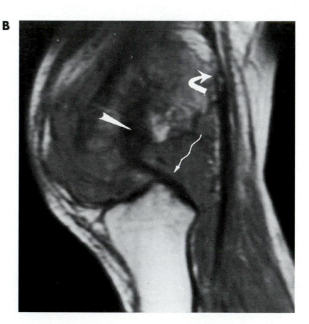

Fig. 4-50   MRI of distal femoral osteosarcoma. **A** and **B,** Sagittal T1-weighted images demonstrate a large mass of predominantly low signal intensity that has invaded knee joint structures including the suprapatellar bursa (*B*), lateral meniscus (*straight arrows*), posterior cruciate ligament (*wavy arrow*), and the Hoffa fat-pad (*H*). There is extensive marrow involvement of the distal femur (*m*), as well as invasion of the neurovascular bundle (*curved arrow*). Low–signal intensity areas representing tumor bone formation (*arrowheads*) are a clue to the correct diagnosis.

Box 4-25   Classical Osteosarcoma: Important Features

- **Peak age:** second decade
- **Male/female ratio:** 1.4:1
- **Most common location:** distal femur or proximal tibia
- **Major radiographic features:** favored site is metaphysis of long bone, especially around knee; may be lytic, blastic, or mixed bone destruction and production; trabecular and cortical destruction; poorly marginated; periosteal new bone is frequent and often takes the form of spiculation or Codman triangles; soft-tissue extension is the rule in larger lesions; MRI and chest CT are essential for pretreatment staging
- **Radiographic differential diagnosis:** Ewing sarcoma, fibrosarcoma, malignant fibrous histiocytoma, chondrosarcoma, osteomyelitis, osteoblastoma, giant cell tumor
- **Treatment:** surgery remains the primary therapy for osteosarcoma; however, in general, preoperative chemotherapy is given in order to facilitate limb salvage procedures

---

Parosteal osteosarcoma   This is a well-differentiated bone-forming neoplasm consisting of densely sclerotic tumor matrix that envelops the underlying bone (Box 4-26).

PEARL

Parosteal osteosarcoma is a low-grade malignancy, most commonly found along the posterior aspect of the distal femur. There is dense bone formation radiographically and mature bone histologically.

---

1. The prognosis for parosteal osteosarcoma is significantly better than that for classical osteosarcoma.
2. The peak age range of incidence for parosteal osteosarcoma is older than for classical osteosarcoma.
3. Approximately 80% of lesions occur between the ages of 20 and 50 years.
4. The lesion occurs over a broad age range, including childhood.
5. The lesion is frequently greater than 5 cm in size at presentation.
6. Approximately 60% of lesions occur in the distal femur.
7. Metaphyseal localization is typical.
8. Extension into the marrow of the affected bone occurs in approximately 10% of cases.
9. Because the lesion is densely ossified, computed tomography, as opposed to magnetic resonance imaging, is generally preferred for the evaluation of medullary extension, which renders prognosis comparable to that of classical osteosarcoma.
10. The most mature tumor bone is found at the center of the lesion, whereas in posttraumatic myositis ossificans it is found peripherally.
11. With inadequate excision, the lesion may recur more aggressively or dedifferentiate.
12. Parosteal osteosarcoma is a monostotic lesion.
13. The differential diagnosis should include consideration of osteochondroma, periosteal osteosarcoma, and posttraumatic myositis ossificans.
14. Except at its origin, the lesion is usually separated from the cortex of the affected bone by a distinct cleft.
15. Pulmonary metastases are less common than with classical osteosarcoma and often occur many years following initial presentation.
16. The 5-year survival rate is approximately 85%.
17. The treatment of choice is wide marginal en bloc excision with limb salvage.
18. Chemotherapy is not a recommended therapeutic strategy for parosteal osteosarcoma.
19. Approximately 40% of lesions occur in the proximal tibia, proximal humerus, and other metaphyseal regions outside the distal femur.

Box 4-26   Parosteal Osteosarcoma: Important Features

- **Peak age:** third decade of life (with the age distribution shifted approximately one decade toward older patients when compared to that of classical osteosarcoma)
- **Male/female ratio:** 1:2
- **Most common location:** distal femur (accounting for approximately 60% of all cases)
- **Major radiographic features:** lobulated and ossified mass arising on the metaphyseal surface of a long bone; posterior lower femur a preferred site; broad attachment to adjacent cortex; thickened and deformed cortex; larger tumors encircle the bone
- **Radiographic differential diagnosis:** myositis ossificans, periosteal osteosarcoma, periosteal chondrosarcoma, high-grade surface osteosarcoma, classical osteosarcoma, osteochondroma
- **Treatment:** surgery with conservative total resection; amputation if the tumor is too large or involves the neurovascular bundle

---

Telangiectatic osteosarcoma   This is a rare, highly vascular variant of osteosarcoma that may contain necrotic tissue and pools of blood centrally with viable tumor limited to the periphery (Box 4-27).

PEARL

Telangiectatic osteosarcoma contains large, blood-filled spaces similar to those of aneurysmal bone cyst. It is osteolytic and often lacks periosteal new bone. The lesion is primarily found in long bones.

1. The peak age range of incidence for telangiectatic osteosarcoma is the same as that for classical osteosarcoma (10 to 25 years).
2. The lesion is usually greater than 5 cm in size at presentation.
3. The metaphyses of long bones are most commonly affected.
4. Radiographically, the lesion tends to be purely lytic and hence difficult to accurately diagnose.
5. Periosteal reaction may or may not be present.
6. The radiographic differential diagnosis should include consideration of Ewing sarcoma, fibrosarcoma, malignant fibrous histiocytoma, giant cell tumor, and aneurysmal bone cyst.
7. Metastatic disease occurs predominantly in the lungs, other bones, and regional lymph nodes.
8. Soft-tissue involvement may or may not be present.
9. The diagnostic imaging workup should include radiography and magnetic resonance imaging of the primary lesion, radiography and computed tomography of the chest, and skeletal scintigraphy.
10. The treatment of choice consists of wide marginal or radical en bloc excision combined with chemotherapy.

Box 4-27 Telangiectatic Osteosarcoma: Important Features

- **Peak age:** second decade (with distribution similar to that of classical osteosarcoma)
- **Male/female ratio:** approximately 2:1
- **Most common location:** distal femur or proximal tibia
- **Major radiographic features:** large size and metaphyseal location; considerable medullary and cortical bone destruction; purely lytic and poorly marginated; periosteal new bone and soft-tissue mass common
- **Radiographic differential diagnosis:** classical osteogenic sarcoma, other sarcomas, aneurysmal bone cyst
- **Treatment:** surgery and chemotherapy

Periosteal osteosarcoma   This is an extremely rare variant of osteosarcoma that arises juxtacortically and initially spares the medullary cavity of the affected bone (Box 4-28).

PEARL

Periosteal osteosarcoma is relatively poorly differentiated and predominantly chondroblastic. The lesion has a better prognosis than does conventional osteosarcoma. It is typically a cortically based lesion with spicules of periosteal new bone.

1. The peak age range of incidence is intermediate between that of classical and parosteal osteosarcoma.
2. Radiographically, amorphous densities or spicules of bone oriented perpendicularly to the cortex may be observed within the soft-tissue component of the lesion.
3. Most lesions are between 2 and 5 cm in size.
4. Long-bone lesions tend to be diaphyseal.
5. The tibia is the most commonly affected bone.
6. Cortical thickening or erosion may occur in association with the lesion.
7. Periosteal reaction, including Codman triangles, is common.
8. The differential diagnosis should include consideration of classical osteosarcoma, juxtacortical chondroma, apophyseal avulsion with early healing response, and parosteal osteosarcoma.
9. Periosteal osteosarcoma metastasizes to the lungs much less frequently than does classical osteosarcoma.
10. The 5-year survival rate is approximately 80%.
11. Magnetic resonance imaging is superior to computed tomography for local staging, particularly with regard to soft-tissue and marrow involvement; detection of the latter would change the diagnosis to classical osteosarcoma.
12. The preferred therapy is wide marginal resection with limb salvage.

PEARL

Typically the prognosis improves for osteosarcomas as one moves from a central location in bone to one that is peripheral. Therefore, central (classical or telangiectatic) osteosarcomas have a worse prognosis than do parosteal lesions. Central osteosarcomas occur in younger patients, parosteal tumors occur in slightly older patients, and periosteal osteosarcomas manifest an intermediate peak age of incidence.

Box 4-28 Periosteal Osteosarcoma: Important Features

- **Peak age:** second decade
- **Male/female ratio:** slight female predominance
- **Most common location:** diaphysis of the tibia or femur
- **Major radiographic features:** diaphyseal location, especially of the tibia; located on the surface of bone, with the medullary canal uninvolved; soft-tissue mass on the bone surface with partial matrix mineralization; periphery of soft-tissue mass free of mineral; adjacent cortex thickened; buttresses or Codman triangles common
- **Radiographic differential diagnosis:** parosteal osteosarcoma, high-grade surface osteosarcoma, classical osteosarcoma, periosteal chondrosarcoma, myositis ossificans
- **Treatment:** surgical resection; chemotherapy may not be indicated in the therapy of this variant of osteosarcoma

Box 4-29 Low-grade Intraosseous Osteosarcoma: Important Features

- **Peak age:** second or third decade
- **Male/female ratio:** approximately 1:1
- **Most common location:** distal femur or proximal tibia
- **Major radiographic features:** medullary lesion extending to bone end; poor margination and large size; most are trabecular or sclerotic; periosteal new bone and soft-tissue mass are usually absent; overall appearance may suggest benignancy with only a small region demonstrating evidence suggesting malignancy
- **Radiographic differential diagnosis:** fibrous dysplasia, giant cell tumor, fibrosarcoma, malignant fibrous histiocytoma
- **Treatment:** surgical resection

Low-grade intraosseous osteosarcoma This is a rare well-differentiated variant of osteosarcoma that lacks soft-tissue involvement (Box 4-29).

1. The lesion constitutes approximately 1% of all osteosarcomas.
2. The peak age range of incidence is the same as that for classical osteosarcoma.
3. The lesion is usually greater than 5 cm in diameter at the time of diagnosis.
4. The metadiaphyseal regions of long bones are most commonly affected.
5. Although many lesions exhibit a mixed pattern of lysis and sclerosis, a significant number are purely lytic and permeative.
6. Recurrent lesions may violate the cortex and induce periosteal reaction.
7. The differential diagnosis should include consideration of Ewing sarcoma, fibrosarcoma, malignant fibrous histiocytoma, osteomyelitis, lymphomas, and fibrous dysplasia.
8. The 5-year survival rate is approximately 85%.
9. Recurrent lesions have a greater tendency to metastasize.
10. Magnetic resonance imaging is the procedure of choice for evaluating the extent of marrow involvement.
11. Wide marginal en bloc excision with limb salvage is the preferred treatment.
12. Aggressive recurrence is usually managed with radical surgery and chemotherapy.

High-grade surface osteosarcoma This is an extremely rare poorly-differentiated variant of classical osteosarcoma that arises on the surface of the bone (Box 4-30).

Box 4-30 High-grade Surface Osteosarcoma: Important Features

- **Peak age:** second or third decade
- **Male/female ratio:** male predominance, but extremely rare tumor
- **Most common location:** femur
- **Major radiographic features:** partially mineralized tumor on the surface of a long bone, most commonly the femur; cortical destruction is usually present; periosteal new bone is frequent
- **Radiographic differential diagnosis:** periosteal osteosarcoma; parosteal osteosarcoma; classical osteosarcoma predominantly affecting the bone surface
- **Treatment:** aggressive surgery (amputation or wide resection) and adjuvant therapy as for classical osteosarcoma

Osteosarcoma occurring after age 60
1. Approximately 30% of lesions occur in the axial skeleton.
2. Approximately 25% of lesions are either craniofacial or soft tissue in origin.
3. Approximately 80% of lesions are purely lytic without radiodense matrix.
4. Approximately half of all lesions arise in a preexisting condition.

5. Fewer than 1% of all patients with Paget disease develop the lesion in a severely affected bone (Box 4-31).

6. Radiation-induced osteosarcoma usually arises in bone that has received more than 3000 rads of cumulative dose (Box 4-32).

7. Common sites of involvement by postirradiation osteosarcoma include the pelvis, shoulder girdle, distal femur, proximal tibia, and distal radius.

8. The interval between radiation therapy and the diagnosis of a radiation-induced osteosarcoma averages 15 years and ranges from 2 to 50 years.

9. Histologic elements of osteosarcoma may arise in a dedifferentiated chondrosarcoma, along with the pathologic features of fibrosarcoma or malignant fibrous histiocytoma.

10. Osteosarcoma may rarely arise from benign lesions including fibrous dysplasia, osteoblastoma, osteochondroma, chronic osteomyelitis, and bone infarcts.

11. Chances for survival are approximately five times better for osteosarcoma arising de novo in an older patient than for osteosarcoma arising in a preexisting lesion.

12. Radical excision is the treatment of choice.

13. Adjunctive chemotherapy has no significant impact on survival with this lesion.

Box 4-31 Paget Sarcoma: Important Features

- **Peak age:** elderly years
- **Male/female ratio:** 2:1
- **Most common location:** pelvis, femur, or humerus
- **Major radiographic features:** most tumors are lytic, though sclerotic and mixed lesions occur; bone of origin shows usual changes of Paget disease; mass or bone production beyond osseous margins; geographic bone destruction; CT or MRI may be useful for detection when soft-tissue component is present; unlike uncomplicated Paget disease, these sarcomas have a predilection for the humerus and only rarely arise in the spine
- **Radiographic differential diagnosis:** Paget disease with marked lysis, evolving benign Paget disease extending into soft tissue
- **Treatment:** aggressive surgical ablation (usually amputation)

Box 4-32 Postirradiation Osteosarcoma: Important Features

- **Peak age:** fourth to seventh decades
- **Male/female ratio:** 1:1.5
- **Most common location:** shoulder or pelvis region

Box 4-32—cont'd

- **Major radiographic features:** tumor may arise from normal bone or area of preexisting lesion within radiation portal; underlying bone often altered by radiation, surgery, the preexisting abnormality, or some combination of these; underlying changes may obscure early diagnosis; latent period ranges from 2 to 50 years, with average of 15 years; postirradiation osteosarcoma similar to classical type; soft-tissue mass, cortical destruction, and extraosseous bone production
- **Radiographic differential diagnosis:** metastasis, radiation changes with preexisting lesion, insufficiency type stress fracture in irradiated or osteopenic bone
- **Treatment:** surgical ablative therapy consistent with the pathologic differentiation of the tumor

---

Osteosarcoma of the jaw   Osteosarcoma of the jaw is a form that occurs later in life than does classical osteosarcoma. Radiographically, osteolysis is evident, sometimes in association with dense tumor matrix. Histologically, the lesion appears less aggressive than classical osteosarcoma. The treatment of choice is wide excision, and the prognosis is much better than that for classical osteosarcoma.

Osteosarcomatosis   Osteosarcomatosis, or multicentric osteosarcoma, is the synchronous development of osteosarcomas at multiple sites. The lesions are always purely osteoblastic, and the distribution is often bilateral and relatively symmetric. Early in the course of disease, the lesions may resemble enostoses radiographically. The condition is rapidly progressive and extremely rare. Most cases occur in children between 5 and 10 years of age. Pulmonary metastases occur early, and the prognosis is extremely poor. The condition is distinguished from a primary osteosarcoma with multiple osseous metastases on the basis of lesion size uniformity (absence of an apparent dominant lesion).

Soft-tissue osteosarcoma

1. The peak age range of incidence for soft-tissue osteosarcoma (40 to 70 years) is older than that for classical osteosarcoma.

2. The lesion most commonly occurs in the thigh.

3. The lesion is occasionally induced by radiation therapy.

4. Tumor matrix occurs at the center of the soft-tissue mass, in contrast to the peripheral pattern of ossification in posttraumatic myositis ossificans.

5. The lesion occasionally contains no radiodense matrix.

6. Metastasis occurs to the regional lymph nodes and lungs.

7. The prognosis is similar to that for classical osteosarcoma.

Cartilage-forming tumors

Medullary chondrosarcoma This is a cartilage-producing sarcoma that arises centrally in the affected bone, either as a primary lesion or secondary to malignant transformation of an enchondroma (Box 4-33).

1. The lesion occurs most commonly during the fourth through sixth decades of life.
2. Secondary lesions tend to arise from malignant transformation of enchondromas in the proximal areas of the skeleton.
3. Soft-tissue involvement occurs with high-grade lesions.
4. Lesions are generally greater than 5 cm in size at the time of diagnosis.
5. Common sites of involvement include the pelvis, shoulder girdle, and long-bone metaphyses (Fig. 4-51).
6. The lesion may be purely lytic or contain calcification ranging from scant flecks to dense aggregates.
7. Periosteal reaction and endosteal thickening may be observed.
8. Histologically, approximately 90% of lesions are low-grade.
9. The lesion is commonly diagnosed incorrectly or late because of its typically nonaggressive radiographic appearance.
10. The differential diagnosis should include consideration of enchondroma, osteosarcoma or fibrosarcoma, malignant fibrous histiocytoma (primary or arising in a bone infarct), lymphoma, and giant cell tumor.
11. Local recurrence is more common than metastatic disease.
12. Pulmonary metastases occur late in the disease course, sometimes up to 20 years after initial diagnosis.
13. The 5-year survival rate is approximately 75%.
14. Local recurrence may occur secondary to tumor seeding during biopsy or resection (Fig. 4-52).
15. Placement of an intramedullary rod may disperse neoplastic cells and predispose to local recurrence.
16. The primary lesion exhibits variable uptake of bone-seeking radiopharmaceuticals, and osseous metastases do not occur; scintigraphy therefore is of no value.
17. Computed tomography is superior to magnetic resonance imaging for the detection of subtle matrix calcification, although the latter method is optimal for staging of marrow and soft-tissue extent.
18. The treatment of choice is en bloc resection with limb salvage or radical excision.
19. Radiation therapy and chemotherapy are effective only as palliative treatment.
20. The lesion is frequently asymptomatic, particularly early in the course of low-grade tumors.

Box 4-33   Medullary Chondrosarcoma: Important Features

- **Peak age:** adulthood (over a broad range through the fourth, fifth, and sixth decades)
- **Male/female ratio:** 1.5:1
- **Most common location:** more than 75% occur in the trunk and proximal humeri and femora, with the acetabular region being a common location
- **Major radiographic features:** predilection for the axial skeleton and the metaphysis or diaphysis of the affected long bone; approximately two thirds of lesions are partially calcified; cortical erosion or destruction is usually present; periosteal reaction is scant or absent; soft-tissue extension is commonly seen in large lesions
- **Radiographic differential diagnosis:** osteosarcoma, fibrosarcoma, enchondroma
- **Treatment:** surgical resection or amputation

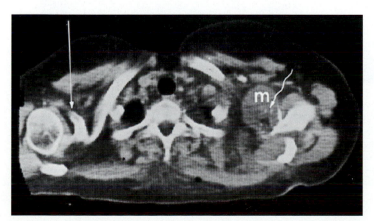

Fig. 4-51   Chondrosarcoma arising from the scapula. CT image reveals a well-defined soft-tissue mass (*m*) with internal calcification (*wavy arrow*) that has destroyed the coracoid process (*straight arrow = normal coracoid for comparison*). CT is superior to MRI for demonstrating calcified or ossified matrix within a lesion.

A

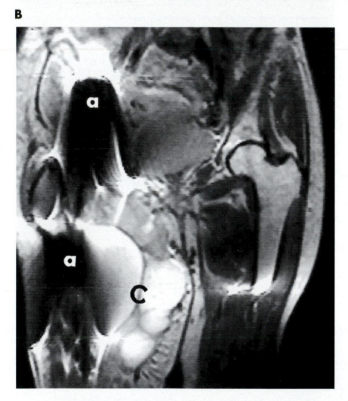

B

Exostotic chondrosarcoma This is a cartilage-producing sarcoma that arises peripherally in the affected bone, either as a primary lesion or secondary to malignant transformation of an osteochondroma (Fig. 4-53; Box 4-34).

1. The lesion tends to occur during the third through fifth decades of life.
2. Most lesions are greater than 5 cm in size at the time of presentation.
3. Common sites of involvement include the pelvis, shoulder girdle, long-bone metaphyses, ribs, and sternum.
4. The cartilaginous cap of the lesion is often densely calcified, with peripheral streaking or a snowstorm-like pattern.
5. A cartilaginous cap greater than 1 cm in thickness suggests malignant degeneration of an osteochondroma.
6. Clinical signs of pain and growth of an exostosis after physeal closure are more important than radiographic appearance in predicting malignant degeneration of an osteochondroma.

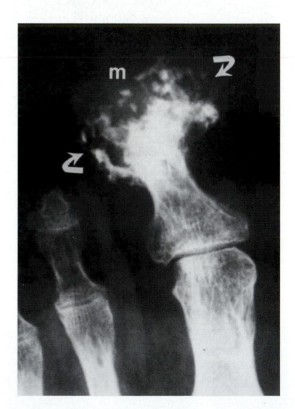

Fig. 4-52   Recurrent chondrosarcoma following surgical resection. **A,** Coronal T1-weighted MR image demonstrates a lobulated mass (*C*) of low signal intensity arising from the femoral diaphysis. **B,** The lesion (*C*) exhibits high signal intensity on a corresponding T2-weighted image (*a = artifact from orthopedic hardware*). Differential diagnosis should include consideration of other osseous and soft-tissue malignancies.

Fig. 4-53   Chondrosarcoma secondary to malignant transformation of a subungual exostosis. A prominent soft-tissue mass (*m*) with extensive calcification (*arrows*) arises from the terminal phalangeal tuft. Subungual exostoses are capped by fibrocartilage as opposed to hyaline cartilage, which distinguishes them from true osteochondromas. The location is typical in this case, but malignant degeneration of subungual lesions is rare.

7. Most exostotic chondrosarcomas are low-grade lesions histologically.
8. Pulmonary metastases are uncommon and generally occur late in the disease course.
9. Radiographic features suggesting malignant degeneration of an osteochondroma include bone destruction, lesion enlargement, and alteration in the pattern of calcification.
10. Magnetic resonance imaging is more reliable than computed tomography for establishing the extent of the soft-tissue mass or cartilaginous cap thickness.
11. Wide or radical excision is the preferred therapy.
12. Neither radiation therapy nor chemotherapy is generally indicated in treating exostotic chondrosarcoma.

Box 4-34 Exostotic Chondrosarcoma: Important Features

- **Peak age:** 20 to 40 years
- **Male/female ratio:** 2:1
- **Most common location:** pelvis or shoulder
- **Major radiographic features:** thick and indistinct cartilaginous cap of a lesion that otherwise has the features of an osteochondroma; radiolucent regions within the lesion; destruction of the underlying osteochondroma or adjacent bone; computed tomography or magnetic resonance imaging is useful for further characterization
- **Radiographic differential diagnosis:** osteochondroma with overlying bursa formation, or soft-tissue impingement causing new-onset pain; asymptomatic osteochondroma
- **Treatment:** surgical resection or amputation

Dedifferentiated chondrosarcoma   Dedifferentiated chondrosarcoma is a highly aggressive neoplasm that arises from degeneration of low-grade chondrosarcoma. The lesion may in turn degenerate into high-grade chondrosarcoma, osteosarcoma, fibrosarcoma, malignant fibrous histiocytoma, or a neoplasm with mixed malignant histologic characteristics. The tumor is not rare, because approximately 10% of the lesions from which it arises undergo dedifferentiation. Histologically, the lesion coexists with its precursor; hence, biopsy sampling error is a potential problem. The 5-year survival rate is approximately 80%, and pulmonary metastases are common. Radical excision and chemotherapy are appropriate in the management of the lesion (Box 4-35).

PEARL

Dedifferentiated chondrosarcoma arises from degeneration of a low-grade chondrosarcoma. Clear cell chondrosarcoma is epiphyseal in location, and may be indistinguishable radiographically from chondroblastoma; the most common locations are the proximal femur, humerus, and tibia.

Box 4-35 Dedifferentiated Chondrosarcoma: Important Features

- **Peak age:** sixth decade
- **Male/female ratio:** approximately 1:1
- **Most common location:** femur or pelvis
- **Major radiographic features:** aggressive-appearing area superimposed on otherwise typical chondrosarcoma; this usually takes the form of a radiolucent or destructive region within the calcified tumor; large size and poor margination; intramedullary calcification of the original cartilaginous tumor is usually present; cortical destruction and an associated soft-tissue mass; many cases are indistinguishable from ordinary medullary chondrosarcomas
- **Radiographic differential diagnosis:** ordinary medullary chondrosarcoma, sarcoma arising in bone infarct, mesenchymal chondrosarcoma
- **Treatment:** radical surgical ablation, sometimes with adjuvant radiation and/or chemotherapy

---

Clear cell chondrosarcoma   This is a very low-grade variant of chondrosarcoma that resembles chondroblastoma (Box 4-36).

1. The lesion typically occurs in the epiphyseal region of long bones.
2. Most patients are affected during the third decade of life and hence are older than those with chondroblastoma.
3. Soft-tissue extension does not usually occur.
4. Most lesions are less than 5 cm in diameter at the time of diagnosis.
5. The lesion most commonly occurs in the proximal femur or humerus.
6. Calcified tumor matrix is usually not present radiographically.
7. Periosteal reaction rarely occurs.
8. After enlarging slowly for several years, the lesion may become more aggressive.
9. The lesion usually manifests a nonaggressive radiographic appearance.

10. The differential diagnosis should include consideration of chondroblastoma, fibrosarcoma, malignant fibrous histiocytoma, and lymphoma.
11. The metastatic potential of clear cell chondrosarcoma is extremely low.
12. Curettage alone is adequate treatment for chondroblastoma but may result in an aggressive local recurrence if used in the management of clear cell chondrosarcoma.
13. The treatment of choice is wide surgical excision.

Box 4-36   Clear Cell Chondrosarcoma: Important Features

- **Peak age:** third decade of life
- **Male/female ratio:** 2:1
- **Most common location:** proximal femur accounts for approximately 50% of cases (the tumor nearly always involves the epiphyseal region of a long bone)
- **Major radiographic features:** epiphyseal location in the proximal femur or humerus; early lesions appear benign with sharp margination, sclerosis at the periphery, and expansion of the affected bone; only about 25% of cases are calcified; larger lesions appear malignant with poor margination and cortical destruction
- **Radiographic differential diagnosis:** chondroblastoma; medullary chondrosarcoma; giant cell tumor
- **Treatment:** surgical resection

Mesenchymal chondrosarcoma   Mesenchymal chondrosarcoma is a rare variant of chondrosarcoma that is highly malignant. The peak age range of incidence is 10 to 40 years, which is younger than that of medullary chondrosarcoma. Osseous involvement occurs most commonly in the ribs and jaw bones, although approximately 30% of lesions arise primarily in soft tissue. Calcification is usually evident radiographically, and the lesion typically manifests aggressive characteristics. Metastatic disease is common and occurs early in the disease course. The treatment of choice is radical resection (Box 4-37).

Box 4-37   Mesenchymal Chondrosarcoma: Important Features

- **Peak age:** third decade
- **Male/female ratio:** slight male predominance
- **Most common location:** pelvis, ribs, maxilla, or mandible

---

Box 4-37—cont'd

- **Major radiographic features:** most lesions have either a nonspecific malignant radiographic appearance or radiographic features suggestive of medullary chondrosarcoma; calcification is usually present; poor margination and cortical destruction occur; frequently there is an associated soft-tissue mass
- **Radiographic differential diagnosis:** medullary chondrosarcoma, osteosarcoma, fibrosarcoma
- **Treatment:** surgical resection

Vascular tumors

Malignant hemangioendothelioma of bone   This is a benign or low-grade malignant vascular tumor that frequently resembles angiosarcoma histologically (Box 4-38).

PEARL

Hemangioendothelioma may be solitary or multifocal; the lesion tends to occur during the fourth and fifth decades with a male predilection. It is typically a lytic lesion with cortical thinning, expansion, or disruption involving the femur, tibia, humerus, pelvis, skull, or spine.

---

1. Polyostotic involvement is more common than monostotic involvement.
2. When polyostotic, lesions tend to involve multiple bones of one extremity.
3. The prognosis is better for polyostotic than for monostotic involvement.
4. Most patients are initially diagnosed during the fourth or fifth decade of life.
5. Typical lesions range in size from 2 to 5 cm.
6. When polyostotic, lesions tend to involve multiple bones in a hand or foot.
7. The majority of lesions are cortically based and metaphyseal or diaphyseal in location.
8. The lesion may be associated with a soft-tissue mass.
9. Major differential diagnostic considerations include enchondromatosis and metastatic disease.
10. Epithelioid hemangioendothelioma is a subtype that is histologically identical to the intravascular bronchoalveolar tumor.
11. Pulmonary metastases occasionally occur.
12. Radiographically, the lesion tends to be nonaggressive or only mildly aggressive in appearance.
13. Resection is the treatment of choice for monostotic lesions.

Box 4-38 Malignant Hemangioendothelioma of Bone: Important Features

- The tumor constitutes 0.1% to 1% of malignant bone tumors.
- There is a slight male predominance.
- 50% of cases occur in the fourth and fifth decades.
- There is a predilection for long bones and vertebrae.
- Clinical findings include pain, local tenderness, and swelling; vertebral lesions may cause spinal cord compression; a history of trauma or radiation therapy for hemangioma may exist.
- Radiolucent zones manifest indistinct margins and absence of sclerotic reaction; variable trabecular destruction and soft-tissue masses occur; periostitis is rare; the lesion may be multifocal or cross joint spaces.
- Pathologic findings include vascular channels with atypical multilayered endothelial cells possessing granular cytoplasm, large nuclei, and nucleoli in a reticulin framework.
- Treatment consists of amputation with or without radiation therapy.
- Prognosis is better when the lesion is multifocal as opposed to solitary.

Hemangiopericytoma   This is an extremely rare vascular lesion that may be benign, locally aggressive, or malignant (Box 4-39).

PEARL

Hemangiopericytoma of bone can be benign or malignant, although the former is more common. It originates from pericytes of Zimmermann, which are cells that surround capillaries, and hence is related to glomus tumor. Electron microscopic identification of pericytes may be required to differentiate hemangiopericytoma from other mesenchymal tumors or sarcomas (fibrosarcoma, chondrosarcoma, synovial sarcoma, malignant fibrous histiocytoma, and Ewing sarcoma).

1. Peak incidence occurs during the fourth and fifth decades of life.
2. Soft-tissue involvement is common.
3. Most lesions are less than 5 cm in size.
4. Typical sites of involvement include the axial skeleton and the metaphyses of the proximal long bones.
5. Periosteal reaction is rare.
6. Hemangiopericytoma generally manifests an aggressive radiographic appearance.
7. Primary soft-tissue lesions are more common than intraosseous tumors and may erode bone secondarily.

8. Malignant lesions may metastasize to lung or bone.
9. Wide resection is the preferred therapy.
10. The 5-year survival rate is approximately 90%.
11. The differential diagnosis should include consideration of chondrosarcoma, lymphoma, fibrosarcoma, and malignant fibrous histiocytoma.
12. Diagnostic imaging workup should include conventional radiography of the lesion and chest, magnetic resonance imaging of the lesion, and radionuclide bone scan.

Box 4-39 Hemangiopericytoma: Important Features

- The lesion is extremely rare.
- There is a slight male predominance.
- The lesion usually occurs in the fourth and fifth decades, particularly in the axial skeleton and long-bone metaphyses.
- Pain with associated swelling is observed clinically.
- Radiographically, the lesion is similar in appearance to hemangioendothelioma.
- Pathologic findings include numerous capillaries surrounded by sheets of spindle cells without frank anaplasia; variable mitotic figures and necrosis are present.
- The treatment of choice is en bloc surgical resection with or without radiation therapy. Radiation alone may be indicated in nonresectable cases.

Angiosarcoma   This is a rare aggressive neoplasm that resembles low-grade malignant hemangioendothelioma histologically.

1. The lesion tends to occur during the fourth and fifth decades of life.
2. Radiographically, the lesion exhibits a permeative appearance, often with an associated soft-tissue mass.
3. The lesion is usually greater than 5 cm in diameter at the time of diagnosis.
4. Angiosarcoma is typically a metaphyseal lesion.
5. The pelvis, femur, humerus, and tibia are the most commonly affected bones.
6. The lesion causes cortical thinning, expansion, and disruption without significant reactive sclerosis or periosteal reaction.
7. Polyostotic involvement occurs in approximately 40% of cases.
8. Patients tend to be younger in the setting of multifocal involvement.
9. Multifocal involvement is often regional in distribution.
10. The differential diagnosis should include consideration of metastatic disease, fibrosarcoma, malignant

fibrous histiocytoma, medullary chondrosarcoma, and lymphoma.

11. Metastases occur most commonly in the lungs and skeleton.

12. Prognosis is better in the setting of multifocal involvement.

13. The recommended diagnostic imaging workup includes radiography and magnetic resonance imaging of the osseous lesion(s), radiography and computed tomography of the chest, and skeletal scintigraphy.

14. The treatment of choice is wide marginal or radical resection.

Angioblastoma   This is a rare low-grade neoplasm that contains elements of vascular, squamous, and alveolar tissue histologically.

PEARL

Angioblastoma or adamantinoma is a malignant lesion of unclear pathogenesis; histologically, it contains mixed elements. Patients are typically in their third or fourth decade of life and frequently have an antecedent history of trauma. The middle third of the tibia is the most common site (90% of cases). Radiographically, the lesion is eccentric, expansile, and lytic with sclerotic margins; satellite lytic foci are sometimes present in contiguity with the main lesion and are characteristic of this tumor. Pulmonary metastases may occur with pneumothorax at presentation (rarely) or following surgical excision with local recurrence.

---

1. The tumor is also commonly known as adamantinoma.

2. The lesion affects the tibia in approximately 90% of cases.

3. The tumor occurs most commonly during the third and fourth decades of life.

4. The lesion may occasionally occur in adolescents.

5. Soft-tissue extension is rare early in the disease course, but a soft-tissue mass may develop as the lesion becomes more locally aggressive.

6. Long-bone lesions are usually middiaphyseal and eccentrically located in the medullary cavity, but some are initially cortically based.

7. Reactive sclerosis is a common radiographic feature.

8. The lesion is generally monostotic but may have satellite foci in the parent and/or adjacent bones.

9. Angioblastoma is a different histologic entity from the adamantinoma or ameloblastoma of the jaw bones.

10. The radiographic appearance ranges from nonaggressive to moderately aggressive.

11. The differential diagnosis should include consideration of other vascular lesions, fibrous dysplasia, and ossifying fibroma.

12. Metastatic involvement of the skeleton, lungs, and regional lymph nodes occurs in approximately 20% of cases.

13. Angioblastoma is usually a low-grade lesion; therefore, metastases may not occur until several years after initial diagnosis.

14. The lesion manifests 5- and 10-year survival rates of approximately 60% and 40%, respectively.

15. Magnetic resonance imaging is the procedure of choice for determining the local extent of the lesion, including possible satellite foci.

16. Curettage with cryosurgery is frequently inadequate as initial therapy.

17. Wide excision should be employed as the initial treatment strategy.

Fibrous tumors

Fibrosarcoma and malignant fibrous histiocytoma of bone   These are two related and relatively common aggressive neoplasms that can be distinguished only on the basis of histologic characteristics.

1. Many lesions previously diagnosed as fibrosarcomas would currently be designated malignant fibrous histiocytomas under the relatively new pathologic classification system for malignant fibrous tumors.

2. Both lesions may arise secondarily in Paget disease, dedifferentiated chondrosarcoma, or irradiated bone.

3. An intramedullary bone infarct may degenerate into a fibrosarcoma, malignant fibrous histiocytoma, or osteosarcoma, although the last is extremely rare (Box 4-40).

PEARL

Malignant degeneration of bone infarcts to fibrosarcoma, malignant fibrous histiocytoma, and (rarely) osteosarcoma has been reported. Bony sclerosis and curvilinear densities are common radiographic findings in bone infarcts.

---

4. Both lesions occur across a broad age range but most commonly arise during the fifth and sixth decades of life.

5. A permeative or moth-eaten appearance with soft-tissue extension is typically evident radiographically.

6. Most lesions exceed 5 cm in size at the time of diagnosis.

7. Central or eccentric diametaphyseal involvement of long bones occurs in approximately 75% of cases.

8. The lesions are most commonly found in the bones around the knee, as well as in the humerus and pelvis.

9. Dystrophic calcification occurs in approximately 15% of lesions, and serpiginous calcification may persist in bone infarcts that have undergone transformation to fibrosarcoma or malignant fibrous histiocytoma.

10. Periosteal reaction is often present.

11. The differential diagnosis should include consideration of metastatic disease, medullary chondrosarcoma, and lymphoma.

12. Very few lesions are low-grade, and the patients with such lesions manifest a 90% 5-year survival rate.

13. Most lesions are high-grade, and affected patients manifest a 25% 5-year survival rate.

14. Metastatic involvement of the skeleton, lungs, regional lymph nodes, and viscera occurs.

15. Magnetic resonance imaging is superior to computed tomography for diagnosing local extent, particularly soft-tissue and bone marrow involvement.

16. Low-grade lesions are best treated with wide excision, and chemotherapy may be a beneficial adjunct.

17. High-grade lesions require wide or radical excision and chemotherapy.

PEARL

Malignant fibrous histiocytoma of bone:
- Male predilection
- Average age 50 years
- Most common in femur, tibia, humerus, ribs, and craniofacial bones
- Variable periosteal or endosteal bone production
- Pathologic fracture in up to 50% of cases

Box 4-40   Malignancy Arising in Bone Infarction: Important Features

- Men are affected more commonly than women.
- Peak age range occurs during the fifth through seventh decades.
- Multiple bone infarcts are commonly present.
- Histologic subtypes include fibrosarcoma, malignant fibrous histiocytoma, and, rarely, osteosarcoma.
- Metastasis is common, and the prognosis is poor.
- Radiographically, the phenomenon may resemble chondrosarcoma arising in an old enchondroma.

Fibrosarcoma and malignant fibrous histiocytoma of soft tissue (Figs. 4-54, 4-55) Fibrosarcoma and malignant fibrous histiocytoma arising primarily in soft tissue together constitute the most common soft-tissue sarcoma occurring in adults. Histologically, both lesions tend to be either low-grade or highly aggressive. Both lesions may incite a reactive pseudocapsule that leads to nonaggressive characteristics on computed tomographic or magnetic resonance imaging studies. At surgery, the lesions are frequently well encapsulated, and tumor cells are always found beyond their apparent margins. Fibrosarcoma can be distinguished from malignant fibrous histiocytoma on the basis of histologic features, with the latter manifesting histiocytes and malignant fibroblasts in a whorled or storiform pattern.

Fatty tumors

Liposarcoma of bone This is an extremely rare aggressive lesion with nonspecific imaging characteristics.

Liposarcoma of soft tissue This is a well-differentiated or high-grade lesion containing variable quantities of fat and occasional dystrophic calcification, cartilage, or bone histologically.

1. The lesion is the second most common soft-tissue sarcoma.

2. Most patients are affected during the third through fifth decades of life.

3. The lesion may be large at the time of diagnosis, because it grows insidiously and is frequently asymptomatic.

4. The buttocks, thigh, lower leg, and retroperitoneum constitute the most common sites of involvement (Fig. 4-56).

5. High-grade lesions manifest abundant histologic cellularity and nonspecific soft-tissue characteristics on cross-sectional imaging studies.

6. Well-differentiated lesions contain variable amounts of fatty tissue; therefore, a specific diagnosis can usually be made on cross-sectional imaging studies.

7. The lesion demonstrates variable signal intensity on T1-weighted magnetic resonance images.

8. The lesion demonstrates high signal intensity on T2-weighted magnetic resonance images (Fig. 4-57).

9. A reactive pseudocapsule is often present, resulting in a nonaggressive appearance on cross-sectional imaging studies.

10. Metastatic disease occurs most commonly in the lungs and viscera.

11. Wide excision is the surgical treatment of choice, regardless of the presence or absence of a reactive pseudocapsule.

12. Chemotherapy is often combined with surgical management.

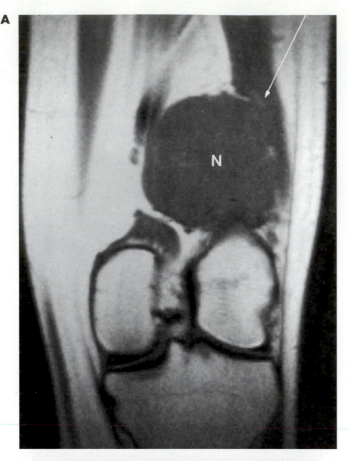

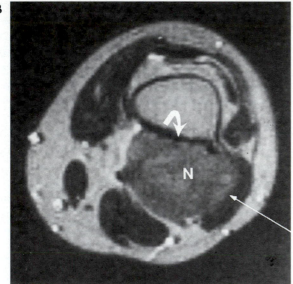

Fig. 4-54 Low-grade malignant fibrous histiocytoma. **A,** Coronal T1-weighted MR image demonstrates a well-defined soft-tissue neoplasm (*N*) of low signal intensity superior to the femoral condyles and with displacement of the adjacent biceps femoris muscle (*arrow*). **B,** Transaxial T2-weighted image reveals intermediate signal intensity within the lesion (*N*), which has invaded the biceps femoris muscle (*straight arrow*) and posterior femoral cortex (*curved arrow*). Differential diagnosis should include consideration of fibrosarcoma, desmoid tumor, and other soft-tissue neoplasms of predominantly fibrous composition.

13. Preoperative or postoperative radiation therapy is often combined with wide excision and chemotherapy as accepted management.

Bone marrow tumors

Plasmacytoma and multiple myeloma These are solitary and disseminated forms of the same disease, characterized by proliferation of malignant plasma cells.

PEARL

Plasma cells can produce immunoglobulins G, A, M, D, and E, and increase in number during infection (especially chronic recurrent multifocal osteomyelitis), as well as in POEMS (polyneuropathy, organomegaly, endocrinopathy, monoclonal gammopathy, and skin disease) syndrome and multiple myeloma. In monoclonal gammopathy (usually involving immunoglobulin G, but also M or A) they are also increased in number.

---

1. Multiple myeloma is the most common primary bone neoplasm (Box 4-41).
2. Patients with multiple myeloma typically have anemia and back pain as early clinical manifestations.
3. Only approximately 5% of affected patients are under 40 years of age.
4. Small soft-tissue masses are commonly associated with myeloma lesions.
5. Generalized osteopenia without focal lesions occurs in approximately 85% of cases.
6. Approximately one third of patients have the presenting syndrome of solitary plasmacytoma, which eventually progresses to multifocal disease in most cases (Fig. 4-58).

PEARL

Solitary plasmacytoma of bone is usually lytic with a sclerotic margin, but it may occasionally be sclerotic. The most common sites of involvement include the spine and pelvis. Despite resection or radiation therapy, the lesion virtually always progresses to disseminated myeloma within 5 years.

---

7. Plasmacytomas generally exceed 5 cm in diameter, whereas multiple myeloma lesions are usually smaller than this (Figs. 4-59, 4-60).
8. Solitary plasmacytoma most commonly occurs in the vertebral bodies, pelvis, femur, and humerus (Fig. 4-61).
9. Many plasmacytomas manifest a nonaggressive, cystic, expansile appearance with septations (Fig. 4-62).

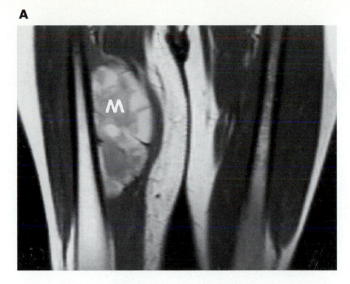

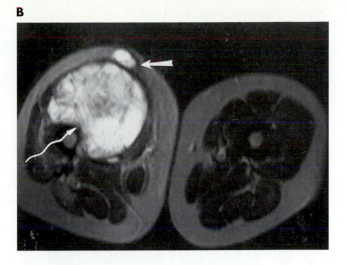

Fig. 4-55   Soft-tissue fibrosarcoma. **A,** Coronal T1-weighted MR image demonstrates a heterogeneous soft-tissue mass (*M*) involving the quadriceps musculature. **B,** Invasion of the subcutaneous fat (*straight arrow*) and femoral cortex (*wavy arrow*) are documented on a transaxial T2-weighted image, where the mass exhibits predominantly high signal intensity. Differential diagnosis should include consideration of other soft-tissue sarcomas.

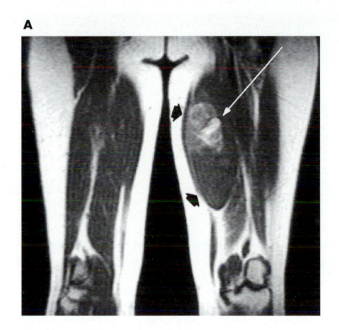

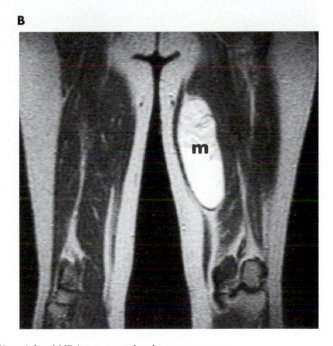

Fig. 4-56   Liposarcoma of the thigh. **A,** Coronal T1-weighted MR image reveals a heterogeneous fat-containing (*white arrow*) mass (*black arrows*) suggesting the correct diagnosis. **B,** The mass (*m*) exhibits uniformly high signal intensity on a corresponding T2-weighted image. Other fat-containing soft-tissue lesions include lipoma, macrodystrophia lipomatosa, hemangioma, and neurofibromatosis.

10. Multiple myeloma most commonly involves the skull, vertebral bodies, ribs, and proximal appendicular skeleton (Fig. 4-63).

11. Associated amyloidosis occurs in approximately 10% to 15% of patients with multiple myeloma.

12. Sclerotic bone lesions occur in approximately 1% of cases of multiple myeloma.

13. Sclerotic plasma cell lesions are a feature of the POEMS (polyneuropathy, organomegaly, endocrinopathy, monoclonal gammopathy, and skin disease) syndrome.

14. Radiographically, multiple myeloma may be simulated by osteoporosis with or without vertebral compression fractures, hyperparathyroidism, or metastatic disease (Fig. 4-64).

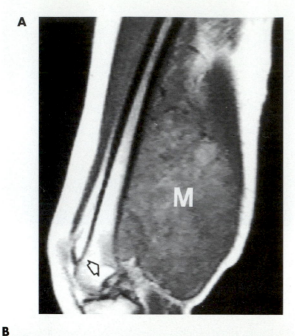

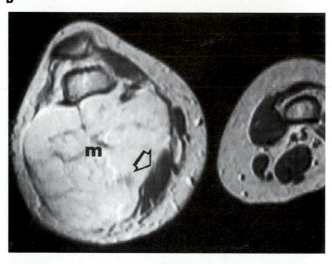

Fig. 4-57 Poorly differentiated liposarcoma of the thigh. **A,** Sagittal T1-weighted MR image demonstrates a large inhomogeneous mass (*M*) containing only sparse fat with early invasion of the distal femur (*arrow*). **B,** Transaxial T2-weighted image reveals predominantly high signal intensity within the mass (*m*), which has invaded and displaced the semimembranosus muscle (*arrow*). Differential diagnosis should include consideration of other soft-tissue sarcomas.

15. The differential diagnosis for solitary plasmacytoma should include consideration of medullary chondrosarcoma and giant cell tumor.
16. The 5-year survival rate for solitary plasmacytoma is approximately 30%.
17. The 5-year survival rate for multiple myeloma is approximately 10%.
18. Radionuclide bone scans are positive in only 25% to 40% of myeloma lesions, whereas radiography is more sensitive than scintigraphy in approximately

40% of lesions; skeletal survey therefore is more reliable for lesion detection.
19. Approximately one out of five lesions can be detected by skeletal scintigraphy but not conventional radiography.
20. Magnetic resonance imaging is more sensitive than either skeletal scintigraphy or conventional radiography for determining disease extent in multiple myeloma (Fig. 4-65).
21. Biopsy is often required because approximately 20% of patients with plasmacytoma do not manifest abnormalities on serum protein electrophoresis or bone marrow aspiration.
22. Radiation therapy and surgery are viable therapeutic options for solitary plasmacytoma.
23. Chemotherapy is the treatment of choice for multiple myeloma.
24. Myeloma lesions that are painful or prone to pathologic fracture are an indication for palliative radiation therapy (Fig. 4-66).

Box 4-41   Multiple Myeloma: Important Features

- **Peak age:** sixth decade of life (rare in patients younger than 40 years of age)
- **Male/female ratio:** approximately 2:1
- **Most common location:** vertebra, ribs, pelvis, or skull
- **Major radiographic features:** multiple small discrete lesions are identifiable involving one or multiple bones; may initially occur as a solitary osseous lesion; purely lytic; surrounding bone does not exhibit reactive sclerosis, nor is there periosteal reaction; endosteal scalloping may be present; expansion of the affected bone and associated soft-tissue mass are common
- **Radiographic differential diagnosis:** metastatic carcinoma, malignant lymphoma, fibrosarcoma
- **Treatment:** radiation and chemotherapy, depending upon the stage of the disease; surgery in cases where there is an impending fracture of the affected bone

---

Non-Hodgkin lymphoma   This is an uncommon primary round cell malignancy of bone (Box 4-42).
  1. The lesion has also been termed *reticulum cell sarcoma.*

PEARL

Reticulum cell sarcoma or histiocytic lymphoma is the third most common primary malignancy of bone in childhood. Occurrence is more common in male individuals. The most common sites are the distal femur and proximal tibia.

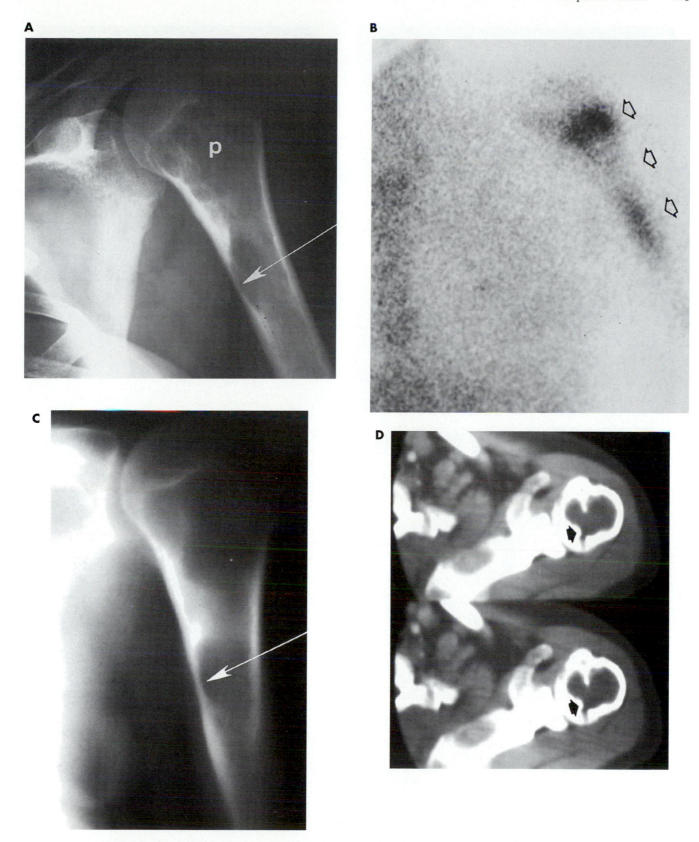

Fig. 4-58   Plasmacytoma of the proximal humerus. **A,** Radiography demonstrates an extensive lytic process (*P*) with endosteal scalloping (*arrow*). **B,** The lesion exhibits increased activity on a radionuclide bone scan (*arrows*). **C** and **D,** Endosteal scalloping (*arrows*) is optimally depicted by conventional or computed tomography. Differential diagnosis should include consideration of metastatic disease, indolent infection, eosinophilic granuloma, and primary fibrous neoplasms of bone.

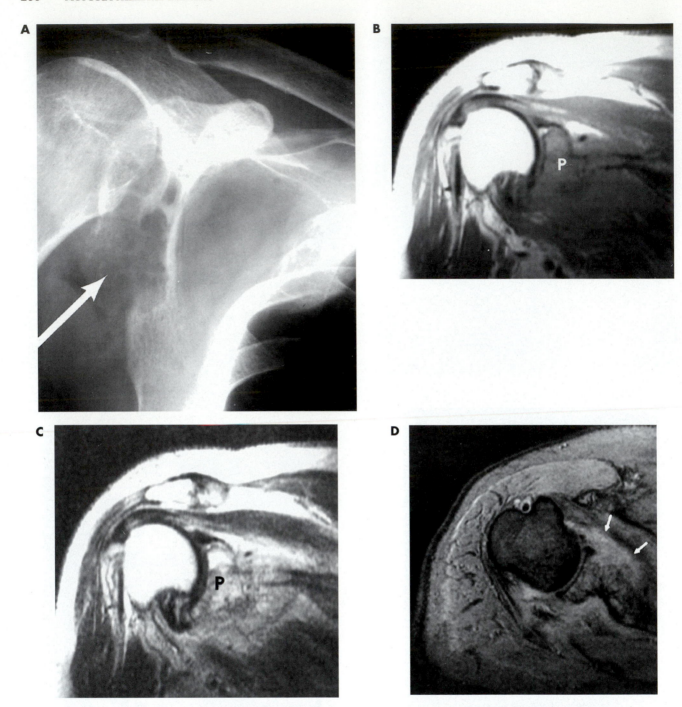

Fig. 4-59 Plasmacytoma of the scapula. **A,** Radiography demonstrates a septate expansile lesion with cortical disruption (*arrow*). **B,** Coronal T1-weighted MR image reveals low signal intensity (*P*) within the marrow of the glenoid. **C,** Corresponding T2-weighted image documents high signal intensity in the lesion (*P*). **D,** Transaxial gradient-echo image demonstrates anterior cortical disruption and soft-tissue extension of the process (*arrows*). Differential diagnosis should include consideration of metastatic disease (renal and thyroid primaries), giant cell tumor, aneurysmal bone cyst, and granulomatous infection.

2. Most patients are between 30 and 60 years of age at initial diagnosis, although the lesion can affect adolescents and young adults.

3. Extensive soft-tissue involvement is frequently observed because of rapid enlargement.

4. Radiographically, the lesion is lytic, permeative, and aggressive in appearance (Fig. 4-67).

5. The lesion is usually asymptomatic in the absence of pathologic fracture.

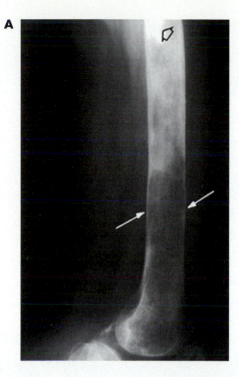

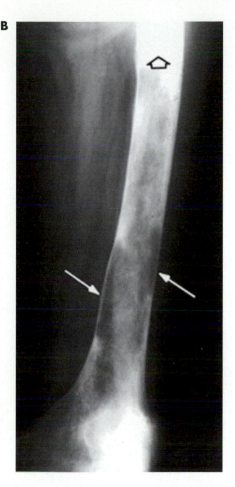

Fig. 4-60 Multiple myeloma. **A** and **B,** Lateral and frontal radiographs of the distal femur reveal multiple areas of osteolysis with endosteal scalloping (*white arrows*) and absence of sclerotic reaction. The patient has undergone hemiarthroplasty (*open arrows*) for pathologic hip fracture. Differential diagnosis should include consideration of metastatic disease and multifocal osteomyelitis.

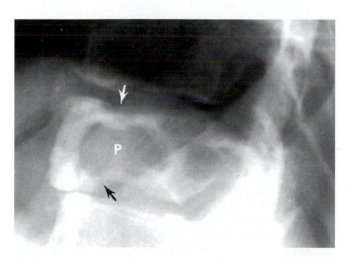

Fig. 4-61 Solitary plasmacytoma of the spine. A well-defined lytic expansile lesion (*P*) with sclerotic margins and pathologic endplate fractures (*arrows*) is evident. Differential diagnosis should include consideration of metastatic disease (particularly from renal or thyroid primaries), granulomatous infection, and lymphoma.

6. Central appendicular sites are more commonly affected than the axial skeleton.
7. Reactive sclerosis occurs in approximately 25% of cases.
8. The femur, pelvis, tibia, humerus, and scapula are the most commonly involved sites.
9. Periosteal reaction is a frequent radiographic finding.
10. The radiographic differential diagnosis should include consideration of osteomyelitis, Ewing sarcoma, fibrosarcoma, malignant fibrous histiocytoma, and Langerhans cell histiocytosis.
11. The lesion most commonly metastasizes to lymph nodes and bone, whereas pulmonary metastases are unusual.
12. The 5-year survival rate is approximately 50%.
13. Pulmonary metastases tend to grow extremely rapidly as compared with other metastatic lung nodules.
14. Skeletal scintigraphy should be performed to determine if the primary lesion has spread to other bones.
15. Magnetic resonance imaging is the optimal technique for determining local extent (Fig. 4-68).
16. Local disease is generally treated by whole bone irradiation.
17. Chemotherapy is indicated for metastatic disease.

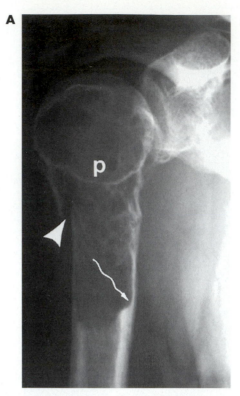

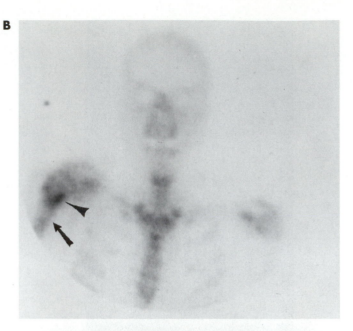

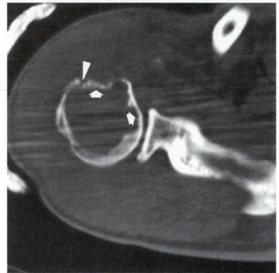

Fig. 4-62 Plasmacytoma of the proximal humerus. **A,** Radiography reveals a septate lytic lesion (*P*) with endosteal scalloping (*wavy arrow*) and pathologic fracture (*arrowhead*). **B,** Increased activity is present within the lesion (*arrow*) and particularly at the site of fracture (*arrowhead*) as seen on a radionuclide bone scan. **C,** CT optimally demonstrates the degree of cortical thinning (*arrows*), as well as the fracture (*arrowhead*).

Box 4-42 Malignant Lymphoma of Bone: Important Features

- **Peak age:** broad age range with peak in the sixth or seventh decade of life
- **Male/female ratio:** 1.5:1
- **Most common location:** may involve any bone and may involve multiple bones, but the femur is involved in approximately 25% of cases
- **Major radiographic features:** the lesion is generally characterized by extensive diaphyseal permeative destruction of the affected bone; the pattern may be one of pure lysis or sclerosis or a mixed pattern of both lysis and sclerosis; there is only sparse periosteal reaction present; the cortex may be thickened, and the lesion exhibits poor margination at the periphery; an associated soft-tissue mass is commonly present; radionuclide bone scan, computed tomography, and magnetic resonance imaging are frequently helpful in defining the extent of the lesion
- **Radiographic differential diagnosis:** Ewing sarcoma, osteosarcoma, osteomyelitis, metastatic carcinoma
- **Treatment:** radiation therapy with or without chemotherapy, depending upon the stage of the disease

---

Hodgkin lymphoma   This is an extremely rare primary neoplasm of bone; the condition usually occurs secondarily in patients with primary lymph node involvement.

1. Approximately 20% of patients with Hodgkin disease have radiographic evidence of osseous involvement.
2. Most patients are affected during the second through fourth decades of life.
3. Associated soft-tissue masses are common.
4. Approximately 75% of lesions occur in the axial skeleton.
5. Hodgkin lymphoma must be included in the differential diagnosis of an "ivory" vertebral body (Fig. 4-69).
6. Approximately 15% of lesions are purely sclerotic, whereas 25% of lesions are purely lytic.

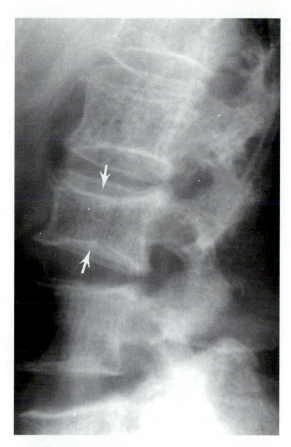

Fig. 4-63   Multiple myeloma. Characteristic abnormalities include generalized osteopenia, multiple small lytic foci of uniform size, and pathologic endplate fractures (*arrows*). Differential diagnosis should include consideration of other causes of severe generalized osteopenia, including postmenopausal or senile osteoporosis, hyperthyroidism, primary hyperparathyroidism, and corticosteroid-induced osteoporosis.

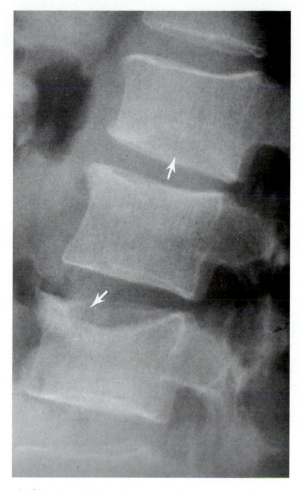

Fig. 4-64   Vertebral endplate fractures in multiple myeloma. Concave and angular deformities (*arrows*) superimposed on diffuse osteopenia are characteristic but nonspecific. Differential diagnosis should include consideration of osteoporosis, primary hyperparathyroidism, hyperthyroidism, and other causes of generalized axial skeletal bone loss.

7. Mixed lytic and sclerotic involvement occurs in approximately 60% of cases (Fig. 4-70).
8. Any identifiable periosteal reaction is minimal.
9. The major differential diagnostic consideration is metastatic disease.
10. Polyostotic involvement occurs in approximately two thirds of cases.

PEARL

Burkitt lymphoma is a B-lymphocyte disease of childhood associated with the Epstein-Barr virus. It is reported to occur in Africa, South America, and the United States. The long-term survival rate is nearly 50%. The skeletal lesions of long bones may simulate Ewing tumor or primary histiocytic lymphoma of bone. The maxilla, mandible, long bones, and pelvis are most frequently involved. Patients are usually under 10 years of age and 10% of cases are multifocal.

Ewing sarcoma   This is a highly malignant round cell neoplasm of childhood (Box 4-43).

1. The lesion is second only to osteosarcoma in frequency as a primary bone tumor of childhood.
2. The peak age range of incidence is 5 to 15 years.
3. Fewer than 5% of lesions occur in patients over the age of 25 years.
4. Radiographically, the lesion tends to be permeative with an associated soft-tissue mass (Fig. 4-71).
5. Approximately 75% of lesions involve the pelvis or long bones (Fig. 4-72).
6. Long-bone lesions are most commonly diaphyseal and based in the medullary canal (Fig. 4-73).
7. The lesion tends to involve long bones among children under age 10, and axial sites in older individuals.
8. Reactive sclerosis occurs in approximately one third of cases.

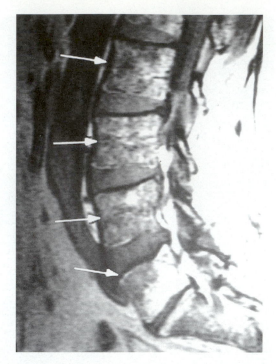

Fig. 4-65 Multiple myeloma. Sagittal T1-weighted MR image demonstrates diffuse patchy replacement of the vertebral bone marrow by areas of low signal intensity (*arrows*). Differential diagnosis should include consideration of leukemia, metastatic disease, Gaucher disease, mastocytosis, and other infiltrative disorders of marrow.

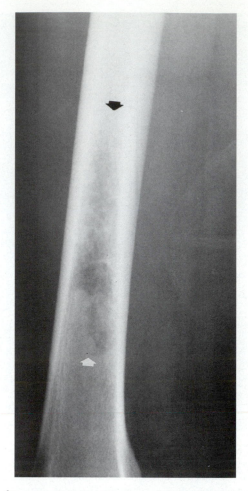

Fig. 4-67 Non-Hodgkin lymphoma of the distal femoral diaphysis in a patient with human immunodeficiency virus infection. A poorly defined lytic lesion (*arrows*) without radiodense tumor matrix or periostitis is present. Differential diagnosis should include consideration of metastatic disease, fibrosarcoma, malignant fibrous histiocytoma, plasmacytoma, and angiosarcoma.

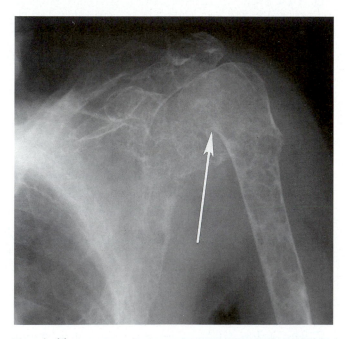

Fig. 4-66 Multiple myeloma. Radiography of the shoulder reveals diffuse osteopenia with multiple lytic lesions, endosteal scalloping, and pathologic fracture (*arrow*). Differential diagnosis should include consideration of cystic angiomatosis and other causes of diffuse longstanding bone marrow infiltration.

9. Clinical symptoms and signs simulate those of infection in up to one third of patients.
10. Ewing sarcoma is extremely rare in black individuals.
11. The differential diagnosis should include consideration of osteomyelitis, neuroblastoma, Langerhans cell histiocytosis, and osteosarcoma.
12. The 5-year survival rate is approximately 50%.
13. Approximately 15% to 30% of patients have metastases at the time of initial diagnosis.
14. Pulmonary and osseous metastases occur with approximately equal frequency.
15. The prognosis is worse for lesions in the axial skeleton than for those in the appendicular skeleton.
16. Ewing sarcoma does not produce radiodense tumor matrix, although reactive bone formation may be found in or adjacent to its associated soft-tissue mass.

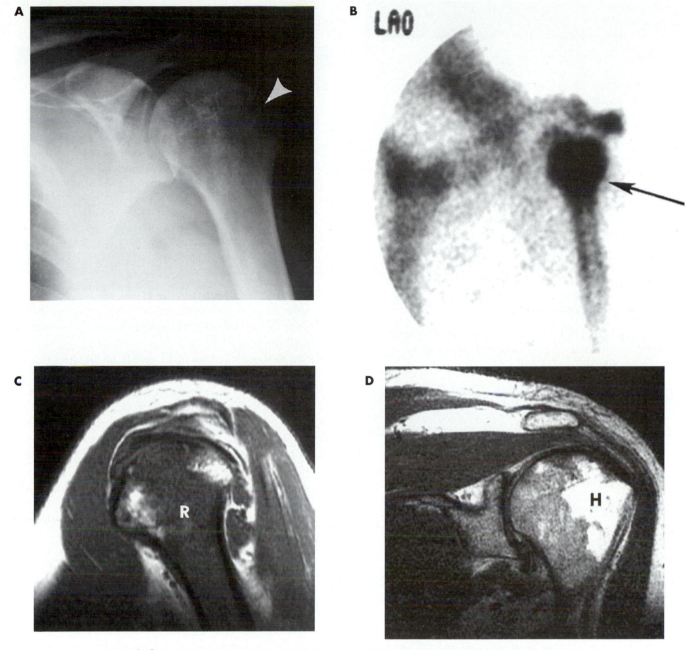

Fig. 4-68 Non-Hodgkin lymphoma of the proximal humerus. **A,** Conventional radiography demonstrates a permeative lytic process (*arrowhead*) involving the epiphysis and metaphysis. **B,** Increased activity (*arrow*) is seen on a radionuclide bone scan. **C,** Sagittal T1-weighted MR image reveals diffuse marrow replacement (*R*) by low signal intensity. **D,** Coronal T2-weighted image documents inhomogeneous high signal intensity (*H*) within the process. Differential diagnosis should include consideration of metastatic disease, hematogenous osteomyelitis, plasmacytoma, and fibrosarcoma.

17. Combined chemotherapy and radiation therapy frequently results in complications, including growth impairment and histologic dedifferentiation.
18. Local recurrence is difficult to distinguish from radiation osteonecrosis on the basis of radiographic criteria.
19. Local recurrence following radiation alone occurs in approximately 25% of cases, suggesting a role for postirradiation resection.

Box 4-43 Ewing Sarcoma: Important Features

- **Peak age:** second decade
- **Male/female ratio:** approximately 1:1
- **Most common location:** pelvic girdle and lower extremity account for approximately 60% of cases

Continued

Box 4-43—cont'd

- **Major radiographic features:** most commonly appears as an extensive diaphyseal lesion; the lesion has a permeative pattern of growth and is poorly marginated; may have regions of both lysis and sclerosis; there is characteristically a prominent periosteal reaction associated with permeation of the cortical bone; a soft-tissue mass most commonly accompanies the osseous lesion; radionuclide bone scan, computed tomography, and magnetic resonance imaging are helpful in defining the extent of the lesion
- **Radiographic differential diagnosis:** malignant lymphoma, osteosarcoma, osteomyelitis, Langerhans cell histiocytosis
- **Treatment:** combination chemotherapy, radiation therapy, and surgery

Miscellaneous tumors

Synovial sarcoma This is a soft-tissue sarcoma of synovial origin that most frequently arises in an extra-articular site.

PEARL

Synovial sarcoma is an uncommon neoplasm that can arise from within a joint but is most frequently extraarticular in location. The most common areas of predilection are the thigh and lower extremity. Patients in the third and fourth decades are most commonly affected. Calcification occurs in 20% to 30% of cases, and osseous erosion or invasion may be seen.

1. The peak age range of incidence is 15 to 35 years.
2. Synovial sarcoma tends to grow relatively slowly.
3. Fewer than 10% of lesions occur within the capsule of an articulation.
4. The lesion may arise from tendons, tendon sheaths, or bursae and hence be remote from an articulation.
5. The lesion most commonly occurs in a lower extremity, particularly around the knee (Fig. 4-74).
6. Calcification occurs in approximately 25% of lesions.
7. Secondary osseous erosion or periosteal reaction occurs in approximately 15% of cases (Fig. 4-75).
8. On magnetic resonance images, islands of tumor may be found peripheral to the primary mass, which may exhibit a nonaggressive-appearing pseudocapsule (Figs. 4-76, 4-77, 4-78).
9. The lesion typically exhibits high signal intensity on T2-weighted magnetic resonance images.
10. Wide excision is the treatment of choice.
11. Local recurrence is frequent, and metastases to the lungs occur.

Chordoma This is a low-grade neoplasm of the spine that arises from notochordal remnants.

PEARL

Chordoma is a rare lesion of notochordal origin composed of highly vacuolated cells. It is locally invasive and rarely metastatic. Male and female patients are affected equally. The lesion usually occurs during the fourth to seventh decades of life. Typical locations include the sacrococcygeal area (50% to 60% of cases), sphenooccipital area (25% to 40% of cases), and remainder of the vertebral column (15% to 20% of cases). Other unusual locations include the paranasal sinuses and nasopharynx. Vertebral lesions occur with lesser frequency in the lumbar and thoracic spine.

1. Sacral lesions tend to appear during the sixth and seventh decades of life, whereas those in the clivus or spine are usually diagnosed during the fourth or fifth decades.
2. Soft-tissue extension may result in an epidural or pelvic mass.
3. Sacral lesions usually exceed 5 cm in size, whereas those in the spine are smaller.
4. Approximately half of all lesions occur in the sacrum.
5. Approximately 35% of all lesions occur in the clivus.
6. Spinal lesions tend to involve the vertebral bodies rather than the posterior elements.
7. Approximately 40% of all primary sacral tumors are chordomas.
8. The lesion tends to grow slowly, hence a sclerotic margin is common.
9. The differential diagnosis of sacral lesions should include consideration of metastatic disease, myeloma, giant cell tumor, and chondrosarcoma.
10. Lesions of the clivus may resemble metastatic disease or chondrosarcoma.
11. The differential diagnosis for vertebral lesions should include consideration of metastatic disease and myeloma.
12. Pulmonary metastases often occur late in the disease course and affect approximately 25% of patients.
13. The 5- and 10-year survival rates are approximately 50% and 25%, respectively.
14. Magnetic resonance imaging is the procedure of choice for establishing local extent.
15. Early wide resection is the preferred therapeutic strategy.
16. Chemotherapy is not generally an effective therapeutic strategy for chordoma.
17. A local recurrence rate of approximately 80% can be anticipated with marginal resection alone.
18. Radiation therapy tends to prevent recurrence but has no effect on survival rate.

A

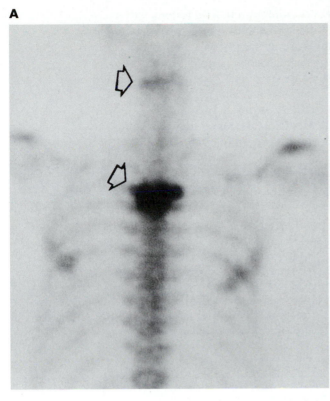

B

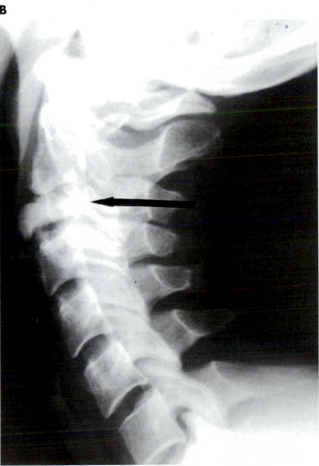

C

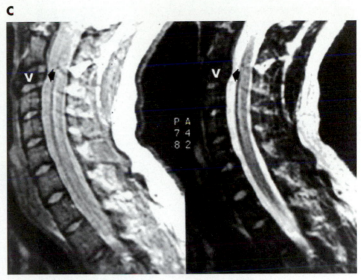

D

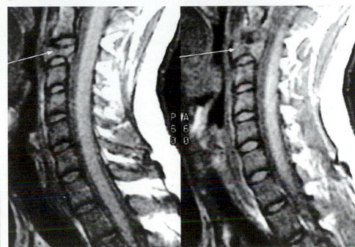

Fig. 4-69 Hodgkin disease of bone. **A,** Skeletal scintigraphy demonstrates abnormal focal areas of increased activity (*arrows*) in the cervical and thoracic spine. **B,** Lateral radiograph reveals osteosclerosis and collapse of the C3 vertebral body (*arrow*). **C,** Sagittal proton density– and T2-weighted MR images demonstrate low signal intensity within the collapsed vertebra (*V*), along with encroachment on the spinal canal (*arrows*). **D,** Gadolinium-enhanced sagittal T1-weighted images document enhancement of the lesion (*arrows*). Differential diagnosis should include consideration of metastatic disease (prostate, breast, medulloblastoma, and carcinoid primaries) and Paget disease. Other causes of the ivory vertebral body are not usually susceptible to pathologic fracture.

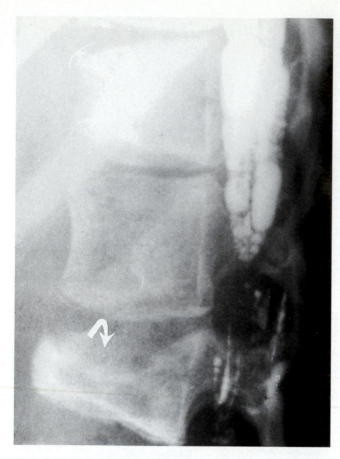

Fig. 4-70 Vertebral compression fracture secondary to Hodgkin lymphoma. Severe concavity of the superior endplate (*arrow*) is associated with a mixed pattern of osteopenia and sclerosis. Differential diagnosis should include consideration of metastatic disease, fungal infection, plasmacytoma, and non-Hodgkin lymphoma.

Rhabdomyosarcoma   Rhabdomyosarcoma is a relatively common soft-tissue neoplasm that arises from muscle. The embryonic histologic type occurs in children and commonly involves the head, neck, and genitourinary tract. The alveolar or pleomorphic histologic type occurs in adults and commonly involves the extremities. Metastatic disease commonly affects the lungs and lymph nodes. The prognosis is poor.

Metastatic disease

Primary sites of origin   Typical radiographic pattern(s) of skeletal involvement by specific primary neoplasms (Table 4-5) are as follows:

1. Lung (adenocarcinoma): purely lytic lesions or mixed lytic and blastic lesions
2. Kidney: purely lytic lesions
3. Bladder: mixed lytic and blastic lesions or purely blastic lesions
4. Gastrointestinal tract (carcinoid): purely blastic lesions
5. Breast: purely lytic lesions, mixed lytic and blastic lesions, or purely blastic lesions

6. Thyroid: purely lytic lesions
7. Lung (oat cell): purely blastic lesions
8. Prostate: mixed lytic and blastic lesions or purely blastic lesions (Fig. 4-79)
9. Neuroblastoma: purely lytic lesions or mixed lytic and blastic lesions
10. Pancreas: purely lytic lesions
11. Medulloblastoma: purely blastic lesions
12. Stomach: purely blastic lesions

PEARL

Metastasis to bone:
- Melanoma and renal cell carcinoma can be indistinguishable radiographically.
- When intestinal carcinoid metastasizes to bone, the lesions are frequently blastic.
- Brain tumors very rarely metastasize outside the cranial vault unless there has been prior surgery; blastic metastases from medulloblastoma occur in children.
- Metastases to long bones do not usually exhibit periosteal reaction, regardless of the primary tumor.

Important concepts

1. Bone metastases eventually occur in 20% to 35% of extraskeletal primary malignancies.
2. Osseous metastases are approximately 25 times more common than primary bone tumors.
3. Approximately 80% of bone metastases arise from primary breast, prostate, lung, or kidney carcinomas.
4. Approximately 10% to 40% of bone metastases are normal on conventional radiographs but positive on skeletal scintigrams.
5. Fewer than 5% of bone metastases are normal on bone scans but detectable on conventional radiographs.
6. Impending pathologic fracture of a bone metastasis is suggested by 50% destruction of cortical bone as seen on conventional radiographs or computed tomography.
7. Only 10% of skeletal metastases are solitary; they usually originate from a primary tumor in the thyroid gland or kidney.
8. Approximately 80% of bone metastases are situated in the axial skeleton, including the vertebrae, pelvis, skull, ribs, proximal femora, and proximal humeri.
9. Vertebral metastases are found in approximately 40% of malignancies at autopsy, although fewer are detected on premortem radiographs.
10. The specificity of skeletal scintigraphy for metastatic disease is poorer than that of conventional radiography, because abnormal foci may occur secondary to tumor, arthritis, trauma, or infection.
11. In patients with widespread prostate or breast metastases, skeletal scintigraphy may reveal a superscan

Table 4-5   Skeletal metastasis: radiographic appearance and incidence

Primary focus	Usual type of skeletal metastasis	Very common	Common	Infrequent	Rare
Breast	Lytic, mixed, or blastic	x			
Lung					
Carcinoma	Predominantly lytic	x			
Carcinoid	Predominantly blastic		x		
Genitourinary system					
Kidney					
Carcinoma	Lytic, expansile	x			
Wilms tumor	Lytic				x
Prostate	Predominantly blastic; lytic in older age group	x			
Urinary bladder	Predominantly lytic; blastic if prostate involved			x	
Adrenal gland					
Pheochromocytoma	Lytic, expansile				x
Carcinoma	Lytic				x
Male and female reproductive system					
Uterus					
Corpus	Lytic			x	
Cervix	Lytic or mixed			x	
Ovary	Predominantly lytic; occasionally blastic				x
Testis	Predominantly lytic; occasionally blastic			x	
Thyroid	Lytic, expansile		x		
Gastrointestinal tract					
Esophagus	Lytic				x
Stomach	Blastic or mixed			x	
Colon	Predominantly lytic; occasionally blastic		x		
Rectum	Predominantly lytic			x	
Biliary tree	Lytic				x
Pancreas	Lytic				x
Liver (hepatoma)	Lytic				x
Head, neck, and central nervous system					
Brain (medulloblastoma)	Lytic or blastic (usually after craniotomy)				x
Neuroblastoma	Lytic, mixed, or blastic	x			
Paranasal sinuses	Lytic				x
Nasopharynx	Lytic or blastic			x	
Salivary glands	Lytic				x
Chordoma	Lytic				x
Skin					
Squamous cell carcinoma	Lytic				x
Melanoma	Lytic, expansile			x	
Malignant neoplasms of bone or soft tissues					
Osteosarcoma	Lytic or blastic				x
Chondrosarcoma	Lytic or mixed				x
Ewing tumor	Lytic, permeative	x			
Fibrosarcoma	Lytic				x
Vascular neoplasms (angiosarcoma, hemangiopericytoma)	Lytic				x

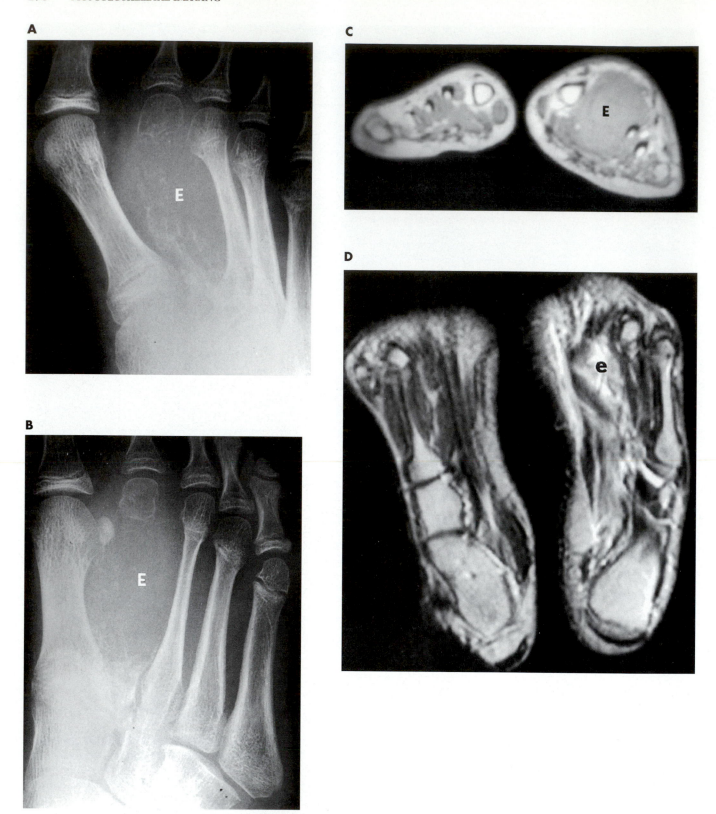

Fig. 4-71 Ewing sarcoma. **A** and **B,** Frontal and oblique radiographs of a child's foot demonstrate expansile destruction of the second metatarsal (*E*) with epiphyseal sparing. **C,** Transaxial T1-weighted MR image documents a large well-circumscribed soft-tissue mass (*E*) of low signal intensity. **D,** T2-weighted image in the plantar plane reveals high signal intensity within the lesion (*e*). Differential diagnosis should include consideration of indolent infection, other small round cell neoplasms, aggressive fibromatosis, and aneurysmal bone cyst.

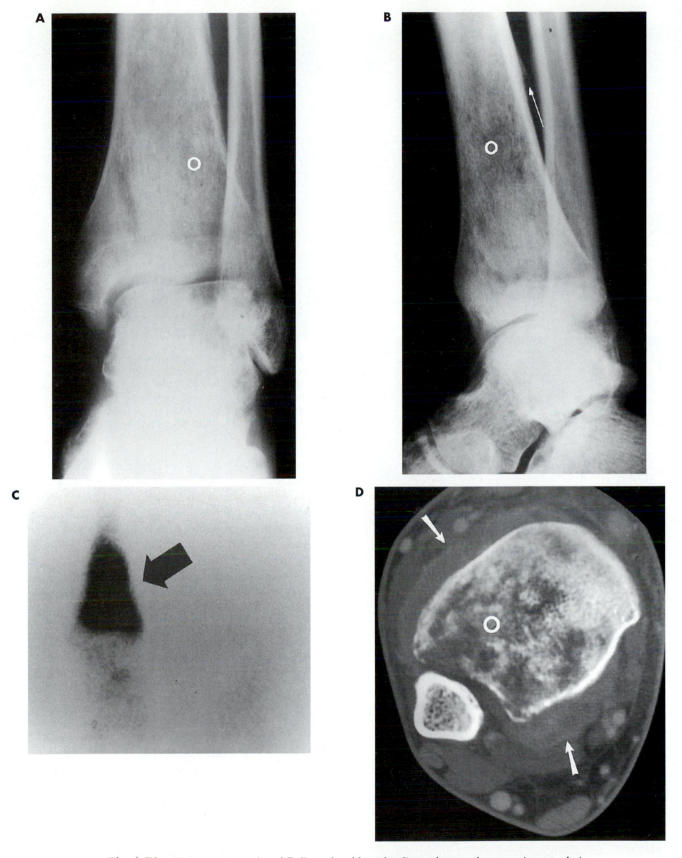

Fig. 4-72   Ewing sarcoma. **A** and **B,** Frontal and lateral radiographs reveal permeative osteolysis (*o*) in the distal tibia with an associated Codman triangle (*arrow*). **C,** The process exhibits increased activity (*arrow*) on a radionuclide bone scan. **D,** CT confirms patchy osteolysis (*O*) and also documents a soft-tissue mass (*arrows*).

Continued

E

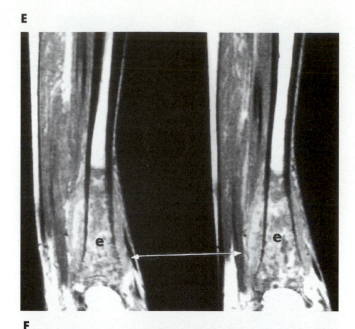

F

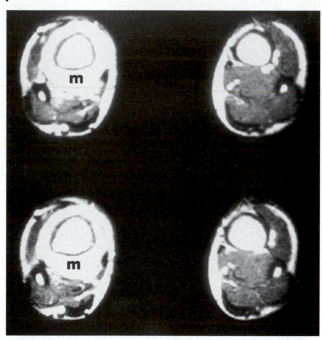

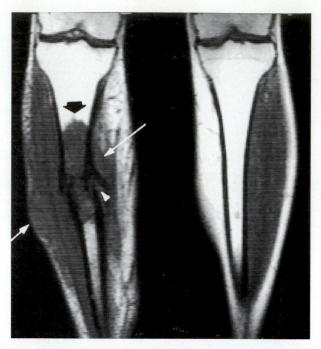

Fig. 4-73 Ewing sarcoma of the tibial diaphysis. Coronal T1-weighted MR image demonstrates a low–signal intensity lesion replacing normal marrow (*black arrow*) with soft-tissue extension (*white arrows*) and pathologic fracture (*arrowhead*). Differential diagnosis should include consideration of osteomyelitis, fibrosarcoma, malignant fibrous histiocytoma, metastatic disease, and lymphoma.

Fig. 4-72, cont'd. **E,** Coronal T1-weighted MR images demonstrate inhomogeneous marrow replacement (*e*) with soft-tissue extension (*double-headed arrow*). **F,** Transaxial T2-weighted images reveal homogeneous high signal intensity within both the marrow and soft-tissue mass (*m*). Differential diagnosis should include consideration of metastatic disease, fibrosarcoma, lymphoma, and other round cell neoplasms.

pattern of excessive diffuse bone uptake with little or no renal excretion of radiopharmaceutical.

12. Computed tomography or magnetic resonance imaging can demonstrate bone marrow infiltration and bone destruction in the setting of scintigram-positive metastases that are negative or equivocal on conventional radiographs (Figs. 4-80, 4-81, 4-82).

13. Angiography with embolization may decrease morbidity from highly vascular metastases, particularly those arising from the kidney.

14. Impending pathologic fracture of a metastasis is suggested by pain or lesion size greater than 2.5 cm; impending fracture is an indication for prophylactic therapy.

15. Unlike the situation with primary bone tumors, periosteal reaction is uncommon with metastatic lesions.

16. Metastases that manifest an expansile, bubbly, geographic appearance on conventional radiographs usually arise from primary tumors of the kidney or thyroid.

17. Changing patterns of radiographic density in bone metastases may reflect disease progression, healing secondary to therapy, or radiation osteonecrosis.

18. Cortically based metastases usually arise from primary tumors of the lung or breast.

19. Osseous metastases that occur distal to the knees or elbows are usually secondary to a primary tumor of the lung.

20. Isolated fractures of the lesser femoral trochanter in adults should be considered pathologic until proven otherwise.

21. Metastatic disease from the breast to the proximal femur usually also involves the acetabulum; hence, total hip arthroplasty should be considered the prophylactic salvage surgery of choice.

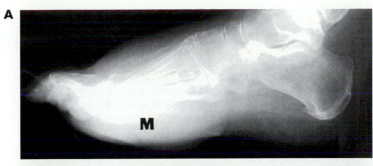

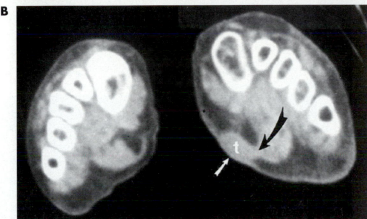

Fig. 4-74 Synovial sarcoma of the foot (clear cell variant). **A,** Lateral radiograph demonstrates a prominent soft-tissue mass (*M*) along the plantar aspect of the foot at the level of the metatarsal shafts. **B,** Computed tomographic image indicates that the tumor (*t*) arises from the plantar aponeurosis (*black arrow*) with deformity of the overlying skin (*white arrow*).

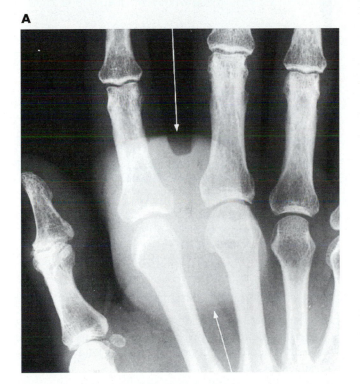

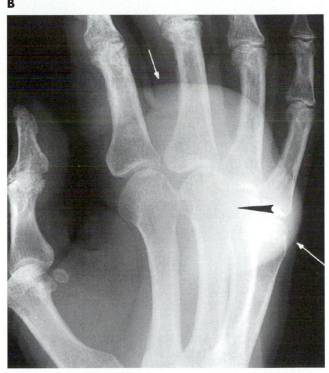

Fig. 4-75 Synovial sarcoma. **A** and **B,** A large soft-tissue mass (*arrows*) arises from the dorsal aspect of the hand and is associated with early erosion of the third metacarpal head (*arrowhead*). Differential diagnosis should include consideration of other soft-tissue sarcomas and human bite or other infection.

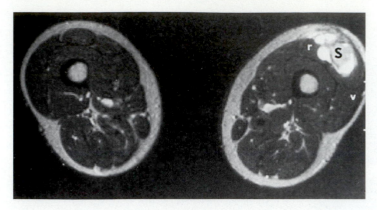

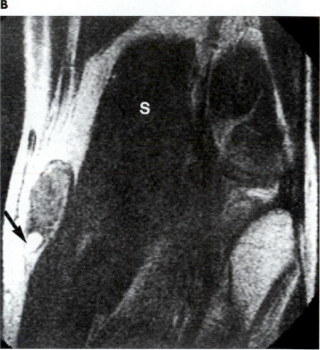

Fig. 4-76   Synovial sarcoma. Transaxial T2-weighted MR image reveals a lobulated mass of high signal intensity (*s*) involving the rectus femoris (*r*), vastus lateralis (*v*), and subcutaneous fat. Differential diagnosis should include consideration of other soft-tissue sarcomas, hematoma, seroma, and abscess.

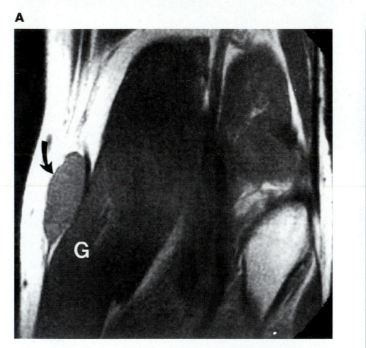

Fig. 4-77   Synovial sarcoma. **A,** Coronal T1-weighted MR image reveals a well-defined mass of low signal intensity (*arrow*) adjacent to the medial head of the gastrocnemius muscle (*G*). **B,** Corresponding T2-weighted image demonstrates an inhomogeneous appearance, with high signal intensity in the inferior portion of the lesion (*arrow*). (*S = semimembranosus muscle.*) Differential diagnosis should include consideration of other benign and malignant soft-tissue neoplasms.

22. Radiographically, vertebral metastases are characterized by involvement of vertebral bodies and pedicles, focal bone destruction, preservation of disk height, and compression fractures as a nonspecific complication (Fig. 4-83).

23. Radiation osteonecrosis is distinguished from metastatic disease by its tendency to involve contiguous bones and its restriction to a region of prior therapy.

24. Posttraumatic osteolysis of the pubic bone may simulate aggressive metastatic disease both radiographically and histologically in an elderly patient.

25. Osteopoikilosis is confidently distinguished from blastic metastatic prostate cancer by its failure to exhibit increased uptake of bone-seeking radiopharmaceuticals.

26. The principal differential diagnostic consideration for metastatic disease in older adults is multiple myeloma, although other bone marrow malignancies can appear similar radiographically (Fig. 4-84).

27. In children, leukemia may be radiographically indistinguishable from metastatic neuroblastoma (Fig. 4-85).

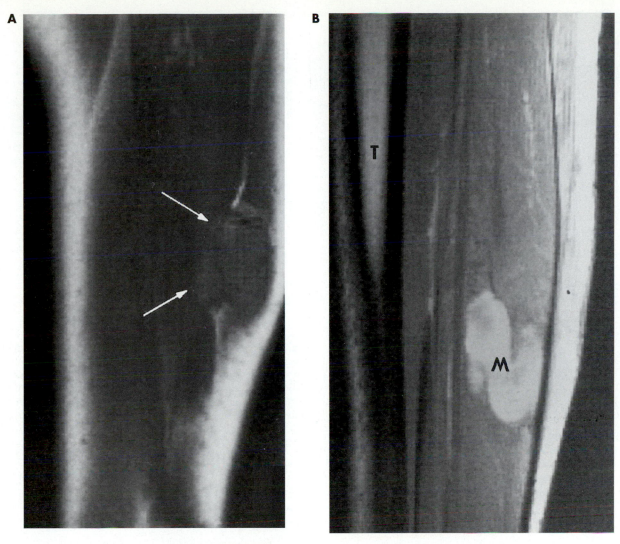

Fig. 4-78 Synovial sarcoma of the calf. **A,** Coronal T1-weighted MR image demonstrates a lobulated soft-tissue mass of low signal intensity (*arrows*) involving both muscle and subcutaneous fat. **B,** Sagittal T2-weighted image reveals predominantly high signal intensity within the mass (*M*). (*T* = *tibia.*) Differential diagnosis should include consideration of other malignant and benign soft-tissue neoplasms, as well as seroma.

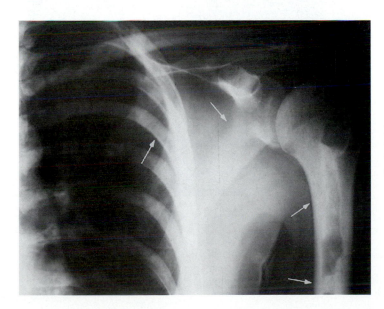

Fig. 4-79 Metastatic prostate carcinoma. Multifocal osteosclerotic lesions (*arrows*) involve the bones of the thorax and shoulder girdle. Differential diagnosis should include consideration of mastocytosis, tuberous sclerosis, sarcoidosis, osteosarcomatosis, and axial melorheostosis.

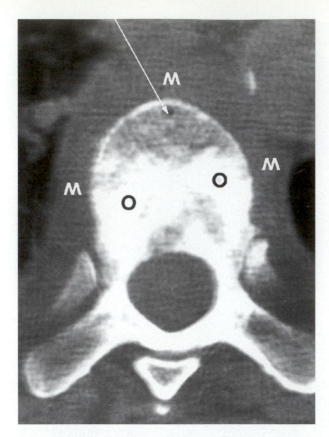

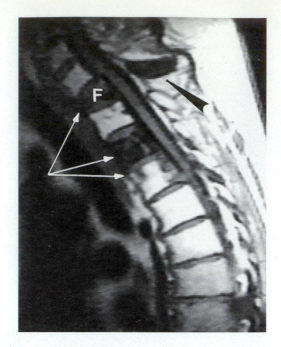

Fig. 4-81 Metastatic prostate carcinoma. Sagittal T1-weighted MR image reveals replacement of normal marrow by multiple low–signal intensity areas, with both vertebral body (*arrows*) and spinous process (*arrowhead*) involvement. A pathologic wedge fracture (*F*) is also present. Differential diagnosis should include consideration of multiple myeloma and indolent infection.

Fig. 4-80 Spinal infection in a patient with metastatic prostate cancer. CT image of a lower thoracic vertebra reveals patchy osteosclerosis (*o*) secondary to marrow involvement by tumor. In addition, a paraspinous soft-tissue mass (*M*) and gas formation (*arrow*) indicate concomitant infection. Other causes of multifocal vertebral osteosclerosis include mastocytosis, tuberous sclerosis, sarcoidosis, and axial melorheostosis.

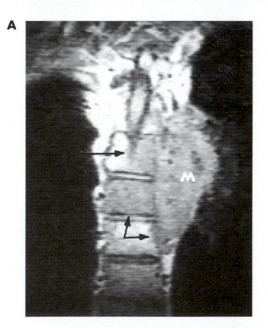

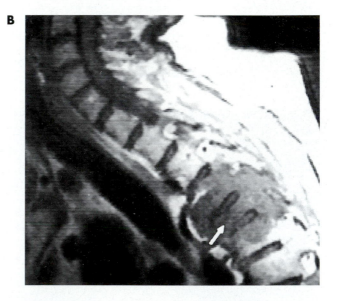

Fig. 4-82 Bronchogenic carcinoma with direct invasion of the spine. **A,** Coronal T1-weighted MR image demonstrates a large pleural-based soft-tissue mass (*M*) with vertebral involvement (*arrows*). **B,** Sagittal T1-weighted image documents involvement of three contiguous levels, with pathologic fracture of the middle segment (*arrow*). Differential diagnosis should include consideration of malignant mesothelioma, metastatic disease, and tuberculosis.

A

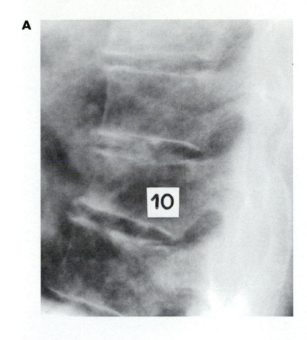

C

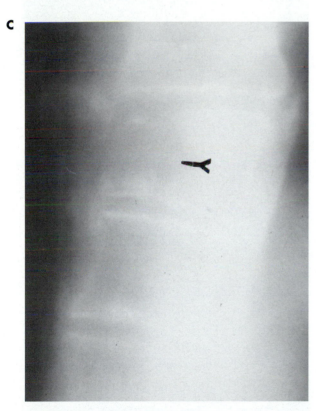

B

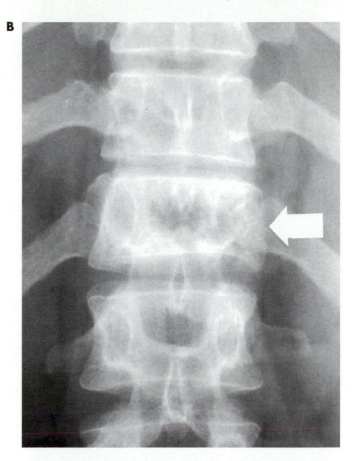

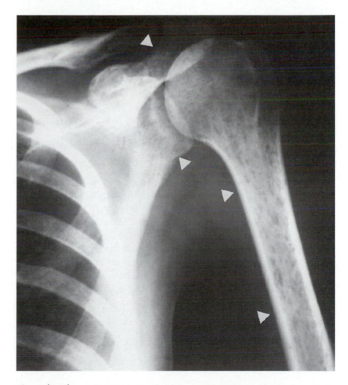

Fig. 4-83 Pathologic vertebral body fractures secondary to metastatic disease. **A,** Renal cell carcinoma involving T10; anterior wedging with superior endplate angulation is indistinguishable from osteoporotic compression fracture. **B,** Melanoma involving T12; deformities of the inferior and lateral vertebral body margins are associated with subtle alterations in trabecular architecture (*arrow*). **C,** Conventional tomography, CT, or MRI may be useful in distinguishing an underlying lesion from osteoporotic collapse of a vertebral body; the lytic focus (*arrow*) in this case represents bronchogenic carcinoma.

Fig. 4-84 Chronic myelogenous leukemia. Multiple well-defined lytic lesions (*arrowheads*) are present in the bony structures of the shoulder. Differential diagnosis should include consideration of multiple myeloma, cystic angiomatosis, reflex sympathetic dystrophy syndrome, and disuse osteoporosis.

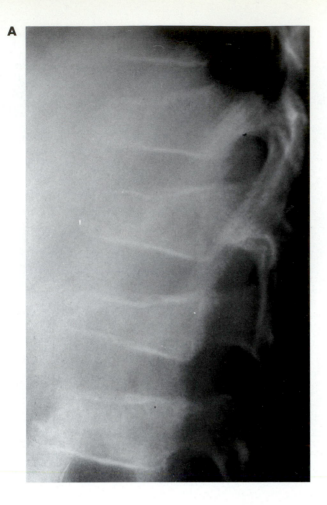

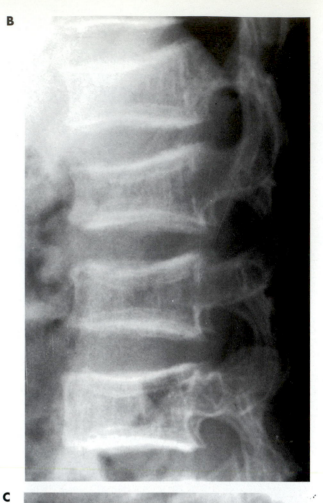

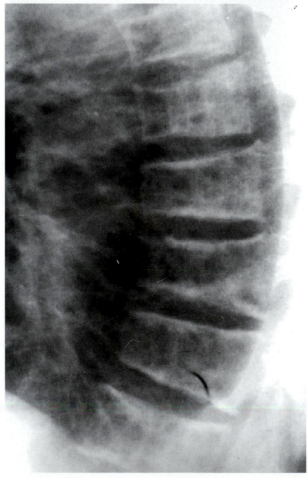

Fig. 4-85 Diffuse vertebral body deformities in infiltrative bone marrow disorders. **A,** Acute lymphocytic leukemia. Anterior wedging of the vertebral bodies is associated with generalized osteopenia. **B,** Neuroblastoma following chemotherapy. Anterior wedging and biconcave endplate deformities are associated with osteopenia and radiodense endplates paralleled by growth arrest-recovery lines. **C,** Chronic myelogenous leukemia. Generalized anterior wedging of the vertebral bodies coexists with osteopenia and accentuated density of the endplates.

PEARL

Lucent metaphyseal bands are a nonspecific finding in leukemia and other chronic childhood illnesses (probably due to a nutritional deficit that interferes with proper osteogenesis) and not a direct sign of leukemic infiltration. Metastatic neuroblastoma demonstrates clumps of extrinsic cells (termed *rosettes*) on bone marrow aspiration. This tumor is usually seen in children under 5 years of age, and long bones are frequently symmetrically involved. The bony lesions are usually lytic with minimal soft-tissue involvement. The urine tests positive for the presence of catecholamines or vanillylmandelic acid.

28. In children, polyostotic fibrous dysplasia and Ollier disease (enchondromatosis) are distinguished from metastatic disease by their tendency to be predominantly unilateral in distribution.

29. In children, metabolic bone disease is distinguished from metastatic disease by its tendency to involve the most rapidly growing areas of the skeleton.

30. The variant of Langerhans cell histiocytosis that is most likely to simulate primary malignancy with osseous metastases is Letterer-Siwe disease.

31. Because it is sometimes disseminated in younger patients, osteomyelitis should be considered in the differential diagnosis of metastatic disease among children.

SUGGESTED READINGS

1. Abdelwahab IF et al: Transarticular invasion of joints by bone tumors: hypothesis, Skeletal Radiol 20:279, 1991.
2. Assoun J et al: Osteoid osteoma: MR imaging versus CT, Radiology 191:217, 1994.
3. Berquist TH: Magnetic resonance imaging of primary skeletal neoplasms, Radiol Clin North Am 31:411, 1993.
4. Bilchik TR et al: Osteoid osteoma: the role of radionuclide bone imaging, conventional radiography, and computed tomography in its management, J Nucl Med 33:269, 1992.
5. Bloem JL and Kroon HM: Osseous lesions, Radiol Clin North Am 31:261, 1993.
6. Brown ML: Bone scintigraphy in benign and malignant tumors, Radiol Clin North Am 31:731, 1993.
7. Cassar-Pullicino VN, McCall IW, and Wan S: Intra-articular osteoid osteoma, Clin Radiol 45:153, 1992.
8. Casselman JW et al: MRI in craniofacial fibrous dysplasia, Neuroradiology 35:234, 1993.
9. Conway WF and Hayes CW: Miscellaneous lesions of bone, Radiol Clin North Am 31:339, 1993.
10. Davies AM, Cassar-Pullicino VN, and Grimer RJ: Incidence and significance of fluid-fluid levels on computed tomography of osseous lesions, Br J Radiol 65:193, 1992.
11. Giudici MAI, Moser RP Jr, and Kransdorf MJ: Cartilaginous bone tumors, Radiol Clin North Am 31:237, 1993.
12. Golfieri R et al: Primary bone tumors: MR morphologic appearance correlated with pathologic examinations, Acta Radiol 32:290, 1991.
13. Greenspan A, Steiner G, and Knutzon R: Bone island (enostosis): clinical significance and radiologic and pathologic correlations, Skeletal Radiol 20:85, 1991.
14. Hayes CW, Conway WF, and Sundaram M: Misleading aggressive MR imaging appearance of some benign musculoskeletal lesions, Radiographics 12:1119, 1992.
15. Hendrix RW, Rogers LF, and Davis TM Jr: Cortical bone metastases, Radiology 181:409, 1991.
16. Hudson TM, Stiles RG, and Monson DK: Fibrous lesions of bone, Radiol Clin North Am 31:279, 1993.
17. Kattapuram SV and Rosenthal DI: Percutaneous biopsy of skeletal lesions, AJR 157:935, 1991.
18. Kelsar PJ, Buck JL, and Suarez ES: Germ cell tumors of the sacrococcygeal region: radiologic-pathologic correlation, Radiographics 14:607, 1994.
19. Kenan S et al: Lesions of juxtacortical origin (surface lesions of bone), Skeletal Radiol 22:337, 1993.
20. Kransdorf MJ and Meis JM: Extraskeletal osseous and cartilaginous tumors of the extremities, Radiographics 13:853, 1993.
21. Kransdorf MJ et al: Osteoid osteoma, Radiographics 11:671, 1991.
22. Kransdorf MJ, Moser RP Jr, and Gilkey GW: Fibrous dysplasia, Radiographics 10:519, 1990.
23. Kumar R et al: Fibrous lesions of bones, Radiographics 10:237, 1990.
24. Manaster BJ: Musculoskeletal oncologic imaging, Int J Radiat Oncol Biol Phys 21:1643, 1991.
25. Manaster BJ and Doyle AJ: Giant cell tumors of bone, Radiol Clin North Am 31:299, 1993.
26. Mohammadi-Araghi H and Haery C: Fibro-osseous lesions of craniofacial bones: role of imaging, Radiol Clin North Am 31:121, 1993.
27. Moore TE et al: Sarcoma in Paget disease of bone: clinical, radiologic, and pathologic features in 22 cases, AJR 156:1199, 1991.
28. Moser RP Jr et al: Primary Ewing sarcoma of rib, Radiographics 10:899, 1990.
29. Murphey MD, Gross TM, and Rosenthal HG: Musculoskeletal malignant fibrous histiocytoma: radiologic-pathologic correlation, Radiographics 14:807, 1994.
30. Stull MA, Kransdorf MJ, and Devaney KO: Langerhans cell histiocytosis of bone, Radiographics 12:801, 1992.
31. Vanel D et al: MR imaging in the follow-up of malignant and aggressive soft tissue tumors: results of 511 examinations, Radiology 190:263, 1994.
32. Varma DGL et al: Chondrosarcoma: MR imaging with pathologic correlation, Radiographics 12:687, 1992.
33. Weatherall PT et al: Chondroblastoma: classic and confusing appearance at MR imaging, Radiology 190:467, 1994.
34. Wilner D: Radiology of bone tumors and allied disorders, Philadelphia, 1982, WB Saunders Co.

CHAPTER 5

Metabolic Bone Disease

OSTEOPOROSIS

1. Osteoporosis is defined as a decreased volume of bone with normal mineralization and histologic characteristics (Box 5-1).

PEARL

Postmenopausal osteoporosis:
- Serum alkaline phosphatase level is usually normal, except during the acute phase of healing fracture.
- The weight-bearing trabeculae are the last to disappear; therefore, vertical trabeculae in the vertebral column persist and appear coarsened radiographically.
- Trabecular bone loss occurs more rapidly than cortical bone loss, and bone loss is not uniform throughout the skeleton.
- Serum calcium and phosphate levels are usually normal.
- Serum estrogen levels are low in the absence of replacement therapy.

2. The radiographic differential diagnosis for osteoporosis should include consideration of osteomalacia, hyperparathyroidism, and multiple myeloma.
3. The distribution of osteoporosis may be either generalized or localized.
4. Potential etiologies for generalized osteoporosis include aging (postmenopausal or senile osteoporosis), hypercorticism, alcoholism, hyperthyroidism, drugs, acromegaly, mastocytosis, ochronosis, amyloidosis, osteogenesis imperfecta, homocystinuria, and idiopathic juvenile osteoporosis.

Causes of osteoporosis include idiopathic juvenile osteoporosis, postmenopausal estrogen deficiency, senility, hypogonadism, thyrotoxicosis, malabsorption, scurvy, calcium deficiency, long-term heparin administration, immobilization, rheumatoid arthritis, malnutrition, alcoholism, osteogenesis imperfecta, Gaucher disease, Cushing syndrome (endogenous or iatrogenic), and homocystinuria.

Cushing syndrome is characterized by diffuse osteoporosis with decreased osteoblastic activity and pseudocallus formation. Reactive new bone formation occurs in adjacent viable bone secondary to osteonecrosis or bone infarction. Mastocytosis exhibits osteoporosis and reactive sclerosis secondary to mast cell infiltration of bone marrow, unlike tuberous sclerosis, where cystlike lesions occur in the terminal phalanges with cortical sclerosis and axial hyperostosis.

5. Bone loss begins at approximately age 35 in both sexes, but the rate is much higher in women, who have an approximately eight-fold greater risk of developing pathologic osteoporosis.
6. The incidence of pathologic osteoporosis is lower in the black race than among whites or those of Asian descent.
7. Risk factors for inappropriate bone loss with age include family history of osteoporosis, estrogen deficiency, calcium deficiency, low body weight, sedentary lifestyle, amenorrhea (secondary to oophorectomy or excessive exercise), alcoholism, and smoking.
8. Age-related osteoporosis is not irreversible, and the rate of bone loss can be altered by therapeutic intervention, particularly using osteoclast inhibitors such as alendronate.
9. The pathophysiology of osteoporosis involves decreased osteoblastic activity and/or increased osteoclastic activity.

PEARL

Calcium metabolism:
- Calcium is constantly added to and removed from bone.
- Between 100 and 400 mg of calcium are excreted in the urine each day.
- The majority of ingested calcium is absorbed in the proximal small bowel under the influence of vitamin D.
- Calcium in the skeleton accounts for 1 to 2 kg of body weight.
- Of total calcium in the body, 98% is present in the skeleton.

10. Cancellous bone loss occurs more rapidly than cortical bone loss, because of the approximately eight-fold greater resorptive surface area of the former.
11. The most common sites of fracture in age-related osteoporosis include vertebral bodies, proximal femur, distal radius, and proximal humerus.
12. Conventional radiography is relatively insensitive for identification of early bone loss; reliable detection requires loss of up to 50% of total mass, especially in purely cancellous bone; hence, quantitative bone densitometry is preferred for early diagnosis and monitoring of osteoporosis (Table 5-1).

PEARL

Osteoporosis:
- Losses of up to 50% of bone mass are required before osteoporosis appears as decreased density on conventional radiographs.
- More sensitive techniques include quantitative CT, dual-energy x-ray absorptiometry, and single-energy x-ray absorptiometry.
- The second metacarpal diaphysis is used to evaluate loss of cortical bone in a technique known as radiogrammetry.
- Osteoporosis results in preferential loss of horizontal trabeculae with relative preservation of the weight-bearing vertical trabeculae.

13. Conventional radiographic signs of osteoporosis include decreased bone density, cortical thinning, and accentuation of weight-bearing trabeculae.
14. Deformities of vertebral bodies occurring secondary to osteoporosis include biconcave endplates, anterior wedging, and acute compression fractures.
15. Drug therapy for age-related osteoporosis may include calcium supplementation, estrogen replacement, vitamin D supplementation, sodium fluoride, bisphosphonates, and calcitonin.
16. Radiographic manifestations of sodium fluoride therapy may include increased bone density, accentuation of trabeculae, and insufficiency fractures (iatrogenic fluorosis).
17. Corticosteroid-induced osteoporosis (Cushing syndrome) may be exogenous or endogenous, and it manifests distinguishing radiographic features including fractures with exuberant callus formation (Fig. 5-1), vertebral collapse with vacuum phenomenon (Fig. 5-2), subchondral ischemic necrosis, and diaphyseal bone infarcts (Box 5-2).

Table 5-1　Quantitative techniques for measuring bone density

Technique (alphabetically)	Bones measured	Examination time (minutes)	Possible accuracy error (%)	Effective radiation dose (μSv)*
Dual-energy x-ray absorptiometry (DXA) *(uses dual-energy photons from an x-ray source)*	Spine, hip, total body, others	1-10	3-9	1
Dual-photon absorptiometry (DPA) *(uses dual-energy photons from a radioactive gadolinium-153 source)*	Spine, hip, total body	20-40	4-10	5
Quantitative computed tomography (QCT) *(uses a conventional CT scanner with specialized calibration hardware and software)*	Spine, hip	20	5-20	60†
Peripheral QCT (PQCT) *(a small-bone version of QCT that measures the bone density of the wrist)*	Forearm	10	4-8	3
Radiographic absorptiometry (RA) *(uses x-rays of the hand and a small metal wedge to calculate bone density)*	Phalanges	1-3	4	1
Single-energy x-ray absorptiometry (SXA) *(uses an x-ray source to measure bone)*	Forearm, calcaneus	4	5	<1
Single-photon absorptiometry (SPA) *(uses single-energy photons from a radioactive isotope source)*	Forearm, calcaneus	15	4-6	<1

*Effective dose refers to radiation that reaches internal organs. For comparison, one chest x-ray gives a radiation dose of about 50 μSv; a lateral spine x-ray, 500 to 1000 μSv; an abdominal CT scan, about 4000 μSv; and natural background radiation is about 2000 to 3000 μSv per year.
†Radiation dose may be up to 600 μSv on older CT scanners.

PEARL

Tumors that elaborate adrenocorticotropic hormone (ACTH) resulting in Cushing syndrome are common and include neuroendocrine tumors (pituitary adenoma, medullary thyroid carcinoma, carcinoid tumor, pancreatic islet cell tumor), oat cell carcinoma, thymoma, adrenal tumors (pheochromocytoma, carcinoma), and other primary neoplasms (breast, liver, prostate, ovary.)

Cushing syndrome is the result of excess circulating adrenal glucocorticoid, either secondary to hyperfunctioning adrenal cortex or administration of corticosteroids. Endogenous Cushing syndrome affects women more often than men by a 3:1 ratio. It is usually associated with adrenal hyperplasia in adults and adrenal tumors in children. Seventy-five percent of endogenous cases are caused by adrenal hyperplasia; the syndrome results less commonly from adenoma or carcinoma and rarely from neuroblastoma or pituitary adenoma. Histopathologically, Cushing syndrome results in osteoporosis and increased callus formation at fracture sites. The classic radiographic abnormalities include osteopenia, multiple vertebral compression fractures, exuberant fracture healing response, and abnormal fat deposits (buffalo hump). Osteonecrosis and bone infarcts are also important complications.

---

18. The association of osteoporosis, scoliosis, and arachnodactyly in a young patient suggests the diagnosis of homocystinuria.

PEARL

Osteoporosis is a frequent and striking feature of homocystinuria; in contrast Marfan syndrome is not associated with osteoporosis. Widespread calcified and narrowed disks, arthropathy, and osteoporosis are pathognomonic of ochronosis (absence of homogentisic acid oxidase). Osteomalacia occurs in Wilson disease, which is an autosomal recessive disturbance of copper metabolism.

---

19. The association of osteoporosis, joint space widening, prominent soft tissues, osseous excrescences, and spadelike tufts suggests the diagnosis of acromegaly.
20. The association of osteoporosis, multiple fractures with exuberant callus formation, bowing of long bones, and blue ocular sclerae suggests the diagnosis of osteogenesis imperfecta.
21. Pharmacologic agents known to induce osteoporosis include corticosteroids, phenytoin (Dilantin), phenobarbital, and heparin (>15,000 U/day).

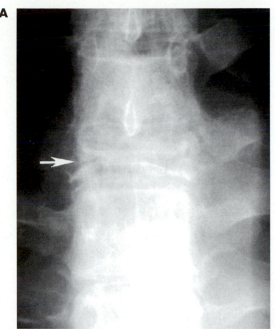

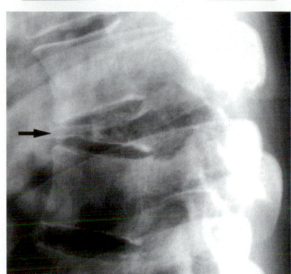

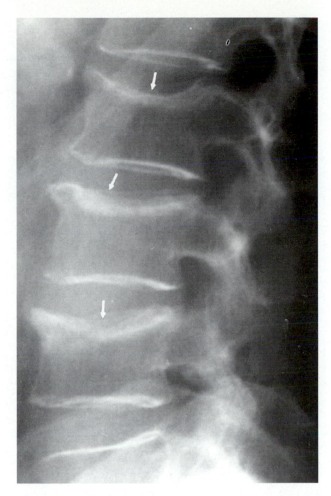

Fig. 5-1 Vertebral compression fractures in steroid-induced osteoporosis. Increased radiodensity of the concave superior endplates (*arrows*) represents the exuberant callus formation associated with fracture healing in the setting of corticosteroid therapy. This phenomenon also occurs in neuroarthropathy, osteogenesis imperfecta, and osteopetrosis.

Fig. 5-2 Vacuum vertebral body secondary to corticosteroid-induced osteonecrosis. **A** and **B,** Frontal and lateral radiographs reveal severe collapse of a thoracic vertebral body, which contains internal gas (*arrows*). Differential diagnosis should include consideration of malignancy with collapse and central necrosis, particularly plasmacytoma and other round cell neoplasms.

22. The acute onset of self-limited but irreversible osteoporosis with vertebral and metaphyseal fractures in a previously healthy 8- to 12-year-old child with no risk factors for the disease suggests the diagnosis of idiopathic juvenile osteoporosis (Fig. 5-3).

PEARL

Idiopathic juvenile osteoporosis is characterized by generalized osteoporosis (predominantly involving the axial skeleton) with multiple fractures, including vertebral body compression fractures.

23. Potential causes of localized osteoporosis include disuse, reflex sympathetic dystrophy syndrome, and transient regional osteoporosis.
24. Radiographically, rapid bone loss or aggressive osteoporosis is characterized by patchy decreased bone density, cortical tunneling or striations, and radiolucent metaphyseal bands.

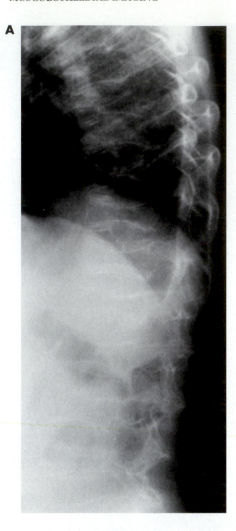

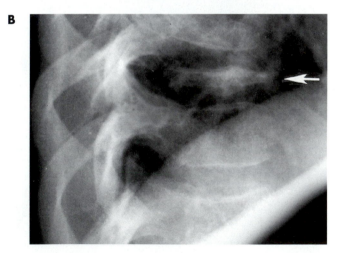

Fig. 5-3   Idiopathic juvenile osteoporosis. **A,** Typical abnormalities include generalized osteopenia, cortical thinning, and widespread biconcave vertebral endplate deformities. Differential diagnosis should include consideration of osteogenesis imperfecta, rickets, and leukemia. **B,** The differential diagnosis for vertebra plana (*arrow*) in a child should include consideration of eosinophilic granuloma, cystic angiomatosis, leukemia, lymphoma, neuroblastoma, fungal infection, and metabolic bone disease with pathologic fracture.

25. Reflex sympathetic dystrophy syndrome or Sudeck atrophy is a form of localized osteoporosis that is mediated by the sympathetic nervous system and is characterized by aggressive osteoporosis, soft-tissue swelling, and hyperesthesia progressing to atrophy and contracture; trauma is the most common of its many causes, and it is occasionally idiopathic (Figs. 5-4, 5-5).

PEARL

Reflex sympathetic dystrophy syndrome:
- Most patients are over 50 years of age at the time of diagnosis.
- Increased accumulation of bone-seeking radiopharmaceuticals is typical at affected sites.
- The shoulder and hand are the most commonly involved sites.
- Soft-tissue swelling and regional osteoporosis are characteristic radiographic findings.
- The disease tends to exhibit bilateral but markedly asymmetric involvement.
- Reported associated conditions have included myocardial infarction; cervical degenerative disease; previous trauma, surgery, or infection; vasculitis; calcific tendinitis; and neoplasm.
- Bone resorption can occur in a cancellous, subperiosteal, intracortical, endosteal, or subchondral location.

26. The two forms of transient regional osteoporosis are transient osteoporosis of the hip and regional migratory osteoporosis; both are characterized by a self-limited course.
27. Magnetic resonance imaging or skeletal scintigraphy is useful for the early diagnosis of transient osteoporosis of the hip; resolution of painful symptomatology and imaging abnormalities generally occurs within 1 year, although contralateral metachronous recurrence is possible.
28. Regional migratory osteoporosis is most common in the male sex during the fourth and fifth decades of life, and tends to involve the knee, ankle, and foot; clinical manifestations of swelling and pain tend to resolve within 1 year but recur metachronously at different sites.

Box 5-1   Osteoporosis: Important Features

- Deficiency of organic bone matrix due to disproportionate osteoclastic and osteoblastic activity
- Common generalized causes: senile, postmenopausal, corticosteroid-induced, nutritional (calcium and/or protein) deficiency

Continued

A

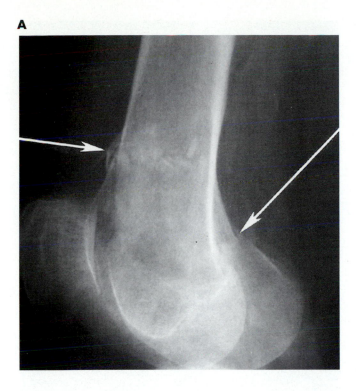

B

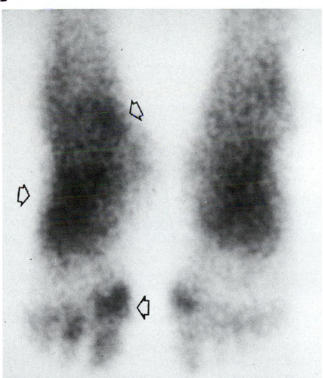

Fig. 5-4   Reflex sympathetic dystrophy syndrome (Sudeck atrophy). **A,** Following a comminuted fracture of the distal femur (*arrows*), the patient developed pain and swelling distal to the injury. **B,** Skeletal scintigraphy reveals diffusely increased activity in the affected foot (*arrows*) as compared to the contralateral side, secondary to hyperemia.

OSTEOMALACIA AND RICKETS

1. The histopathology of rickets and osteomalacia is characterized by incomplete mineralization of normal osteoid tissue.
2. Both disorders are characterized by decreased serum levels of calcium and phosphate and an increased serum level of alkaline phosphatase.

PEARL

Osteomalacia is characterized by elevated serum alkaline phosphatase levels with low serum levels of calcium and phosphate. Familial hypophosphatemia is X-linked vitamin D–resistant osteomalacia, and involves defective phosphate absorption by the kidney. Osteomalacia may occasionally occur in neurofibromatosis secondary to defective renal function. Rickets is osteomalacia occurring in the immature skeleton.

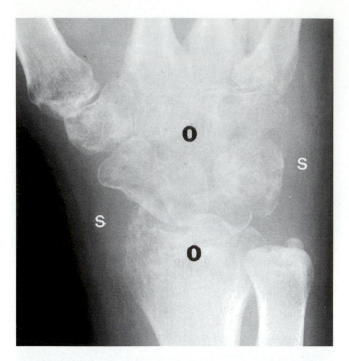

Fig. 5-5   Reflex sympathetic dystrophy (Sudeck atrophy). Characteristic findings include severe patchy osteopenia (*o*) and diffuse soft-tissue swelling (*s*). Disuse osteoporosis is the major differential diagnostic consideration.

3. Both disorders are characterized by decreased levels of calcium and phosphate in the urine.
4. Generalized osteopenia is a major radiographic sign of rickets and osteomalacia, and it is caused by a reduction in the number of trabeculae (Box 5-3).

PEARL

Rickets usually results in osteopenia. However, with renal rickets, osteosclerosis can occur as a result of associated secondary hyperparathyroidism. Scurvy results in osteoporosis whereas pyknodysostosis and tuberous sclerosis result in osteosclerosis. Pyknodysostosis is an autosomal recessive dysplastic condition that resembles osteopetrosis but has wormian bones and acro-osteolysis. Systemic mastocytosis can produce either increased, decreased, or mixed bone density.

---

5. Indistinctness of trabeculae is a radiographic sign of both disorders that occurs secondary to incomplete mineralization of trabeculae (Fig. 5-6).
6. Looser zones or pseudofractures are wide, often incomplete, transverse radiolucent bands that contain unmineralized osteoid; these lesions are often symmetric in distribution and tend to involve the medial aspects of the proximal femora, posterior

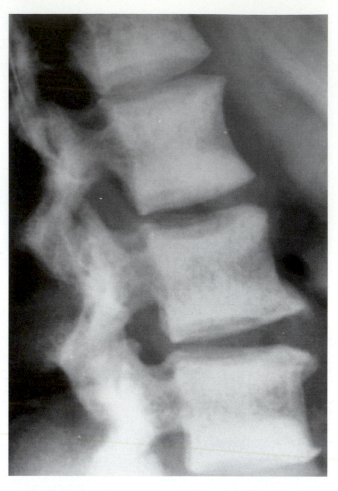

Fig. 5-6   Atypical axial osteomalacia. Characteristic findings in the spine include indistinctness of the trabeculae and generalized osteosclerosis, particularly affecting the endplate regions. Differential diagnosis should include consideration of renal osteodystrophy, fluorosis, and other causes of diffuse osteosclerosis.

aspects of the proximal ulnae, axillary margins of the scapulae, pubic rami, and ribs.

PEARL

Pseudofractures occur in rickets, osteomalacia, and Paget disease. In the last, they involve tensile areas as opposed to the compressive sites typically affected in vitamin D deficiency states.

---

7. Radiographic abnormalities in rickets most often occur at sites of rapid growth, such as the distal femur, proximal tibia, distal radius, distal tibia, and proximal humerus.
8. Radiographically, rachitic alterations are predominantly metaphyseal, with a wide radiolucent zone of provisional calcification composed of unmineralized osteoid.

In children with rickets, the failure of normal ossification is most pronounced at the growth plate. The zone of provisional calcification does not mineralize but continues to expand, leading to marked elongation and widening of the metaphyseal portion of the growth plate.

Vitamin D is a steroid that requires a two-step hydroxylation in order to become the most active form. The first hydroxylation, which occurs at the 25 position, occurs in the liver, followed by hydroxylation at the 1 position in the kidney. 1,25-dihydroxy-D3 is the active form of vitamin D. Parathyroid hormone increases the rate of renal hydroxylation of vitamin D.

9. The metaphyses in rickets are typically wide, irregular, and cupped; important potential complications of the disease include vertebral endplate deformities, bowing of long bones, and slipped epiphyses (Fig. 5-7).
10. Vitamin D or cholecalciferol is a prohormone produced endogenously by skin exposure to ultraviolet radiation and derived exogenously from dietary intake.
11. Vitamin D undergoes two hydroxylations to become an active metabolite; in the liver, a hydroxyl group is added at the 25 position, and in the kidney, a second hydroxyl group is added at the 1 position.

12. The two primary functions of vitamin D are to maintain (1) normal serum calcium and phosphate levels and (2) bone mineralization.
13. Vitamin D has direct effects on bone, the intestine, the kidneys, and the parathyroid glands.
14. The effects of vitamin D on bone are mobilization of calcium and phosphate via osteoclastic stimulation (in conjunction with parathyroid hormone) and stimulation of bone mineralization.
15. The effect of vitamin D on the intestine is stimulation of calcium and phosphate absorption.

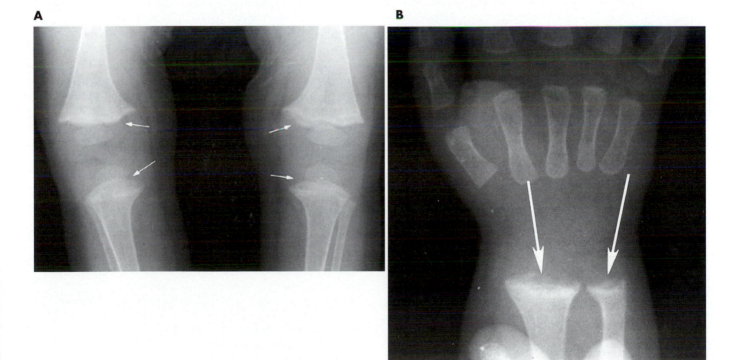

A   **B**

Fig. 5-7   Rickets. Frontal radiographs of the knees (**A**) and wrist (**B**) demonstrate physeal widening with irregular cupping of the metaphyseal margins (*arrows*). Differential diagnosis should include consideration of hypophosphatasia and metaphyseal chondrodysplasias.

16. The effect of vitamin D on the parathyroid glands is suppression of parathyroid hormone production.
17. The effect of vitamin D on the kidney is stimulation of tubular phosphate resorption.
18. Vitamin D levels are increased by dietary calcium deficiency and hypophosphatemia and decreased by high vitamin D levels (self-regulation) and low parathyroid hormone levels.

PEARL

Vitamin D is used for homeostatic maintenance of serum calcium and phosphorus levels and mineralization of bone. The two main target areas for action are the intestine and the bone, although the kidneys and the parathyroids are also affected. There are three hormones that mediate this balance: calcitonin, parathyroid hormone, and vitamin D. Bone acts as the calcium reservoir. The effect of vitamin D on the intestine is to increase absorption of calcium and phosphorus. Vitamin D acts at the cellular level by influencing cellular deoxyribonucleic acid (DNA). The action of vitamin D on bone is twofold: (1) mobilization of calcium and phosphate from already mineralized bone and (2) stimulation of maturation and mineralization of the organic matrix. The first process requires vitamin D and parathyroid hormone. This process is necessary for the maintenance of normal serum calcium and phosphate levels. The second function is mineralization of the matrix resulting from a direct effect on the bone cells and osteoid. The action of vitamin D on the kidney increases renal tubular absorption of phosphate.

---

19. Potential causes of osteomalacia and rickets include dietary calcium deficiency, gastrointestinal malabsorption, hepatobiliary disease, renal disease, drug therapy, neoplastic disease, and neonatal factors; amyloidosis is also associated with osteomalacia, but not osteoporosis.
20. Vitamin D deficiency results from hepatocellular disease because of interference with hydroxylation of prohormone vitamin D, and from biliary disease because of malabsorption of vitamin D.
21. Pharmacologic agents implicated in the development of osteomalacia and rickets include phenobarbital, phenytoin, and certain bisphosphonates.
22. Osteomalacia or rickets occurs in renal osteodystrophy secondary to interference with the second hydroxylation step of vitamin D.
23. The renal tubular disorders induce vitamin D–resistant or refractory rickets and include cystinosis and familial X-linked hypophosphatemia; in these conditions, rickets occurs secondary to renal tubular phosphate loss, resulting in elevated levels of urinary calcium and phosphate.

PEARL

Osteomalacia and rickets:
- In the tumor-associated variety, hemangiopericytoma has been the most frequent histologic diagnosis, although nonossifying fibroma, giant cell tumor, osteoblastoma, and fibrous dysplasia have also been implicated.
- In the X-linked hypophosphatemic variety, spinal alterations resembling diffuse idiopathic skeletal hyperostosis occur in association with a degenerative-like enthesopathy.
- Atypical axial osteomalacia occurs exclusively in adult men, spares appendicular sites, does not produce Looser zones, and does not respond to vitamin D therapy.
- Among low–birth weight premature infants, rachitic alterations usually appear at about 12 weeks of age.
- The conditions may develop in a wide variety of small bowel malabsorptive states, including sprue, celiac disease, regional enteritis, scleroderma, multiple jejunal diverticula, blind loop syndromes, and subsequent to gastrectomy or small bowel resection.
- Looser zones or pseudofractures occur most frequently in the axillary aspects of the scapulae, ribs, superior and inferior pubic rami, inner margins of the proximal femora, and posterior margins of the proximal ulnae, commonly in a symmetric distribution.
- Calvarial changes in rickets include basilar invagination and a squared configuration resulting from posterior flattening and excessive osteoid deposition in the frontal and parietal regions (craniotabes).

---

24. The radiographic differential diagnosis of rickets should include consideration of hypophosphatasia and the Schmid type of metaphyseal chondrodysplasia.

Box 5-3   Osteomalacia (Rickets): Important Features

Causes:
- Malabsorption, inadequate intake of vitamin D; liver disease; renal tubular dystrophies; chronic renal failure; anticonvulsants; tumor-associated
- Generalized osteopenia
- Coarsened, indistinct trabecular pattern secondary to incomplete osteoid mineralization
- Pseudofractures (medial femoral neck, inferolateral scapulae, ischiopubic rami)
- Vertebral compression fractures, biconcavity
- Bowing, fracture of long bones
- Basilar invagination
- Physeal widening, metaphyseal irregularity
- Rachitic rosary
- If renal, occurs with secondary hyperparathyroidism

Neonatal Rickets

1. Radiographically, rachitic alterations usually do not become apparent before the age of 6 months.
2. The condition occurs secondary to dietary calcium, phosphate, and vitamin D levels that are insufficient to meet the demands of rapid skeletal growth.
3. Low-birth weight premature infants requiring long periods of parenteral nutrition may develop the condition.
4. Neonatal hepatobiliary or renal disease may contribute to the pathogenesis of the condition.
5. Copper deficiency in premature infants simulates neonatal rickets radiographically.

Tumor-associated Rickets and Osteomalacia

1. Tumor-associated rickets and osteomalacia are vitamin D resistant or refractory.
2. The pathogenesis of both disorders may involve tumor secretion of a humoral substance that inhibits renal tubular resorption of phosphate.
3. The conditions are associated with small benign bone and soft-tissue neoplasms, including hemangiopericytoma, hemangioma, nonossifying fibroma, and giant cell tumor.
4. Autosomal dominant inheritance is observed.
5. Serum calcium levels are normal.
6. Decreased serum phosphate levels are typical.
7. Urinary levels of phosphate and alkaline phosphatase are increased.
8. The metabolic bone disease responds dramatically to resection of the associated neoplasm.

PEARL

Osteomalacia can be seen in Wilson disease, fibrous dysplasia, neurofibromatosis, hemangiopericytoma, and as a result of anticonvulsant therapy. Hemangiopericytoma of soft tissue is the most common tumor associated with osteomalacia. Heparin therapy results in osteoporosis but not osteomalacia (osteoporosis is decreased skeletal mass with increased bone porosity, whereas osteomalacia is decreased mineralization of bone with incompletely mineralized osteoid seams).

HYPERPARATHYROIDISM

1. The condition may be primary (usually secondary to parathyroid adenoma) or secondary (usually secondary to chronic renal disease); in the latter case, rickets or osteomalacia coexists (Box 5-4).

PEARL

Hyperparathyroidism:
- Lung tumors can secrete a peptide that is similar to parathyroid hormone, causing a paraneoplastic syndrome.
- Parathyroid hormone levels are normal to slightly decreased in lung cancer.
- Genetic disorders such as the multiple endocrine neoplasia syndromes (types 1 and 2) can result in primary hyperparathyroidism.
- Chronic renal failure results in secondary hyperparathyroidism.
- Type 1 multiple endocrine neoplasia is composed of pituitary adenoma, primary hyperparathyroidism, and pancreatic adenomas.
- Type 2 is defined as medullary carcinoma of the thyroid, primary hyperparathyroidism, and pheochromocytoma.
- Type 2B (also known as Type 3) is composed of medullary carcinoma of the thyroid, pheochromocytoma, and multiple mucosal neuromas.
- Four familial conditions are associated with pheochromocytoma: familial pheochromocytoma, multiple endocrine neoplasia type 2 syndrome, neurofibromatosis, and von Hippel–Lindau syndrome.

PEARL

Primary hyperparathyroidism is a group of disorders characterized by uncontrolled secretion of parathyroid hormone. It most commonly occurs in patients over 50 years of age with a female-to-male ratio of 2:1. Lung cancer can produce a paraneoplastic syndrome in which the tumor secretes a parathyroid hormone–like peptide. Chondrocalcinosis is a finding of primary hyperparathyroidism. Primary hyperparathyroidism is found in multiple endocrine neoplasia type 1 and 2 syndromes.

2. Serum levels of calcium and alkaline phosphatase are increased, whereas serum levels of phosphate are decreased.
3. Urinary levels of calcium and phosphate are increased.

PEARL

Calcitonin is a polypeptide secreted by the perifollicular cells of the thyroid gland. Increased serum calcium levels cause the perifollicular cells to release calcitonin, which acts upon the kidney and bone to lower the calcium levels. It is thus a physiologic antagonist to parathyroid hormone. Calcitonin inhibits osteoclasts and increases renal excretion of calcium.

4. Clinical manifestations of hyperparathyroidism include bone pain, weakness, nephrolithiasis, and peptic ulcer disease.
5. Musculoskeletal abnormalities are evident radiographically in approximately 50% of patients.
6. The imaging procedure of choice for diagnosis and follow-up is high-detail frontal radiography of the hands and wrists.
7. Hyperparathyroidism may induce either bone resorption or formation.

PEARL

Hyperparathyroidism (primary or secondary) can cause diffuse widespread osteoporosis, subperiosteal bone resorption, absence of the lamina dura of the teeth, and arterial calcification. Primary hyperparathyroidism is most commonly caused by adenoma (90% of cases) or hyperplasia (10% of cases) and rarely by carcinoma. Secondary hyperparathyroidism is most frequently seen in renal insufficiency and occasionally occurs with calcium deficiency. Chondrocalcinosis (hyaline cartilage and fibrocartilage) is more common in the primary form, whereas soft-tissue calcifications and osteosclerosis predominate in the secondary form. Brown tumors were once more common in the primary form, although the prevalence is currently greater in the secondary form because of the increased longevity of patients with chronic renal failure. Pseudohyperparathyroidism is a paraneoplastic syndrome involving ectopic parathyroid hormone secreted by pulmonary or renal tumors.

8. Radiographic manifestations of hyperparathyroidism include generalized osteopenia, bone resorption, osteosclerosis, brown tumors, chondrocalcinosis, and soft-tissue and vascular calcification.

PEARL

Primary and secondary hyperparathyroidism:
- Brown tumors occur with greater frequency in the secondary form.
- Chondrocalcinosis occurs more commonly in the primary form.
- Soft-tissue calcification, including interstitial calcification of the lung and vascular calcification, occurs more commonly in the secondary form.
- Cortical tunneling occurs with equal frequency in both conditions.
- Of patients with hyperparathyroidism, approximately 30% to 40% show bone involvement.

9. Specific forms of bone resorption occurring in hyperparathyroidism include subperiosteal, subchondral, subligamentous, trabecular, intracortical, and endosteal, the most common of which is subperiosteal (Fig. 5-8).

PEARL

Osteoclasts (which are responsible for bone resorption) originate from primitive mesenchymal osteoprogenitor cells, have multiple nuclei, work on the bone surface, and are stimulated by parathyroid hormone.

10. Common sites of subperiosteal bone resorption in hyperparathyroidism include the radial aspects of phalanges (particularly the middle phalanges of the index, middle fingers), terminal tufts, marginal areas of small joints, proximal medial aspects of long

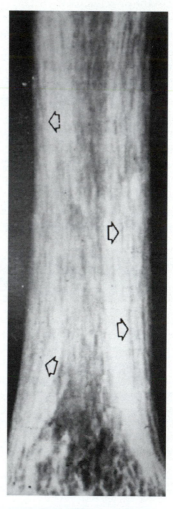

Fig. 5-8   Intracortical tunneling or striations (*arrows*) in hyperparathyroidism. This phenomenon may occur in any situation of rapid or severe bone loss, including reflex sympathetic dystrophy syndrome, hyperthyroidism, and advanced postmenopausal or corticosteroid-induced osteoporosis.

bones (tibia, femur, humerus), and lamina dura of teeth.

11. The salt-and-pepper skull represents trabecular bone resorption in the diploic space.

12. Common sites of subchondral bone resorption in hyperparathyroidism include the sacroiliac joints, acromioclavicular joints, and symphysis pubis.

13. Common sites of subligamentous bone resorption in hyperparathyroidism include the calcaneus, inferior aspect of the clavicle, and pelvis.

14. Bone resorption in hyperparathyroidism may simulate inflammatory arthritis in its radiographic appearance (sacroiliitis, marginal and subchondral erosions).

PEARL

Hyperparathyroidism exhibits erosions at the site of origin of the plantar aponeurosis and the site of insertion of the Achilles tendon, especially along its superior aspect. There is often significant associated osteoporosis (including cortical tunneling), as well as vascular calcification. Differential diagnosis of erosions at entheses should include consideration of plantar fasciitis, rheumatoid arthritis, Reiter syndrome, psoriatic arthritis, gout, and infection.

15. Osteosclerosis in hyperparathyroidism may be localized or diffuse, and it is more common in the secondary form (Box 5-5).

16. The rugger jersey spine is an example of osteosclerosis occurring in hyperparathyroidism.

17. A brown tumor is a localized collection of osteoclasts, fibrous tissue, and giant cells that appears as an osteolytic lesion; it is radiographically and pathologically similar to giant cell tumor and currently more common in the secondary form of hyperparathyroidism because of the increasing longevity of patients with chronic renal disease.

18. Following resection of a parathyroid adenoma, brown tumors may assume an osteosclerotic appearance simulating blastic metastatic disease.

19. Chondrocalcinosis in hyperparathyroidism is more common in the primary form.

PEARL

Laboratory abnormalities associated with hyperparathyroidism include elevated serum calcium levels, decreased serum phosphate levels, increased serum alkaline phosphatase levels, increased urinary calcium levels, and increased urinary phosphate levels.

20. Soft-tissue calcification in hyperparathyroidism is more common in the secondary form.

21. Vascular calcification in hyperparathyroidism is more common in the secondary form and resembles that observed in diabetes mellitus.

RENAL OSTEODYSTROPHY

1. The major components of renal osteodystrophy are rickets or osteomalacia and secondary hyperparathyroidism; aluminum intoxication may coexist in patients on dialysis therapy.

2. Renal osteodystrophy in young patients is primarily manifested as rickets, with alterations of hyperparathyroidism occurring later.

3. Renal osteodystrophy in adults is dominated by the manifestations of secondary hyperparathyroidism.

4. Chronic renal failure results in retention of phosphate, which induces hypocalcemia and subsequent parathyroid hyperplasia with secondary hyperparathyroidism.

5. Chronic renal failure is associated with decreased intestinal absorption of calcium and skeletal resistance to parathyroid hormone.

6. In chronic renal failure, the serum calcium level is normal or low, whereas serum phosphate and alkaline phosphatase levels are elevated.

7. The radiographic manifestations of renal osteodystrophy include the combined features of rickets or osteomalacia and secondary hyperparathyroidism, as well as periosteal enostosis, a finding that does not occur in primary hyperparathyroidism.

8. In the periosteal enostosis of severe chronic renal osteodystrophy, a radiolucent zone may be interposed between the periostitis and host bone; the femora, pubic rami, and metatarsals are most commonly affected.

Box 5-4 Hyperparathyroidism (Primary, Secondary): Important Features

- Elevated serum parathyroid hormone levels, hypercalcemia, hypophosphatemia
- Subperiosteal bone resorption (radial aspect of middle phalanges), intracortical tunneling
- Subchondral, subligamentous, subtendinous bone resorption
- Generalized osteopenia (primary)
- Brown tumors: lytic, expansile, well marginated (higher frequency in primary, but more common in secondary)
- Acro-osteolysis

Continued

Box 5-4—cont'd

- Resorption of dental lamina dura
- Grainy demineralization of skull (salt-and-pepper appearance)
- Generalized osteosclerosis, rugger jersey spine (secondary)
- Pathologic fractures, especially through brown tumors
- Metastatic soft-tissue calcification (periarticular, Mönckeberg-type medial vascular)
- Chondrocalcinosis (primary)

Box 5-5   Generalized Osteosclerosis: Differential Diagnosis

- Renal osteodystrophy (secondary hyperparathyroidism)
- Metastatic disease: breast, prostate
- Paget disease
- Myelofibrosis
- Fluorosis
- Mastocytosis
- Sickle cell anemia
- Pyknodysostosis
- Osteopetrosis
- Von Buchem disease

Dialysis Therapy

1. Depending upon the calcium level in dialysate and the degree of phosphate level control, the radiographic manifestations of secondary hyperparathyroidism may resolve in response to dialysis.
2. Soft-tissue calcification is common in dialysis patients, and large tumoral deposits occur with highest frequency in this setting.
3. Dialysis cysts tend to develop in the phalanges and carpal bones.
4. The radiographic manifestations of dialysis spondyloarthropathy are identical to those of spinal infection; percutaneous biopsy is mandatory because dialysis patients are at risk for both conditions.
5. Potential complications of arteriovenous dialysis shunts include aneurysm formation, carpal tunnel syndrome, and bacteremia with secondary hematogenous osteomyelitis or septic arthritis.
6. Aluminum intoxication is responsible for osteomalacia and stress fractures in dialysis patients.
7. Ischemic necrosis in dialysis patients is induced by corticosteroid therapy, as well as by subchondral resorption with collapse secondary to hyperparathyroidism.
8. Crystal deposits and amyloid may be found in biopsy material obtained from areas of spondyloarthropathy in dialysis patients.

Aluminum Intoxication Occurring as a Complication of Dialysis

1. Radiographic manifestations include rachitic or osteomalacic alterations and stress fractures.
2. Clinical features include bone pain, proximal muscle weakness, and encephalopathy.
3. Vitamin D therapy has no effect on the osteomalacia induced by aluminum intoxication.
4. Histologically, osteomalacia is observed, with aluminum deposits at the interface between mineralized and unmineralized osteoid.
5. Aluminum hydroxide antacids administered for control of hyperphosphatemia also contribute to the development of aluminum intoxication.

HYPOPARATHYROIDISM

1. Potential causes of hypoparathyroidism include surgical excision of or trauma to the parathyroid glands, or the disorder may be idiopathic (Box 5-6).
2. In hypoparathyroidism, decreased production of parathyroid hormone results in hypocalcemia and neuromuscular dysfunction.
3. The major radiographic manifestations of hypoparathyroidism are osteosclerosis, soft-tissue calcification, basal ganglia calcification, and hypoplastic dentition; of these, osteosclerosis is the most common and may be localized or generalized.
4. Less constant radiographic features of hypoparathyroidism include osteoporosis, premature physeal closure, vertebral hyperostosis, enthesopathy, and radiodense bands in metaphyses and vertebral endplates.

Box 5-6   Hypoparathyroidism: Important Features

- Decreased serum parathyroid hormone levels, hypocalcemia, hyperphosphatemia
- Short stature
- Generalized osteosclerosis or osteoporosis
- Premature physeal closure
- Ligamentous, subcutaneous calcification
- Dental abnormalities (delayed eruption, hypoplasia, thickened lamina dura)
- Calvarial thickening, basal ganglia calcification

PSEUDOHYPOPARATHYROIDISM

1. The pathophysiology of pseudohypoparathyroidism involves end organ resistance to parathyroid hormone (Box 5-7).
2. Pseudohypoparathyroidism manifests a characteristic somatotype that includes obesity, short stature, and brachydactyly.

PEARL

Pseudohypoparathyroidism and pseudo-pseudohypoparathyroidism are believed to be related genetic abnormalities. Both have similar radiographic appearances, although the former has decreased calcium and increased phosphate in the serum whereas the latter has normal blood chemistries. Both may exhibit basal ganglia calcification. Clinically, presenting symptoms include brachydactyly, obesity, short stature, exostoses, and mental retardation. The exostoses are small and may arise from the diaphyses. Muscular tetany occurs as a result of decreased serum calcium levels.

3. Radiographic abnormalities common to both hypoparathyroidism and pseudohypoparathyroidism include osteosclerosis, soft-tissue calcification, and basal ganglia calcification.

PEARL

Hypoparathyroidism, pseudohypoparathyroidism (PHP), and pseudo-pseudohypoparathyroidism (PPHP):
- In PHP, the renal tubules and other end organs do not respond to parathyroid hormone.
- PPHP is an incomplete genetic manifestation of PHP, with patients having normal blood calcium and phosphate levels.
- The physical and radiologic appearances may be identical.
- Slipped capital femoral epiphyses may occur.
- Secondary hyperparathyroidism occurs in 10% of patients with PHP but does not occur in PPHP.
- Dense metaphyseal bands, generalized osteosclerosis, and soft-tissue calcification are potential features of all three disorders.

4. Hypocalcemia is a laboratory abnormality common to both hypoparathyroidism and pseudohypoparathyroidism.
5. Pseudohypoparathyroidism is associated with shortening of metacarpals, metatarsals, and phalanges; involvement most commonly affects the first, fourth, and fifth rays.
6. Pseudohypoparathyroidism should be included in the differential diagnosis of a positive metacarpal sign, along with Turner syndrome (Fig. 5-9).
7. Radiographically, patients with pseudohypoparathyroidism may demonstrate small exostoses that extend perpendicularly from the surface of the bone.
8. Additional features of pseudohypoparathyroidism include growth deformities, bowing of long bones, and cone-shaped epiphyses.

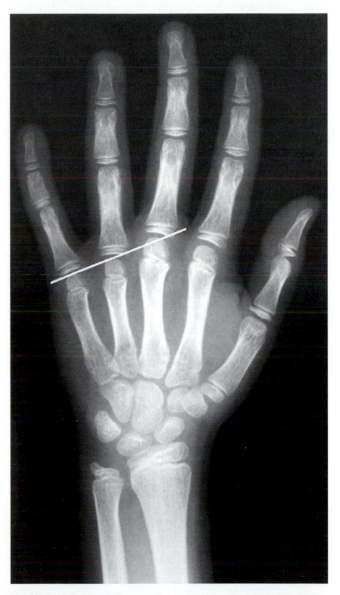

Fig. 5-9   Short fourth metacarpal with positive metacarpal sign (*line*) in pseudohypoparathyroidism. Differential diagnosis should include consideration of isolated developmental variation, growth plate injury, Turner syndrome, and Noonan syndrome.

Box 5-7 Pseudohypoparathyroidism and Pseudo-pseudohypoparathyroidism: Important Features

- Different clinical expressions of same genetic disorder
- Tissue insensitivity to parathyroid hormone
- PHP: elevated serum parathyroid hormone levels; hypocalcemia, hyperphosphatemia
- PPHP: normal blood chemistries
- Cone-shaped epiphyses
- Short metacarpals, metatarsals, and phalanges secondary to premature physeal closure
- Metastatic soft-tissue and ligamentous calcification
- Basal ganglia calcification
- Short stature
- Small diaphyseal exostoses
- Calvarial thickening
- Degenerative enthesopathy
- Abnormal dentition

PSEUDO-PSEUDOHYPOPARATHYROIDISM

Pseudo-pseudohypoparathyroidism is the normocalcemic form or stage of pseudohypoparathyroidism and is also caused by failure of end organ response to parathyroid hormone (Box 5-8). The clinical manifestations of the disorder are identical to those of pseudohypoparathyroidism.

PEARL

PHP and PPHP are more common in women and appear to represent an X-linked dominant trait. In PHP, the serum calcium level is decreased and the serum phosphorus level is increased. Both entities may have presenting symptoms of short stature, short digits (predominantly the fourth and fifth metacarpals), basal ganglia calcification, mental retardation, obesity, and diaphyseal exostoses.

Radiographic features of the disease are identical to those of pseudohypoparathyroidism, except that basal ganglia calcification is relatively rare.

Box 5-8 Pseudohypoparathyroidism and Pseudo-pseudohypoparathyroidism: Additional Important Features

- Short stature, round face, obesity, brachydactyly, abnormal dentition, mental retardation, strabismus, dermatoglyphic abnormalities, impaired taste and olfaction

Box 5-8—cont'd

- Serum parathyroid hormone levels are normal or increased (failure of end organ response)
- More frequent in women; X-linked dominant trait
- Positive metacarpal sign
- Serum calcium levels low in PHP and normal in PPHP
- Radiographic changes secondary to hyperparathyroidism can occur in PHP
- Diagnosis: second decade
- PHP and PPHP can occur in the same family and may represent the same disease
- Incidence of phalangeal shortening higher in PHP than in PPHP
- Degree of metacarpal shortening higher in PPHP than PHP

HYPERTHYROIDISM

1. In children, the primary musculoskeletal effect of hyperthyroidism is accelerated skeletal maturation resulting in advanced bone age.
2. In adults, hyperthyroidism induces increased bone turnover rate resulting in generalized osteoporosis.
3. Potential complications of hyperthyroidism in the spine include vertebral compression fractures and accentuated thoracic kyphosis.
4. Hyperthyroidism is also associated with a myopathy that may clinically simulate arthritis.
5. Thyroid acropachy is a rare complication of hyperthyroidism that occurs following therapy; patients are usually euthyroid but may be hypothyroid or hyperthyroid (Box 5-9).
6. Soft-tissue manifestations of thyroid acropachy include localized swelling and digital clubbing.
7. Radiographically, thyroid acropachy is characterized by dense feathery periostitis that is frequently asymmetric in distribution (Fig. 5-10).
8. Involvement of the metacarpals and phalanges in thyroid acropachy is much more common than that of the long bones.

Box 5-9 Thyroid Acropachy: Important Features

- Affects 0.5% to 1.0% of patients with thyrotoxicosis
- Usually follows treatment of hyperthyroidism, when patient is euthyroid or hypothyroid
- Affects any age, with equal sex incidence
- Periostitis more prominent on radial aspect of bone

Continued

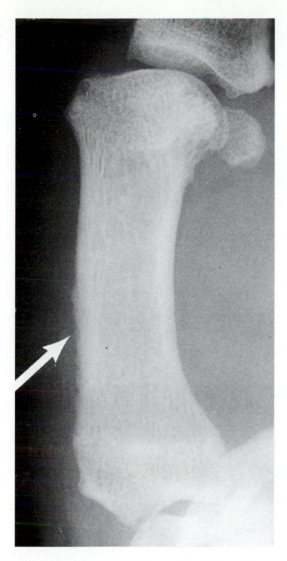

Fig. 5-10 Irregular periostitis (*arrow*) involving a short tubular bone in thyroid acropachy. Differential diagnosis should include consideration of secondary hypertrophic osteoarthropathy, pachydermoperiostosis, and florid reactive periostitis.

Box 5-9—cont'd

- Progressive; resolution poorly correlated with correction of thyroid function
- Usually occurs several years after onset of hyperthyroidism

---

HYPOTHYROIDISM

1. In children, the primary musculoskeletal effect of hypothyroidism is delayed skeletal maturation resulting in retarded bone age (Fig. 5-11; Box 5-10).

PEARL

Early closure of the physes is associated with hypergonadism and hyperpituitarism. Cushing syndrome may cause delayed closure of physes. Craniopharyngioma, by virtue of its compression on the hypothalamus and pituitary, may also delay closure of the physes. Cretinism (hypothyroidism) is also associated with delayed physeal closure.

2. Epiphyseal dysgenesis is a feature of hypothyroidism that may simulate Legg-Calvé-Perthes disease or an epiphyseal dysplasia during childhood.
3. The epiphyses in hypothyroidism tend to be stippled during infancy, fragmented during childhood, and cone shaped during adolescence.

PEARL

Hypothyroidism results in delayed closure of physes leading to a characteristic fragmented epiphyseal appearance. Other findings include wormian bones, well-differentiated inner and outer tables of the skull resulting from an osteopenic diploic space, prolonged separation of sutures and fontanelles, and shortened slender long bones with dense transverse bands at the metaphyseal ends.

4. In the skull, wormian bones or sutural ossicles are a characteristic radiographic feature of hypothyroidism.

PEARL

Wormian bones are associated with cretinism, pyknodysostosis, osteogenesis imperfecta, and cleidocranial dysostosis, among numerous other conditions.

5. A characteristic manifestation of hypothyroidism in the spine is short, anteriorly wedged vertebral bodies (bullet-shaped) at the thoracolumbar junction, resulting in kyphosis.
6. Untreated infantile hypothyroidism or cretinism results in characteristic clinical manifestations including short stature, obesity, and mental retardation.

A

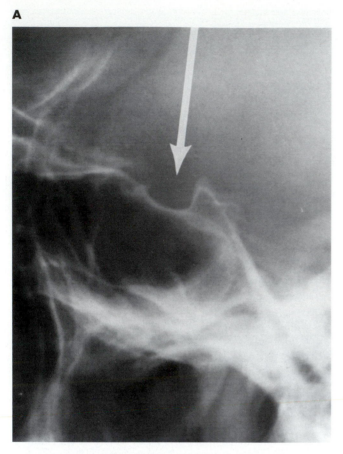

B

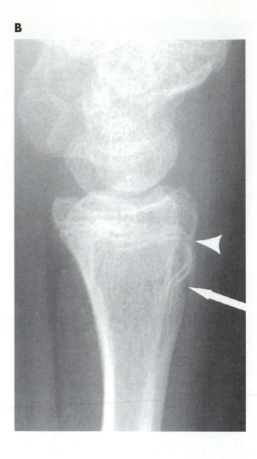

Fig. 5-11 Panhypopituitarism in a 28-year-old patient. **A,** The sella turcica is extremely hypoplastic (*arrow*) (normal dimensions are approximately 14 by 17 mm on a lateral radiograph). **B,** Abnormal findings at the wrist include generalized osteopenia with predisposition to fracture (*arrow*; buckle type) and delayed skeletal maturation (*arrowhead*).

Box 5-10 Hypothyroidism (Cretinism): Important Features

- Delayed skeletal maturation
- Fragmented, irregular epiphyses
- Wide sutures and fontanelles with delayed closure, wormian bones
- Delayed dentition
- Delayed pneumatization of sinuses, mastoid air cells
- Short stature
- Osteoporosis
- Hypoplastic phalanges of fifth finger
- Thoracolumbar vertebral wedging with kyphosis
- Coxa vara with flattened femoral head in adults

ACROMEGALY

1. Excess production of growth hormone leads to pituitary gigantism in the immature skeleton and acromegaly in the mature skeleton (Box 5-11).

2. Soft-tissue thickening in acromegaly is most readily recognized in the heel pad and digits.

3. Radiographic measurements useful in the diagnosis of acromegaly include heel pad thickness, sesamoid index of the first metacarpophalangeal joint, tuftal width of the third finger, second metacarpophalangeal joint thickness, and soft-tissue thickness at the level of the second proximal phalangeal midshaft.

PEARL

Acromegaly is caused by excessive growth hormone production after physeal closure. The hallmarks of the disease are increased size of the sella turcica, mandibular prognathism, increased heel pad thickness, widened joint spaces, spadelike tufts of the distal phalanges, premature osteoarthritis, increased sesamoid index (flexor pollicis brevis), and osteoporosis. Posterior scalloping of the vertebral bodies also occurs.

4. Radiographic features of acromegaly in the skull include sella turcica enlargement, calvarial thickening, sinus enlargement, and prominent facial bones.
5. Radiographic manifestations of acromegaly in the spine include increased vertebral body dimensions, intervertebral disk space widening, posterior vertebral body scalloping, and exaggerated spinal curvature (Fig. 5-12).
6. Bone density in acromegaly tends to be decreased, particularly late in the disease.
7. Radiographic manifestations of acromegaly in the hands and feet include spadelike tufts (occasionally with pseudoforamina), joint space widening, osseous enlargement, and excrescences at tendon and ligament attachment sites (enthesopathy).
8. The most important complication of acromegaly is premature degeneration of thickened cartilage with secondary osteoarthritis.

9. Bone proliferation at entheses (enthesopathy) in acromegaly simulates the peripheral skeletal manifestations of diffuse idiopathic skeletal hyperostosis.
10. Chondrocalcinosis is a radiographic feature of acromegaly that also occurs in hyperparathyroidism, calcium pyrophosphate dihydrate crystal deposition disease, ochronosis, and hemochromatosis.

Box 5-11 Acromegaly: Important Features

- Excessive growth hormone production, usually secondary to eosinophilic adenoma
- Sellar enlargement, deformity
- Thickening of articular cartilage
- Soft-tissue thickening (heel pad thickness greater than 23 mm)
- Premature osteoarthritis and degenerative disk disease
- Large frontal sinuses, calvarial hyperostosis, mandibular prognathism
- Generalized osseous enlargement (reflected in sesamoid index), phalangeal tuftal proliferation
- Posterior vertebral body scalloping
- Degenerative enthesopathy
- Osteoporosis
- Gigantism in children: sellar enlargement, large skeleton with normal bone age

---

PAGET DISEASE

Paget disease is a disorder of unknown etiology, although both genetic and viral factors have been suggested (Box 5-12). The disease usually occurs in patients older than 50 years, although it may occur at a younger age. The disease is relatively common in Europe, with the exception of Scandinavia. In England and Wales it is estimated to occur in 7% of men and 4% of women older than 50 years. It is rare in Africa and Asia. In the United States, the disorder is not uncommon but shows considerable regional and ethnic variation. In all areas, incidence increases with age.

Approximately 80% of cases occur in asymptomatic patients and are discovered during routine radiographic examination or following the evaluation of an elevated serum alkaline phosphatase level. Symptoms may include bone or joint pain, bowing or increased warmth of an extremity, calvarial enlargement, kyphosis, cranial nerve or spinal cord compression, or congestive heart failure. Frequently the symptoms result from complications of Paget disease. The osseous involvement may be polyostotic or monostotic. In the latter instance, the vertebral bodies (including the sacrum), skull, pelvis, femur, and tibia are the bones most frequently involved.

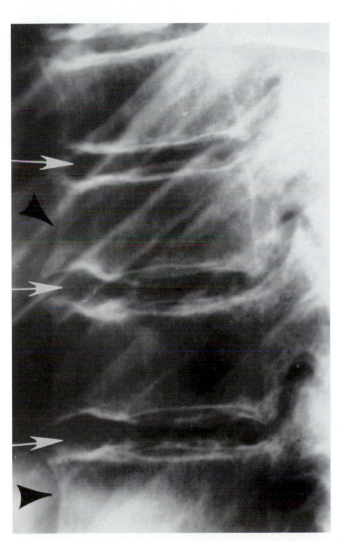

Fig. 5-12 Increased anteroposterior diameter of the thoracic vertebral bodies (*arrowheads*) in association with thickening of the intervertebral disks (*arrows*) is characteristic of acromegaly.

The essential pathologic abnormality is disordered and extremely active bone remodeling. The radiographic appearances and the biochemical abnormalities reflect the histologic changes; osteoblastic activity being associated with an elevated serum alkaline phosphatase level and osteoclastic activity with an increase in the urinary excretion of hydroxyproline. In the acute phase, there is active and unbalanced osteoclastic bone resorption.

Radiographically, this results in an area of lytic destruction. In the calvarium this originates in either the frontal or occipital region and is known as osteoporosis circumscripta. In a long bone, the lesion originates at the articular margin and extends into the diaphysis in a flamelike manner. Histologically, the osteoclastic process is accompanied by some osteoblastic activity, which results in augmented uptake of isotope during radionuclide bone scans.

In the intermediate stage, there is increased osteoblastic activity resulting in thickening of the cortex, coarsening of the trabecular pattern, loss of corticomedullary differentiation, and generalized bony overgrowth. In long bones, the bony overgrowth often results in bowing. In the skull, the new bone formation takes the form of focal patchy density within areas of osteoporosis. In addition, there is thickening of the calvarium associated with loss of differentiation between the inner and outer tables. Basilar invagination may be noted. In the spine, there may be generalized enlargement of an involved vertebral body, which is frequently associated with thickening of the cortex and coarsening of the trabecular pattern. Finally, in the inactive phase, there is a diffuse increase in density of the involved bone. Two or more phases of Paget disease may coexist in the same patient or bone.

Fractures, either through osteolytic or osteoblastic lesions, may be the presenting feature of Paget disease. Insufficiency fractures of the tensile cortex of long bones may progress to complete fractures. Degenerative arthritis, particularly involving the hip joints in association with protrusio acetabuli, may occur.

Malignant transformation is rare, but osteosarcoma is the most frequently associated histologic type. Fibrosarcoma and chondrosarcoma occur less frequently. Giant cell tumor may occur in association with pagetic involvement, but may also occur in a normal bone of a patient with Paget disease. Most commonly the skull, spine, and innominate bones are affected. Multiple lesions may occur. Intranuclear inclusion bodies have been noted in giant cells associated with Paget disease.

Not all cases of soft-tissue swelling in Paget disease are due to malignant transformation. Juxtacortical masses with spiculated new bone may result solely from florid Paget disease extending into the soft tissues. Extramedullary hematopoiesis has also been noted in Paget disease. Both metastases and myeloma may occur

in bones involved by Paget disease, but the significance of these associations is controversial.

Calcitonin, bisphosphonates, and mithramycin, which inhibit osteoclastic activity, may arrest the disease process or even partially restore normal bone architecture.

The following is a summary of important features of Paget disease (Box 5-13):

1. The disease is more common in the male sex and usually appears after the age of 40 years.
2. Histologically, abnormal bone remodeling is observed, with both osteoclastic and osteoblastic activity involved.

PEARL

Conditions producing rapid turnover of bone include Paget disease, hyperparathyroidism, and familial hyperphosphatasia. The lytic phase, followed by the sclerotic phase, of Paget disease is representative of the rapid turnover of bone associated with this condition. Hyperparathyroidism results in bone resorption and remodeling. Familial hyperphosphatasia (also known as juvenile Paget disease) has an appearance similar to that of Paget disease. Patients have an elevated serum alkaline phosphatase level indicative of an increased bone turnover rate. In contrast, hypophosphatasia results in generalized deficient or absent bone mineralization; patients typically have multiple fractures.

---

3. The disease is most commonly encountered in temperate climates and is rare among Asian and African individuals.
4. The serum alkaline phosphatase level is usually elevated, secondary to bone formation.
5. Serum and urinary hydroxyproline levels are usually elevated, secondary to bone resorption.

PEARL

Elevation of urinary hydroxyproline levels is a sensitive indicator of Paget disease. Urinary calcium, acid phosphatase, and phosphate levels may be increased in Paget disease. Urinary phosphoethanolamine levels are increased in hypophosphatasia.

---

6. Clinical presentations referable to the musculoskeletal system include bone pain, progressive osseous enlargement, bowing of long bones, and pathologic fracture.

7. Deafness may result from cranial nerve compression secondary to skull base enlargement or middle ear ossicle involvement.

8. Spinal cord compression may occur secondary to basilar invagination or vertebral enlargement.

9. High-output congestive heart failure is rare and results from increased blood flow in pagetic bone.

10. Because the disease is frequently asymptomatic, it may be an incidental radiographic finding.

11. The three sequential stages of Paget disease may coexist and include lytic, mixed, and sclerotic radiographic patterns.

12. The characteristic radiographic feature of early Paget disease is well-defined osteolysis.

13. The characteristic radiographic features of advanced Paget disease include osseous enlargement, increased bone density, cortical thickening, and trabecular thickening with distortion (Fig. 5-13).

14. Vertebral involvement occurs in approximately 75% of patients, with a predilection for the lumbar spine.

15. Patterns of spinal involvement include the picture frame and ivory vertebrae, and pathologic compression fracture is a potential complication (Fig. 5-14).

16. The skull and pelvis are each affected in approximately two thirds of cases (Fig. 5-15).

17. The lytic phase of Paget disease in the calvarium is known as osteoporosis circumscripta, and usually begins in the frontal or occipital bones.

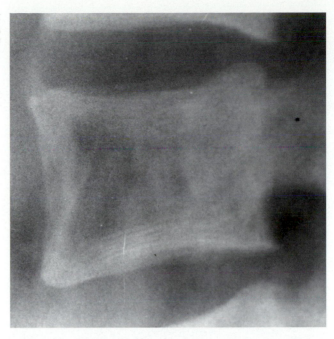

Fig. 5-14 Picture frame vertebral body in Paget disease. Characteristic accentuated density and thickness of the vertebral body margins with relative radiolucency of the center is associated with osseous enlargement and a coarse trabecular pattern.

18. The cotton wool skull consists of focal osteosclerotic areas superimposed on a background of relative osteolysis.

19. Basilar invagination may complicate involvement of the skull base by Paget disease in approximately one third of patients.

20. The earliest radiographic sign of Paget disease in the pelvis is cortical thickening along the iliopectineal line (Fig. 5-15).

21. Protrusio acetabuli is a potential complication of involvement in the innominate bone.

22. Tubular bone involvement occurs in approximately one third of cases and is initially characterized by subarticular osteolysis that progresses into the diaphysis with a well-defined flame- or blade of grass–shaped margin.

23. Initial osteolysis that begins in the diaphysis of a tubular bone is a phenomenon that occurs almost exclusively in the tibia.

24. Paget disease may be monostotic or polyostotic, and skeletal scintigraphy is the procedure of choice for staging the extent of involvement (Fig. 5-15).

25. Pseudofractures are a potential complication of Paget disease, and they tend to occur on the tensile side of a bone in contrast to those of osteomalacia, which occur on the compressive side.

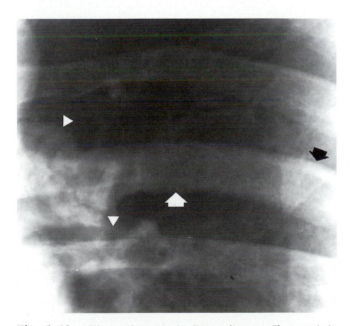

Fig. 5-13 Rib involvement in Paget disease. Characteristic findings include osseous enlargement, cortical thickening (*arrows*), and premature osteoarthritic change at the costovertebral and costotransverse articulations (*arrowheads*).

A

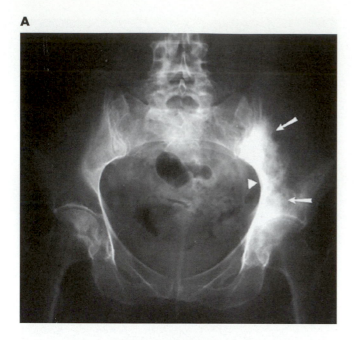

B

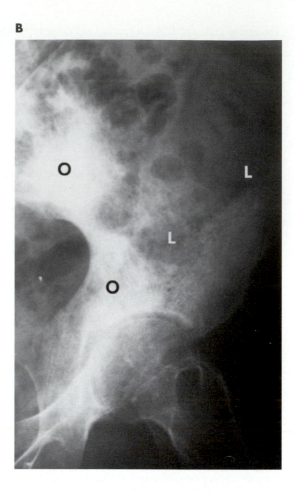

C

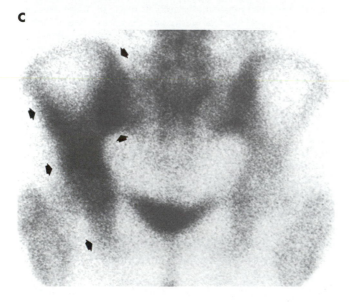

D

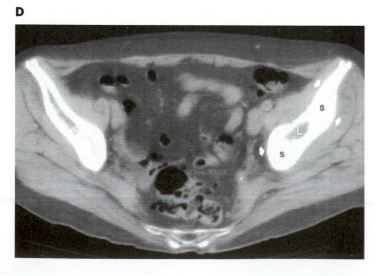

Fig. 5-15   Paget disease of the innominate bone.
A, Frontal radiograph of the pelvis reveals unilateral sclerosis
of the pelvic inlet (*arrows*) with thickening of the
iliopectineal line (*arrowhead*). **B,** Oblique view demonstrates
a mixed pattern of osteosclerosis (*O*) and osteolysis (*L*).
C, Radionuclide bone scan exhibits more extensive
involvement (*arrows*) than that suggested by conventional
radiography. **D,** CT documents mild osseous enlargement,
cortical thickening (*arrows*), and mixed areas of lysis (*L*) and
sclerosis (*S*). Differential diagnosis should include
consideration of metastatic disease (particularly from prostate
and breast primaries) and active chronic osteomyelitis.

PEARL

Pseudofractures affect tensile areas of bone in Paget disease. Increased levels of serum alkaline phosphatase and hydroxyproline and of urinary hydroxyproline occur. Malignant degeneration is a complication of Paget disease (dominant tissue type: osteosarcoma, 50% to 60% of cases; fibrosarcoma, 20% to 25% of cases; chondrosarcoma, 10% of cases). Rarely, pagetic bone will degenerate to a giant cell tumor; this usually occurs with skull and facial bone involvement.

26. The most serious potential complication of Paget disease is sarcomatous degeneration, which occurs in fewer than 1% of all patients but up to 10% of those with widespread involvement.

PEARL

Complications of Paget disease include basilar invagination, pathologic fractures, sarcomatous degeneration, protrusio acetabuli, spinal stenosis, narrowing of cranial foramina, vertebral collapse, premature degenerative joint disease, and high-output congestive heart failure.

27. The most common histologic type present in Paget sarcoma is osteosarcoma.
28. Giant cell tumors arising in Paget disease are usually benign and exhibit a predilection for the skull and facial bones.
29. Involvement of subchondral bone by Paget disease usually leads to premature degenerative joint disease (Fig. 5-13).
30. Therapeutic options in Paget disease include administration of calcitonin, bisphosphonates, and mithramycin, and skeletal scintigraphy is the preferred imaging method for monitoring response.

Box 5-12 Paget Disease (Osteitis Deformans): Important Features

- Onset in fifth through sixth decades; male predilection
- Lytic and sclerotic phases, usually coexistent
- Osseous enlargement, deformity, bowing
- Blade of grass (V-shaped configuration of advancing lytic front in long bones)
- Osteoporosis circumscripta (sharply demarcated zone of decreased density in skull)
- Cotton wool skull (mixed pattern in calvarium)
- Basilar invagination, vertebral expansion resulting in spinal cord compression

Box 5-12—cont'd

- Pseudofractures at tensile sites
- Thick, coarse trabeculae; cortical thickening
- Pelvis most common (often initial) site, especially iliopectineal line
- Pathologic fractures in flat and tubular bones (bone is weak despite increased density), less commonly in spine
- Sarcomatous degeneration (osteosarcoma, fibrosarcoma, chondrosarcoma elements comprising Paget sarcoma), giant cell tumor
- Slow progression, usually polyostotic (pelvis, proximal femora, skull, tibiae, vertebrae)
- Premature osteoarthritis secondary to deficient support of articular cartilage

Box 5-13 Paget Disease (Osteitis Deformans): Additional Important Features

- Affects approximately 3% of population over age 40
- High-output congestive heart failure due to vascularity of bone rather than arteriovenous shunting
- Elevated serum and urinary hydroxyproline levels
- Elevated serum alkaline phosphatase level
- Axial skeleton or proximal femur affected in 75% to 80% of cases
- Basilar invagination and neurologic compromise occur in approximately one third of cases with skull involvement
- Sarcomatous transformation to predominantly osteosarcoma (50% to 60% of cases), fibrosarcoma (20% to 25% of cases), chondrosarcoma (10% of cases); malignancy occurs in 5% to 10% of cases with widespread disease, 1% to 3% with less extensive involvement
- Giant cell tumor in involved skull, facial, or pelvic bone; this lesion in face or multicentric sites suggests underlying Paget disease

Hyperphosphatasia (Juvenile Paget Disease [Box 5-14])

1. Calvarial thickening with interspersed areas of increased and decreased density replacing normal bone architecture
2. Generalized demineralization, expansion, and bowing of long tubular bones; widening of short tubular bones
3. Dissolution of normal cortical architecture, replacement by strands of longitudinal trabeculae
4. Cortical thickening on inner side of bowed tubular bones
5. Short stature
6. Premature tooth loss

Box 5-14   Hyperphosphatasia (Juvenile Paget Disease): Important Features

- Cortical thickening, bowing of long bones
- Calvarial hyperostosis
- Elevated levels of serum alkaline and acid phosphatase, uric acid, and aminopeptidase
- Osseous enlargement, deformity
- Coarse trabecular pattern
- Pseudofractures, pathologic fractures
- Cystic areas in metaphyses

HYPOPHOSPHATASIA

1. The radiographic differential diagnosis of hypophosphatasia should include consideration of rickets and the Schmid type of metaphyseal chondrodysplasia (Box 5-15).
2. Serum levels of alkaline phosphatase are severely decreased.

PEARL

Hypophosphatasia is an inborn error of metabolism resulting in a decreased level of serum alkaline phosphatase. This results in osteomalacia and formation of abnormal amounts of osteoid. The hallmark of Paget disease is increased osteoid formation in the sclerotic phase. Fibrous dysplasia presents a spectrum of manifestations, one of which is increased osteoid formation. Osteomalacia and rickets are associated with increased amounts of unmineralized osteoid; causes include X-linked hypophosphatemia (vitamin D-resistant rickets), renal tubular acidosis, and anticonvulsant therapy.

3. Serum and urinary levels of phosphoethanolamine are elevated.

PEARL

Hypophosphatasia exhibits an elevated level of urinary phosphoethanolamine.

	Serum Levels	
	Calcium	Phosphorus
Hyperparathyroidism	Increased	Decreased
Hypoparathyroidism	Decreased	Increased
Pseudohypoparathyroidism	Decreased	Increased
Pseudo-pseudohypoparathyroidism	Normal	Normal
Hypophosphatasia	Normal	Normal

4. The most severe form of the disease is associated with death during infancy, and the clinical presentation is that of a "tar baby."

PEARL

There are several forms of hypophosphatasia, including lethal and tarda autosomal recessive variants. Both are disorders of the serum enzyme alkaline phosphatase with increased urinary excretion of phosphoethanolamine. The radiographic appearance is one of rickets or osteomalacia. Affected patients suffer from mental retardation and dental abnormalities.

5. Radiographically, the most severe form of hypophosphatasia is characterized by absent cranial mineralization, short bowed bones with fractures and deformities, and metaphyseal alterations resembling those of rickets.
6. Radiographically, the least severe form of hypophosphatasia demonstrates mild bowing deformities and delayed fracture healing.
7. Radiographically, the intermediate form of hypophosphatasia manifests generalized osteopenia, metaphyseal alterations resembling those of rickets, craniosynostosis, and wormian bones; the clinical course is one of progressive improvement.

PEARL

Hypophosphatasia is a disorder characterized by low serum alkaline phosphatase levels resulting in osteomalacia and increased urinary excretion of phosphoethanolamine. Wilson disease is characterized by increased serum and urinary copper levels with decreased serum ceruloplasmin levels resulting in central nervous system, hepatic, and corneal (Kayser-Fleischer rings) abnormalities. Homocystinuria is an inborn error of methionine metabolism caused by deficiency of cystathionine synthetase.

Box 5-15   Hypophosphatasia: Important Features

- Autosomal recessive deficiency of serum alkaline phosphatase with elevated level of urinary phosphoethanolamine (lethal and tarda forms)
- Short, thin, coarsely trabeculated, or unmineralized bones
- Metaphyseal growth deficiency or fracture with irregular callus, marked deformity

Box 5-15—cont'd

- Submetaphyseal demineralization with irregularity, bowing
- Deficient calvarial ossification with craniosynostosis
- Wide, irregular physeal plates
- Dental anomalies, joint pain, ectopic calcification, gait disturbances, mental retardation

GAUCHER DISEASE

1. Gaucher disease is a storage disorder, with accumulation of sphingolipid in the reticuloendothelial system and bone marrow infiltration by Gaucher cells.
2. The infantile and juvenile forms of the disease are associated with mental retardation and early death.
3. Gaucher disease is a familial condition that occurs with high frequency among the Ashkenazic Jewish people.
4. The majority of patients first experience symptoms during childhood or young adulthood.
5. Bone marrow infiltration by Gaucher cells results in expansion of the distal femur, known as the Erlenmeyer flask deformity, in up to 50% of patients; the differential diagnosis of this finding should include consideration of Niemann-Pick disease, anemias, fibrous dysplasia, metaphyseal dysplasia of Pyle, osteopetrosis, and heavy metal poisoning.
6. Ischemic necrosis is an important complication of the disease in approximately half of cases, and it exhibits a predilection for the femoral head; intramedullary bone infarcts also occur, particularly in metadiaphyseal areas of long bones.
7. Bone density in Gaucher disease tends to be decreased in a generalized distribution (Fig. 5-16).
8. In the spine, H-shaped vertebral bodies are a characteristic finding, the differential diagnosis of which should include consideration of sickle cell anemia and sickle-thalassemia (Figs. 5-17, 5-18).
9. Localized lytic bone lesions are an occasional manifestation of Gaucher disease and represent focal accumulations of Gaucher cells (Fig. 5-19).
10. Systemic manifestations of Gaucher disease include enlargement of the liver and spleen.
11. Patients with Gaucher disease have an increased incidence of osteomyelitis as compared to the normal population.
12. Niemann-Pick disease is a storage disorder that involves abnormal accumulation of sphingomyelin; the radiographic and clinical manifestations of the disease are similar to those of Gaucher disease, except that patients are not prone to development of osteonecrosis and localized lytic lesions.

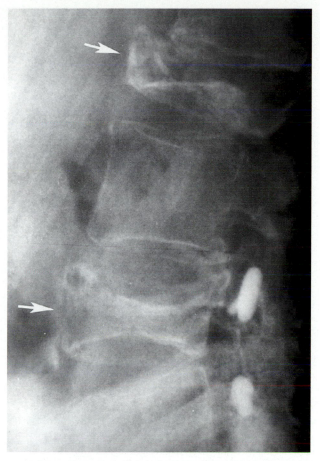

Fig. 5-16 Vertebral compression fractures in Gaucher disease. Severe height loss is evident at L1, and biconcave endplate deformities involve the L3 vertebral body (*arrows*); generalized osteopenia is present. The findings are nonspecific and could occur in advanced osteoporosis, primary hyperparathyroidism, hyperthyroidism, multiple myeloma, metastatic disease, and other conditions associated with diminished bone mass.

13. Magnetic resonance imaging is the procedure of choice for determining the extent of marrow involvement in Gaucher disease, as well as for the early diagnosis of its most important complication (Fig. 5-20).

MUCOPOLYSACCHARIDOSES

1. The mucopolysaccharidoses are a group of hereditary disorders that manifest a common pattern of radiographic abnormalities known as dysostosis multiplex (Box 5-16).
2. Distinction among the various subtypes of diseases in the group is best accomplished on the basis of clinical presentation and specific mucopolysaccharides excreted in the urine (Box 5-17).

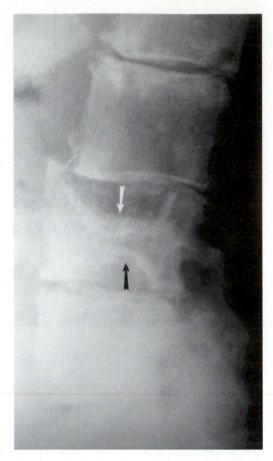

Fig. 5-17   H-shaped vertebra in Gaucher disease. Central depression of the endplates (*arrows*) with sparing of the periphery is characteristic and occurs secondary to bone infarction. Differential diagnosis should include consideration of sickle cell anemia and sickle-thalassemia.

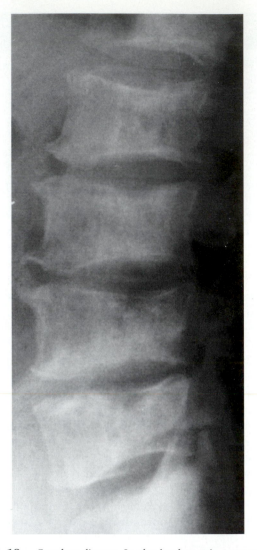

Fig. 5-18   Gaucher disease. In the lumbar spine, generalized central endplate deformities and intervertebral disk space narrowing are associated with patchy areas of osteosclerosis superimposed on diffuse osteopenia. Differential diagnosis should include consideration of sickle cell anemia and sickle-thalassemia.

PEARL

Gargoylism (Hurler syndrome) is characterized by mental retardation, early clouding of the cornea, heart disease, and coarse facies; the findings are manifested in the first few years of life. Morquio syndrome is characterized by short stature, corneal changes, and normal intelligence, and it manifests itself in early infancy. Achondroplasia and asphyxiating thoracic dystrophy of Jeune can be diagnosed in utero, and are not mucopolysaccharidoses. Achondroplasia can be recognized sonographically after 27 weeks of gestation and is associated with advanced paternal age.

---

3. Mental retardation is an important clinical feature of all mucopolysaccharidoses except Morquio, Scheie, and Maroteaux-Lamy syndromes.
4. Radiographic abnormalities in the skull include macrocephaly and J-shaped sella turcica.

5. In the spine, focal kyphosis occurs at the thoracolumbar junction, in association with one or more small, oval, and retrolisthesed vertebral bodies with anterior central or inferior beaking.
6. Radiographic abnormalities in the pelvis include flared iliac wings and constricted superior acetabular regions, with increased acetabular angles.
7. The metacarpal bones are typically short, broad, proximally tapered, and arranged in a fan-shaped configuration.
8. The carpal angle in mucopolysaccharidoses is decreased, a finding also present in Madelung deformity, dyschondrosteosis, and Turner syndrome.
9. The long bones are short and undertubulated.

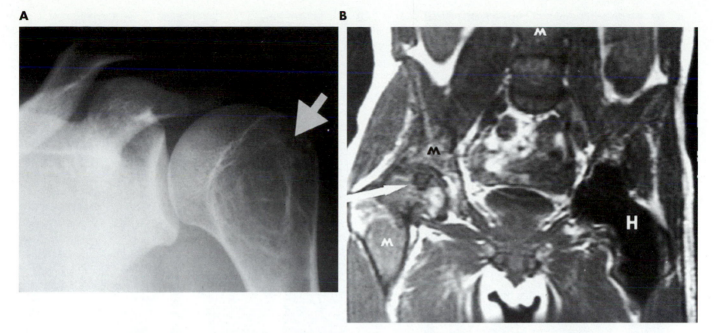

Fig. 5-19   Gaucher disease. **A,** A lytic lesion in the proximal humerus (*arrow*) has resulted from a focal accumulation of lipid-laden cells. **B,** Coronal T1-weighted MR image demonstrates diffuse low signal intensity within the bone marrow (*M*), along with osteonecrosis of the right femoral head (*arrow*) and contralateral hip replacement (*H*) for the same process. The combination of findings is virtually diagnostic of the disease.

10. In the thorax, oar-shaped ribs are characteristically observed.
11. In addition to the radiographic features of dysostosis multiplex, patients with Morquio syndrome also demonstrate odontoid hypoplasia, atlantoaxial instability, and severe platyspondylia in the spine.
12. Buckling of the sternum in Morquio syndrome may result in restricted motion of the thoracic cage.
13. Compression and fragmentation of the capital femoral epiphyses in Morquio syndrome resembles the radiographic findings of an epiphyseal dysplasia.

Box 5-16   Mucopolysaccharidoses (Dysostosis Multiplex): Important Features

- Flat, irregular, wedged, and/or oval vertebral bodies with central or inferior beaking anteriorly
- Kyphosis
- Shortening of long and short tubular bones
- Irregular epiphyses, metaphyseal flaring
- Coxa valga, deep acetabula
- Narrow pelvis (goblet shape) and flared iliac wings
- Tapered metacarpal bases, decreased carpal angle
- Morquio syndrome (type IV): hypoplastic or absent odontoid process, atlantoaxial subluxation
- Large skull, J-shaped sella turcica

Box 5-17   Features of Specific Mucopolysaccharidoses

MAJOR TYPES OF MUCOPOLYSACCHARIDOSIS

A. Type I (Hurler syndrome)
B. Type II (Hunter syndrome)
C. Type III (Sanfilippo syndrome)
D. Type IV (Morquio syndrome)
E. Type V or IS (Scheie syndrome)
F. Type VI (Maroteaux-Lamy syndrome)
G. Type VII (Sly syndrome)

MAJOR FEATURES OF THE VARIOUS TYPES

- Keratan sulfaturia: D
- Aortic regurgitation: D, E, G
- Normal intelligence: D, E, F
- Odontoid hypoplasia with atlantoaxial subluxation: D
- No corneal opacities: B
- Proximal tapering of second through fifth metacarpals: All
- No hepatosplenomegaly: E
- No joint stiffness, flexion contractures: D
- X-linked recessive inheritance: B
- Dysostosis multiplex (osteopenia, macrocephaly, thick calvarium, oar-shaped ribs, anterior beaking of vertebral bodies, goblet-shaped pelvis, coxa valga, increased acetabular angles, epiphyseal dysplasia, metaphyseal widening, short limbs): All

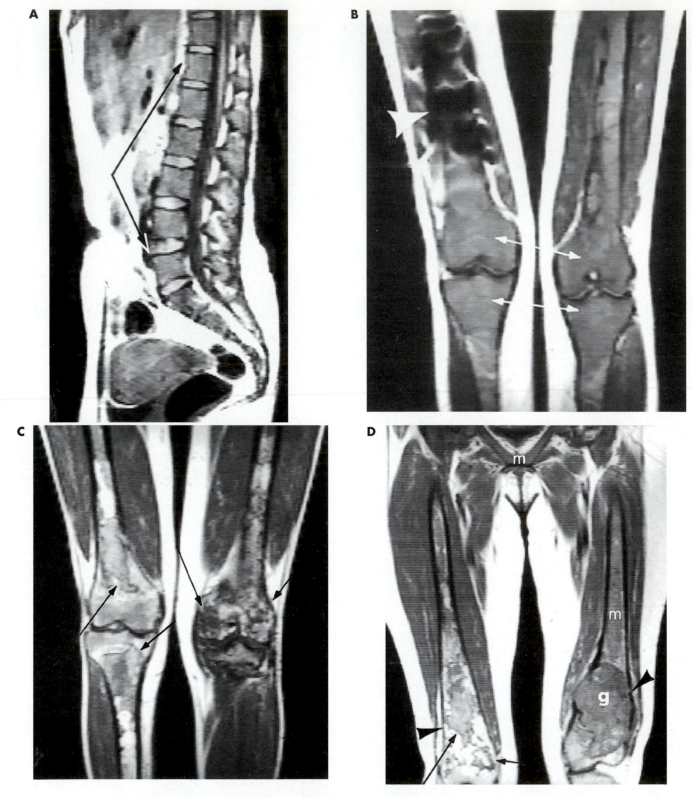

Fig. 5-20   MRI in Gaucher disease. **A,** Sagittal T1-weighted image demonstrates abnormally low signal intensity within the vertebral marrow (*arrows*). **B,** Similar findings (*arrows*) are evident on a coronal T1-weighted image (*arrowhead = artifact from orthopedic hardware*). **C,** Multiple intramedullary and subchondral areas of infarction (*arrows*) are evident on a coronal T1-weighted image. **D,** Coronal T1-weighted image reveals diffuse low signal intensity within the marrow (*m*), multiple intramedullary bone infarcts (*arrows*), Erlenmeyer flask deformity of both distal femora (*arrowheads*), and an expansile focal accumulation of Gaucher cells (*g*).

HOMOCYSTINURIA

1. Homocystinuria is a familial disorder with autosomal recessive inheritance, characterized by deficiency of the enzyme cystathionine synthetase (Box 5-18).

PEARL

Homocystinuria is an autosomal recessive disorder associated with cystathionine synthetase enzyme deficiency. Affected patients have the tendency for development of thrombosis, malar flush, and a high arched palate. Mental retardation is common. Arachnodactyly and large secondary ossification centers with some irregularity are present. The bone age may be advanced. Osteoporosis is usually prominent, especially in the spine. The presence of osteoporosis, metaphyseal flaring, and multiple contractures helps to differentiate homocystinuria from Marfan syndrome.

2. Because of an associated defect in collagen synthesis that affects multiple organ systems, the clinical and radiographic features of homocystinuria resemble those of Marfan syndrome.
3. Homocystinuria most commonly occurs among patients of northern European descent.
4. Clinical manifestations of the disease in the central nervous system include seizures, mental retardation, and spontaneous ocular lens dislocation.
5. Cardiovascular abnormalities include venous and arterial thrombosis secondary to abnormal platelet aggregation, cystic medial necrosis in elastic arteries, and spontaneous dissections; the last complication is rare in homocystinuria as compared to its incidence in Marfan syndrome.
6. Generalized osteoporosis with vertebral manifestations including biconcave endplate deformities and compression fractures is a characteristic finding in homocystinuria that distinguishes it from Marfan syndrome.

PEARL

Homocystinuria is an inborn error of methionine metabolism that has a clinical presentation of mental retardation, sternal deformities, scoliosis, and arachnodactyly. Radiographically, affected patients can appear marfanoid. However, osteoporosis that is inappropriate for age is a sign of homocystinuria that is not seen in Marfan syndrome.

7. Additional abnormalities of homocystinuria in the spine include scoliosis, posterior scalloping of vertebral bodies, and increased sagittal diameter of vertebral bodies.
8. Arachnodactyly is a characteristic finding in the hands and feet.
9. Articular manifestations of homocystinuria include joint laxity and flexion contractures.
10. In the bony thorax, pectus excavatum is a typical finding

PEARL

Homocystinuria is an autosomal recessive deficiency of cystathionine synthesis. Radiographic findings include osteoporosis, biconcave vertebrae, scoliosis, kyphosis, arachnodactyly, varus deformity of the humeral neck, enlarged carpal bones, accelerated skeletal maturation, and microcephaly. Clinical features include marfanoid habitus, cutaneous malar flush, fine sparse dry hair, ectopic lens, hepatosplenomegaly, and mental retardation.

Box 5-18   Homocystinuria: Important Features

- Cystathionine synthetase deficiency affecting collagen, elastin
- Osteoporosis, fractures
- Scoliosis
- Pectus excavatum or carinatum
- Coxa valga, genu valgum, pes cavus
- Physeal calcifications (particularly at wrist)
- Metaphyseal cupping, widening
- Carpal bone enlargement (especially capitate, hamate)
- Vascular complications

IDIOPATHIC TUMORAL CALCINOSIS

1. Radiographically, idiopathic tumoral calcinosis is characterized by large periarticular masses of soft-tissue calcification (Fig. 5-21).
2. The most common sites of involvement are the hip, shoulder, and elbow.
3. The disorder exhibits no sex predilection.
4. The frequency of idiopathic tumoral calcinosis is highest among blacks.
5. The serum calcium level is normal.
6. The serum phosphate level is elevated.

A

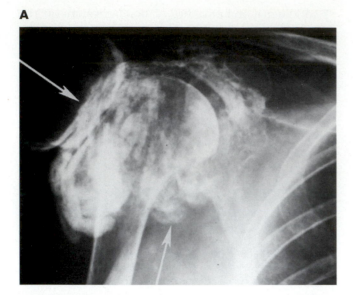

B

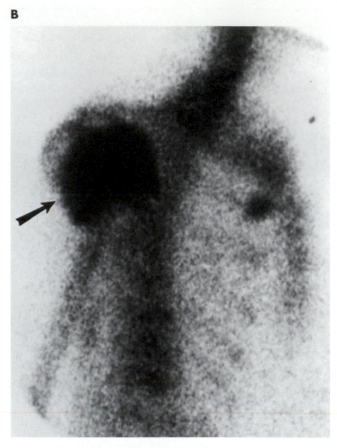

Fig. 5-21   Tumoral calcinosis involving the shoulder. **A**, Extensive cloudlike metastatic soft-tissue calcification (*arrows*) is evident in a periarticular distribution. **B**, The process exhibits increased activity (*arrow*) on a radionuclide bone scan. This phenomenon occurs in renal osteodystrophy and as an idiopathic inherited defect in phosphate metabolism.

7. The serum erythrocyte sedimentation rate is elevated.
8. The disorder may be caused by enhanced renal tubular phosphate resorption.
9. Idiopathic tumoral calcinosis is a form of metastatic soft-tissue calcification.
10. Computed tomography or dependent radiographs may reveal fluid levels within the calcified soft-tissue masses.
11. The calcified soft-tissue masses may be complicated by restricted joint function, cutaneous ulceration, and secondary infection.
12. Similar calcified soft-tissue deposits can occur in patients with chronic renal disease.
13. The recurrence rate is extremely high following surgical resection.
14. Dietary restriction of calcium and phosphate, along with administration of phosphate-binding antacids, constitutes accepted management.

PEARL

Idiopathic tumoral calcinosis:
- It appears during the second and third decades of life.
- The periarticular calcific masses are separated from each other by intervening radiolucent areas and may demonstrate fluid levels on dependent radiographs.
- The hip region is most commonly affected.
- It exhibits a predilection for black men, approximately one third of whom have a family history of the disease.
- Positive findings have been documented on skeletal scintigraphic studies.

NUTRITIONAL DISORDERS (Boxes 5-19, 5-20, 5-21)

Major radiographic features of specific nutritional disorders are summarized as follows:

1. Generalized osteopenia: hypovitaminosis C (scurvy), hypervitaminosis D
2. Generalized osteosclerosis: hypervitaminosis D, fluorosis
3. Abnormal collagen formation: hypovitaminosis C
4. Painful periostitis of long bones: hypervitaminosis A
5. Metaphyseal corner fractures of Pelkin: hypovitaminosis C
6. Ligamentous ossification, particularly pelvic: fluorosis, hypervitaminosis A in adults
7. Subperiosteal hemorrhage: hypovitaminosis C
8. Scorbutic rosary at costochondral junctions: hypovitaminosis C
9. Cone-shaped epiphyses: hypervitaminosis A
10. Sclerotic epiphyseal rim (Wimberger line): hypovitaminosis C
11. Extensive subperiosteal bone formation: hypovitaminosis C
12. Prominent soft-tissue calcification: hypervitaminosis D
13. Decreased bone production, most evident at rapid growth sites: hypovitaminosis C
14. Insufficiency-type stress fractures: fluorosis
15. Dense metaphyseal line of Fränkel: hypovitaminosis C
16. Periostitis: hypervitaminosis A, hypervitaminosis D, fluorosis
17. Radiographically occult until after 6 months of age: hypervitaminosis A, hypovitaminosis C
18. Radiolucent metaphyseal bands: hypovitaminosis C
19. Growth disturbance: hypervitaminosis A, hypovitaminosis C
20. Fractures from minor trauma: hypovitaminosis C, fluorosis

Box 5-19 Fluorosis: Important Features

- Endemic, occupational, iatrogenic
- Diffuse osteosclerosis (particularly axial)
- Degenerative enthesopathy
- Insufficiency fractures
- Ligamentous calcification (especially sacrospinous, sacrotuberous)
- Periosteal proliferation in long bones
- Differential diagnosis: osteopetrosis, osteoblastic metastases, myelofibrosis, renal osteodystrophy, hypoparathyroidism and its variants

Box 5-20 Hypervitaminosis D: Important Features

- Elevated serum calcium, phosphorus, alkaline phosphatase levels
- Metastatic soft-tissue calcification (usually periarticular)
- Cortical, trabecular thickening
- Radiodense metaphyseal bands (zones of provisional calcification) with diminished density in adjacent bone
- Radiodense vertebral endplates
- Calcification of falx cerebri

Box 5-21 Hypervitaminosis A: Important Features

- Cranial sutural widening
- Diaphyseal periostitis
- Premature physeal closure
- Enthesopathy with tendon, ligament calcification

PEARL

Hypervitaminosis A results in pseudotumor cerebri with widening of the sutures and bulging of the fontanelles secondary to increased intracranial pressure; a possible presenting symptom is drowsiness. Clinical findings include scaly skin, photophobia, and anemia. Hypervitaminosis D results in regions of cortical thickening and metastatic soft-tissue calcification, in association with elevated serum calcium and phosphate levels.

DRUG- AND TOXIN-INDUCED DISORDERS

Radiographic features of specific drug- and toxin-induced disorders are as follows:

1. Osteoporosis: methotrexate, alcohol, corticosteroids, heparin
2. Ischemic necrosis: alcohol, corticosteroids

PEARL

Metadiaphyseal infarcts are less common than ischemic necrosis of the femoral head in chronic alcoholism.

3. Rickets, osteomalacia: phenytoin, phenobarbital, bisphosphonates, ifosfamide
4. Broad metaphyses with radiodense bands: lead, bismuth
5. Stippled epiphyses: Coumadin
6. Developmental delay with variable skeletal anomalies: alcohol
7. Periostitis in infants: prostaglandins
8. Axial and appendicular hyperostosis: retinoids (isotretinoin, etretinate)
9. Calvarial thickening: phenytoin
10. Scurvylike metaphyseal and epiphyseal alterations in children: methotrexate
11. Heel pad thickening: phenytoin
12. Bandlike acro-osteolysis: polyvinyl chloride

MISCELLANEOUS METABOLIC DISORDERS

Important aspects of additional metabolic disorders affecting the musculoskeletal system are summarized in Boxes 5-22 through 5-24.

Box 5-22   Diabetes Mellitus: Important Features

- Mönckeberg-type medial vascular calcification
- Osteomyelitis secondary to spread from contiguous soft-tissue infection
- Neuropathic bone and joint disease
- Most common sites: foot, ankle
- Lipoatrophic variant affects soft tissues and bone marrow (Fig. 5-22)

Box 5-23   Oxalosis: Important Features

- Primary hyperoxaluria with increased urinary excretion of oxalic and glycolic acids
- Oxalate deposits in soft tissues
- Osteoporosis
- Pathologic fractures
- Metaphyseal and subperiosteal rarefaction
- Secondary hyperparathyroidism

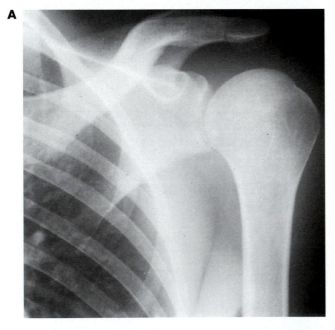
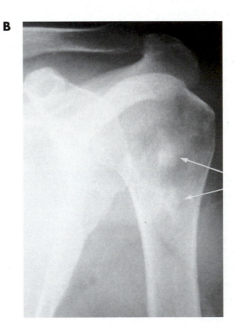

Fig. 5-22   Lipoatrophic diabetes. **A,** Radiography of the shoulder reveals complete absence of subcutaneous and intermuscular fat. **B,** Atrophy of bone marrow fat in another patient is associated with multiple osteosclerotic foci (*arrows*) in metaphyseal areas.

Box 5-24   Phenylketonuria: Important Features

- Metaphyseal cupping of long bones (especially at wrist)
- Calcific spicules extending vertically from metaphyses into physeal cartilage, evolving into striations with maturation
- Sclerotic metaphyseal margins
- Osteoporosis
- Delayed skeletal maturation

Differential diagnostic considerations for specific metabolic disorders affecting the sella turcica and spine are summarized in Boxes 5-25 and 5-26.

Box 5-25   Metabolic Disorders Affecting the Sella Turcica

- Sphenooccipital synchondrosis normally fuses between 14 and 20 years of age
- Subarachnoid space extension into sella turcica occurs in approximately 20% of people
- Small sella: primary hypopituitarism, growth hormone deficiency, congenital disorders, Sheehan syndrome, myotonic dystrophy, Cushing syndrome secondary to adrenal adenoma
- Chromophobe adenomas are more likely to cause sellar enlargement than basophilic adenomas
- Mucopolysaccharidoses: arachnoid cysts produce an elongated, enlarged chiasmatic sulcus (J-shape)
- Large sella: hypogonadism, neurofibromatosis, pituitary neoplasms, craniopharyngioma, intrasellar aneurysm, oxycephaly, cretinism, "empty sella" syndrome
- Nelson syndrome: rapidly growing ACTH-producing adenoma following bilateral adrenalectomy for Cushing disease; more common in women
- Demineralization of dorsum sellae: increased intracranial pressure, aging

Box 5-26   Metabolic Disorders Affecting the Spine

DIABETES MELLITUS

- DISH: 13% overall incidence,
      21% incidence in older patients
- Spinal neuroarthropathy
- Osteomyelitis, disk infection

LIPOATROPHIC DIABETES

- Osteosclerosis of vertebral bodies

Box 5-26—cont'd

HYPERTHYROIDISM

- Vertebral osteopenia, accentuation of primary trabeculae
- Biconcave endplate deformities
- Kyphosis or kyphoscoliosis
- Thoracolumbar involvement predominates over cervical

ACROMEGALY

- Elongation, widening of vertebral bodies; thoracolumbar involvement predominates over cervical
- Extensive osteophytosis
- Increased disk space height, especially lumbar
- Accentuated thoracic kyphosis, lumbar lordosis
- Hypertrophic apophyseal joint changes
- Posterior scalloping of vertebral bodies

HYPOTHYROIDISM

- T12, L1 vertebral bodies: short, bullet-shaped with gibbus
- Osteoporosis
- Delayed development, fusion of apophyses
- Irregular vertebral body contours
- Relative disk space widening
- Increased anterior atlantoaxial distance

HYPERPARATHYROIDISM

- Subchondral bone resorption at discovertebral junctions
- Schmorl nodes
- Generalized vertebral osteopenia
- Brown tumors

HYPOPARATHYROIDISM

- Generalized vertebral osteosclerosis
- Calcification of spinal ligaments with osteophytosis resembling DISH or ankylosing spondylitis

CUSHING SYNDROME

- Vertebral osteoporosis
- Biconcave endplate deformities, compression fractures with excessive callus
- Exaggerated thoracic kyphosis
- Osteonecrosis of vertebral bodies with vacuum phenomena

PAGET DISEASE

- Lumbar spine predominant
- Vertebral body, posterior element enlargement with neurologic encroachment; extension into osteophytes possible
- Enlarged, coarsened trabeculae

Continued

Box 5-26—cont'd

- Condensation along contours of vertebral bodies (picture frame)
- Occasional ivory vertebrae, dense pedicles
- Biconcave endplates, disk space narrowing, osseous bridging
- Vertebral collapse
- Sarcomatous degeneration

RICKETS

- Scoliosis, decreased height
- Expansion of intervertebral disks, biconcave vertebral endplates
- Equivalent involvement throughout spine

OSTEOMALACIA

- Coarsened, indistinct trabeculae
- Diffuse vertebral osteopenia
- Biconcave endplates, uniform and symmetric involvement throughout spine
- Atypical axial form: dense, coarse trabecular pattern, cervical involvement predominates over lumbar

RENAL OSTEODYSTROPHY

- Osteomalacic alterations
- Osteosclerotic bands adjacent to vertebral endplates (rugger jersey spine)
- Diffuse osteosclerosis

HYPOPHOSPHATASIA

- Paravertebral calcification
- Resemblance to rickets, osteomalacia

X-LINKED HYPOPHOSPHATEMIC VITAMIN D–RESISTANT OSTEOMALACIA

- Generalized increase in vertebral density
- Spinal changes resembling those of ankylosing spondylitis
- Calcification of paravertebral ligaments, annulus fibrosus, apophyseal joint capsules

SUGGESTED READINGS

1. Genant JK, Gluer CC, and Lotz JC: Gender differences in bone density, skeletal geometry, and fracture biomechanics, Radiology 190:636, 1994.
2. Holder LE, Cole LA, and Myerson MS: Reflex sympathetic dystrophy in the foot: clinical and scintigraphic criteria, Radiology 184:531, 1992.
3. Karantanas AH, Kalef-Ezra JA, and Glaros DC: Quantitative computed tomography for bone mineral measurement: technical aspects, dosimetry, normal data and clinical applications, Br J Radiol 64:298, 1991.
4. Kaufmann GA, Sundaram M, and McDonald DJ: Magnetic resonance imaging in symptomatic Paget's disease, Skeletal Radiol 20:413, 1991.
5. Kunin JR and Strouse PJ: "Yarmulke" sign of Paget's disease, Clin Nucl Med 16:788, 1991.
6. Lang P et al: Osteoporosis: current techniques and recent developments in quantitative bone densitometry, Radiol Clin North Am 29:49, 1991.
7. Mayo-Smith W and Rosenthal DI: Radiographic appearance of osteopenia, Radiol Clin North Am 29:37, 1991.
8. Mithal A et al: Radiological spectrum of endemic fluorosis: relationship with calcium intake, Skeletal Radiol 22:257, 1993.
9. Murphey MD et al: Musculoskeletal manifestations of chronic renal insufficiency, Radiographics 13:357, 1993.
10. O'Keeffe D: Morphometry, Radiol Clin North Am 29:165, 1991.
11. Rosenberg AE: Pathology of metabolic bone disease, Radiol Clin North Am 29:19, 1991.
12. Ryan PJ and Fogelman I: Osteoporotic vertebral fractures: diagnosis with radiography and bone scintigraphy, Radiology 190:669, 1994.
13. Sundarum M, Dessner D, and Ballal S: Solitary, spontaneous cervical and large bone fractures in aluminum osteodystrophy, Skeletal Radiol 20:91, 1991.
14. Verlooy H et al: Common bone features in osteomalacia, secondary hyperparathyroidism, and renal osteodystrophy, Clin Nucl Med 16:372, 1991.
15. Wahner H and Fogelman I: Dual-energy x-ray absorptiometry in clinical practice, New York, 1994, Scovill Paterson.
16. Wang Y et al: Endemic fluorosis of the skeleton: radiographic features in 127 patients, AJR 162:93, 1994.
17. Young W, Sevcik M, and Tallroth K: Metaphyseal sclerosis in patients with chronic renal failure, Skeletal Radiol 20:197, 1991.

CHAPTER 6

Congenital Disorders

DEVELOPMENTAL DYSPLASIA OF THE HIP (Box 6-1)

Recognized Characteristics

1. The highest incidence of the condition occurs in the white race and female sex by a ratio of approximately 5:1.
2. Intrauterine factors predisposing to the disorder include oligohydramnios and breech presentation.
3. The incidence of developmental hip dysplasia with clinical signs that persist beyond the initial few weeks of life is approximately 1 per 1000 births.
4. Conventional radiography performed before approximately 6 weeks of age is associated with a high rate of false negative diagnoses.
5. On frontal radiographs of the pelvis, the Hilgenreiner line is drawn tangent to the inferior aspects

319

A

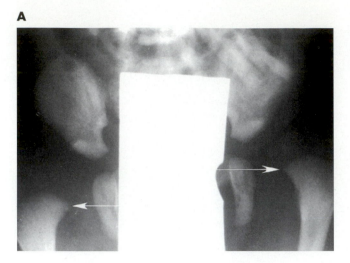

B

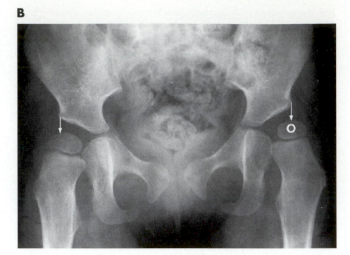

Fig. 6-1   Bilateral developmental hip dysplasia. **A,** Infant with severe lateral displacement of both femora (*arrows*). **B,** Older child with diminished lateral coverage (*arrows*) of both femoral heads, particularly on the left where the femoral head ossification center (*o*) is hypoplastic.

of the ossified iliac bones, and the Perkin lines are constructed perpendicular to it through the lateral edges of the ossified acetabula; the femoral head ossification centers should normally lie in the inferomedial quadrants formed by these lines.

6. Lateral femoral subluxation can be quantitated by measuring (1) the horizontal distance from the acetabular teardrop to either the medial corner of the proximal femoral metaphysis or (2) the center-edge angle, formed by the Perkin line and a line joining the center of the femoral head ossification center to the lateral edge of the ossified acetabulum (Fig. 6-1).

7. Superior femoral subluxation can be quantitated by measuring the vertical distance from the proximal femoral metaphysis to the Hilgenreiner line.

8. The femur is considered dislocated if it is displaced both laterally and superiorly.

9. The Shenton line forms a smooth continuous curve from the obturator foramen to the medial aspect of the femoral neck.

10. The acetabular angle is formed by a line drawn parallel to the ossified acetabular roof and the Hilgenreiner line, and it normally measures less than 30 degrees.

11. Femoral anteversion normally measures approximately 30 degrees at birth and 10 degrees at skeletal maturity; in developmental hip dysplasia, this parameter is often greater than is normal for the patient's age.

12. Ossification of the proximal femoral epiphysis is usually delayed in the setting of developmental hip dysplasia (Fig. 6-2).

13. The soft-tissue components of the neonatal hip can be assessed by contrast arthrography, ultrasonography, or magnetic resonance imaging (Fig. 6-3).

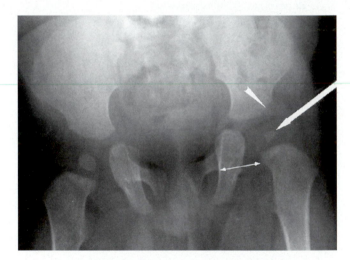

Fig. 6-2   Developmental dysplasia of the hip. On the left, there is delayed development of the femoral head ossification center (*arrow*), a shallow acetabulum (*arrowhead*), and disruption of the Shenton line (*double-headed arrow*).

14. Partial or complete inversion of the fibrocartilaginous superior acetabular labrum or limbus can prevent closed reduction of a dislocated hip; additional characteristics of the soft tissues in developmental hip dysplasia include elongation of the ligamentum teres, hourglass constriction of the joint capsule, tightening of the iliopsoas muscle, and hypertrophy of the intracapsular soft tissues (pulvinar).

15. Use of computed tomography with three-dimensional image reconstruction is a useful preoperative strategy for characterizing osseous structure in older patients with untreated developmental hip dysplasia (Box 6-2).

16. Potential complications of developmental hip dysplasia include ischemic necrosis of the femoral head

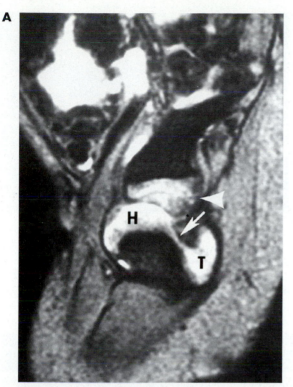

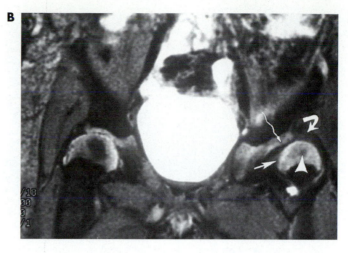

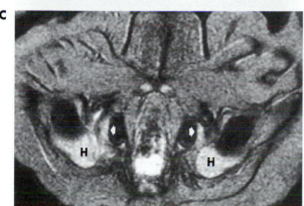

Fig. 6-3 MRI in developmental hip dysplasia. **A,** Sagittal gradient-echo image of normal anatomy. Note continuity of growth cartilage (*arrow*) between the femoral head (*H*) and greater trochanter (*T*) (*arrowhead = triradiate cartilage*). **B,** Coronal gradient-echo image in unilateral dysplasia. The left femoral head is hypoplastic and laterally dislocated (*straight arrow*), and its secondary ossification center is delayed in development (*arrowhead*). The acetabular labrum remains in the normal everted position (*curved arrow*), although the triradiate cartilage is shallow (*wavy arrow*). **C,** Transaxial gradient-echo image in bilateral dysplasia. Both femoral heads (*H*) are posteriorly dislocated with respect to the triradiate cartilages (*arrows*).

(secondary to manipulation), pseudoacetabulum formation in the iliac wing with limb shortening (untreated severe dysplasia), and premature degenerative joint disease in the hip (Fig. 6-4).

17. Therapeutic management of mild developmental hip dysplasia includes closed reduction and abduction splinting.

18. Surgical procedures applied to the soft tissues in moderate to severe developmental hip dysplasia include open reduction with removal of obstructing soft tissue, iliopsoas tendon release, and abductor tenotomy.

19. To improve coverage of the femoral head by the acetabulum in developmental hip dysplasia, varus derotational osteotomy may be performed on the proximal femur.

Box 6-1 Developmental Hip Dysplasia: Important Features

- Teratologic: caused by external factors (neuromuscular diseases such as meningomyelocele, arthrogryposis multiplex congenita)
- Typical: secondary to hormonally induced capsular laxity, intrauterine position

SOFT-TISSUE ALTERATIONS

- Elongation, hypertrophy of ligamentum capitis femoris (also pulvinar); displacement of transverse acetabular ligament; impedes reduction
- Inversion of labrum (limbus); slow, gradual, impedes reduction

Continued

A

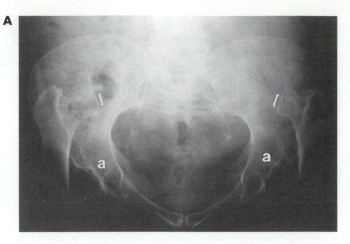

B

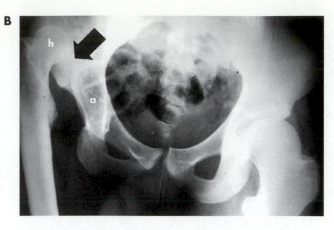

Fig. 6-4   Severe untreated developmental dysplasia of the hip. **A,** The flattened femoral heads (*brackets*) are superiorly dislocated and articulate with pseudoacetabula in the iliac wings posteriorly. The true acetabula (*a*) are hypoplastic. **B,** There is posterolateral dislocation (*arrow*) of the hypoplastic femoral head (*h*) with respect to the underdeveloped true acetabulum (*a*).

Box 6-1—cont'd

- Transposition of iliopsoas tendon leading to "hourglass" deformity of capsule; impedes reduction
- Capsular contractures

In teratologic cases, the hip is often severely or completely dislocated at birth. In typical cases, the hip is rarely truly dislocated at birth.

TYPE I: DISLOCATABLE OR POSITIONALLY UNSTABLE
- Mild marginal acetabular alterations
- Femoral anteversion without asphericality
- Mild adduction-flexion contractures (capsule, muscle)
- Negative Orotolani sign, decreased abduction

TYPE II: SUBLUXED
- Femoral anteversion with asphericality
- Shallow acetabulum
- Narrow anterior acetabular margin
- Superior and posterior acetabular marginal deformities (eversion of labrum)
- Increased acetabular acclivity (bone)
- Increased flexion-adduction contractures (capsule, muscle)
- Early limbus inversion secondary to superior-posterior displacement of femur by extensor-abductors (gluteus medius and minimus)
- Positive Orotolani sign, decreased abduction

TYPE III: DISLOCATED
- Marked deformity of acetabular margin and femoral head
- Posterosuperior displacement of femoral head
- False acetabulum with labral inversion
- Inverted limbus; contractures (capsule, muscle)
- Elongated ligamentum capitis femoris

Box 6-1—cont'd

- Transverse acetabular ligament pulled into acetabulum
- Positive Orotolani sign for 24 to 48 hours, positive Barlow sign (dislocation)

PRENATAL CONTRIBUTORS
- Maternal: spinal configuration, uterine and abdominal muscle tone, parity, restrictive clothing, cesarean delivery, hormonal factors
- Fetal: oligohydramnios, breech presentation, prematurity and low birth weight, muscle tone, genotype
- Postnatal: swaddling, adduction-extension positioning

CLINICAL DIAGNOSIS
- Signs: limb shortening, unequal creases, limited abduction, Trendelenburg gait
- Provocative maneuvers: Ortolani test (abduct to relocate), Barlow test (adduct to dislocate)

INCIDENCE
- Hip laxity: 1.5%; established DDH: 0.15%; female>male; white>black; left>right>bilateral (70%, 25%, 5% of cases)
- Types I, II account for 95% of unstable hips in neonates

TREATMENT
- Follow-up for neonatal laxity
- Splinting or casting for dislocatable or dislocated-relocatable cases
- Surgery for irreducibility, contractures, failed conservative therapy; options: osteotomy (Salter, Pemberton, Chiari, femoral), acetabuloplasty, limbectomy, adductor and iliopsoas releases; complications: ischemic necrosis, secondary coxa vara, premature osteoarthritis, pseudoacetabulum formation

Box 6-2 Imaging in Developmental Hip Dysplasia

RADIOGRAPHY

- Putti triad: delayed epiphyseal ossification, increased acetabular acclivity, and superolateral femoral head displacement
- Normal radiograph does not exclude developmental hip dysplasia
- Screening film: anteroposterior, legs together, neutral rotation
- At 4 to 6 weeks: proximal and lateral migration of femoral neck relative to ilium; shallow acetabulum, especially superiorly; false acetabulum; delayed ossification of femoral head center (abnormal side may occasionally have accelerated ossification)

Acetabular Index

- 27.5 degrees: neonatal average
- 30 degrees: normal
- 7.5 degrees of asymmetry: abnormal

Lateral Migration

Quadrants:
IU = inner upper
IL = inner lower
OU = outer upper
OL = outer lower

Perkin line

IU	OU
IL	OL

← Hilgenreiner line

- If head or medial metaphysis:
  OL: subluxation
  OU: dislocation
  IL: stability

Superior Migration

- Vertical distance of femoral ossification center or metaphysis from Hilgenreiner line

Shenton Line

- Smooth curve in normal case
- Medial femoral metaphysis to superior border of obturator foramen

Marginal Lateral Acclivity of Acetabulum

- Appears after 1 month of age, indicates incongruity

ULTRASONOGRAPHY

- For patients up to approximately 10-12 months of age
- Structures of interest: femoral head (size and shape), acetabulum (depth and development), acetabulum-head relationship, cartilage, synovial fluid, ligamentum teres

Echo Characteristics

- High: fibrocartilage, capsule, ligamentum teres
- Intermediate-low: femoral head, muscles
- Anechoic: effusion, articular cartilage, triradiate cartilage

Box 6-2—cont'd

Ultrasonographic Grading Scheme

- I: normal
- II: acetabular deficiency
- III: subluxation with normal or abnormal cartilage
- IV: dislocation

ARTHROGRAPHY

- Indications: late discovery, incomplete reduction, recurrent dislocation, preoperative assessment
- Follows preparatory traction in most cases and is often performed under anesthesia
- Structures of interest: femoral head, labrum, iliopsoas tendon, pulvinar, capsule, ligamentum teres

COMPUTED TOMOGRAPHY

- Useful in assessing the acetabulum-femoral head relationship, impediments to reduction, acetabular and femoral torsion, preoperative planning in cases of late diagnosis

MAGNETIC RESONANCE IMAGING

- Favored for its multiplanar imaging capabilities and superb delineation of nonossified soft-tissue structures among children older than 12 months of age

Surgical Procedures on the Bony Pelvis for Treatment

1. Steele triple innominate osteotomy is a reconstructive procedure for skeletally mature patients.
2. Salter opening wedge osteotomy and Pemberton acetabuloplasty are reconstructive procedures for skeletally immature patients.
3. Chiari medial displacement osteotomy is a salvage procedure for older patients.
4. Salter opening wedge osteotomy is an osteotomy that extends horizontally from the anteroinferior iliac spine to the sacrosciatic notch.
5. Chiari medial displacement osteotomy is an intraarticular osteotomy extending across the superior aspect of the acetabulum.
6. Salter opening wedge osteotomy and Pemberton acetabuloplasty require placement of an opening bone wedge.
7. Salter opening wedge osteotomy and Pemberton acetabuloplasty are procedures in which the acetabular roof is shifted anterolaterally to cover the femoral head.
8. Pemberton acetabuloplasty is an osteotomy that extends from the anteroinferior iliac spine to the triradiate cartilage.

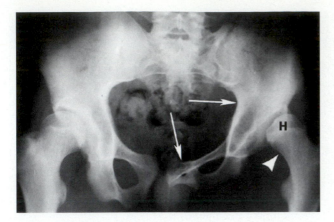

Fig. 6-5   Developmental dysplasia of the hip following inadequate treatment. Left-sided pelvic deformity (*arrows*) is secondary to prior osteotomies aimed at improving acetabular coverage. The femur is displaced superolaterally (*arrowhead*) and its head is hypoplastic (*H*).

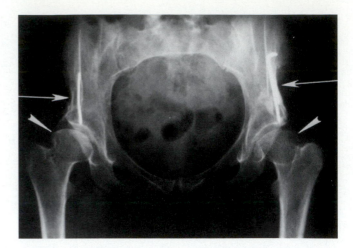

Fig. 6-6   Residual deformity of the femoral heads (*arrowheads*) following surgical treatment (circumacetabular osteotomies) (*arrows*) of developmental hip dysplasia. Differential diagnosis should include consideration of Legg-Calvé-Perthes disease, other causes of osteonecrosis in childhood, and old slipped capital femoral epiphysis.

9. Chiari medial displacement osteotomy is an osteotomy in which bone lying inferior to the osteotomy is displaced medially.
10. Steele triple innominate osteotomy is an osteotomy that extends across the iliac neck, ischium, and pubis.
11. Chiari medial displacement osteotomy, Salter opening wedge osteotomy, Pemberton acetabuloplasty, and Steele triple innominate osteotomy increase the degree of femoral head coverage by the acetabulum (Fig. 6-5).
12. Salter opening wedge osteotomy is an osteotomy in which the symphysis pubis acts as a hinge.
13. Pemberton acetabuloplasty is an osteotomy in which the triradiate cartilage acts as a hinge.
14. Chiari medial displacement osteotomy is an osteotomy in which the femoral head is displaced medially.
15. Salter opening wedge osteotomy and Pemberton acetabuloplasty allow for congruent growth of the femoral head and acetabulum.
16. Steele triple innominate osteotomy involves rotation of a free-floating acetabulum (Fig. 6-6).

CONGENITAL FOOT DEFORMITIES

Fundamental Concepts

1. Valgus (abductus): eversion of the hindfoot or forefoot (Fig. 6-7).
2. Varus (adductus): inversion of the hindfoot or forefoot.
3. Planus: depression of the longitudinal arch of the foot (Fig. 6-7).
4. Cavus: elevation of the longitudinal arch of the foot.
5. Equinus: fixed plantar flexion of the hindfoot. (Fig. 6-7).
6. Forefoot pronation: outward rotation of the forefoot relative to the hindfoot.
7. Forefoot supination: inward rotation of the forefoot relative to the hindfoot.

PEARL

Pes is used to describe *acquired* foot deformities, whereas talipes is used to describe *congenital* foot deformities. In this context, equinus refers to abnormal elevation of the posterior part of the calcaneus, with plantar flexion of the hindfoot. Adduction refers to medial deviation of the forefoot toward the sagittal axis of the body. Cavus refers to elevation of the longitudinal arch of the foot. Planus refers to flattening of the longitudinal arch of the foot.

Imaging

1. Qualitative and quantitative assessment of congenital foot deformities is best made based on weight-bearing radiographs.
2. In the evaluation of hindfoot varus or valgus, the talus is the assumed bone of reference.

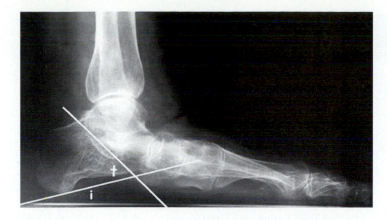

Fig. 6-7   Equinovalgus hindfoot with pes planus and generalized osteopenia. The talocalcaneal angle (*t*) is greater than 45 degrees. The calcaneal inclination angle (*i*) is less than 15 degrees.

3. On lateral radiographs, the talocalcaneal or Kite angle normally measures between 25 and 45 degrees, but it can be up to 50 degrees in newborns (Fig. 6-7).

4. On frontal radiographs, the talocalcaneal angle normally measures between 30 and 50 degrees in newborns and between 15 and 40 degrees in older children.

5. On frontal radiographs, the midtalar line normally passes through or slightly medial to the first metatarsal base.

6. On frontal radiographs, the midcalcaneal line normally passes through the fourth metatarsal base.

7. Hindfoot varus occurs in congenital clubfoot and paralytic deformities.

8. On lateral radiographs in hindfoot varus, the talocalcaneal angle is decreased to less than 25 degrees.

9. On frontal radiographs in hindfoot varus, the talocalcaneal angle is decreased to less than 15 degrees and the midtalar line passes lateral to the first metatarsal base.

10. Hindfoot valgus occurs in flexible flatfoot deformity, congenital vertical talus, and neurologic deformities.

11. On lateral radiographs in hindfoot valgus, the talocalcaneal angle is increased to greater than 50 degrees in newborns and greater than 45 degrees in older children.

12. On frontal radiographs in hindfoot valgus, the talocalcaneal angle is increased to greater than 50 degrees in newborns and greater than 40 degrees in older children; the midtalar line passes medial to the first metatarsal base.

13. On lateral radiographs, the angle formed by the long axes of the calcaneus and tibia normally measures between 60 and 90 degrees, and there is dorsiflexion of the calcaneus.

14. Equinus hindfoot occurs in congenital clubfoot and congenital vertical talus.

15. On lateral radiographs in equinus hindfoot, the calcaneotibial angle is increased to greater than 90 degrees, and there is plantar flexion of the calcaneus.

16. Calcaneus hindfoot occurs in cavus deformities and spastic deformities.

17. On lateral radiographs in calcaneus hindfoot, the calcaneotibial angle is decreased to less than 60 degrees, and there is excessive dorsiflexion of the calcaneus.

18. On normal lateral radiographs, the fifth metatarsal lies in the most plantar position, with superimposition of the remaining metatarsals; on normal frontal radiographs, proximal convergence of the metatarsals is observed, with slight overlap of the bases.

19. Forefoot varus occurs in congenital clubfoot and spastic deformities and is associated with inversion and supination.

20. On lateral radiographs in forefoot varus, the fifth metatarsal lies in the most plantar position and the first metatarsal lies in the most dorsal position, with decreased superimposition of the metatarsals.

21. On frontal radiographs in forefoot varus, the forefoot is narrowed with increased overlap of the metatarsal bases.

22. Forefoot valgus occurs in flexible flatfoot deformity and is associated with congenital vertical talus and spastic deformities.

23. On lateral radiographs in forefoot valgus, the first metatarsal lies in the most plantar position.

24. On frontal radiographs in forefoot valgus, the forefoot is broadened with decreased overlap of the metatarsal bases.

PEARL

Congenital foot deformities:
- Cavus foot deformity occurs in neuromuscular disorders such as poliomyelitis and meningomyelocele.
- Congenital vertical talus includes a talonavicular joint dislocation.
- Peroneal spastic flatfoot is frequently associated with tarsal coalition.
- Flexible flatfoot deformity is associated with an increase in the talocalcaneal angle, hindfoot valgus, and forefoot abduction.
- Congenital vertical talus is frequently associated with myelomeningocele or arthrogryposis multiplex congenita.
- In some cases of congenital clubfoot, the talus is small, with a dome that is less convex than normal and medioplantar dislocation of its neck and head.
- Following inadequate or improper treatment of congenital clubfoot, the rocker-bottom deformity or the flattop talus deformity can be observed.

Congenital Clubfoot

1. The condition is also known as talipes equinovarus.
2. Congenital clubfoot is more common in male individuals by a ratio of 2 to 3:1.
3. Potential causes include persistence of normal early fetal anatomy, abnormal intrauterine position, ligamentous laxity secondary to defective connective tissue, and muscular imbalance.
4. Hindfoot varus is one component of the deformity.
5. Equinus hindfoot is one component of the deformity.
6. Varus configuration of the forefoot occurs in congenital clubfoot.
7. The incidence of the condition is approximately 1 in 1000 births.

Metatarsus Adductus

1. The condition is the most common structural foot deformity of infancy.
2. Metatarsus adductus is more common in female individuals.
3. Metatarsus adductus is usually bilateral.
4. The incidence of the condition is approximately 1 in 100 births.
5. Forefoot varus is one component of the deformity.
6. The calcaneotibial angle on lateral radiographs is normal in metatarsus adductus.
7. Hindfoot valgus may be a component of the deformity.

Pes Cavus

1. Calcaneus hindfoot occurs in pes cavus.
2. Compensatory plantar flexion of the forefoot is one component of the deformity.
3. The condition occurs in lower motor neuron disease such as poliomyelitis.
4. Upper motor neuron lesions such as Friedreich ataxia may cause pes cavus.
5. The condition occurs in peroneal-type muscular dystrophy, including Charcot-Marie-Tooth disease.
6. Vascular ischemia as in the Volkmann contracture may cause pes cavus.

Congenital Vertical Talus

1. The condition is commonly referred to as the rocker-bottom foot.
2. Dorsal dislocation of the navicular locks the talus in extreme plantar flexion.
3. Hindfoot valgus is one component of the deformity.
4. Equinus hindfoot occurs in congenital vertical talus.
5. Forefoot valgus occurs in congenital vertical talus.
6. On lateral radiographs, dorsiflexion of the forefoot is observed.
7. Clinically, a rigid flatfoot deformity is evident.
8. The condition is frequently associated with meningomyelocele.
9. Congenital vertical talus may occur either in isolation or in association with other congenital anomalies.

Flexible Flatfoot Deformity

1. The condition is also known as pes planovalgus.
2. Because the deformity is flexible, non–weight-bearing radiographs are normal.
3. The condition is hereditary and affects approximately 4% of the population.
4. Hindfoot valgus is one component of the deformity.
5. The calcaneotibial angle on lateral radiographs is normal in flexible flatfoot deformity.
6. Forefoot valgus occurs in flexible flatfoot deformity.
7. Ligamentous laxity is the most likely cause of the deformity.
8. The condition is distinguished from congenital vertical talus by the absence of talonavicular joint dislocation.
9. Clinically, flattening of the midtarsal arch is observed.

TARSAL COALITION

1. Although most cases of tarsal coalition are congenital, the condition may also be caused by trauma, surgery, inflammatory arthritis, or infection.

PEARL

Tarsal coalition represents abnormal fusion of the tarsal bones. The union may be fibrous, cartilaginous, or osseous, and it can be congenital or acquired. Acquired forms result from infection, trauma, articular disorders, or postsurgical change. In order of decreasing frequency, the predominant sites are talocalcaneal, calcaneonavicular, talonavicular, and calcaneocuboid. A penetrated axial radiograph (Harris-Beath projection) and computed tomography are used in assessing talocalcaneal coalition.

2. The frequency of tarsal coalition in the general population is approximately 1%.
3. Tarsal coalition may be familial or a component of congenital malformation syndromes including the Apert form of acrocephalosyndactyly, hereditary symphalangism, and hand-foot-uterus syndrome.
4. Tarsal coalition is more common in male individuals.
5. Failure of normal segmentation is the most likely cause of congenital cases.
6. Bilateral involvement is observed in approximately 25% of patients.
7. The sites are listed in decreasing order of frequency: talocalcaneal, calcaneonavicular, talonavicular, and calcaneocuboid.
8. Most patients are affected during the second or third decade of life.
9. Peroneal spastic flatfoot is the typical clinical presentation, and limited subtalar motion may also be observed.

PEARL

Talocalcaneal and calcaneonavicular coalitions are the most common tarsal coalitions. A coalition leads to a painful flatfoot (sometimes called peroneal spastic flatfoot), usually appearing in the second or third decade of life.

10. Tarsal coalitions may be osseous, cartilaginous, or fibrous.
11. On conventional radiographs, nonosseous coalitions manifest abnormally close approximation between the involved bones, with cortical irregularity and sclerosis; elongation of the anterior calcaneal process in calcaneonavicular coalition produces the anteater sign on the lateral view (Fig. 6-8).
12. The best tool for the diagnosis of calcaneonavicular coalition is a 45-degree medial oblique conventional radiograph; false negative diagnoses are possible with computed tomography or magnetic resonance

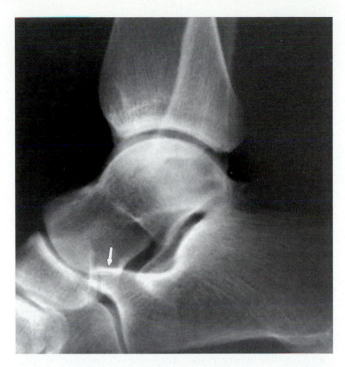

Fig. 6-8 Elongation and squaring of the anterior calcaneal process (*arrow*) in calcaneonavicular coalition (anteater sign). An oblique radiograph would optimally depict the abnormality, which does not require cross-sectional imaging for diagnosis (unlike talocalcaneal coalition).

imaging because of the oblique orientation of coalitions at this site.

13. Talocalcaneal coalition produces more marked symptoms and more secondary radiographic signs than does calcaneonavicular coalition.
14. Talocalcaneal coalition most commonly occurs at the level of the sustentaculum tali of the calcaneus.
15. Dorsal beaking of the talar head occurs in diffuse idiopathic skeletal hyperostosis (Forrestier disease), acromegaly, and rheumatoid arthritis, as well as tarsal coalition.
16. The ball-and-socket ankle mortise is a secondary radiographic sign of talocalcaneal coalition that results from obligatory transfer of inversion-eversion function from the subtalar joints to the tibiotalar joint.
17. Additional secondary radiographic signs of talocalcaneal coalition include close approximation between the talus and calcaneus, asymmetry of the talar necks inferiorly, and widening of the lateral talar process.
18. Computed tomography is the best procedure for the diagnosis of talocalcaneal coalition; false positive and false negative results may be obtained with the axial (Harris-Beath) calcaneal view because of suboptimal positioning or radiographic technique.
19. On bone scans, increased activity is seen both at the site of coalition and at joints that are secondarily

subjected to abnormal stress, such as the tibiotalar and talonavicular articulations.

20. The CT gantry should either be in neutral position or angled away from the knees, because a false positive diagnosis of subtalar coalition may be obtained if it is angled toward the knees.

21. For definitive diagnosis of typical nonosseous subtalar coalitions, contrast material may be injected into the talonavicular joint, which communicates with the articulation between the talus and sustentaculum tali (the posterior subtalar joint does not); failure of the contrast material to opacify this space indicates interposed cartilage or fibrous tissue.

22. Magnetic resonance imaging or conventional tomography can also be used to diagnose talocalcaneal coalition; the lateral projection is optimal for the latter method.

SCLEROSING BONE DYSPLASIAS

1. The sclerosing bone dysplasias constitute a spectrum of disorders caused by failure of osteoclastic bone resorption at sites of either endochondral or intramembranous ossification.

2. Osteopoikilosis is an asymptomatic condition characterized by multiple enostoses or bone islands located predominantly in the epiphyses and metaphyses; it exhibits normal findings on radionuclide bone scans (Box 6-3).

3. Osteopathia striata is also known as Voorhoeve disease, and it is characterized by linear striations in the metaphyseal regions.

PEARL

Osteopathia striata:
- Linear, regular bands of increased density extending from metaphyses through variable portions of long bones
- Fanlike striation of the iliac bones
- Multiple areas of rarefaction interspersed between striations
- Occasionally multiple small exostoses

Box 6-3 Osteopoikilosis: Important Features

- Hereditary benign sclerosing bone dystrophy
- Small round or oval foci of osteosclerosis, usually less than 1 cm in size
- Most common in areas of high trabecular bone content
- Not progressive with age

Box 6-3—cont'd

- Cutaneous abnormalities in 25% of cases (dermatofibrosis lenticularis disseminata, keloids, scleroderma-like lesions)
- Most commonly affects long-bone epiphyses and metaphyses, pelvis, scapulae, carpus, tarsus
- No increased activity seen on radionuclide bone scans
- May coexist with osteopathia striata (Voorhoeve disease), melorheostosis, hyperostosis frontalis interna
- Enostoses are symmetrical in distribution, well-defined, and uniform in size

Melorheostosis (Box 6-4)

1. The condition is characterized by wavy hyperostosis that resembles wax dripping from a burning candle.
2. Both the endosteal and periosteal surfaces of the cortex are affected.
3. The bones of the lower extremities are most commonly involved.
4. Involvement of axial skeletal sites is predominantly endosteal and appears as osteosclerotic foci that resemble blastic metastases.
5. The condition tends to involve multiple bones in the same extremity or ray.
6. Clinical manifestations of melorheostosis include bone pain and joint stiffness.
7. Soft-tissue calcification and ossification are relatively common and exhibit a predilection for periarticular regions (Fig. 6-9).

PEARL

Melorheostosis:
- Childhood: irregular linear areas of increased density along major axis of tubular bones and, less frequently, in other parts of skeleton; monostotic, monomelic, or polyostotic
- Advanced: linear radiodensities with osteophytic periosteal excrescences (flowing candle wax), ectopic bone formation
- Shortening (less frequently, lengthening) and abnormal curvature of involved bones

8. Melorheostosis may be associated with osteopoikilosis and osteopathia striata in mixed sclerosing bone dysplasia.
9. The osteosclerotic lesions of melorheostosis exhibit increased uptake of bone-seeking radiopharmaceuticals, unlike the situation in osteopoikilosis and osteopathia striata.

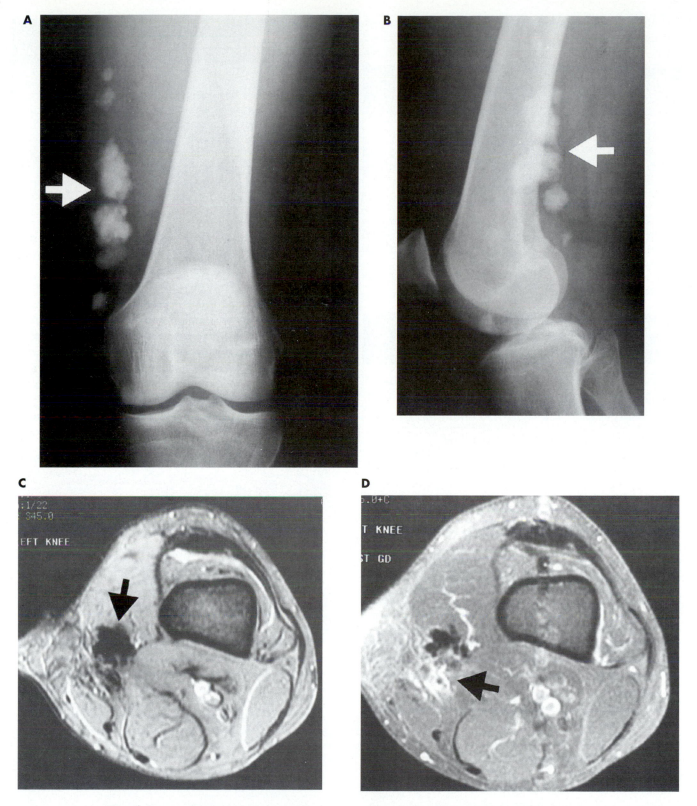

Fig. 6-9   **A** and **B,** Soft-tissue calcification (*arrows*) in melorheostosis. **C,** Gradient-echo transaxial MR image reveals signal void (*arrow*) characteristic of mineralized tissue. **D,** Transaxial T1-weighted fat-suppression image following intravenous administration of gadolinium-DTPA reveals enhancement (*arrow*) in the vicinity of the evolving calcification, indicating inflammation. Differential diagnosis should include consideration of other causes of dystrophic, metastatic, and idiopathic soft-tissue calcification.

10. Melorheostosis is also known as Leri disease or osteosis eburnisans monomelica.

Box 6-4 Melorheostosis: Important Features

- Linear irregular corticomedullary osteosclerosis resembling wax drippings
- Unilateral distribution, multiple long bones, confinement to one side of bone typical
- Soft-tissue calcification with contractures
- Deformities of hands, feet
- Associated with osteopoikilosis, osteopathia striata (benign sclerosing bone dystrophies)

See Table 6-1 for diseases associated with various hyperostotic conditions.

Table 6-1 Diseases associated with various hyperostotic conditions

Lesion	Possible associated diseases
Osteoma	Gardner syndrome
Osteopoikilosis	Osteopathia striata
	Melorheostosis
	Hyperostosis frontalis interna
Osteopathia striata	Osteopoikilosis
	Melorheostosis
	Osteopetrosis
	Focal dermal hypoplasia
Melorheostosis	Linear scleroderma
	Osteopoikilosis
	Osteopathia striata
	Neurofibromatosis
	Tuberous sclerosis
	Hemangiomas

Osteopetrosis (Box 6-5)

1. The condition is also known as Albers-Schönberg disease.
2. The disorder may cause a bone-within-bone appearance radiographically.
3. The disease should be considered in the differential diagnosis of the Erlenmeyer flask deformity of long bones.
4. The infantile form of osteopetrosis is severe and often fatal, and it exhibits autosomal recessive inheritance.
5. The adult form of osteopetrosis may be inherited as either an autosomal recessive or autosomal dominant trait.
6. Typical clinical manifestations include anemia and increased susceptibility to infection.

PEARL

Osteopetrosis:
- Generalized homogeneous increase in bone density, poor corticomedullary differentiation, loss of trabecular structure (infancy)
- Alternating transverse bands of greater and lesser density in tubular bones, arcuate bands in iliac bones, sandwich vertebrae, bone-within-bone appearance
- Thickening and sclerosis of cranial base and vault, underpneumatization of mastoids and paranasal sinuses

---

7. As in fluorosis, the osteosclerotic bone of osteopetrosis is fragile and frequently fractures, particularly with transverse orientation.
8. Radiodense metaphyseal bands in osteopetrosis are oriented parallel to the physis.
9. Bone marrow signal intensity on MRI studies is decreased in osteopetrosis, by an amount commensurate with the severity of the disease.
10. Magnetic resonance imaging is useful in monitoring the success of bone marrow transplantation in the infantile form of the disease.
11. Involvement of the skull base may cause cranial nerve damage with secondary blindness and/or deafness.

Box 6-5 Osteopetrosis: Important Features

- Hereditary disease characterized by failure of osteoclastic resorption
- Generalized osteosclerosis
- Cortical thickening with medullary encroachment
- "Bone within bone" appearance
- Trabecular thickening
- Metaphyseal undermodeling (Erlenmeyer flask deformity)
- Radiolucent and radiodense bands near ends of long bones
- Obliteration of mastoid air cells, paranasal sinuses, basal foramina by osteosclerosis with secondary cranial nerve deficits
- Crowding of bone marrow space with pancytopenia and extramedullary hematopoiesis
- Less severe forms: onset in late infancy or childhood; symmetric pattern predominantly affecting axial skeleton and long bones; discrete foci, linear streaks, concentric layers, or diffuse sclerosis, especially at growing bone ends

---

Pyknodysostosis (Box 6-6)

The following are recognized characteristics of pyknodysostosis:

1. Inherited as an autosomal recessive trait

PEARL

Pyknodysostosis:

- Persistence of anterior fontanelle and cranial sutures into adulthood, wormian bones, small facial bones, hypoplasia of paranasal sinuses, obtuse mandibular angle
- Generalized sclerosis with moderate metaphyseal undermodeling of tubular bones, multiple fractures
- Partial (rarely complete) aplasia of distal phalanges of fingers and toes
- Hypoplasia of acromial end (rarely of major portions) of the clavicle; occasional failure of complete segmentation in upper cervical and lower lumbar spine

2. Short-limbed dwarfism
3. Generalized osteosclerosis
4. Absence of the cranial nerve damage that occurs in osteopetrosis and progressive diaphyseal dysplasia
5. Decreased bone strength with increased fracture susceptibility
6. Hypoplasia or resorption of the distal clavicle
7. Obtuse mandibular angle
8. Absence of cortical thickening that occurs in progressive diaphyseal dysplasia and melorheostosis
9. Wormian bones

PEARL

Wormian bones occur in pyknodysostosis, osteogenesis imperfecta, rickets, kinky-hair syndrome of Menkes, cleidocranial dysostosis, hypothyroidism, hypophosphatasia, otopalatodigital syndrome, pachydermoperiostosis, and Down syndrome, among other conditions.

PEARL

Sutural widening may be actual, when the calvarial segments are separated by increased intracranial pressure, or apparent, when they are deficiently ossified or infiltrated by cells. Causes include the following:

- Normal variation (related to prematurity or growth spurt)
- Increased intracranial pressure (tumor, hydrocephalus)
- Infiltration by tumors (neuroblastoma, leukemia, lymphoma)
- Congenitally deficient ossification (osteogenesis imperfecta, pyknodysostosis, cleidocranial dysostosis)
- Metabolic disorders (hypoparathyroidism, lead intoxication, hypophosphatasia, hypothyroidism, rickets, hypervitaminoses, protein or calorie malnutrition)

10. Acro-osteolysis
11. Delayed closure of sutures and fontanelles
12. Afflicted impressionist painter Toulouse-Lautrec

Box 6-6 Pyknodysostosis: Important Features

- Osteosclerosis, medullary encroachment in long bones
- Sclerosis of skull base
- Wide sutures, wormian bones
- Short-limbed dwarfism (acromelic type)
- Distal phalangeal and clavicular dysplasia

Progressive Diaphyseal Dysplasia (Engelmann Disease) (Box 6-7)

The following are recognized characteristics of progressive diaphyseal dysplasia:

1. Autosomal dominant inheritance
2. Cranial nerve damage
3. Sclerosis of skull base
4. Normal metaphyses (absence of Erlenmeyer flask deformity that occurs in osteopetrosis)
5. Symmetric endosteal and periosteal cortical thickening of long and short tubular bone diaphyses, with narrowing of medullary cavities
6. Muscular weakness
7. Appearance usually during childhood
8. Bone pain
9. Waddling gait
10. Increased uptake on skeletal scintigraphy
11. Exophthalmos
12. Craniodiaphyseal dysplasia is a severe variant of this condition (Box 6-8)

Box 6-7 Progressive Diaphyseal Dysplasia (Engelmann Disease): Important Features

- Symmetric endosteal and periosteal cortical thickening, with progressive broadening of long-bone diaphyses
- Sparing of metaphyses and epiphyses
- Clavicles, cervical vertebrae, calvarium may be affected

Continued

Box 6-7—cont'd

- Autosomal dominant inheritance
- Calvarial hyperostosis, sclerosis of skull base, increased intracranial pressure, cranial nerve encroachment
- Clinical features: decreased muscle mass and subcutaneous fat, weakness, gait disorder, bony enlargement, leg pain, and elevated erythrocyte sedimentation rate with onset in second through fourth decades
- Differs from Ribbing disease in symmetric distribution and pattern of inheritance

Box 6-8   Craniodiaphyseal Dysplasia: Important Features

- Progressive facial distortion, small stature, mental retardation, deafness, visual disturbance, and death prior to age 20
- Severe sclerosis and hyperostosis of skull and facial bones
- Widened ribs and clavicles
- Diaphyseal expansion of tubular bones with lack of metaphyseal flare

CHROMOSOMAL DISORDERS

Trisomy 21 or Down Syndrome (Box 6-9)

The following are recognized characteristics of Down syndrome:

1. Atlantoaxial instability

PEARL

Causes of atlantoaxial subluxation include Down syndrome, pharyngitis in children (Griesl syndrome), rheumatoid arthritis, psoriatic arthritis and other seronegative spondyloarthropathies, trauma, septic arthritis, Ehlers-Danlos syndrome, and Morquio syndrome.

2. Microcephaly
3. Eleven pairs of ribs
4. Decreased iliac index
5. Short stature
6. Craniofacial dysmorphism
7. Congenital heart disease
8. Absence of the positive metacarpal sign that is a feature of Turner syndrome

9. Two manubrial ossification centers
10. Flared iliac wings
11. Mental and motor retardation
12. Decreased acetabular angles
13. Clinodactyly
14. Brachydactyly

Box 6-9   Down Syndrome: Important Features

- Developmental hip dysplasia in 40% of cases
- Atlantoaxial instability
- Flared iliac wings, decreased acetabular angles
- Duodenal atresia is most common gastrointestinal tract abnormality
- Accessory epiphyses (manubrium), cuboid vertebral bodies, high arched palate, delayed suture closure, sinus hypoplasia, microcephaly, 11 rib pairs, hypoplastic fifth middle phalanx with clinodactyly and short, irregular metacarpals

Chromosome X Monosomy or Turner Syndrome (Box 6-10)

The following are recognized characteristics of Turner syndrome:

1. Short stature
2. Short fourth metacarpal (Fig. 6-10)

PEARL

A short fourth metacarpal occurs in Turner syndrome, pseudohypoparathyroidism, and pseudo-pseudohypoparathyroidism, among other conditions predisposing to premature physeal closure.

3. Aortic coarctation
4. Horseshoe kidney
5. Cubitus valgus
6. Transient lymphedema
7. Overgrowth of medial femoral condyle
8. Ovarian dysgenesis
9. Foot deformities including pes cavus, tarsal coalition, and short fourth metatarsals
10. Flattening of medial tibial plateau
11. Shield chest
12. Decreased bone density (Fig. 6-10)
13. Medial proximal tibial exostoses
14. Decreased carpal angle (Fig. 6-10)
15. Pelvic deformities including protrusio acetabuli and a male configuration of the pelvic inlet

Box 6-10 Turner Syndrome (Gonadal Dysgenesis): Important Features

- Short stature, webbed neck
- Short fourth metacarpal (positive metacarpal sign)
- Large medial femoral condyle, medial tibial plateau exostoses
- Cubitus valgus
- Decreased carpal angle
- Lymphedema
- Osteopenia secondary to estrogen deficiency
- Short fourth metatarsals
- Delayed skeletal maturation
- Thin ribs, clavicles
- Vertebral body irregularities
- Atlantoaxial anomalies

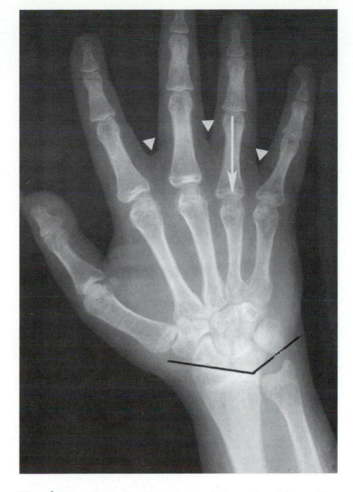

Fig. 6-10 Short fourth metacarpal (*arrow*) with positive metacarpal sign, diffuse osteopenia, decreased carpal angle (*lines*), and interdigital webbing (*arrowheads*) in Turner syndrome. The only viable alternative diagnosis for this constellation of findings is Noonan syndrome.

Trisomy 18 or Edwards Syndrome

The following are recognized characteristics of Edwards syndrome:

1. Physical and mental retardation
2. Congenital foot deformities
3. Eleven pairs of ribs
4. Sternal aplasia or hypoplasia
5. Flexion deformities of fingers
6. Small pelvis
7. Increased iliac index
8. Prominent occiput
9. Hypoplasia of mandible and maxilla
10. Absence of the atlantoaxial instability that occurs in Trisomy 21
11. Hypoplasia of ribs and clavicles
12. Facial dysmorphism
13. Low birth weight

OSTEOGENESIS IMPERFECTA (Box 6-11)

Recognized Characteristics

1. Severe osteoporosis (Figs. 6-11, 6-12)
2. Wormian bones
3. Multiple fractures with exuberant callus formation and malunion
4. Blue sclerae
5. Elevated serum alkaline phosphatase level secondary to multiple fractures
6. Normal fracture healing, although malunion is common
7. Otosclerosis
8. Articular laxity, as opposed to flexion contractures
9. Dentinogenesis imperfecta
10. Bleeding diathesis secondary to abnormal vasculature and platelet function
11. Basilar invagination
12. Kyphoscoliosis
13. Vertebral endplate deformities (Figs. 6-11, 6-12)
14. Defective type I collagen
15. Thin skin
16. Popcornlike calcification of physes
17. Limb shortening and bowing
18. In utero diagnosis by second-trimester ultrasonography
19. Protrusio acetabuli

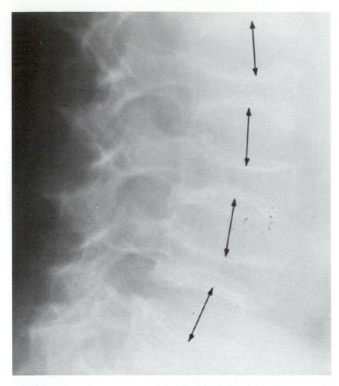

Fig. 6-11   Osteogenesis imperfecta in a child. Typical abnormalities in the spine include generalized osteopenia, cortical thinning, and biconcave vertebral bodies at all levels (*arrows*). Differential diagnosis should include consideration of idiopathic juvenile osteoporosis, rickets, and leukemia.

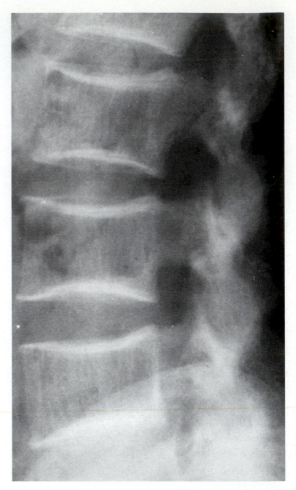

Fig. 6-12   Fibrogenesis imperfecta osseum. Typical radiographic findings of this uncommon disorder include generalized osteopenia, coarse trabeculation, and biconcave endplate deformities with accentuated density. Differential diagnosis should include consideration of osteogenesis imperfecta tarda and corticosteroid-induced osteoporosis.

Box 6-11   Osteogenesis Imperfecta (Lobstein Disease): Important Features

- Deficient formation of bone matrix, collagen
- Affects bones, skin, sclerae, periarticular tissues
- Osteoporosis, cortical thinning, deficient trabecular structure
- Multiple fractures with exuberant callus formation
- Congenital forms: radiolucent fetal skeleton, deformity secondary to multiple fractures
- Bowing of long bones, epiphyseal enlargement secondary to repetitive physeal fracture
- Thin calvarium, wormian bones, basilar invagination with platybasia
- Tardive forms (appearing later in childhood or in adulthood): biconcave vertebrae, slender long bones, bowing, cortical thinning, osteopenia, fractures
- Protrusio acetabuli

Characteristics of Specific Subtypes of Osteogenesis Imperfecta (Box 6-12)

PEARL

Osteogenesis imperfecta is a heterogeneous group of disorders caused by faulty collagen formation. The most severe form results in multiple fractures in utero. Patients with the mildest form have only occasional fractures at a later age. Findings include shortening of bones secondary to multiple telescoping fractures, exuberant callus formation, wormian bones, cystic and/or calcific alterations in the growth plates, pseudoarthroses, and bone thinning. Deafness, dentinogenesis imperfecta, and blue sclerae are clinical findings.

1. Autosomal dominant inheritance: osteogenesis imperfecta tarda type I
2. Autosomal recessive inheritance: osteogenesis imperfecta congenita
3. Death in utero or shortened life span: osteogenesis imperfecta congenita
4. Multiple fractures in utero: osteogenesis imperfecta congenita
5. Fracture rate decreasing at puberty: osteogenesis imperfecta tarda type I
6. Short, thick, and bowed long bones: osteogenesis imperfecta congenita
7. Gracile long bones of normal length: osteogenesis imperfecta tarda type I, osteogenesis imperfecta tarda type II
8. Large skull: osteogenesis imperfecta congenita
9. Form with lowest incidence of fracture: osteogenesis imperfecta tarda type II
10. Multiple fractures beginning after birth: osteogenesis imperfecta tarda type I, osteogenesis imperfecta tarda type II
11. Popcornlike calcification of physes: osteogenesis imperfecta congenita
12. Form with latest age of presentation: osteogenesis imperfecta tarda type II

Box 6-12   Osteogenesis Imperfecta: Summary

- Congenital (thick, short) and tardive (thin, gracile) patterns of bone deformity
- Blue sclerae (over 90% of cases)
- Dentinogenesis imperfecta
- Thin skin, hyperplastic scars, premature vascular calcification, joint laxity, hernias, platelet dysfunction
- Otosclerosis, platybasia, basilar invagination (with or without hydrocephalus), intracranial hemorrhage, wormian bones
- Kyphoscoliosis, narrow triradiate pelvis, protrusio acetabuli, shepherd crook femoral deformity, traumatic physeal fragmentation
- Fracture healing normal with exuberant callus
- Sofield procedure (multiple osteotomies and intramedullary rodding) used to correct long-bone bowing from multiple fractures

---

MARFAN SYNDROME (Box 6-13)

The following are recognized characteristics of Marfan syndrome:
1. Arachnodactyly
2. Normal bone density (as opposed to the decreased density that occurs in homocystinuria)
3. Kyphoscoliosis

PEARL

Marfan syndrome is an ocular, skeletal, and connective tissue disorder. Scoliosis is a prominent feature, and spontaneous pneumothoraces may occur. Aortic dilatation occurs as a result of weakness of the media. Tubular bones are increased in length, and the metacarpal index quantifies arachnodactyly in the second through fifth metacarpals. Marfan syndrome is transmitted as an autosomal dominant trait. Mental retardation and osteoporosis occur in homocystinuria, but not in Marfan syndrome. Pectus carinatum and excavatum are commonly associated with both disorders.

---

4. Hypermobile joints
5. Autosomal dominant inheritance
6. Posterior scalloping of vertebral bodies
7. Cardiac valvular insufficiency
8. Pectus excavatum or carinatum
9. Tall, thin stature
10. Ectopia lentis
11. Joint dislocations
12. Cystic medial necrosis with spontaneous rupture or dissection of great vessels
13. Disproportionately long extremities
14. Pes planus
15. Protrusio acetabuli
16. Dolichocephaly
17. Increased metacarpal index
18. Pulmonary dysaeration, spontaneous pneumothorax

Box 6-13   Marfan Syndrome: Important Features

- Hereditary multisystem disorder of collagen and elastin synthesis
- Arachnodactyly
- Slender elongation of long tubular bones, cortical thinning
- Pectus excavatum or carinatum
- Pes planus
- Muscular hypoplasia, hypotonia
- Bone density appropriate for age
- Kyphoscoliosis

EHLERS-DANLOS SYNDROME (Box 6-14)

The following are recognized characteristics of Ehlers-Danlos syndrome:
1. Thin, hyperelastic skin
2. Hypermobile joints secondary to ligamentous laxity

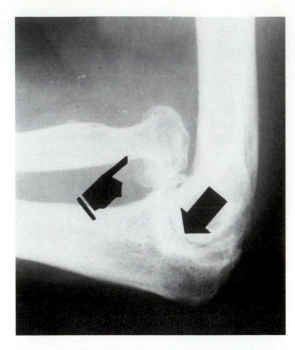

Fig. 6-13 Chronic radial head dislocation (*pointing finger*) with secondary osteoarthritis (*arrow*) in the Ehlers-Danlos syndrome. Differential diagnosis should include consideration of previous trauma (particularly the Monteggia injury), osteo-onychodysostosis, arthrogryposis multiplex congenita, and other causes of ligamentous laxity.

3. Connective tissue fragility
4. Easy bruisability
5. Predominant occurrence among male individuals
6. Defective collagen synthesis
7. Subluxation or dislocation of joints, which is also a major feature of Larsen syndrome (Fig. 6-13)

PEARL

Larsen syndrome:
- Multiple dislocations (especially hips, knees, elbows) with secondary epiphyseal deformities
- Vertebral malformations (especially cervical) with abnormal spinal curvature
- Supernumerary calcaneal ossification center (appearing in late infancy or later, fusing with main center at about age 8 years)
- Supernumerary carpal bones; short, sometimes broad and irregular metacarpals; lack of distal tapering of proximal and middle phalanges; short, broad first through fourth distal phalanges; premature fusion of first distal phalangeal epiphysis
- Pes equinovarus

---

8. Spherical subcutaneous calcifications in forearms and shins, secondary to hemorrhage and fat necrosis
9. Existence of at least 10 distinct types

10. Premature osteoarthritis
11. Autosomal dominant, autosomal recessive, and X-linked recessive patterns of inheritance
12. Developmental hip dysplasia
13. Kyphoscoliosis
14. Varicose veins
15. Gastrointestinal and genitourinary diverticula
16. Cardiac valvular insufficiency
17. Vascular fragility with aneurysms and dissection
18. Normal bone strength in all types (patients not predisposed to fractures)
19. Pulmonary dysaeration

Box 6-14 Ehlers-Danlos Syndrome: Important Features

- In approximately 30% of cases, multiple small round or oval densities with calcified margins are present in subcutaneous tissues of the extremities, varying from 2 to 15 mm in diameter and random in distribution (representing necrosis of fatty nodules); especially forearms, shins
- Hypermobility of joints, subluxations, osteoarthritis related to tissue hyperelasticity
- Affecting tall white male individuals of European descent
- Pes planus, pectus excavatum or carinatum, kyphoscoliosis, steep downward sloping of ribs, Raynaud phenomenon, posterior vertebral body scalloping
- Spontaneous aortic dissection, large vessel rupture, bowel or bronchial rupture, hemorrhage in type IV (Sachs syndrome)
- Skin: hyperelasticity, easy bruisability, cigarette-paper scars, molluscoid tumors (pressure points), discoloration (subcutaneous hemorrhage), impaired suture holding
- Typed according to specific defects in collagen metabolism
- Mitral valve prolapse

NEUROFIBROMATOSIS (Box 6-15)

1. Musculoskeletal involvement occurs in approximately 85% of patients.
2. The most common musculoskeletal manifestation of neurofibromatosis is kyphoscoliosis, and it may be complicated by paraplegia if rapidly progressive.
3. Kyphoscoliosis in neurofibromatosis most commonly affects the midthoracic spine, and the deformity is usually angular (Fig. 6-14).
4. Mesodermal dysplasia is responsible for kyphoscoliosis, as well as for many of the other musculoskeletal abnormalities in the disease.
5. Posterior scalloping of vertebral bodies in neurofibromatosis is usually secondary to dural ectasia.

Posterior scalloping of vertebral bodies occurs in neurofibromatosis (secondary to dural ectasia or neurofibroma), Marfan or Ehlers-Danlos syndrome, achondroplasia, acromegaly, mucopolysaccharidoses, intraspinal neoplasm, and elevated cerebrospinal fluid pressure.

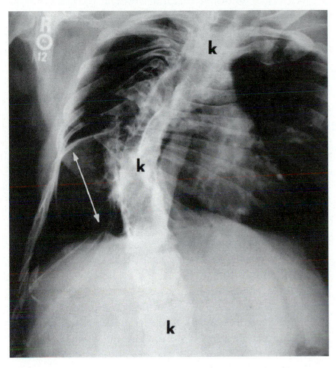

Fig. 6-14 Severe kyphoscoliosis (*k*) with associated rib deformities (*arrows*) in neurofibromatosis. Although this is the most common musculoskeletal manifestation of the disease, the differential diagnosis should include consideration of idiopathic scoliosis, Marfan syndrome, Ehlers-Danlos syndrome, spondyloepiphyseal and spondylometaphyseal dysplasias, and a large number of other congenital malformation syndromes.

6. Neural foraminal widening with pedicle erosion and paraspinal soft-tissue mass may be caused by dumbbell neurofibroma or lateral meningocele (Fig. 6-15).
7. The tibia, fibula, or clavicle may exhibit bowing, narrowing, or pseudoarthrosis in neurofibromatosis secondary to mesodermal dysplasia (Fig. 6-16; Box 6-16).
8. Neurofibromatosis should be considered in the differential diagnosis of rib notching, resulting from pressure erosion by neurofibromas of the intercostal nerves.

Rib notching is associated with neurofibromatosis, as well as with congenital heart disease. In coarctation of the aorta, the rib notching occurs as a late finding and typically involves the third through ninth ribs. Both before and after repair of tetralogy of Fallot, rib notching can occur secondary to pulmonary outflow tract obstruction with systemic collaterals.

9. Irregular twisting and thinning of ribs in the disease occurs secondary to mesodermal dysplasia.
10. The periosteum is loosely attached in neurofibromatosis, resulting in ossified subperiosteal hemorrhages, the radiographic appearance of which may simulate that of scurvy or battered child syndrome.
11. Neurofibromatosis resembles arteriovenous malformation, lymphangiomatosis, macrodystrophia lipomatosa, and Klippel-Trenaunay-Weber syndrome when it causes localized gigantism.

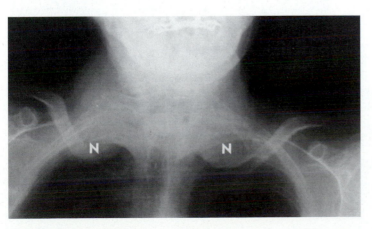

Fig. 6-15 Bilateral apical neurofibromas (*N*) in a patient with neurofibromatosis. In this disease, lateral thoracic meningomyeloceles related to dural ectasia could have a similar appearance.

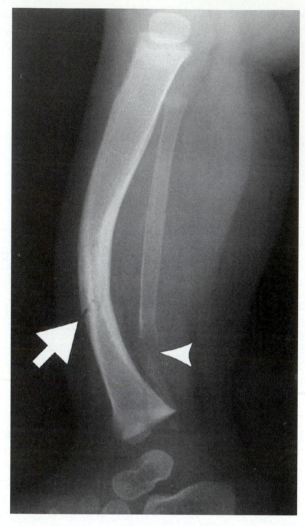

Fig. 6-16   Congenital pseudoarthrosis of the tibia (*arrow*) and fibula (*arrowhead*), secondary to mesodermal dysplasia, is pathognomonic of neurofibromatosis, although it may also occur as an isolated developmental aberration.

13. Intraosseous cystic lesions in neurofibromatosis may be caused by direct invasion from neurofibromatous soft tissue.

14. Small exostoses may arise adjacent to soft-tissue neurofibromas.

15. On computed tomographic studies, soft-tissue neurofibromas are usually characterized by density slightly lower than that of normal muscle.

16. On T2-weighted magnetic resonance images, soft-tissue neurofibromas manifest high signal intensity (Fig. 6-17).

17. On cross-sectional imaging studies, plexiform neurofibromas exhibit a serpiginous appearance that may simulate that of vascular malformation (Fig. 6-18).

18. The most common manifestation of neurofibromatosis in the skull is macrocranium.

19. The sphenoid wing may be hypoplastic or absent in the disease (Fig. 6-19).

20. The cause of exophthalmos in neurofibromatosis is brain herniation into the posterior bony orbit secondary to hypoplasia of the posterosuperior orbital wall.

21. Defects in the calvarium may be accompanied by ipsilateral mastoid hypoplasia, occur most commonly in the left lambdoid suture, and are caused by mesodermal dysplasia (Fig. 6-20).

22. Neuromas or gliomas may cause enlargement of cranial foramina, particularly the optic canal and internal auditory canal.

23. Fracture healing response in neurofibromatosis is poor secondary to mesodermal dysplasia, which may result in nonunion and pseudoarthrosis.

24. The most significant potential complication of soft-tissue neurofibroma is malignant transformation to neurofibrosarcoma.

25. Cutaneous manifestations of neurofibromatosis include tumors (neurofibromas, schwannomas), café-au-lait spots, and molluscum fibrosum.

PEARL

Major causes of regional or focal gigantism include Klippel-Trenaunay-Weber syndrome, arteriovenous malformations, neurofibromatosis, and macrodystrophia lipomatosa. All of these conditions may cause changes in long bones including focal gigantism or dwarfism. This may be secondary to chronic hyperemia from soft-tissue proliferation.

12. A benign bone neoplasm that may be encountered in the disease is nonossifying fibroma or fibrous cortical defect; multiple lesions frequently occur.

PEARL

Neurofibromatosis:
- Sharply angular kyphoscoliosis of the thoracolumbar spine is a hallmark of neurofibromatosis.
- Absence of the lesser or greater sphenoid wings is also characteristic.
- Scalloping of the posterior aspect of the vertebral bodies is a common finding.
- Soft-tissue overgrowth with osseous hyperplasia is frequent.
- Pseudoarthrosis of the tibia is a common finding.
- Aplasia or hypoplasia of the fibula and focal gigantism may occur.

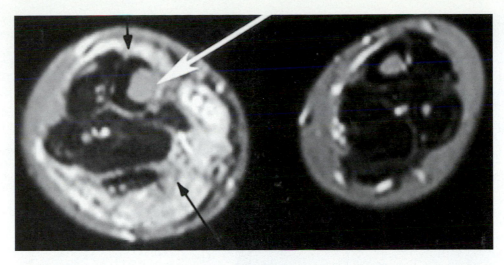

Fig. 6-17   Neurofibromatosis with plexiform lesion. Transaxial T2-weighted MR image reveals a serpiginous and infiltrative soft-tissue mass of high signal intensity (*black arrows*) in the calf, with osseous erosion (*white arrow*). Differential diagnosis should include consideration of hemangioma and soft-tissue sarcoma.

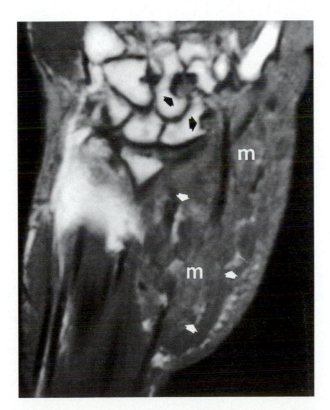

Fig. 6-18   Plexiform neurofibroma of the wrist with osseous involvement (*black arrows*). T1-weighted coronal MR image reveals a convoluted low–signal intensity mass (*m*) containing sparse areas of fat (*white arrows*). Diffuse increased signal intensity on T2-weighted images is characteristic of such lesions.

Box 6-15   Neurofibromatosis (von Recklinghausen Disease): Important Features

- Bone changes in about 50% of cases, caused by erosion from soft-tissue lesions or intrinsic osseous dysplasia
- Erosive deformities most common in posterior vertebral bodies, ribs
- Mesodermal dysplastic deformities more common and occur earlier
- Scoliosis or kyphoscoliosis most frequently localized and often an isolated abnormality
- Infrequently, bowing of one or more long bones, or absence of posterior and superior orbital walls (associated with pulsating exophthalmos)
- Dysplastic scalloping of multiple posterior vertebral body margins
- Ununited fractures possibly leading to pseudoarthrosis
- Other disorders of bone growth associated with minimal to massive hypertrophy or overlying soft tissues
- Rarely, cystic-appearing intramedullary lytic lesions
- Soft-tissue opacities on radiographs from individual subcutaneous neurofibromas

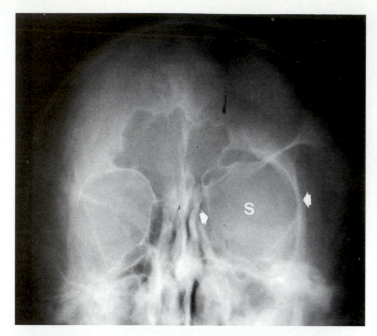

Fig. 6-19 Unilateral absence of the sphenoid wing (*s*) with orbital enlargement (*arrows*) in neurofibromatosis. The findings are virtually specific for this condition and occur secondary to mesodermal dysplasia.

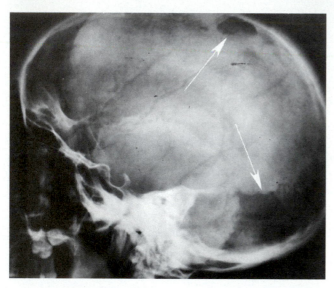

Fig. 6-20 Sutural defects (*arrows*) resulting from mesodermal dysplasia in neurofibromatosis; lambdoid involvement on the left side of the skull is particularly characteristic of this condition.

Box 6-16   Congenital or Developmental Pseudoarthrosis: Important Features

- Associated with neurofibromatosis or (less commonly) fibrous dysplasia
- Clavicular involvement almost always on right side, except in patients with dextrocardia
- Bone grafting leads to healing in only 25% to 35% of cases
- Prognosis for healing better for fractures developing at an older age
- Most commonly affects tibia and/or fibula
- Represents mesodermal dysplasia
- Abnormal bone is congenital; fracture and pseudoarthrosis occur after birth (developmental)

FIBRODYSPLASIA OSSIFICANS PROGRESSIVA

1. The disease is a hereditary disorder of mesodermal tissue.
2. Progressive ossification of muscles, tendons, and ligaments occurs.
3. Inheritance is autosomal dominant with variable penetrance, and spontaneous mutations frequently occur.
4. Pathologically, the condition resembles posttraumatic myositis ossificans.
5. Electromyography studies demonstrate myopathic abnormalities.
6. Ossification at entheses results in excrescences that are attached to the underlying bone.

Fibrodysplasia ossificans progressiva:
- Osteophyte-like excrescences arise from tubular bones.
- Joint contractures are a characteristic clinical feature.
- Radiographs of the hands or feet are useful in distinguishing the condition from posttraumatic myositis ossificans.
- There is a high frequency of associated congenital anomalies, including joint ankylosis.
- Spinal alterations may resemble those of diffuse idiopathic skeletal hyperostosis, seronegative spondyloarthropathies, or juvenile chronic arthritis.

7. Most affected patients have acute painful torticollis secondary to involvement of the sternocleidomastoid muscle.
8. Bridging ossification between adjacent bones in the spine, shoulder girdle, upper arms, and pelvis eventually results in severely restricted motion and/or respiratory compromise.
9. Approximately 75% of patients manifest bilateral hallux microdactyly and/or phalangeal synostosis, along with hallux valgus; similar congenital abnormalities of the thumbs occur somewhat less frequently (Box 6-17).

Box 6-17 Differential Diagnosis of Radial Ray Anomalies

- Fibrodysplasia ossificans progressiva
- Vertebral defects, anal atresia, tracheoesophageal fistula with esophageal atresia, and radial and/or renal anomalies (VATER complex)
- Vertebral, anal, cardiovascular, tracheal, esophageal, renal, and limb bud anomalies (VACTERL complex)
- Thrombocytopenia–absent radius (TAR) syndrome
- Fanconi anemia
- Cornelia de Lange syndrome
- Holt-Oram syndrome
- Diastrophic dwarfism (hitch-hiker thumb)

10. Minor trauma may precipitate exacerbations and remissions of the process.
11. The condition is also known as myositis ossificans progressiva.

ARTHROGRYPOSIS MULTIPLEX CONGENITA

The following are recognized characteristics of arthrogryposis multiplex congenita:
1. Long neuromuscular-type scoliosis
2. Decreased bone density, resulting predominantly from disuse

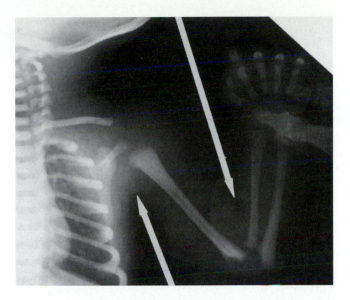

Fig. 6-21 Arthrogryposis multiplex congenita. Flexion contractures are present at the shoulder and elbow (*arrows*). Other causes of flexion contractures include scleroderma, burns, epidermolysis bullosa, long-term immobilization of an articulation, fibrodysplasia ossificans progressiva, and melorheostosis.

3. Congenital clubfoot or valgus foot deformities
4. Generalized muscle atrophy
5. Decreased size and number of anterior horn cells in the spinal cord
6. Symmetric flexion contractures with cutaneous webbing (Fig. 6-21)
7. Developmental hip dysplasia
8. Hypoplastic or absent patellae
9. Thickened, relatively radiodense joint capsules
10. Clubhand deformity with ulnar deviation
11. Fibrofatty replacement of muscle fibers
12. Multiple joint dislocations
13. Presentation in utero (diagnosis possible with ultrasonography) or at birth
14. Frequent fractures
15. Adduction of thumbs
16. Carpal and tarsal coalition
17. Overtubulated or gracile long bones, secondary to disuse
18. Most common distribution of the disorder: both upper and lower extremities (approximately 50% of patients)
19. Least common distribution of the disorder: lower extremities only (approximately 10% of patients)

CLEIDOCRANIAL DYSOSTOSIS (Box 6-18)

The following are recognized characteristics of cleidocranial dysostosis:
1. Autosomal dominant inheritance
2. Brachycephaly

3. Wormian bones (Box 6-19)
4. Delayed cranial suture and fontanelle closure
5. Broad mandible with prognathism
6. Posterior wedging of thoracic vertebrae
7. Bilateral clavicular aplasia or hypoplasia
8. Hypoplastic scapulae
9. Absent or delayed ossification of pubic bones
10. Relatively long second and fifth metacarpals
11. Short middle phalanges
12. Tapered distal phalanges
13. Pseudoepiphyses and cone-shaped epiphyses
14. Hypoplastic iliac wings
15. Dental anomalies
16. Coxa vara or valga
17. Abnormal modeling of long bones
18. Retarded development of membranous bones
19. In utero diagnosis possible with ultrasonography

Box 6-18 Cleidocranial Dysostosis: Important Features

- Aplasia or hypoplasia of clavicles
- Delayed closure of cranial sutures, wormian bones
- Defective ossification of pubic, ischial bones
- Hemivertebrae
- Pointed terminal tufts in hands

Box 6-19 Major Differential Diagnosis of Wormian Bones

- Cleidocranial dysostosis
- Cretinism, hypothyroidism
- Idiopathic normal variation
- Osteogenesis imperfecta
- Down syndrome
- Hypophosphatasia
- Progeria
- Pyknodysostosis
- Rickets

HEREDITARY OSTEO-ONYCHODYSOSTOSIS (HOOD) SYNDROME (Box 6-20)

The following are recognized characteristics of hereditary osteo-onychodysostosis:
1. Also known as nail-patella syndrome or Fong disease
2. Autosomal dominant inheritance

PEARL

Osteo-onychodysostosis:
- Autosomal dominant inheritance
- Proteinuria, hematuria leading to chronic renal failure, renal osteodystrophy
- Flexion contractures of hip, knee, elbow; digital web formation; hypoplasia of deltoid, triceps, quadriceps muscles
- Elbow: asymmetric development of humeral condyles, capitellar hypoplasia, radial head subluxation or dislocation
- Absent or hypoplastic patellae, hypoplastic lateral femoral condyles, sloping tibial plateaus with prominent tubercles
- Clinical features: cubitus valgus, abnormal iris pigmentation, abnormal gait, genu valgum

---

3. Dysplastic nails
4. Hypoplastic or absent patellae (Fig. 6-22)
5. Posterior iliac horns (Fig. 6-23)
6. Hypoplasia of radial head and capitellum with dislocation (Fig. 6-24)
7. Hypoplasia of the lateral femoral condyles with genu valgum
8. Renal disease
9. Flaring of iliac wings with decreased iliac angles
10. Renal osteodystrophy
11. Foot deformities
12. Short stature
13. Ocular abnormalities

Box 6-20 Hereditary Osteo-onychodysostosis (HOOD or Nail-patella Syndrome): Important Features

- Hypoplastic or absent patellae
- Dysplastic nails
- Posterior iliac horns (80% of cases, pathognomonic, usually bilateral)
- Flared iliac crests with prominent anterosuperior spines
- Hypoplastic radial head and capitellum resulting in elbow deformities
- Renal osteodystrophy in later life secondary to glomerulonephritis

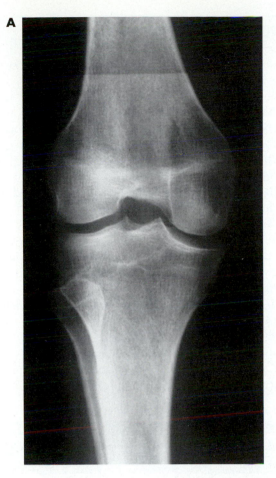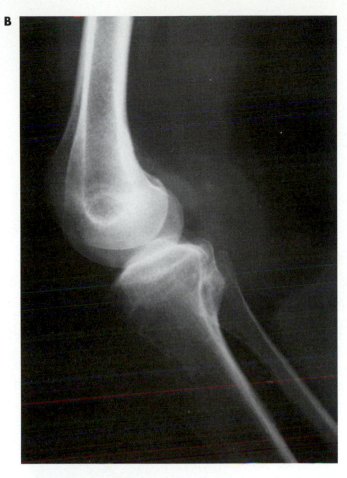

Fig. 6-22   A and **B,** Congenital absence of the patella in osteo-onychodysostosis. Differential diagnosis is limited to surgical resection and arthrogryposis multiplex congenita.

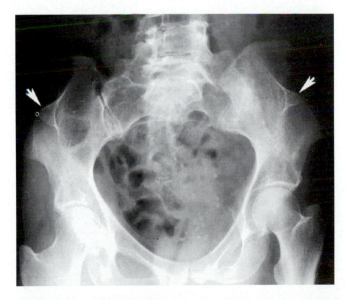

Fig. 6-23 Bilateral iliac horns (*arrows*) in osteo-onychodysostosis. These excrescences are pathognomonic of this condition and are not true osteochondromas because they lack a cartilaginous cap.

KLIPPEL-FEIL SYNDROME

The following are recognized characteristics of Klippel-Feil syndrome:

1. Fused cervical or cervicothoracic vertebrae
2. Absence of posterior element fusion is usually a feature of acquired, as opposed to congenital, vertebral synostoses
3. Omovertebral bone
4. Also termed *congenital brevicollis*
5. Genitourinary anomalies
6. Sprengel deformity of scapula
7. Low occipital hairline
8. Aural anomalies
9. Decreased mobility of the head and neck
10. Spina bifida
11. Webbed neck (pterygium colli)
12. Cervical ribs
13. Hemivertebrae
14. Short neck
15. Deafness

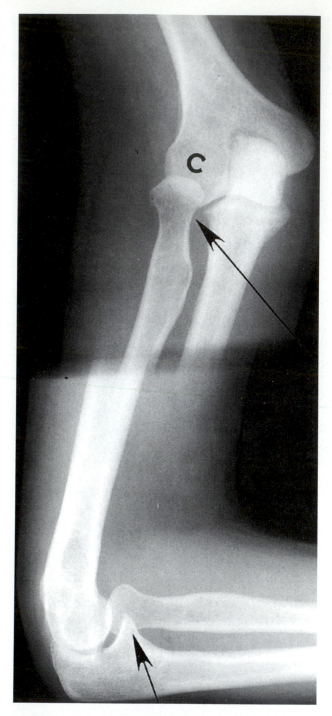

Fig. 6-24   Elbow alterations in osteo-onychodysostosis. Deformity of the capitellum (*c*) is associated with a hypoplastic and dislocated radial head (*arrows*). Differential diagnosis should include consideration of the Ehlers-Danlos syndrome, trauma during childhood, and other causes of ligamentous laxity with predisposition to joint subluxation.

16. Absence of constriction at levels of fused intervertebral disks is usually a feature of acquired, as opposed to congenital, vertebral synostoses
17. May occur in VATER or VACTERL complexes

MADELUNG DEFORMITY (Box 6-21)

1. Madelung deformity exhibits a predilection for women.
2. The distal radius is bowed in a volar and ulnar direction.
3. The ulna is relatively long and may be dislocated dorsally.
4. The carpal angle is decreased in the Madelung deformity, as in the mucopolysaccharidoses.
5. Madelung deformity may occur as an isolated finding.
6. Decreased carpal angle also occurs in the Turner syndrome.
7. The deformity occurs in a form of short-limbed dwarfism known as dyschondrosteosis or Leri-Weill disease.
8. Decreased carpal angle also may occur in enchondromatosis.
9. Decreased carpal angle also may occur in the multiple cartilaginous exostosis syndrome.
10. A Madelung-type deformity may be a consequence of trauma (distal radial fracture).

Box 6-21   Madelung Deformity: Important Features

- Decreased carpal angle (normally 130 to 137 degrees)
- Isolated or associated with dyschondrosteosis (a form of mesomelic dwarfism)
- Volar curvature of radius with ulnar elongation
- Three to five times more common in women than in men
- Radiographic differential diagnosis: previous trauma, Turner syndrome, multiple cartilaginous exostoses, enchondromatosis, mucopolysaccharidoses
- Bilateral asymmetric involvement more common than unilateral
- Clinical features: visible forearm deformity; pain, muscle fatigue; limitation of dorsal extension, ulnar deviation, supination at wrist

---

CAUDAL REGRESSION SYNDROME

1. The mildest form is characterized by partial agenesis of the sacrum.
2. The most severe cases involve total vertebral agenesis below the lower thoracic spine.
3. Developmental hip dysplasia is commonly associated with caudal regression syndrome.
4. Talipes equinovarus is commonly associated with caudal regression syndrome.

5. Approximately 20% of affected patients have mothers with diabetes mellitus.
6. Neurogenic bowel and bladder dysfunction is an associated finding.
7. Osseous fusion between the iliac bones may be observed in severe cases of caudal regression syndrome.
8. Intraspinal abnormalities coexist with the osseous alterations.
9. The condition is amenable to diagnosis by intrauterine ultrasonography.

PRIMARY PROTRUSIO ACETABULI

1. The disorder is also known as the Otto pelvis.
2. Primary protrusio acetabuli is usually bilateral in distribution.
3. The disorder is often familial.
4. The majority of affected patients are women.
5. The disorder may be secondary to a congenital defect in the development of the acetabular triradiate cartilage.
6. Premature degenerative joint disease is a potential complication, with axial migration of the femoral heads.
7. The differential diagnosis of protrusio acetabuli should include consideration of rheumatoid arthritis, septic arthritis, seronegative spondyloarthropathies, osteomalacia, osteogenesis imperfecta, and Paget disease.

PROXIMAL FEMORAL FOCAL DEFICIENCY

1. Proximal femoral focal deficiency is usually an isolated abnormality.
2. Approximately 90% of cases are unilateral.
3. Partial or complete agenesis of the proximal femur occurs.
4. Radiographically, the femur is typically short and displaced superiorly, laterally, and posteriorly with respect to the pelvis.
5. The distal femur is usually normal in proximal femoral focal deficiency.
6. Secondary abnormalities of the acetabulum and pelvis are common.
7. Delayed ossification of the capital femoral epiphysis is characteristic.
8. At skeletal maturity, subtrochanteric pseudoarthrosis or varus deformity, prominent spacing between the femoral head and dysplastic shaft, or ossification limited to the distal femur may be observed.
9. In younger patients, the severity of the deformity may be established by visualization of unossified cartilage using contrast arthrography or magnetic resonance imaging.

10. The major differential diagnostic consideration is infantile coxa vara.
11. The deformity is amenable to diagnosis by intrauterine ultrasonography.

INFANTILE COXA VARA

1. The disorder usually manifests itself with limb length discrepancy and gait abnormality during the first few years of life.
2. The normal neck-shaft angle of the proximal femur is approximately 150 degrees at birth and between 120 and 130 degrees in adults; the neck-shaft angle of the proximal femur is less than 120 degrees in all forms of coxa vara, including the infantile type.
3. Infantile coxa vara is unilateral in 60% to 75% of cases.
4. The proximal femoral physis is wide and vertically oriented in infantile coxa vara.
5. The disorder exhibits no sex predilection.
6. Radiolucent bands forming an inverted V shape delineate a triangular segment of bone in the superomedial aspect of the femoral neck.
7. The etiology of the deformity is unknown.
8. The degree of varus deformity frequently progresses with growth.
9. The differential diagnosis of coxa vara should include consideration of mild proximal femoral focal deficiency, slipped capital femoral epiphysis, cleidocranial dysostosis, osteogenesis imperfecta, rickets, renal osteodystrophy, and fibrous dysplasia.

PEARL

Infantile coxa vara is a form of proximal femoral dysplasia that becomes apparent at the age when the child first walks. The patient typically has a painless, lurching gait. Radiographs demonstrate varus positioning of the femoral neck and a widened, vertically oriented growth plate. The condition occurs with equal incidence among male and female individuals, and 60% to 75% of cases are unilateral.

SHORT-LIMBED DYSPLASIAS

1. Rhizomelia refers to short-limbed dysplasia in which the proximal segments (humeri and femora) are most severely affected.
2. Mesomelia refers to short-limbed dysplasia in which the middle segments (radius-ulna and tibia-fibula) are most severely affected.

3. Acromelia refers to short-limbed dysplasia in which the distal segments (metacarpals-phalanges and metatarsals-phalanges) are most severely affected.

Classification of Specific Forms as Rhizomelic, Mesomelic, or Acromelic

1. Achondroplasia: rhizomelic

PEARL

Achondroplasia:

- The primary abnormality in achondroplasia is a defect in endochondral bone formation.
- Although the disorder is transmitted as a mendelian dominant trait, the vast majority of cases are sporadic.
- It is the most common form of short-limbed dysplasia.
- It is classified as rhizomelic dysplasia because the predominant shortening is of the proximal segments (femora and humeri).
- Neonatal death is due primarily to brainstem compression associated with a small foramen magnum.
- The skull has a characteristic prominence of the calvarium.
- In the spine, there is hypoplasia of the vertebral bodies with scalloping of their posterior margins; the interpediculate distances progressively decrease, and bony spinal stenosis is present.
- Narrowing of the sacrosciatic notches in the pelvis creates a characteristic "champagne glass" configuration.
- In the hand, there is shortening of all digits, which are uniform in length; separation of the ring and middle fingers creates a trident hand configuration.

---

2. Asphyxiating thoracic dysplasia: acromelic
3. Thanatophoric dysplasia: rhizomelic
4. Pseudoachondroplasia: rhizomelic
5. Dyschondrosteosis: mesomelic
6. Chondrodysplasia punctata (recessive): rhizomelic
7. Acrodysostosis: acromelic
8. Achondrogenesis: rhizomelic
9. Hypochondroplasia: rhizomelic
10. Chondroectodermal dysplasia: mesomelic

Other Fundamental Concepts

The following are various specific forms and the associated predominant involvement of epiphyses, metaphyses, or both

1. Spondyloepiphyseal dysplasia: epiphyseal
2. Asphyxiating thoracic dysplasia: metaphyseal
3. Chondrodysplasia punctata: epiphyseal
4. Achondroplasia: metaphyseal
5. Kniest disease: both
6. Thanatophoric dysplasia: metaphyseal

7. Pseudoachondroplasia: both
8. Metaphyseal chondrodysplasias: metaphyseal
9. Metatropic dysplasia: both
10. Hypochondroplasia: metaphyseal
11. Chondroectodermal dysplasia: epiphyseal
12. Spondylometaphyseal dysplasia: metaphyseal
13. Diastrophic dysplasia: both
14. Multiple epiphyseal dysplasia: epiphyseal
15. Achondrogenesis: metaphyseal

Type II achondrogenesis is less severe than type I and is characterized by more complete ossification and absence of rib fractures. Asphyxiating thoracic dysplasia is also known as the Jeune syndrome. Chondroectodermal dysplasia is also known as the Ellis–van Creveld syndrome. The autosomal dominant form of chondrodysplasia punctata is also known as Conradi-Hünerman disease. The differential diagnosis of stippled epiphyses should include consideration of chondrodysplasia punctata, fetal coumadin syndrome, and Zellweger syndrome. Dyschondrosteosis is also known as Leri-Weill disease.

Classification of Specific Forms as Autosomal Dominant, Autosomal Recessive, or Both

1. Achondroplasia: autosomal dominant (Box 6-22)
2. Hypochondroplasia: autosomal dominant
3. Pseudoachondroplasia: both
4. Thanatophoric dysplasia: autosomal recessive
5. Achondrogenesis: autosomal recessive
6. Asphyxiating thoracic dysplasia: autosomal recessive
7. Chondroectodermal dysplasia: autosomal recessive
8. Diastrophic dysplasia: autosomal recessive
9. Chondrodysplasia punctata: both
10. Dyschondrosteosis: autosomal dominant
11. Spondyloepiphyseal dysplasia: autosomal dominant

Box 6-22 Achondroplasia: Important Features

- Autosomal dominant defect in endochondral bone formation; most cases are spontaneous mutations
- Rhizomelic dysplasia
- Bony spinal stenosis with or without gibbus, hypoplastic wedged lumbar vertebral bodies, posterior scalloping, decreasing interpediculate distance from L1 to L5, short pedicles
- Extremities disproportionately short relative to trunk
- Metaphyseal flaring, ball-in-socket epiphyses
- Short broad pelvis (champagne glass), squared iliac wings, decreased acetabular angles
- Large skull, small foramen magnum, trident hand (uniform length of short bones)

Major Clinical Features

1. Achondroplasia: neurologic symptoms secondary to spinal stenosis (Fig. 6-25)
2. Hypochondroplasia, pseudoachondroplasia: articular laxity with sparing of the elbow
3. Achondrogenesis: hydropic, edematous appearance
4. Chondrodysplasia punctata: ichthyotic cutaneous alterations
5. Diastrophic dysplasia, chondrodysplasia punctata: multiple joint contractures
6. Dyschondrosteosis: Madelung deformity
7. Spondyloepiphyseal dysplasia: coxa vara
8. Thanatophoric dysplasia: death in early infancy
9. Achondroplasia, hypochondroplasia, dyschondrosteosis: limited elbow extension
10. Asphyxiating thoracic dysplasia: respiratory difficulties secondary to hypoplasia of the thoracic cage
11. Chondroectodermal dysplasia: polydactyly of the hands and feet
12. Diastrophic dysplasia: hitch-hiker thumb
13. Spondyloepiphyseal dysplasia: pectus carinatum
14. Achondroplasia, hypochondroplasia: accentuated lumbar lordosis with protuberant abdomen
15. Achondrogenesis: incompatibility with life
16. Diastrophic dysplasia, spondyloepiphyseal dysplasia: congenital clubfoot
17. Chondroectodermal dysplasia, chondrodysplasia punctata: congenital heart disease
18. Chondroectodermal dysplasia: hypoplastic nails
19. Achondroplasia: trident hand (Fig. 6-26)
20. Thanatophoric dysplasia, asphyxiating thoracic dysplasia, chondroectodermal dysplasia: narrow thorax
21. Asphyxiating thoracic dysplasia: progressive renal disease
22. Achondroplasia, thanatophoric dysplasia, chondrodysplasia punctata: facial dysmorphism
23. Spondyloepiphyseal dysplasia: decreased muscle tone with secondary waddling gait
24. Achondroplasia, thanatophoric dysplasia, achondrogenesis: macrocranium (Fig. 6-26)
25. Chondroectodermal dysplasia: dental abnormalities
26. Diastrophic dysplasia: cystic swelling of the outer ear
27. Chondrodysplasia punctata: cataracts

A

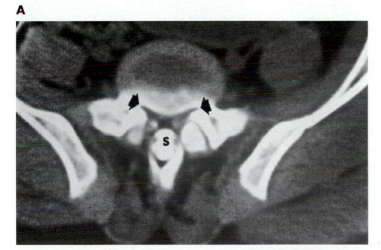

B

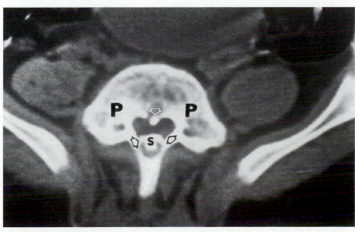

Fig. 6-25   Spinal stenosis in achondroplasia. **A** and **B**, Sequential CT images reveal short pedicles (*P*) along with central (*open arrows*) and lateral recess (*closed arrows*) narrowing (*s = opacified thecal sac*).

A

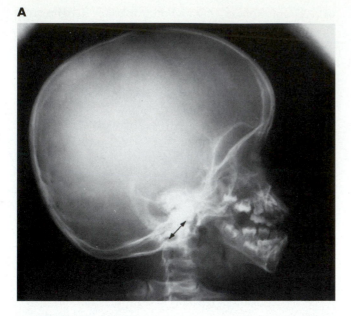

B

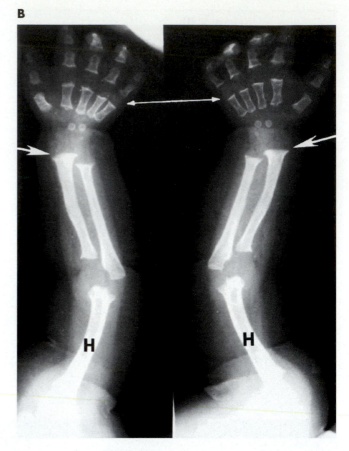

Fig. 6-26   Achondroplasia. **A,** A large skull with craniofacial disproportion and a small foramen magnum (*arrows*) is characteristic. **B,** Predominant shortening of the humeri (*H*) (rhizomelic dysplasia), metaphyseal flaring (*arrows*), and a trident appearance of the hands (*double-headed arrow*) are typical.

28. Thanatophoric dysplasia, achondrogenesis: protuberant abdomen without accentuated lumbar lordosis
29. Chondroectodermal dysplasia: sparse hair
30. Spondyloepiphyseal dysplasia: retinal detachment with myopia
31. Achondroplasia: genu varum
32. Achondroplasia: hydrocephalus
33. Chondroectodermal dysplasia: genu valgum

Radiographic Findings in Specific Forms

1. Platyspondyly: thanatophoric dysplasia, pseudoachondroplasia, spondyloepiphyseal dysplasia, spondylometaphyseal dysplasia, diastrophic dysplasia, Kniest disease, metatropic dysplasia

---

┌PEARL┐

Spondyloepiphyseal dysplasia tarda:
- Childhood: mild epiphyseal dysplasia (especially proximal femora, distal tibiae), widened ends of tubular bones (especially proximal femora), mild to moderate platyspondyly with occasional endplate irregularity
- Early adulthood: irregular articular surfaces of tubular bones with preserved articular spaces
- Later adulthood: secondary degenerative arthropathy

---

2. Stippled or punctate epiphyses: chondrodysplasia punctata (recessive), chondrodysplasia punctata (dominant), Kniest disease

┌PEARL┐

Kniest disease:
- Platyspondyly with anterior vertebral body wedging, occasional coronal clefting in infancy
- Broad iliac bones with basilar hypoplasia
- Very broad and short femoral necks, delayed ossification of capital femoral epiphyses (as late as adolescence), which are eventually large and flattened
- Short tubular bones with broad metaphyses and large, deformed epiphyses in childhood and adulthood

3. Talipes equinovarus: spondyloepiphyseal dysplasia, diastrophic dysplasia
4. Scoliosis or kyphoscoliosis: spondyloepiphyseal dysplasia, chondrodysplasia punctata (dominant), diastrophic dysplasia, metatropic dysplasia

PEARL

Spondyloepiphyseal dysplasia congenita:
- Infancy and early childhood: platyspondyly, which may persist into adulthood
- Childhood: notching in superior and inferior endplates
- Short broad iliac bones with basilar hypoplasia and lacelike crests, horizontal proximal femoral growth plates with prominent medial spurs in neck region (childhood), delayed femoral epiphyseal ossification
- Short tubular bones with irregular epiphyseal and metaphyseal ossification

5. Coxa vara: spondyloepiphyseal dysplasia, spondylometaphyseal dysplasia, metaphyseal chondrodysplasias

PEARL

Spondylometaphyseal dysplasia:
- Childhood: flat, biconvex vertebral bodies with endplate irregularities and tonguelike anterior protrusion centrally
- Adolescence: partial restoration to normal form
- Childhood: small, irregular capital femoral epiphyses with irregularity of subchondral acetabula
- Adulthood: marked femoral head dysplasia
- Short tubular bones with expanded, markedly irregular metaphyses and small, deformed epiphyses

6. Postaxial (ulnar-sided, fibular-sided) polydactyly: chondroectodermal dysplasia (Box 6-23)
7. Small foramen magnum: achondroplasia (Fig. 6-26)
8. Large skull: achondroplasia, achondrogenesis, Kniest disease (Fig. 6-26)
9. Madelung deformity: dyschondrosteosis
10. Coronal clefts in vertebral bodies: chondrodysplasia punctata (recessive), chondrodysplasia punctata (dominant)
11. Short ribs with narrow thorax: thanatophoric dysplasia, chondroectodermal dysplasia, asphyxiating thoracic dysplasia

PEARL

Thanatophoric dysplasia:
- Narrow anteroposterior and lateral thoracic diameters due to short ribs
- Marked platyspondyly with notchlike ossification defects of central portions of superior and inferior endplates

Continued

PEARL—cont'd

- Decreased vertical diameter and horizontal inferior margins of iliac bones, small sacrosciatic notches, short and broad pubic and ischial bones
- Shortening and relative broadening, bowing, and metaphyseal flaring of long tubular bones; extreme shortening and broadening of short tubular bones
- Small facial bones, relatively large calvarium, sometimes kleeblattschädel (cloverleaf) skull

12. Nasal hypoplasia: acrodysostosis
13. Odontoid hypoplasia; spondyloepiphyseal dysplasia, Kniest disease
14. Diaphyseal overtubulation: metatropic dysplasia

PEARL

Metatropic dysplasia:
- Early infancy: defective vertebral body ossification with small, flattened, or diamond-shaped appearance
- Later: platyspondyly, anterior wedging (frequently with humplike build-up of bone in central and dorsal portions of endplates, particularly at thoracolumbar junction), kyphoscoliosis
- Narrow thorax
- Hypoplasia of iliac bases with crescent-shaped crests and low-set anterosuperior spines, small and deformed capital femoral epiphyses, hyperplastic proximal femoral metaphyses
- Shortening, marked metaphyseal flaring, and epiphyseal dysplasia of tubular bones

15. Asymmetric shortening of limbs: chondrodysplasia punctata (dominant)
16. Vertebral endplate irregularities: pseudoachondroplasia, spondylometaphyseal dysplasia, chondrodysplasia punctata (dominant), multiple epiphyseal dysplasia

PEARL

Pseudoachondroplasia:
- Shortening and metaphyseal widening of tubular bones
- Flattened epiphyses, delayed ossification of capital femoral epiphyses, underossification of lateral portions of distal femoral epiphyses
- Short, broad femoral necks; broad trochanteric regions

Continued

PEARL—cont'd

- Irregular deformity and shortening of metacarpals, metatarsals, and phalanges; oval-shaped first metacarpal bone
- Delta-shaped deformity of distal femoral and radial metaphyses
- Pes equinovarus
- Cervical kyphosis, vertebral body irregularity, moderate lower lumbar interpediculate narrowing, often progressive thoracolumbar kyphoscoliosis

17. Decreasing interpediculate distance from upper to lower lumbar spine: achondroplasia, hypochondroplasia, thanatophoric dysplasia

PEARL

Hypochondroplasia:
- Moderate narrowing of interpediculate distances, anteroposterior shortening of pedicles, increased dorsal concavity of lumbar vertebral bodies
- Shortening and relative squaring of tubular bones
- Short, broad femoral necks
- Elongated distal fibulae relative to tibiae
- Occasional moderate increase in size of calvarium

18. Decreased carpal angle: dyschondrosteosis
19. Marked micromelia: achondrogenesis

PEARL

Achondrogenesis, type I:
- Absent or severely retarded ossification of vertebral bodies, sacrum
- Barrel-shaped thorax, short horizontal ribs with splayed ends and multiple fractures
- Small iliac bones with decreased vertical diameter, retarded ossification of pubic and ischial bones
- Extremely short tubular bones with concave ends, longitudinally projecting bone spurs at periphery of metaphyses, bowing of undertubulated shafts (especially femur, radius, ulna)

20. Capitate-hamate fusion: chondroectodermal dysplasia

PEARL

Chondroectodermal dysplasia (Ellis–van Creveld syndrome):
- Postaxial or axial hexadactyly with or without metacarpal and/or phalangeal fusion
- Infancy: low iliac wings, hooklike downward protrusion of medial and often lateral acetabular aspects; often premature ossification of capital femoral epiphyses; sometimes narrow thorax, short ribs
- Childhood: normalization of pelvis and thorax
- Progressive distal tubular bone shortening with short, broad middle phalanges
- Late infancy, childhood: capitate-hamate fusion, cone-shaped epiphyses of middle and distal phalanges
- Slanting proximal tibial metaphyses with apex located centrally; epiphyseal ossification centers adjacent to medial portions of metaphyses

21. Squared iliac wings with narrow sacrosciatic notches: achondroplasia, thanatophoric dysplasia (Fig. 6-27)
22. Cloverleaf deformity of the skull: thanatophoric dysplasia
23. Abnormal origin of thumbs with short first metacarpals: diastrophic dysplasia
24. Increased fibular length: hypochondroplasia
25. Telephone receiver–shaped long bones: thanatophoric dysplasia
26. Ossification absent in lower lumbar spine and sacrum: achondrogenesis

PEARL

Achondrogenesis, type II:
- Absent or severely retarded ossification of vertebral bodies, absent sacral ossification
- Barrel-shaped thorax with short ribs
- Small iliac bones with crescent-shaped inner and inferior margins, absent or severely delayed ossification of pubic and ischial bones
- Very short tubular bones with metaphyseal flaring and cupping

27. H- or U-shaped vertebral bodies: thanatophoric dysplasia
28. Posterior scalloping of vertebral bodies: achondroplasia, diastrophic dysplasia (Fig. 6-28)
29. Delayed ossification of pubic rami: spondyloepiphyseal dysplasia
30. Barrel-shaped thorax with multiple fractures of short, horizontal ribs: achondrogenesis

A

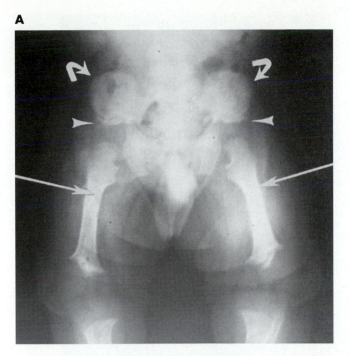

B

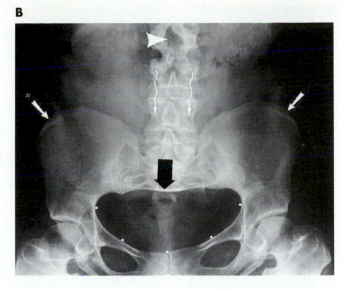

Fig. 6-27   Pelvic alterations in achondroplasia. **A,** In a child, rhizomelic shortening of the femora (*straight arrows*) with metaphyseal flaring, horizontal acetabular roofs (*arrowheads*), and squaring of the iliac wings (*curved arrows*) are evident. **B,** In an adult, a champagne glass appearance to the pelvic inlet (*dots*), squaring of the iliac wings (*white arrows*), sacral horizontality (*black arrow*), and progressive inferior narrowing of the interpediculate distances (*wavy arrows)* are seen. The patient has undergone laminectomies (*arrowhead*) for spinal stenosis.

31. Horizontal acetabular roofs: achondroplasia, thanatophoric dysplasia, spondyloepiphyseal dysplasia (Fig. 6-27)
32. Flared iliac wings: asphyxiating thoracic dysplasia
33. Deficient mineralization of skull and tarsal bones: achondrogenesis
34. Brachycephaly with thickened calvarium and skull base: acrodysostosis
35. Wedged vertebrae at thoracolumbar junction: achondroplasia (Box 6-24)

Box 6-23   Chondroectodermal Dysplasia (Ellis–van Creveld Syndrome): Important Features

- Short, deformed tubular bones (mesomelic dysplasia)
- Ectodermal dysplasia
- Polydactyly
- Carpal coalition
- Hypoplastic dentition
- Narrow, elongated thorax

Box 6-24   Wedged Vertebrae at Thoracolumbar Junction in Children: Differential Diagnosis

- Achondroplasia
- Cretinism, hypothyroidism
- Down syndrome
- Mucopolysaccharidoses, mucolipidoses
- Normal variation
- Trauma (acute); battered child syndrome
- Eosinophilic granuloma
- Neoplasm
- Granulomatous infection

Clinical Course in Specific Forms

1. Normal life span; pain and limited motion of wrist: dyschondrosteosis
2. Normal life span; premature degenerative joint disease: pseudoachondroplasia

A

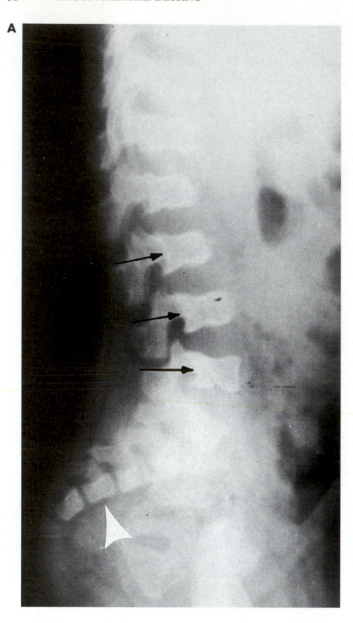

B

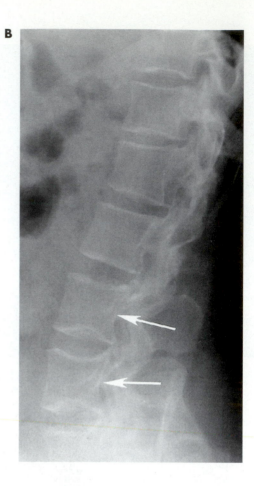

Fig. 6-28   Posterior scalloping of lumbar vertebral bodies (*arrows*) in achondroplasia. **A,** Child. Horizontal sacrum (*arrowhead*) is also evident. **B,** Adult. Differential diagnosis should include consideration of neurofibromatosis, acromegaly, Marfan syndrome, and intraspinal neoplasm.

3. Normal life span; premature degenerative joint disease; scoliosis with occasional paraplegia: spondyloepiphyseal dysplasia
4. Normal life span; nerve root compression; paraparesis or paraplegia; hydrocephalus: achondroplasia
5. Stillbirth or death in neonatal period: achondrogenesis
6. Death in neonatal period or normal life span; restricted activity secondary to scoliosis and clubfoot; premature degenerative joint disease: diastrophic dysplasia
7. Stillbirth, death in neonatal period, or normal life span; normal intelligence; orthopedic problems including joint contractures; cataracts: chondrodysplasia punctata (dominant)

8. Normal life span; vaginal delivery of offspring precluded by small pelvis: hypochondroplasia
9. Death in neonatal period from respiratory distress and cardiac failure: thanatophoric dysplasia
10. High probability of death in infancy from congenital heart disease; pulmonary insufficiency resulting from narrow thorax; disability secondary to genu valgum: chondroectodermal dysplasia
11. Failure to thrive, retarded psychomotor development, and recurrent infections with death in infancy: chondrodysplasia punctata (recessive)
12. Death in first year of life from respiratory insufficiency resulting from small thorax, or normal life span; renal failure in late childhood or adulthood: asphyxiating thoracic dysplasia

Metaphyseal Chondrodysplasia (Schmid Type)

1. Bone density and mineralization are normal (in contrast to the situation in rickets).
2. Serum levels of calcium and phosphate are normal.
3. The serum alkaline phosphatase level is normal (in contrast to the situation in rickets).
4. The condition is characterized by abnormal endochondral bone formation (in contrast to the mineralization defect of rickets).
5. The long bones are short and bowed.
6. Widened growth plates with osseous projections arising from the metaphyseal margins are evident radiographically.

PEARL

Metaphyseal chondrodysplasia (Schmid type):
- Short tubular bones
- Cupping, fraying, splaying of metaphyses (especially lower extremities, affecting proximal femora more severely than distal)
- Coxa vara, short femoral necks, genu varum

7. The radiographic differential diagnosis of the condition should include consideration of rickets, hypophosphatasia, and other metaphyseal chondrodysplasias.

PEARL

Other metaphyseal chondrodysplasias:
McKusick type:
- Shortening, metaphyseal dysplasia (more severe in knee region than in proximal femora) of tubular bones
- Disproportionately long fibulae, predominantly at distal ends
- Childhood: flat, biconvex vertebral bodies
- Adulthood: small vertebral bodies

Jansen type:
- Infancy: generalized demineralization with occasional fractures, metaphyseal changes simulating rickets, cortical erosion and subperiosteal new bone formation, reticular pattern of calvarium
- Childhood: severe shortening and curvature of tubular bones; severe widening, splaying, and fragmentation of metaphyses, with wide separation from epiphyseal ossification centers; hyperostosis of calvarium and skull base
- Adulthood: shortening and abnormal alignment of tubular bones, wide and irregular metaphyses, sclerosis of skull base and occasional calvarial thickening

MISCELLANEOUS CONGENITAL DISORDERS

1. Rib deformities and vertebral abnormalities occur in association with meningomyelocele.
2. Williams syndrome is a potential cause of diffuse osteosclerosis.

PEARL

Williams syndrome consists of idiopathic hypercalcemia, mental retardation, and unusual facies. The skeleton manifests increased radiographic density. Vascular anomalies include supravalvular aortic stenosis, infantile-type coarctation of the aorta, pulmonary valvular stenosis, peripheral pulmonary artery stenosis, and stenosis of the carotid arteries.

3. Wilms tumor may be associated with musculoskeletal abnormalities.

PEARL

Wilms tumor is associated with congenital hemihypertrophy, a situation possibly due to incomplete twinning. Male-to-female ratio is 1:2. Bones are longer and thicker on the affected side (usually the right). Associated malformations include aniridia, polydactyly, hypospadias, cryptorchidism, nevi, and hemangiomas. Tissue overgrowth may also be a feature of Beckwith-Wiedemann syndrome, which is also associated with Wilms tumor.

4. Several congenital disorders primarily affect the metaphyses and skull (Box 6-25).

PEARL

Metaphyseal dysplasia (Pyle disease):
- Autosomal recessive inheritance
- Erlenmeyer flask deformity of tubular bone metaphyses, bowing
- Supraorbital prominence, obtuse mandibular angle, cranial vault hyperostosis
- Wide nasal bridge, high palate, hearing and visual disturbances, short trunk, long extremities, contractures
- Joint pain, muscular weakness, scoliosis, genu valgum, dental malocclusion, bone fragility

Continued

PEARL—cont'd

Craniometaphyseal dysplasia:
- Transverse metaphyseal bands of increased density, with or without generalized sclerosis of tubular bones
- Moderate metaphyseal undertubulation
- Sclerotic bands framing the vertebral bodies; sclerotic areas in pelvis, scapulae, clavicles, ribs
- Sclerosis of skull base, less frequently calvarium; occasionally craniosynostosis
- Sclerotic changes occur in severely affected infants and may be reversible

5. Many congenital spinal abnormalities are isolated and considered normal variations if asymptomatic (Fig. 6-29).
6. Progeria is one of several congenital syndromes associated with premature senescence (Box 6-26).

Box 6-25   Metaphyseal Dysplasia (Pyle Disease): Important Features

- Metaphyseal undermodeling (Erlenmeyer flask deformity) extending into diaphyses (especially long bones)
- Decreased bone density
- Craniofacial hyperostosis
- Genu valgum

Box 6-26   Progeria: Important Features

- Congenital short-limbed dwarfism with premature aging
- Short thin clavicles
- Hypoplastic facial bones
- Delayed sutural closure
- Coxa valga
- Acro-osteolysis of terminal phalanges
- Slender long bones
- Generalized osteoporosis
- Normal skeletal age
- Sparse subcutaneous fat
- Premature atherosclerotic vascular calcification
- Survival rare beyond 20 years of age

7. Hypoplasia of the glenoid neck is an uncommon congenital malformation that may have a similar pathogenesis to developmental hip dysplasia (Fig. 6-30).
8. The acrocephalosyndactyly syndromes represent an apparent genetic linkage between craniosynostosis (Box 6-27; Fig. 6-31) and digital syndactyly involving both soft tissue and bone (Box 6-28; Fig. 6-32).

PEARL

Progeria:
- Acro-osteolysis or osseous dissolution resembling hyperparathyroidism or massive osteolysis of Gorham
- Hypoplastic facial bones and mandible, delayed closure of cranial sutures, coxa valga
- Dwarfism, alopecia, brown trunk pigmentation, skin and subcutaneous fat atrophy, limited hip and knee extension, receding chin, beaked nose, exophthalmos
- Pathologic fractures

Box 6-27   Craniosynostosis: Important Features

A. Differential Diagnosis
- Idiopathic
- Acrocephalopolysyndactyly (Carpenter syndrome)
- Acrocephalosyndactyly syndromes
- Craniofacial dysostosis (Crouzon disease)
- Hyperostotic diseases (Pyle, Engelmann, Van Buchem diseases; osteopetrosis)
- Hyperthyroidism
- Hypophosphatasia
- Microcephaly
- Rickets, treated
- Arrested hydrocephalus

B. Types of Calvarial Deformity
- Brachycephaly (short head): premature bilateral closure of coronal and/or lambdoid sutures
- Oxycephaly, acrocephaly, turricephaly (pointed head): premature closure of all sutures
- Plagiocephaly (oblique head): unilateral premature closure of coronal and/or lambdoid sutures
- Scaphocephaly (boat head), dolichocephaly (long head): premature closure of sagittal suture
- Trigonocephaly (triangular head): premature closure of metopic suture

Box 6-28   Acrocephalosyndactyly Syndromes: Important Features

- Carpenter syndrome includes polydactyly of the feet.
- Craniosynostosis is most severe in Apert and Carpenter syndromes.
- Associated features of Chotzen syndrome include short stature, cryptorchidism, and seizures.
- Short stature, pericardial cysts, contractures, rectal prolapse, and deformed ears occur in Waardenburg syndrome.

Continued

Box 6-28—cont'd

- Obesity and autosomal recessive inheritance characterize Summit and Carpenter syndromes.
- Pfeiffer syndrome is characterized by mild craniosynostosis, occasional hallux duplication, and autosomal dominant inheritance.

9. Condylar hyperplasia is an uncommon developmental disorder associated with unilateral overgrowth of the mandibular condyle (Box 6-29).

Box 6-29   Condylar Hyperplasia: Important Features

- Presentation: age 15 to 19 years
- Unilateral progressive hyperplasia of mandibular condyle, secondary to persistent activity of growth zone
- Headache, jaw pain, malocclusion
- Female predilection
- Pandimensional condylar enlargement, normal or elongated condylar neck
- Self-limited, but may progress for over 20 years
- Treatment: orthodontic, surgical correction

10. The endosteal hyperostosis syndromes include the Van Buchem and Worth types (Box 6-30).

Box 6-30   Endosteal Hyperostosis: Important Features

Van Buchem type:
- Cortical hyperostosis, usually polyostotic; most commonly affecting mandible, clavicles, scapulae, ribs, long bones
- Protracted cases: resorption of original cortex, widened medullary canal; diaphyseal expansion; longitudinal overgrowth and bowing deformities of long bones, sometimes with interosseous bony ridges

Worth type:
- Hyperostosis, increased density of calvarium; sclerosis of skull base; underpneumatization of paranasal sinuses
- Hyperostosis, increased density of mandible after puberty
- Endosteal thickening of diaphyseal cortex of tubular bones
- Widened medial clavicles

11. Dysosteosclerosis is a potential cause of diffuse osteosclerosis affecting the axial skeleton (Box 6-31).

Box 6-31   Dysosteosclerosis: Important Features

- Thickening and sclerosis of calvarium and skull base with decreased or absent pneumatization of paranasal sinuses and mastoids
- Sclerosis of ribs, clavicles, scapulae, and pelvic bones with hypoplasia of the iliac bones
- Flattening, dorsal wedging, and sclerosis of vertebral bodies
- Tubular bones: wide metaphyseal flaring with sclerosis of epiphyses, metaphyseal margins, and central portions; cortical thinning and radiolucency in expanded submetaphyseal areas

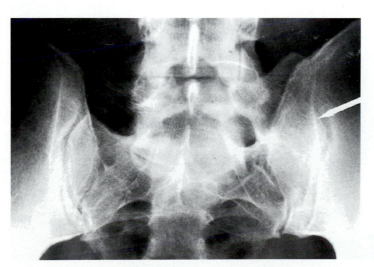

Fig. 6-29   A transitional segment (*arrow*) is a common congenital variation at the lumbosacral junction and may be associated with a pseudoarthrosis.

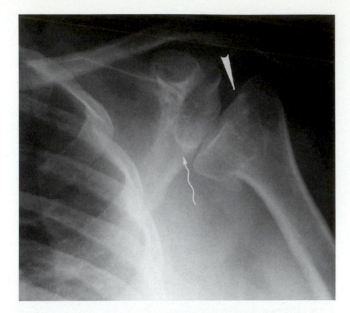

Fig. 6-30 Developmental dysplasia of the glenohumeral joint. In utero incongruity of the articulation has resulted in hypoplasia of the glenoid neck (*arrow*) and deformity of the humeral head (*arrowhead*). Differential diagnosis should include consideration of multidirectional instability with trough fracture of the humeral head and osseous Bankart lesion of the glenoid, previous trauma, and osteonecrosis with secondary osteoarthritis.

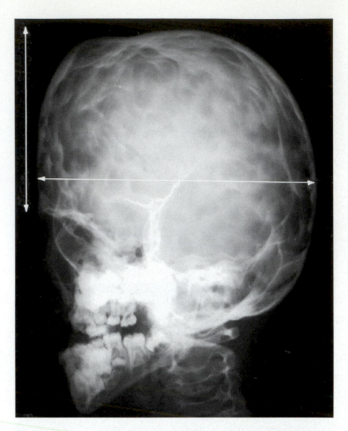

Fig. 6-31 Craniosynostosis with brachycephaly (*long arrow*) and turricephaly (*short arrow*), as well as accentuated convolutional markings of the calvarium. Differential diagnosis of premature sutural closure should include consideration of isolated developmental anomaly, acrocephalosyndactyly syndromes, arrested hydrocephalus, treated rickets, accelerated skeletal maturation, and other congenital malformation syndromes.

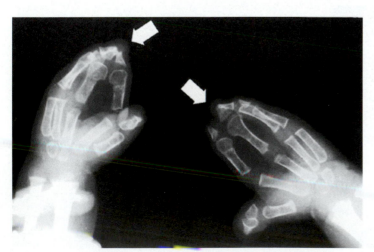

Fig. 6-32 Cutaneous and osseous syndactyly (*arrows*) in Apert syndrome. Differential diagnosis should include consideration of other acrocephalosyndactyly syndromes and isolated congenital malformation.

12. Irregular ossification is the most striking radiographic feature of the Dyggve-Melchior-Clausen syndrome (Box 6-32).

Box 6-32 Dyggve-Melchior-Clausen Syndrome: Important Features

- Platyspondyly, radiolucent vertebral bodies with irregular endplates
- Small iliac wings bordered by a wide lacelike zone of irregular ossification, small or unossified femoral heads and necks
- Severe shortening and distortion of tubular bones with markedly irregular epiphyseal and metaphyseal ossification
- Disappearance of irregular ossification in adolescence

---

13. Holt-Oram syndrome is characterized by an association between radial ray anomalies and congenital heart disease (Box 6-33).

Box 6-33 Holt-Oram Syndrome: Important Features

- Triphalangeal, clinodactylic, absent, or proximally placed thumb with cutaneous syndactyly
- Congenital heart disease (atrial or ventricular septal defects)
- Radial hypoplasia and/or malformation, preaxial (radial-sided) carpal bone anomalies, radioulnar synostosis
- Autosomal dominant inheritance, variable penetrance
- Included in VACTERL complex

---

14. The mesomelic dysplasias predominantly affect the bones of the forearm and calf, and include the Langer and Nievergelt types (Box 6-34).

Box 6-34 Mesomelic Dysplasia: Important Features

Langer type:
- Distal ulnar shortening with delayed or absent distal epiphyseal ossification and hyperplastic interosseous ridges
- Radial bowing of radii with radiovolar angulation of distal articular surface, radiovolar dislocation of radial heads
- Proximal shortening of fibulae with lateral angulation of shafts and expansion of midshafts, resulting in triangular configuration (as seen on frontal projection)
- Short and plump tibiae with hypoplasia of lateral aspects of distal epiphyses (childhood), lateral angulation of distal tibial articular surfaces with fibular deviation of tali

---

Box 6-34—cont'd

Nievergelt type:
- Shortening and triangular or rhomboid deformity of tibiae and sometimes radii, less marked shortening and widening of fibulae and sometimes ulnae
- Older children and adults: tarsal bone synostosis
- Radioulnar synostoses with radial head dislocation (and rarely proximal ulnar subluxation)

---

15. Osseous irregularity and constriction are the predominant radiographic manifestations of osteodysplasty (Melnick-Needles syndrome) (Box 6-35).

Box 6-35 Osteodysplasty (Melnick-Needles Syndrome): Important Features

- Bowing, metaphyseal flaring, irregularities of cortical density and thickness; distorted diaphyseal contours of long tubular bones; severe coxa valga
- Irregular contours, multiple constrictions, and distorted appearance of ribs, clavicles
- Lateral constriction of supraacetabular regions, flared iliac wings, caudal tapering of ischia, attenuation of pubic and ischial bones
- Micrognathia, delayed anterior fontanelle closure, sclerosis of skull base and sometimes calvarium
- Anterior vertebral body concavity, often with increased height, kyphosis, and/or scoliosis

---

16. The otopalatodigital syndrome predominantly affects the bones of the skull, hands, and feet (Box 6-36).

Box 6-36 Otopalatodigital Syndrome: Important Features

- Skull: prominent supraorbital ridge and posterior parietal and occipital regions; vertical clivus; small facial bones with decreased mastoid and paranasal sinus pneumatization; small mandible with increased angle
- Hand: accessory carpals with partial or complete coalescence (especially capitate-hamate); short and broad distal phalanges with premature epiphyseal fusion; short radial side of fifth middle phalanx; elongated ulnar side of second metacarpal bone with accessory ossification center that often fuses with trapezoid

Continued

Box 6-36—cont'd

- Foot: short metatarsals and phalanges of hallux; multiple fusions between metatarsals, cuneiforms, and navicular; accessory ossification center of proximally elongated fifth metatarsal bone
- Other: hypoplasia and posterior dislocation of proximal radius; narrow lumbar pedicles with wide interpediculate distances; failed ossification of some neural arches; small iliac wings with diminished flaring; variable hip dislocation and coxa valga
- Male individuals affected more severely than female individuals

---

17. Sclerosteosis is an uncommon condition associated primarily with hyperostosis of the skull and long bones (Box 6-37).

Box 6-37   Sclerosteosis: Important Features

- Hyperostosis of calvarium, base of skull, and mandible
- Lack of normal diaphyseal constriction of tubular bones
- Radial deviation of second and third fingers, cutaneous syndactyly of fingers
- Cortical sclerosis and hyperostosis of tubular bones

---

18. An association between short ribs and polydactyly occurs in the Saldino-Noonan syndrome (Box 6-38).

Box 6-38   Short Rib–Polydactyly Syndrome (Saldino-Noonan Syndrome): Important Features

- Short, horizontally oriented ribs
- Small iliac bones with flat acetabular roofs
- Marked shortening of long tubular bones with pointed or ragged appearance of ends
- Incomplete ossification of short tubular bones, postaxial (ulnar-sided or fibular-sided) polydactyly

---

19. Shortening of short tubular bones and epiphyseal deformities are the major radiographic features of the trichorhinophalangeal dysplasias, which also affect the hair and nose (Box 6-39).

Box 6-39   Trichorhinophalangeal Dysplasia (Type I): Important Features

- Shortening of one or more phalanges and/or metacarpals
- Characteristic epiphyseal deformity of affected tubular bones (Giedion type 12), cone-shaped epiphyses, premature epiphyseal fusion
- Small capital femoral epiphyses, sometimes with Legg-Calvé-Perthes-like changes

---

20. Narrowing or constriction of the medullary spaces in the calvarium and tubular bones is characteristic of tubular stenosis (Kenny-Caffey disease) (Box 6-40).

Box 6-40   Tubular Stenosis (Kenny-Caffey Disease): Important Features

- Narrow medullary cavities of tubular bones, normal or increased cortical thickness
- Diaphyseal constriction of tubular bones with relative metaphyseal flaring
- Absent diploic space of calvarium, wide metopic suture, and delayed anterior fontanelle closure

---

SUGGESTED READINGS

1. Castillo M: Congenital abnormalities of the nose: CT and MR findings, AJR 162:1211, 1994.
2. Cupta SK et al: Cleido-cranial dysostosis-skeletal abnormalities, Australas Radiol 36:238, 1992.
3. Demirci A and Sze G: Cranial osteopetrosis: MR findings, AJNR 12:781, 1991.
4. Eich GF, Babyu P, and Giedion A: Pediatric pelvis: radiographic appearance in various congenital disorders, Radiographics 12:467, 1992.
5. Elster AD et al: Autosomal recessive osteopetrosis: bone marrow imaging, Radiology 182:507, 1992.
6. Elster AD et al: Cranial imaging in autosomal recessive osteopetrosis. I. Facial bones and calvarium, Radiology 183:129, 1992.
7. Greenspan A: Sclerosing bone dysplasias: a target site approach, Skeletal Radiol 20:561, 1991.
8. Herman TE and McAlister WH: Inherited diseases of bone density in children, Radiol Clin North Am 29:149, 1991.
9. Ibis E et al: Three-phase bone scintigraphy in Maffucci's syndrome, Clin Nucl Med 17:505, 1992.
10. Jones KL: Smith's recognizable patterns of human malformation, ed 4, Philadelphia, 1988, Saunders.

11. Kleiman PK: Schmid-like metaphyseal chondrodysplasia simulating child abuse, AJR 156:576, 1991.

12. Lachman RS et al: Collagen, genes and the skeletal dysplasias on the edge of a new era: a review and update, Eur J Radiol 14:1, 1992.

13. Lawson JP: Clinically significant radiologic anatomic variants of the skeleton, AJR 163:249, 1994.

14. Manns RA and Davies AM: Glenoid hypoplasia: assessment of computed tomographic arthrography, Clin Radiol 43:316, 1991.

15. Maroteaux P and Spranger J: Spondylometaphyseal dysplasias: a tentative classification, Pediatr Radiol 21:293, 1991.

16. Smoker WRK: Craniovertebral junction: normal anatomy, craniometry, and congenital anomalies, Radiographics 14:255, 1994

17. Snow RD and Lecklitner ML: Musculoskeletal findings in Klippel-Trenaunay syndrome, Clin Nucl Med 16:928, 1991.

18. Spirt BA et al: Prenatal sonographic evaluation of short-limbed dwarfism: an algorithmic approach, Radiographics 10:217, 1990.

19. Taybi H and Lachman RS: Radiology of syndromes, metabolic disorders, and skeletal dysplasias, ed 3, Chicago, 1990, Year Book.

20. Westcott MA et al: Congenital and acquired orthopedic abnormalities in patients with myelomeningocele, Radiographics 12:1155, 1992.

21. Williams JW, Monaghan D, and Barrington NA: Craniofacial melorheostosis: case report and review of the literature, Br J Radiol 64:60, 1991.

Hematologic Disorders

HEMOPHILIA (Box 7-1)

1. Hemophilia is a bleeding disorder that results from clotting factor deficiency; the major consequences are hemarthroses, arthropathy, pseudotumors, and soft-tissue hemorrhage (Fig. 7-1).

2. Hemophilia A results from factor VIII deficiency and occurs in male individuals as a result of X-linked recessive inheritance.

3. Hemophilia B or Christmas disease results from factor IX deficiency and occurs in male individuals as a result of X-linked recessive inheritance.

4. Von Willebrand disease results from a combination of factor VIII deficiency and platelet dysfunction, and it occurs in both sexes via autosomal dominant inheritance.

5. The most common sites of hemarthrosis in hemophilia are the knee, elbow, ankle, hip, and shoulder.

6. Potential complications of hemarthrosis, which tends to affect multiple joints in an asymmetric distribution, are arthropathy and flexion contractures.

7. Hemarthroses appear as radiodense effusions because of hemosiderin deposition in hypertrophied synovium; the same phenomenon occurs in pigmented villonodular synovitis, and hence both diseases may manifest low signal intensity on magnetic resonance images owing to the paramagnetic effect of iron deposition (Fig. 7-2).

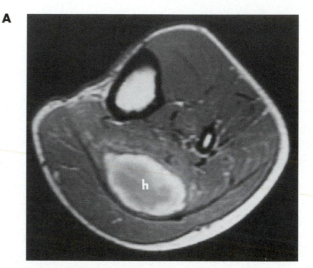

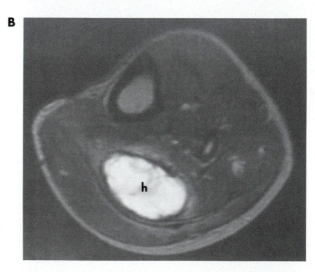

Fig. 7-1 Soleus hematoma in hemophilia. The well-encapsulated mass (*h*) exhibits areas of high signal intensity on both T1-weighted (**A**) and T2-weighted (**B**) MR images, characteristic of subacute hemorrhage. No hemosiderin effect has yet developed.

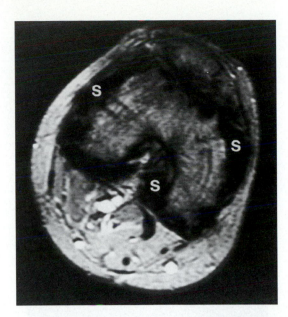

Fig. 7-2 Hemosiderin-laden synovium (*S*) in hemophilia of the knee. Differential diagnosis should include consideration of pigmented villonodular synovitis and other causes of recurrent hemarthrosis.

PEARL

In hemophilia, hemarthrosis leads to effusion, and hyperemia results in periarticular osteopenia. Repeated episodes of intraarticular hemorrhage lead to joint space narrowing, irregular bone erosion, and subchondral cyst formation. Chronic disease results in enlargement of the ends of the long bones. Bleeding around the cruciate ligaments of the knee produces widening of the intercondylar notch of the femur. Pseudotumors of bone may occur.

The radiographic changes of hemophilia are very similar to those of juvenile chronic arthritis. Increased radiographic density and decreased MR signal intensity of synovial tissue are features of hemophilia but not juvenile chronic arthritis. The distribution of disease tends to be different; hemophilia most commonly involves the knees, ankles, and elbows, whereas juvenile chronic arthritis frequently involves the hands, wrists, larger joints, and spine.

8. Synovial inflammation in hemophilia causes hyperemia, which is responsible for radiographic findings that include periarticular osteoporosis, epiphyseal overgrowth, and premature physeal closure.
9. The long-bone diaphyses in hemophilia are characterized by shortening and overtubulation.
10. The arthropathy of hemophilia demonstrates joint space narrowing, subchondral cysts, osseous erosions, and premature degenerative joint disease radiographically.

11. In the knee, squaring of the inferior patellar pole and widening of the intercondylar notch are characteristic findings that also occur in juvenile chronic arthritis.
12. The hemophilic elbow typically manifests radial head enlargement and trochlear notch enlargement.
13. In the ankle, hemophilia produces tibiotalar slant, which is also observed in trauma, juvenile chronic arthritis, epiphyseal dysplasias, and sickle cell anemia.
14. Pseudotumors in hemophilia occur most commonly in the femur, pelvis, and tibia and result from intraosseous, subperiosteal, or soft-tissue hemorrhage (Fig. 7-3).

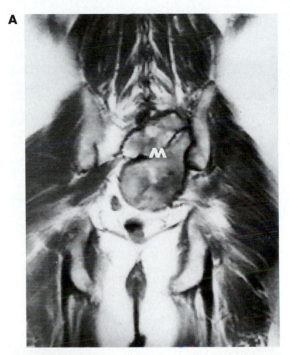

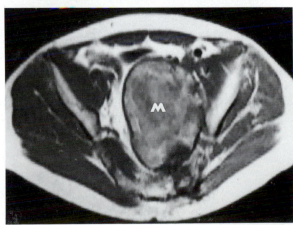

Fig. 7-3 Hemophilic pseudotumor. Coronal (**A**) and transaxial (**B**) T1-weighted MR images demonstrate a large mass of inhomogeneous signal intensity (*M*) arising from the anterior aspect of the sacrum. Differential diagnosis should include consideration of soft-tissue sarcoma and abscess.

A

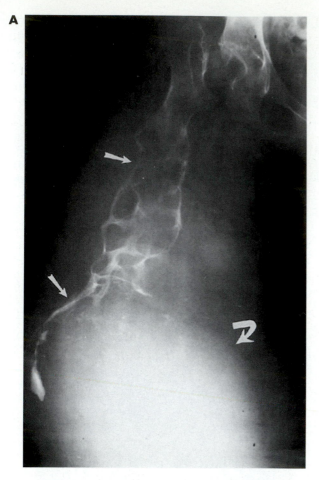

B

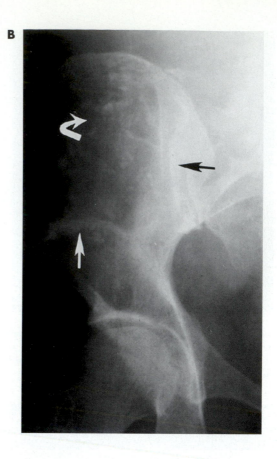

Fig. 7-4 Hemophilic pseudotumors. **A,** An expansile, septated long-bone lesion (*straight arrows*) is associated with calcification and massive soft-tissue extension (*curved arrow*) at the level of the knee. **B,** A well-defined lytic flat-bone lesion (*straight arrows*) with numerous internal calcifications (*curved arrow*) arises from the iliac wing. Differential diagnosis should include consideration of chondrosarcoma, as well as other benign and malignant neoplasms with dystrophic calcification.

15. Radiographic manifestations of hemophilic pseudo-tumor include bone destruction with sharp sclerotic margins, endosteal or periosteal scalloping, and soft-tissue mass (Fig. 7-4).

16. Characteristics of hemophilic pseudotumors that are generally absent in malignant neoplasms include sclerotic margins and cortical scalloping.

Box 7-1   Hemophilia: Important Features

- Hemophilia A: deficient factor VIII
- Hemophilia B (Christmas disease): deficient factor IX
- Hemophilia C: deficient factor XI
- Synovial hemorrhage leading to hemosiderin deposits with fibrosis and adhesions, resulting in cartilage and subchondral bone erosions
- Soft-tissue swelling, radiodense effusions, joint space narrowing, erosions
- Epiphyseal overgrowth, premature physeal fusion
- Weight-bearing joints (knee most common) predominantly affected
- Squared inferior patellar pole

SICKLE CELL ANEMIA (Box 7-2)

1. Sickle cell anemia occurs in approximately 1% of North American black individuals and manifests hemoglobin S.
2. The major consequences of sickle cell anemia include hematopoietic marrow hyperplasia, infection, and infarction (Fig. 7-5).

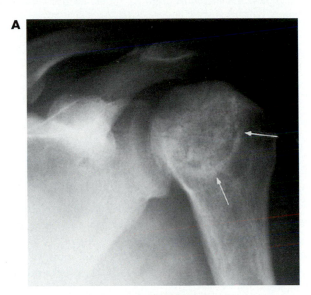

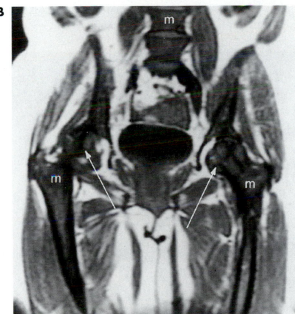

Fig. 7-5   Sickle cell anemia. **A,** Osteonecrosis of the humeral head (*arrows*) is a recognized complication of the disease. **B,** Coronal T1-weighted MR image reveals diffuse low signal intensity within the bone marrow (*m*) secondary to hematopoietic expansion and hemosiderin deposition. Inhomogeneity of both femoral heads (*arrows*) indicates osteonecrosis. Differential diagnosis should include consideration of Gaucher disease.

3. Dactylitis or the hand-foot syndrome occurs in up to 20% of children with sickle cell anemia and is usually caused by infarction of the short tubular bones in the hands and feet; radiographic features include soft-tissue swelling and periosteal reaction, simulating osteomyelitis.

PEARL

Sickle cell dactylitis is usually caused by bone infarcts secondary to thrombosis, although secondary osteomyelitis may occur.

4. Subchondral ischemic necrosis in sickle cell anemia is frequently bilateral and exhibits a predilection for the femoral head and humeral head (Fig. 7-5).
5. In the spine, H-shaped vertebral bodies occur secondary to preferential sludging and infarction in the relatively poorly vascularized central portion of the vertebral endplates; this deformity is also a characteristic radiographic feature of sickle-thalassemia and Gaucher disease (Fig. 7-6).
6. Radiographically, diaphyseal bone infarcts in sickle cell anemia may exhibit serpiginous radiodensities, patchy osteosclerosis, and/or periosteal reaction.
7. Bone density in sickle cell anemia is usually decreased, although this may be difficult to appreciate because of the genetically higher bone density of the targeted ethnic population.
8. The most common organism responsible for osteomyelitis in sickle cell anemia is *Staphylococcus aureus.*
9. Osteomyelitis caused by *Salmonella* organisms occurs with greater frequency among individuals affected by sickle cell anemia than in the general population.
10. Outside the musculoskeletal system, patients with sickle cell anemia may demonstrate cardiomegaly, pulmonary infarction, cholelithiasis, splenic infarction, and renal papillary necrosis radiographically.
11. Hyperplasia of hematopoietic marrow in sickle cell anemia, as well as the early diagnosis of its major musculoskeletal complications (infection, infarction), is best established by the use of magnetic resonance imaging (Fig. 7-5).
12. The predominant manifestions of hemoglobin SC disease are ischemic necrosis, hematopoietic marrow hyperplasia in the skull, and splenomegaly.
13. Patients with sickle cell trait have hemoglobin AS and, aside from occasional bone infarction, are generally free of musculoskeletal complications.

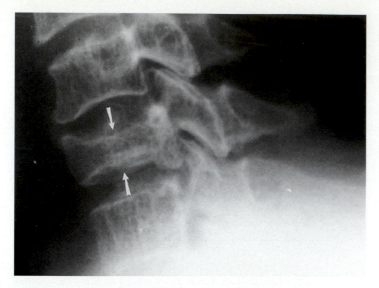

Fig. 7-6 H-shaped vertebral body in sickle cell anemia. Lateral radiograph of the cervical spine reveals characteristic central endplate depression (*arrows*) with relative sparing of the periphery, localized to the C5 level; generalized osteopenia is also present. Differential diagnosis should include consideration of Gaucher disease and sickle-thalassemia.

Box 7-2 Sickle Cell Anemia: Important Features

- Dactylitis: periosteal new bone formation, diaphyseal expansion, cortical and trabecular thinning, widening of intertrabecular spaces; ages 6 months to 8 years (peak age 2 years)
- Acute bone infarction may be indistinguishable from osteomyelitis on bone-gallium scintigraphy or magnetic resonance imaging
- Disease affects 1 in 500 black children in the United States, and a total of more than 50,000 Americans
- More than two million black American individuals (8% to 10%) have sickle cell trait
- A twenty-five–fold greater incidence of *Salmonella* osteomyelitis occurs among affected patients than in the general population
- Clinical findings: colic, irritability, bacterial pneumonia, meningitis, pallor, jaundice, hepatosplenomegaly, scleral icterus, anemic cardiac murmurs
- Symptoms appear between the ages of 6 months and 2 years
- Skeletal manifestations of marrow hyperplasia are less severe than in thalassemia major

---

THALASSEMIA

1. Thalassemia is also known as Cooley or Mediterranean anemia.
2. Because the hemolytic anemia of thalassemia major (homozygous beta-thalassemia) is more severe than that of sickle cell anemia, radiographic features of hematopoietic marrow hyperplasia are much more striking in the former (Fig. 7-7).
3. The hair-on-end appearance of the calvarium in thalassemia is caused by radiodense striations in a markedly widened diploic space.
4. Facial bone involvement with obliteration of the paranasal sinuses results in the characteristic rodent facies.
5. The Erlenmeyer flask deformity is a common radiographic feature of thalassemia (Box 7-3).

Box 7-3 Differential Diagnosis of Metaphyseal Undermodeling

- Anemias (thalassemia, sickle cell anemia)
- Fibrous dysplasia
- Gaucher disease, Niemann-Pick disease
- Healing fracture, metaphyseal injury
- Enchondromatosis (Ollier disease)
- Heavy metal poisoning (lead, bismuth)
- Metaphyseal dysplasias (Pyle disease, craniometaphyseal dysplasia)
- Mucopolysaccharidoses
- Multiple cartilaginous exostosis syndrome
- Neoplasms (benign, expansile)
- Osteopetrosis

---

6. Ischemic necrosis is extremely rare in thalassemia; H-shaped vertebral bodies have occasionally been described.
7. The tubular bones manifest widening of the medullary cavity, cortical thinning, and coarse trabeculation (Fig. 7-7).
8. Extramedullary hematopoiesis is a well-recognized manifestation of thalassemia and may appear as

A

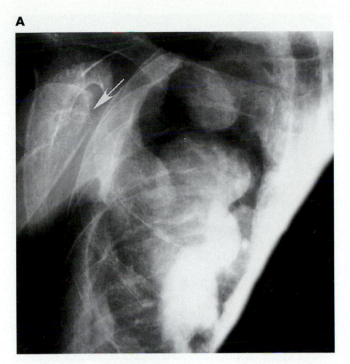

B

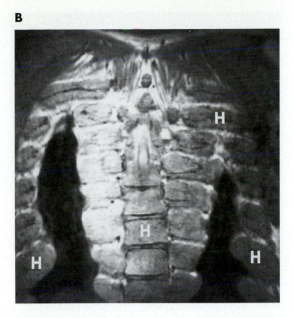

Fig. 7-7   Thalassemia major. **A,** Pathognomonic radiographic findings include generalized osteopenia with tendency toward fracture (*arrow*), coarse trabecular pattern, and osseous enlargement secondary to bone marrow expansion. **B,** Coronal T1-weighted MR image reveals diffuse low signal intensity (*H*) in areas of hematopoietic marrow expansion.

soft-tissue masses in the pelvis, retroperitoneum, or paravertebral region.

9. Spontaneous fracture is a potential complication of thalassemia (Fig. 7-7).

10. Clinically, patients with thalassemia major experience presenting symptoms early in childhood and usually die by young adulthood.

11. Ischemic necrosis is common in sickle-thalassemia; signs of marrow hyperplasia also occur, and musculoskeletal infection is a variable feature. In general, the severity of each complication is less than that of the homozygous form of the anemia (sickle cell anemia versus thalassemia major) in which it predominates.

12. Computed tomography or magnetic resonance imaging can document tissue iron overload secondary to multiple blood transfusions in patients with thalassemia and other anemias.

SYSTEMIC MASTOCYTOSIS (Box 7-4)

1. Systemic mastocytosis usually appears after puberty.
2. Clinical symptoms include flushing, nausea, vomiting, and diarrhea.
3. Cutaneous or mucous membrane involvement resembles urticaria pigmentosa of childhood.
4. Osteoporosis may occur in the disease secondary to histamine or prostaglandin release by mast cells.

5. Localized lytic lesions are most commonly encountered in the ribs, pelvis, skull, and tubular bones.
6. Generalized, as opposed to localized, osteoporosis is more commonly observed.
7. Osteosclerosis in the disease is a manifestation of host bone reaction to mast cell infiltration of marrow (Fig. 7-8).

PEARL

Mastocytosis is a recognized cause of diffuse increased bone density. It is a rare proliferative disorder that begins in adult life, affecting both men and women. Mast cell proliferation in bone marrow causes the reactive sclerosis, as in metastatic carcinomatosis. Osteoblastic metastases include prostate, breast, and bowel (carcinoid, stomach) tumors; medulloblastoma; and Hodgkin lymphoma. Tuberous sclerosis produces focal areas of osteosclerosis (axial skeleton in particular) and osteolysis (appendicular skeleton, especially metacarpals and metatarsals).

8. Focal sclerotic lesions are more common than diffuse osteosclerosis.
9. Localized osteoblastic areas superimposed on a background of diffuse osteoporosis is the most common radiographic pattern.

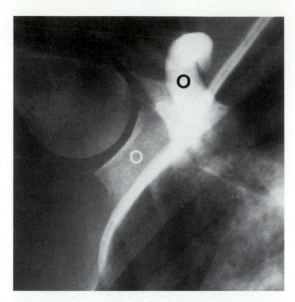

Fig. 7-8   Systemic mastocytosis. Generalized osteosclerosis (*O*) is secondary to widespread marrow replacement by mast cells. Differential diagnosis should include consideration of renal osteodystrophy, metastatic prostate carcinoma, fluorosis, osteopetrosis, and myelofibrosis.

10. The radiographic differential diagnosis of systemic mastocytosis may include consideration of metastatic disease, myelofibrosis, fluorosis, sickle cell anemia, and other causes of generalized osteosclerosis.

PEARL

Systemic mastocytosis:
- Osseous changes in 50% of cases, usually in axial skeleton; resorption and thickening of bone trabeculae by diffuse deposits of marrow mast cells, leading to mixed osteosclerotic and osteolytic pattern
- Osteolytic areas common in ribs, pelvis, skull, and tubular bones, producing cystlike appearance
- Sclerosis resembles myelofibrosis, and may be diffuse or multifocal
- Diploic space of skull often obliterated by thickened sclerotic tables

---

11. Mast cell infiltration of the liver, spleen, lymph nodes, bone marrow, and skin may occur, and splenomegaly is a common clinical and radiographic finding.
12. Involvement of the axial skeleton is more pronounced than that of appendicular areas.

Box 7-4   Systemic Mastocytosis: Important Features

- Dermatologic manifestations: confluent macules, papules, nodules; chronic dermatitis; skin pigmentation; oral, nasal, rectal mucosal involvement
- Discrete lytic lesions under 5 cm in diameter with sclerotic margins, most commonly in ribs, pelvis, skull, tubular bones
- Generalized osteoporosis (affecting axial more than appendicular skeleton) secondary to calcium malabsorption, mast cell histamine and heparin production
- Intramedullary bone infarction, osteonecrosis
- Focal or diffuse osteosclerosis
- Pancytopenia, hepatosplenomegaly, peptic ulcers, diarrhea, vomiting, shocklike episodes, weight loss, lymphadenopathy

---

MYELOFIBROSIS

1. The condition is also known as myelosclerosis and agnogenic myeloid metaplasia.
2. In myelofibrosis, fibrosis occurs in areas of bone marrow that are normally involved in hematopoiesis.
3. Compensatory hematopoiesis occurs in areas of bone normally occupied by fatty marrow, which may in turn become fibrotic.
4. Myelofibrosis generally has an insidious onset in middle-aged to elderly patients of both sexes.
5. Radiographic patterns of the disease include either generalized osteosclerosis or patchy areas of increased bone density with cortical thickening.
6. Bone density may be normal or decreased, and osteolytic lesions may occur in a minority of patients.
7. Subperiosteal hemorrhage may occur secondary to thrombocytopenia.
8. Bone pain and arthralgias can occur in both spinal and extraspinal sites.
9. Extramedullary hematopoiesis is manifested as hepatosplenomegaly and paravertebral soft-tissue masses; spinal cord compression may occur secondarily.

PEARL

Extramedullary hematopoiesis occurs most often between the T8 and T12 levels and can cause spinal cord compression. It may occur with myelofibrosis and congenital hemolytic anemias, and is hypervascular on contrast-enhanced CT or MR examinations as well as angiography.

10. Up to 80% of patients with myelofibrosis have elevated serum or urinary uric acid levels, and secondary gout occurs in up to 20% of cases; the latter may antedate the diagnosis of myelofibrosis.
11. The radiographic differential diagnosis of myelofibrosis should include consideration of metastatic prostate carcinoma, fluorosis, and systemic mastocytosis.

PEARL

In cases of longstanding leukemia or lymphoma, a sclerotic phase can occur with associated myelofibrosis. The therapy for thalassemia may lead to hemosiderosis. Mast cells in systemic mastocytosis may produce heparin, histamine, and other vasoactive substances that stimulate bone resorption. Dactylitis noted in sickle cell disease can be due to either infection and/or infarction, although the latter is usually the primary process. The most common organism to cause osteomyelitis in the disease is *S. aureus*.

In general, hematologic and other bone marrow disorders frequently have their most striking and/or earliest manifestations in the axial skeleton, because of the normal concentration of hematopoietic marrow at this location in adults (Box 7-5).

Box 7-5   Hematologic and Bone Marrow Disorders Affecting the Spine

SICKLE CELL ANEMIA

Vertebral osteoporosis, coarse trabeculae
H-shaped or biconcave endplate deformities
Vertebral collapse
Accentuated thoracic kyphosis, lumbar lordosis
Osteosclerosis due to infarction
Osteomyelitis: disk space narrowing, vertebral body destruction, paravertebral abscess
Paravertebral extramedullary hematopoiesis

THALASSEMIA

Vertebral osteoporosis
Biconcave, rarely H-shaped, endplate deformities
Compression fractures
Paravertebral extramedullary hematopoiesis
Enlarged hemosiderin-laden lymph nodes

HEREDITARY SPHEROCYTOSIS

Occasional paravertebral extramedullary hematopoiesis

SICKLE CELL–THALASSEMIA DISEASE

H-shaped or biconcave vertebral bodies
Incidence of spinal abnormalities: less than in SS disease, greater than in SC disease

Box 7-5—cont'd

PLASMA CELL MYELOMA

Over half of solitary lesions involve spine: multicystic and expansile with thick trabeculae or purely osteolytic with or without extension across disk to adjacent level
Multiple focal lytic (rarely sclerotic) lesions; relative sparing of posterior elements
Vertebral osteopenia, biconcave deformities, collapse resembling that of osteoporosis
Paraspinous, extradural soft-tissue masses
Diffuse osteosclerosis: fewer than 3% of cases, usually after treatment
Anterior scalloping of vertebral bodies

POEMS SYNDROME

Polyneuropathy, organomegaly, endocrinopathy, monoclonal gammopathy, skin changes
Benign sclerotic vertebral plasmacytomas
Irregular osseous excrescences involving posterior elements

WALDENSTRÖM MACROGLOBULINEMIA

Resembles multiple myeloma
Diffuse vertebral osteopenia
Osteolytic lesions: 10% to 15% of cases

AMYLOIDOSIS

Vertebral osteoporosis, collapse

LEUKEMIA (ACUTE)

Diffuse vertebral osteopenia
Rare spinal osteosclerosis
Osteolytic lesions in fewer than 3% of cases
Paravertebral soft-tissue masses

POLYCYTHEMIA VERA

Patchy radiolucent vertebral lesions

LYMPHOMA

Osteolytic lesions
Vertebral compression fractures
Localized or diffuse osteosclerosis, especially in Hodgkin disease (ivory vertebra)
Mixed lytic and blastic involvement

MULTICENTRIC RETICULOHISTIOCYTOSIS

Early, severe atlantoaxial involvement: subluxation, osseous erosion

MYELOFIBROSIS

Osteosclerosis (40% to 50% of cases), diffuse or with focal areas of radiolucency
Condensation of bone at superior, inferior endplates ("sandwich" vertebra)
Paravertebral extramedullary hematopoiesis

Continued

Box 7-5—cont'd

LANGERHANS CELL HISTIOCYTOSIS

Vertebra plana
Bubbly, lytic, expansile lesions of vertebral bodies,
    posterior elements
Paraspinous masses

FABRY DISEASE

Osteoporosis of thoracolumbar vertebrae

GAUCHER DISEASE, NIEMANN-PICK DISEASE

Vertebral osteopenia, loss of secondary trabeculae with
    accentuation of primary trabeculae
Compression fractures, Schmorl nodes, vertebra plana
Kyphosis, gibbus, occasional ankylosis across disk
    space
H-shaped endplate deformities

SJÖGREN SYNDROME

Atlantoaxial subluxation
Apophyseal joint involvement

SYSTEMIC MASTOCYTOSIS

Diffuse vertebral osteopenia
Circumscribed lytic lesions, less than 5 cm in size with
    sclerotic margins
Localized sclerotic lesions
Diffuse vertebral osteosclerosis

SUGGESTED READINGS

1. Murayama S et al: MR imaging of pediatric hematologic disorders, Acta Radiol 32:267, 1991.
2. Steiner RM et al: Magnetic resonance imaging of diffuse bone marrow disease, Radiol Clin North Am 31:383, 1993.

CHAPTER 8

Miscellaneous Conditions and Techniques

INFANTILE CORTICAL HYPEROSTOSIS

1. The condition is also known as Caffey disease.
2. Clinical manifestations include tender soft-tissue swelling, hyperirritability, and fever.
3. The disease usually appears before the age of 6 months and is occasionally evident at birth.
4. A self-limited clinical course is typical, but occasional cases are prolonged.
5. Radiographically, marked cortical hyperostosis is characteristic, and osseous bridging may be observed (Fig. 8-1).

369

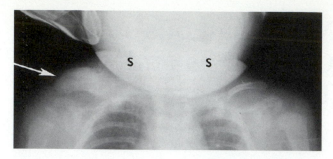

Fig. 8-1 Cortical thickening and expansion of the clavicle (*arrow*) in infantile cortical hyperostosis (Caffey disease). Differential diagnosis should include consideration of chronic recurrent multifocal osteomyelitis and small round cell neoplasms of childhood. Diffuse swelling of the lower jaw (*s*) is also evident.

6. Involvement of the mandible and clavicles is most characteristic of infantile cortical hyperostosis and generally distinguishes it from other causes of periosteal reaction in infants; the disease may also affect the ribs, scapulae, tubular bones, and/or calvarium (Fig. 8-2).

7. Different osseous sites may be affected sequentially in the disease.

8. Tubular bone involvement is predominantly diaphyseal in infantile cortical hyperostosis, and the metaphyses tend to be unaffected.

9. The disease must be distinguished from physiologic periostitis, which affects long tubular bones symmetrically in approximately 35% of newborns and resolves by the age of 6 months.

10. Because rickets and scurvy appear before the age of 6 months only in severely stressed premature infants, they are not likely to be confused with the disease.

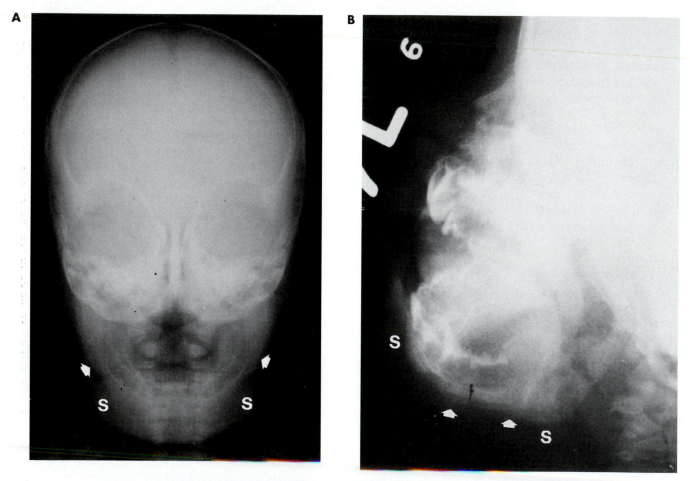

Fig. 8-2 **A** and **B,** Immature periosteal reaction involving the mandible (*arrows*) with associated soft-tissue swelling (*s*) in infantile cortical hyperostosis (Caffey disease). Differential diagnosis should include consideration of osteomyelitis, leukemia, and other small round cell neoplasms of childhood.

DIFFERENTIAL DIAGNOSIS OF PERIOSTEAL REACTION IN INFANTS

The differential diagnosis of periosteal new bone formation in the neonate and infant includes:
1. Infantile cortical hyperostosis
2. Physiologic periostitis of infancy
3. Subperiosteal hemorrhage in scurvy
4. Healing fractures in rickets
5. Congenital syphilis
6. Acute hematogenous osteomyelitis
7. Battered child syndrome
8. Acute leukemia
9. Metastatic neuroblastoma
10. Sickle cell dactylitis
11. Hypervitaminosis A
12. Prostaglandin therapy
13. Cyanotic congenital heart disease

Physiologic periostitis is seen in normal infants after 1 month of age. It is usually symmetric in distribution.

In congenital syphilis, the initial stage involves bony destructive changes in the metaphyses. The characteristic metaphyseal destructive lesion along the medial aspect of the proximal tibia is known as the Wimberger sign. This is followed by a second stage of diffuse diaphyseal periosteal reaction. The amount of periosteal reaction may be considerable and may simulate that of Caffey disease. However, the involvement of the metaphyses in congenital syphilis is not seen in Caffey disease. The third stage involves diaphyseal osteomyelitis with multiple radiolucent defects within the cortex. Any or all of these three stages may be observed in the same patient.

Caffey disease or infantile cortical hyperostosis almost always occurs before 6 months of age and is manifested clinically by fever, irritability, soft-tissue swelling, anemia, and pseudoparalysis. Radiographically there is exuberant periosteal reaction with a characteristic distribution. This disease particularly involves the mandible, clavicles, and ribs. The long bones, scapulae, and skull are also affected. The vertebral column and bones of the hands and feet are almost never involved. Recovery is usually complete within several weeks to several months.

Hypervitaminosis A almost always occurs after the age of 6 months to 1 year. There is usually no fever, although irritability may be present. The distribution of the periosteal reaction primarily includes the ulnae, metatarsals, clavicles, tibiae, and fibulae. The mandible is almost never involved, in contrast to the situation in Caffey disease.

In the battered child syndrome, the periosteal reaction relates to subperiosteal hemorrhage and healing of traumatic lesions, including characteristic metaphyseal corner fractures. Trauma, scurvy, and hemophilia are important causes of subperiosteal hemorrhage.

PEARL

Metaphyseal reactive alterations with microfractures accompany subperiosteal hematomas found in battered children.

Cortical hyperostosis occurs in hypervitaminosis A, Caffey disease, and tuberous sclerosis. Caffey disease is also known as infantile cortical hyperostosis, and causes diaphyseal cortical bone proliferation with relative sparing of the epiphyses and metaphyses. In contrast, osteopoikilosis involves hyperostosis of the spongiosa.

Syphilis demonstrates periosteal reaction in both the congenital and acquired forms. Sickle cell disease may also manifest secondary periosteal reaction after bone infarction, especially in the setting of dactylitis. Hypervitaminosis A in its chronic form may also cause periosteal reaction. In contrast, hypervitaminosis D usually causes metastatic soft-tissue calcification and cortical thickening. Scurvy is characterized by subperiosteal hemorrhage with exuberant periostitis. Eosinophilic granuloma can also cause considerable periosteal reaction.

PERIOSTITIS OF ACUTE CHILDHOOD LEUKEMIA

1. It occurs in up to 35% of cases.
2. It represents subperiosteal infiltration by malignant cells.
3. It may be clinically occult or cause arthritic complaints.
4. It is particularly prominent in the terminal phalanges.
5. Acute childhood leukemia is radiographically indistinguishable from metastatic neuroblastoma, because both diseases may manifest decreased bone density, focal osteolytic lesions, and periostitis.

PERIOSTITIS OF HYPERVITAMINOSIS A

1. It tends to manifest an undulating or wavy pattern.
2. The periostitis of hypervitaminosis A most commonly affects the ulna.
3. It may be associated with metaphyseal cupping and cone-shaped epiphyses, particularly in the distal femur.
4. The periostitis of hypervitaminosis A usually appears after the age of 6 months, in contrast to that of infantile cortical hyperostosis.

5. It may be associated with hyperirritability and elevated intracranial pressure.

SARCOIDOSIS (Box 8-1)

1. Bone involvement occurs in up to 15% of patients.

Box 8-1   Sarcoidosis: Important Features

- Erythema nodosum, cutaneous granulomas
- Osseous involvement rare without skin lesions
- Hand: predominant osseous site, honeycombed or lacelike septated lytic lesions, acro-osteosclerosis
- Arthritis in 10% to 35% of cases, with women affected more often than men
- Scintigraphy useful for staging bone involvement
- Localized or generalized osteosclerosis, affecting axial skeleton more than appendicular skeleton

2. The majority of patients with bone involvement also have skin lesions and intrathoracic disease, including hilar adenopathy, pulmonary infiltrates or fibrosis, and apical bullae.
3. The disease is more common in black individuals than among white individuals or those of Asian descent.
4. Extraskeletal manifestations of the disease include hepatosplenomegaly and ocular abnormalities.
5. The disease usually appears during young adulthood and manifests no sex predilection.
6. Generalized osteopenia and generalized osteosclerosis are recognized patterns of bone involvement by the disease.
7. Osteolytic lesions with lacy septations are characteristic and tend to involve the short tubular bones of the hands.
8. Acro-osteosclerosis of the terminal phalangeal tufts can occur in sarcoidosis.
9. Localized areas of osteosclerosis are part of the radiographic spectrum of the disease.
10. Polyarticular arthralgia occurs in the disease and is occasionally associated with nonspecific radiographic abnormalities.
11. Soft-tissue calcification in sarcoidosis is not idiopathic (calcinosis) but occurs secondary to hypercalcemia and hence is of the metastatic type.
12. Soft-tissue swelling commonly accompanies involvement of acral skeletal sites.
13. The differential diagnosis of osteolytic phalangeal lesions should include consideration of enchondroma, tuberous sclerosis, and brown tumor of hyperparathyroidism.

PEARL

In tuberous sclerosis, osseous abnormalities include areas of sclerosis (typically in the axial skeleton), lytic lesions (especially in the metacarpals and metatarsals), and (occasionally) diffuse sclerosis of a single rib. Growth disturbances do not typically occur. In contrast, enchondromatosis, osteoid osteoma, arteriovenous malformations, fractures in children, and juvenile chronic arthritis can cause limb shortening.

14. The differential diagnosis of focal osteosclerotic lesions in the axial skeleton should include consideration of metastatic disease, tuberous sclerosis, systemic mastocytosis, and myelofibrosis (Box 8-2).

Box 8-2   Tuberous Sclerosis: Important Features

- Localized multifocal areas of increased bone density, characteristically in posterior vertebral bodies and pedicles
- Two thirds of cases: cystlike changes in phalanges accompanied by irregular periosteal new bone formation in metacarpals and subungual fibromas (which may cause pressure erosion)
- Thickened, undulating cortical contours in tubular bones
- Medullary osteoblastic deposits tend to occur after puberty
- Intracranial calcifications
- Adenoma sebaceum in up to 15% of cases; especially during childhood, puberty, pregnancy
- Earliest cutaneous sign: hypopigmented macules (leukoderma)
- Shagreen patches, café-au-lait spots

ENVIRONMENTAL BONE DISEASE

Radiographic Features Associated with Specific Environmental Insult(s) (Box 8-3)

1. Scoliosis: radiation
2. Acro-osteolysis: thermal and electrical burns, frostbite, polyvinyl chloride exposure (Fig. 8-3)
3. Increased susceptibility to osteomyelitis: radiation, thermal and electrical burns, frostbite
4. Periarticular calcification and ossification: thermal and electrical burns, frostbite
5. Increased density, resorption, or premature fusion of secondary ossification centers: frostbite
6. Joint contractures: radiation, thermal and electrical burns

A

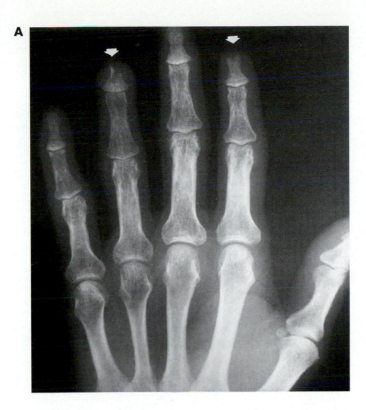

B

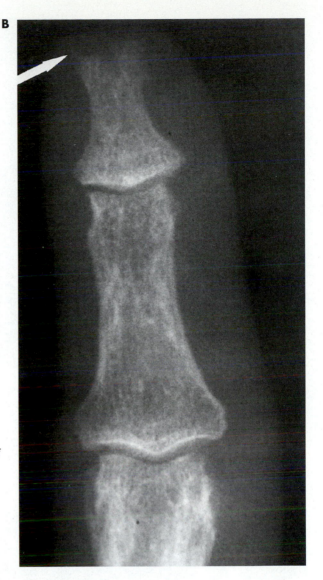

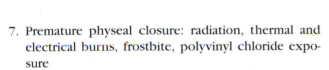

Fig. 8-3 **A** and **B,** Acro-osteolysis and osteopenia in frostbite. Soft-tissue and osseous resorption involves the second and fourth digits of the hand (*arrows*). Differential diagnosis should include consideration of thermal and electrical burns, leprosy, scleroderma, thromboangiitis obliterans (Buerger disease), and neuroarthropathy.

7. Premature physeal closure: radiation, thermal and electrical burns, frostbite, polyvinyl chloride exposure
8. Benign osteochondroma: radiation
9. Osteonecrosis: radiation, thermal and electrical burns, frostbite, caisson disease (Figs. 8-4, 8-5, 8-6; Box 8-4)
10. Leukemia: radiation
11. Iliac wing hypoplasia: radiation
12. Cutaneous interdigital webbing: thermal and electrical burns
13. Malignant neoplasm (osteosarcoma, chondrosarcoma, fibrosarcoma, malignant fibrous histiocytoma): radiation
14. Limb length discrepancy: radiation, thermal and electrical burns
15. Pathologic fracture: radiation, thermal and electrical burns
16. Sparing of thumbs: frostbite
17. Decreased bone density: radiation, thermal and electrical burns, frostbite (Fig. 8-3)

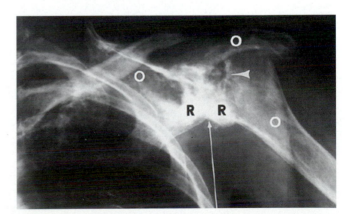

Fig. 8-4 Radiation osteonecrosis of the shoulder. Characteristic findings include patchy osteolysis (*o*) and reactive sclerosis (*R*) with cartilage loss (*arrow*) and fragmentation of subchondral bone (*arrowhead*). The major alternative diagnosis is neuroarthropathy.

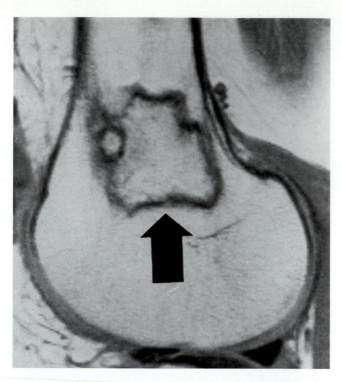

Fig. 8-5   Intramedullary bone infarct. Sagittal T1-weighted MR image demonstrates a well-defined lesion (*arrow*) with central fat and serpiginous low–signal intensity margins representing bone. Differential diagnosis should include consideration of corticosteroid therapy, sickle cell anemia, Gaucher disease, alcoholism, and caisson disease.

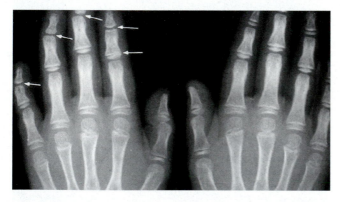

Fig. 8-6   Thermal injury in childhood. Osteonecrosis of the epiphyseal ossification centers of the phalanges has resulted in their dissolution and/or premature fusion (*arrows*).

18. Periostitis: radiation, thermal and electrical burns, frostbite
19. Soft-tissue swelling: radiation, thermal and electrical burns, frostbite
20. Calcification and ossification of aural pinnae: frostbite

21. Articular cartilage damage with degenerative joint disease: radiation, thermal and electrical burns, frostbite (Fig. 8-7)

PEARL

Immersion foot:
- Immersion foot is a condition found among individuals who have stood for prolonged periods in cold water.
- The combination of dependent position with associated edema, cold, moisture, and maintenance of a static posture for long periods of time is responsible.
- Tight footgear or leggings may also contribute to development of the condition.
- The initial stage of the process consists of prolonged arteriolar spasm with secondary edema.
- Later there is intense hyperemia and formation of small thrombi with resulting ischemia.
- Ultimately, dry gangrene, atrophy, contractures, fibrosis, and scarring occur.

Box 8-3   Radiation Effects on the Growing Skeleton: Important Features

- Two types of scoliosis (lateral flexion, rotary); usually affecting children irradiated for abdominal neoplasms (Wilms tumor, neuroblastoma); severity determined by age, dose, size-distribution of field
- Vertebral growth arrest lines (bone-in-bone appearance, 1000 to 2000 rads; gross endplate irregularities, 2000 to 3000 rads)
- Iliac hypoplasia, acetabular dysplasia, coxa valga, coxa vara, hip dislocation, slipped capital femoral epiphysis, leg shortening
- Maxillary, nasal, temporal bone growth arrest (retinoblastoma therapy)
- Postirradiation osteocartilaginous exostoses without malignancy
- Potential effects (1600 to 6500 rads) in ribs, ilia, clavicles, scapulae, posterior elements of spine, humeri, femora, tibiae
- Dose for induction of malignancy: 1200 rads over 2 weeks to 24,000 rads over 2 years
- Estimated risk of malignancy induction for megavoltage therapy: 1.5%; three to four times higher for retinoblastoma therapy alone
- Most radiation-induced malignancies are osteosarcomas, arising up to 20 years after treatment (latent period)

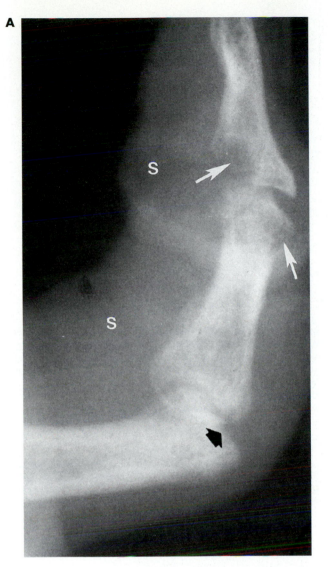

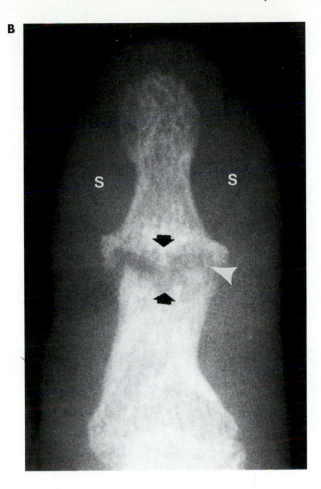

Fig. 8-7   Erosive arthropathy of the interphalangeal joints in frostbite. **A** and **B**, Radiographic findings include periarticular osteopenia, marginal (*white arrows*) and central (*black arrows*) osseous erosions, soft-tissue swelling (*S*), and evolving ankylosis (*arrowhead*). Differential diagnosis should include consideration of inflammatory osteoarthritis, seronegative spondyloarthropathies, and multicentric reticulohistiocytosis (lipoid dermatoarthritis).

Box 8-4   Dysbaric Osteonecrosis (Caisson Disease): Important Features

- Correlates poorly with decompression sickness
- Incidence higher with repeated exposures, exposures to higher pressures, rapid rates of exposure, obesity
- Usually 4 to 12 months elapse between time of exposure and positive radiographs
- Incidence in compressed air workers: 10% to 12%
- Incidence in divers: 15% to 65%

Differential Diagnosis of Acro-osteolysis (Fig. 8-8)

1. Frostbite
2. Thermal and electrical burns
3. Polyvinyl chloride exposure
4. Pyknodysostosis
5. Neuroarthropathy (Fig. 8-9)
6. Hyperparathyroidism
7. Hajdu-Cheney syndrome (familial) (Box 8-5)
8. Psoriatic arthritis
9. Mixed connective tissue disease

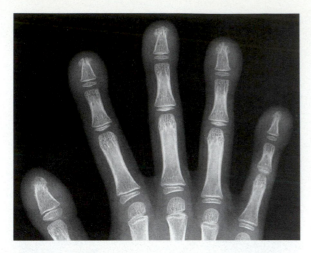

Fig. 8-8 Digital clubbing associated with acro-osteolysis. Differential diagnosis in a pediatric patient should include consideration of cyanotic congenital heart disease, cystic fibrosis, biliary atresia, and pachydermoperiostosis.

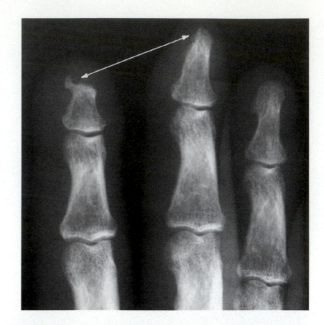

Fig. 8-9 Acro-osteolysis (*arrows*) in neuropathic disease of the hand. Differential diagnosis should include consideration of hyperparathyroidism, scleroderma, thermal injury, leprosy, psoriatic arthritis, polyvinyl chloride exposure, and familial forms of acro-osteolysis.

PEARL

Idiopathic osteolysis:
- Idiopathic forms of osteolysis include phalangeal acro-osteolysis, tarsocarpal osteolysis, and multicentric osteolysis (Gorham disease)
- Initially, localized rarefaction and irregular trabeculation of affected area
- Progressive loss of osseous structures with partial or complete disappearance of bones
- Phalangeal involvement most common
- Associated conditions: osteoporosis, wormian bones, alveolar resorption, absent frontal sinuses, hypermobile joints

10. Leprosy
11. Scleroderma
12. Lesch-Nyhan syndrome
13. Epidermolysis bullosa (Box 8-6)
14. Ainhum (Box 8-7)
15. Conditions associated with digital clubbing (Fig. 8-8)

PEARL

Ainhum:
- Ainhum is a disease process of unknown cause found among individuals who live in South Africa and occasionally in residents of the southern United States.
- It is almost always found only in the fifth and/or fourth toe.
- The condition rarely occurs in the hand.
- It is characterized by a constricting ring in the soft tissues, which may erode bone and lead to autoamputation.

PEARL

Acro-osteolysis occurs in scleroderma, mixed connective tissue disease, psoriasis, epidermolysis bullosa, frostbite, thermal and electrical injuries, neuropathic disease, leprosy, pyknodysostosis, and conditions associated with clubbing. The phenomenon may be bandlike in hyperparathyroidism, familial forms of acro-osteolysis such as Hajdu-Cheney syndrome, and exposure to polyvinyl chloride and snake or scorpion venom.

Box 8-5   Familial Acro-osteolysis: Important Features

- Painful, shortened, clubbed digits with cutaneous ulcerations, beginning in second decade
- Lytic destruction, resorption of terminal phalangeal tufts (hands, feet)
- Resorption of alveolar processes, loss of teeth (mandible, maxilla)
- Craniosynostosis
- Basilar invagination
- Spinal osteoporosis with fracture
- Kyphoscoliosis

Box 8-6   Epidermolysis Bullosa: Important Features

- Four types: simple (autosomal dominant), hyperplastic dystrophic (autosomal dominant), dystrophic polydysplastic (recessive), lethal (recessive)
- Poor dermal-epidermal adherence
- Metacarpophalangeal and interphalangeal flexion contractures, interdigital webbing, distal phalangeal tapering, osteoporosis, slender tubular bone diaphyses, maxillary and mandibular underdevelopment, mandibular prognathism, caries, periapical abscesses, tooth loss
- Mucous membrane involvement (eyes, nose, oropharynx, anus, genital tract, esophagus) in recessive types
- Skin: vesicles, bullae, ulcerations, scars, keloids, milia, pigmentation
- Webbing between fingers is caused by epithelial bridging across opposing cutaneous bullae
- Claw hand is caused by flexion deformities of fingers
- Soft-tissue atrophy evolves into shortening and tapering of distal phalanges, which become pointed and cone shaped in advanced cases

Box 8-7   Ainhum: Important Features

- Predominantly affects middle-aged black Africans and descendants in North and South America, Panama, West Indies; also reported in India
- Not caused by a parasite; pathogenesis involves traumatic rotational strain from prolonged barefooted ambulation
- Fifth toe most commonly involved
- Digits of hand involved rarely
- Frequently bilateral
- Evolution of process: up to 20 years (mean, 5 years)
- Bony alterations follow soft-tissue disease: focal osteoporosis, cortical resorption of proximal or middle phalanx, pathologic fracture
- Begins along medial aspect of involved digit
- Autoamputation can occur

Differential Diagnosis of Soft-tissue Calcification and Classification as Dystrophic, Metastatic, or Idiopathic

1. Frostbite: dystrophic
2. Thermal and electrical burns: dystrophic (Fig. 8-10)
3. Scleroderma: idiopathic
4. Sarcoidosis: metastatic

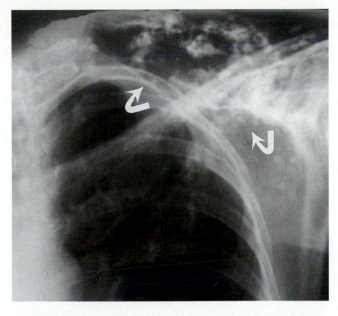

Fig. 8-10   Dystrophic soft-tissue calcification (*arrows*) secondary to thermal injury. Differential diagnosis should include consideration of dermatomyositis and various causes of metastatic soft-tissue calcification.

PEARL

Elevated serum phosphate level is more critical than elevated serum calcium level for the development of metastatic soft-tissue calcification. A calcium-phosphate solubility product of greater than 60 to 70 will cause tumoral or metastatic calcification.

5. Tumoral calcinosis: metastatic
6. Hyperparathyroidism: metastatic
7. Synovial sarcoma: dystrophic
8. Calcium hydroxyapatite crystal deposition disease: dystrophic
9. Mixed connective tissue disease: idiopathic
10. Gout: metastatic
11. Milk-alkali syndrome: metastatic
12. Cysticercosis: dystrophic
13. Pseudohypoparathyroidism: metastatic
14. Renal osteodystrophy: metastatic

PEARL

Causes of metastatic soft-tissue calcification include hyperparathyroidism, hypoparathyroidism, renal osteodystrophy, hypervitaminosis D, milk-alkali syndrome, sarcoidosis, and disseminated neoplasms (metastatic disease, leukemia, and multiple myeloma). Tumoral calcinosis is a form of metastatic soft-tissue calcification seen in renal osteodystrophy and an inborn error of phosphate metabolism.

15. Systemic lupus erythematosus: idiopathic
16. Ehlers-Danlos syndrome: dystrophic
17. Soft-tissue chondrosarcoma: dystrophic
18. Dermatomyositis: idiopathic (Fig. 8-11)
19. Hypoparathyroidism: metastatic
20. Tuberculosis: dystrophic
21. Hypervitaminosis D: metastatic
22. Synovial osteochondromatosis: dystrophic
23. CREST (calcinosis, Raynaud phenomenon, esophageal motility disorders, sclerodactyly, and telangiectasia) syndrome: idiopathic
24. Weber-Christian disease: dystrophic
25. Chondrodysplasia punctata: dystrophic
26. Dracunculiasis: dystrophic
27. Pseudoxanthoma elasticum: dystrophic (Box 8-8)
28. Multiple myeloma: metastatic
29. Idiopathic calcinosis universalis: idiopathic

PEARL

Idiopathic calcinosis universalis:
- Radiographically, its soft-tissue calcifications cannot be confidently distinguished from those of hyperparathyroidism and dermatomyositis.
- It is exclusively a disease of infants and children.
- Calcific deposits occur in tendons, ligaments, muscles, and subcutaneous fat.
- Involvement is more extensive in children than in infants.
- It is not a form of metastatic soft-tissue calcification.

---

30. Neonatal subcutaneous emphysema: dystrophic
31. Calcium gluconate extravasation: dystrophic
32. Subcutaneous pitressin tannate injection: dystrophic
33. Leprosy: dystrophic

Box 8-8   Pseudoxanthoma Elasticum: Important Features

- Ocular abnormalities (angioid retinal streaks, chorioretinitis)
- Calcifications in skin, tendons, ligaments, periarticular regions (especially metacarpophalangeal joints, hips, elbows), large peripheral arteries and veins
- Skin thickening, redundancy, hyperextensibility; yellowish papules
- Calcification of falx, tentorium, choroid plexus, petroclinoid ligaments
- Life-threatening complications: upper and lower extremity claudication, coronary insufficiency, abdominal angina, hypertension, hemorrhage
- Autosomal recessive inheritance; female predominance
- Primary defect in elastic fibers
- Most common manifestation is calcification in soft tissues (calcinosis cutis consists of linear calcification of dermis, or extensive soft-tissue and periarticular deposits) and blood vessels (medial or intimal calcification)

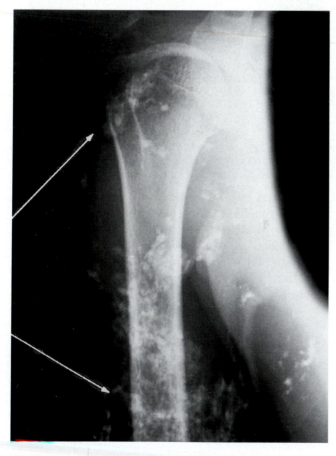

Fig. 8-11 Dermatomyositis with calcinosis universalis. Frontal radiograph of the shoulder demonstrates multiple soft-tissue calcifications (*arrows*) with a sheetlike distribution. Differential diagnosis should include consideration of other collagen-vascular diseases, as well as various causes of metastatic and dystrophic soft-tissue calcification.

Major Sites of Soft-tissue Calcification in Specific Disorders

1. Leprosy: nerves
2. Endocrine diseases, thermal or physical trauma, relapsing polychondritis: pinnae of ears
3. Granulomatous infections: lymph nodes
4. Hyperparathyroidism, renal osteodystrophy, milk-alkali syndrome, hypervitaminosis D, collagen-vascular disorders: periarticular sites

5. CPPD crystal deposition disease, hemochromatosis, hyperparathyroidism, acromegaly: articular cartilage
6. Calcium hydroxyapatite and CPPD crystal deposition diseases: tendons, bursae
7. Scleroderma, other collagen-vascular disorders: fingertips
8. Renal osteodystrophy, diabetes mellitus, progeria: arteries
9. Alkaptonuria, CPPD crystal deposition disease, hyperparathyroidism, immobilization, trauma: intervertebral disks

Differential Diagnosis of Periarticular Ossification

1. Frostbite
2. Thermal and electrical burns
3. Posttraumatic myositis ossificans
4. Paraplegia or quadriplegia (Fig. 8-12)
5. Soft-tissue osteosarcoma
6. Melorheostosis
7. Synovial osteochondromatosis
8. Fibrodysplasia ossificans progressiva
9. Carbon monoxide poisoning
10. Tetanus
11. Soft-tissue implants of giant cell tumor

PEARL

Potential causes of *both* soft-tissue calcification and ossification include:
- Venous insufficiency in the lower extremities
- Neurologic injury resulting in immobilization
- Thermal or physical trauma
- Neoplastic disease

Differential Diagnosis of Calcification or Ossification of the Aural Pinnae

1. Frostbite
2. Mechanical trauma
3. Addison disease
4. Hyperparathyroidism
5. Gout
6. Acromegaly
7. Calcium pyrophosphate dihydrate crystal deposition disease
8. Alkaptonuria
9. Dermatomyositis

A

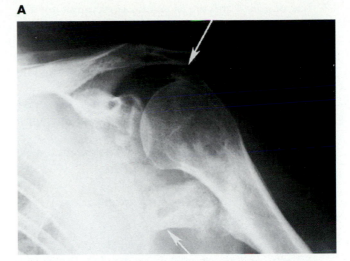

B

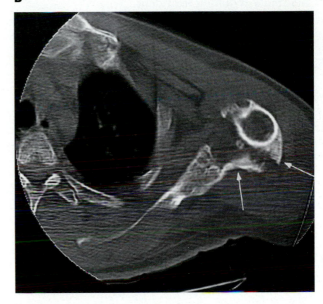

Fig. 8-12 Periarticular ossification in quadriplegia. **A,** Radiography demonstrates mature bone formation (*arrows*) in the soft tissues around the glenohumeral joint. **B,** CT documents nearly complete osseous bridging (*arrows*) across the posterior aspect of the joint. Differential diagnosis should include consideration of closed head injury, burns, posttraumatic myositis ossificans, and fibrodysplasia ossificans progressiva.

SCOLIOSIS

Radiographic Technique

1. A high-quality standing anteroposterior film should be obtained initially for evaluation of possible vertebral structural abnormalities.
2. Follow-up radiographs for assessment of progression should be obtained in the posteroanterior projection.

3. Use of gonad and breast shields is recommended in scoliosis radiography.
4. The x-ray beam should be tightly collimated to the smallest possible field size.
5. Use of an intensifying screen, preferably of the rare earth type, is mandatory.
6. With proper technique, the breast dose from posteroanterior radiography may be as low as one tenth of that for anteroposterior films.
7. The most radiosensitive tissue in adolescent female patients is the breast.

Idiopathic Scoliosis

1. Female individuals are preferentially affected by a 7:1 ratio and are more likely to develop progressive deformity requiring treatment.

PEARL

Scoliosis is more common in female individuals and tends to increase in severity during bone growth and after the accelerated development of spinal degenerative disk and joint disease. It may be the presenting sign in osteoid osteoma of the spine.

2. It is the most common form of scoliosis, accounting for approximately 85% of cases.
3. Thoracic, lumbar, thoracolumbar, or double curves may occur.
4. Surgical treatment is indicated for progressive curves or those exceeding 50 degrees.
5. Infantile idiopathic scoliosis exhibits thoracic predilection in male individuals, is usually left convex, and is refractory to treatment.
6. The condition is frequently associated with spondylolysis at the L5 level.

Congenital Scoliosis

1. Congenital scoliosis is usually secondary to hemivertebrae and/or intervertebral bars, either of which may induce unbalanced growth.
2. Congenital scoliosis may be associated with tethered spinal cord or diastematomyelia, either of which is best diagnosed by magnetic resonance imaging.
3. The recommended treatment for congenital scoliosis is early surgical spinal fusion, because external bracing is not effective.
4. Diseases of collagen synthesis that may be associated with scoliosis include Ehlers-Danlos syndrome, Marfan syndrome, and homocystinuria.

5. Neuromuscular scoliosis is characterized by a single long curve and may be caused by paralysis, spasticity, arthrogryposis multiplex congenita, or muscular dystrophy.
6. Neurofibromatosis is associated with a short angular curve most commonly located in the upper thoracic spine, caused by mesodermal dysplasia.
7. The scoliosis of neurofibromatosis may be associated with posterior scalloping of vertebral bodies or with rib deformities, and it can be complicated by rapid progression or paralysis.
8. The most common neoplasm responsible for scoliosis is osteoid osteoma, and the curve is typically long and concave toward the lesion; the lesion is usually located in the posterior elements and may be radiographically occult, requiring skeletal scintigraphy to suggest the diagnosis.
9. Radiation therapy is a potential iatrogenic cause of scoliosis in skeletally immature patients; although rare with modern therapy, it was most frequently observed in patients with Wilms tumor (nephroblastoma).
10. Acute trauma is an additional cause of acquired scoliosis that nearly always requires surgical spinal fusion for therapy owing to the severity of associated findings.
11. Therapeutic options for scoliosis include external bracing, electrical stimulation, and surgical spinal fusion.

PEARL

Congenital scoliosis may be convex right or left and occurs secondary to a congenital anomaly of the spine. Adolescent scoliosis usually involves multifactorial etiologies including inheritance. Adolescent scoliosis is the most frequently occurring type of idiopathic scoliosis in the United States. Girls are affected approximately seven times as frequently as boys. Idiopathic scoliosis is found in up to 15% of patients with congenital heart disease.

ORTHOPEDIC HARDWARE

Posterior Spinal Instrumentation

1. Reduces scoliosis by correcting rotary deformity: Cotrel-Dubousset instrumentation
2. Must be accompanied by sufficient bone graft for a fusion mass to form: Harrington instrumentation, Luque instrumentation, Cotrel-Dubousset instrumentation
3. Secured most tightly by sublaminar wiring: Luque instrumentation

4. Consists of superior and inferior shoes that differ in size and shape: Harrington instrumentation
5. Corrects associated rib hump deformity of scoliosis: Cotrel-Dubousset instrumentation
6. Failure tends to occur at junction of ratcheted and smooth portions of hardware: Harrington instrumentation
7. Secured most safely by Wisconsin segmental spinal instruments (WSSI) wiring through the thick portions of the spinous processes with buttons: Luque instrumentation
8. Preserves the thoracic kyphosis: Cotrel-Dubousset instrumentation
9. Hardware may be in the form of rods, U shapes, or rectangles: Luque instrumentation
10. Ratcheted portion of hardware is placed cephalad: Harrington instrumentation
11. Most complicated apparatus for correction of scoliosis: Cotrel-Dubousset instrumentation
12. May be complicated by detachment of shoes from laminae or pedicles: Harrington instrumentation
13. May be complicated by dural tear during removal: Luque instrumentation

Anterior Spinal Instrumentation

1. Screw breakage is a potential complication of cervical and lumbar fusion.
2. The Dwyer, Dunn, and Zielke devices are occasionally employed for the treatment of nontraumatic scoliosis.
3. Following cervical diskectomy, displacement of the bone block placed in the intervertebral disk space is a potential complication.
4. Aortic complications may result from left-sided hardware placed in the lumbar spine.
5. Screws employed for cervical or lumbar fusion may back out or cut out through osteoporotic bone.
6. Flexion-extension radiography is the procedure of choice for detecting postoperative instability related to failure of graft incorporation and/or pseudoarthrosis.

Total Hip Arthroplasty (Box 8-9)

1. The acetabular cup angle on frontal radiographs should measure 30 to 45 degrees from horizontal; if it exceeds 50 degrees, prosthesis dislocation is more likely to occur as a complication.
2. On lateral groin radiographs, the acetabular cup should be anteverted by 10 to 15 degrees; if less, it should be compensated by a corresponding degree of femoral component anteversion, or prosthesis dislocation is more likely to occur as a complication.
3. Lateralization of the acetabular cup is a potential complication among patients with osteoarthritis.

4. Protrusio migration of the acetabular cup is a potential complication in patients with rheumatoid arthritis, Paget disease, or osteomalacia.
5. The level of the femoral trochanters should be the same as in the contralateral hip to optimize function of the gluteus medius and iliopsoas muscles.
6. The position of the femoral component should be neutral or slightly valgus, with the prosthesis contacting the lateral cortex proximally and the medial cortex distally; varus positioning predisposes to loosening of the femoral component.
7. Cemented femoral components should be covered by polymethyl methacrylate in zones of maximal stress, including the tip, medial aspect of proximal stem, and lateral aspect of distal stem.
8. An unequivocal sign of loosening for cemented arthroplasties, which can be seen on a single radiograph, is cement fracture.
9. Radiolucency at the cement-bone interface that exceeds 2 mm in width is suggestive of but not diagnostic of component loosening, whereas progressive widening of the radiolucency on sequential radiographs is unequivocal plain-film evidence of this complication.
10. Reactive fibrous tissue (histiocytosis) may occasionally result in extensive scalloped bone resorption around a femoral or acetabular component (particle disease) that can be asymptomatic and unassociated with loosening.
11. Radiographic signs of loosening for porous ingrowth and other noncemented components include prosthesis motion or toggling within a normal radiolucent interface zone of up to 2 mm in width, and progressive widening of the radiolucent interface zone on sequential films.
12. Interval change in position of a prosthetic component, or subsidence, is an unequivocal radiographic sign of loosening; acetabular components tend to migrate superiorly and femoral components tend to migrate inferiorly, resulting in limb shortening as a complication.
13. During the first 5 years following arthroplasty, the femoral component is most likely to undergo loosening.
14. Beyond the first 5 years following arthroplasty, the acetabular component is most likely to undergo loosening.
15. Radiographic signs suggestive of infection in a total hip arthroplasty include periosteal reaction and heterotopic ossification, but the definitive diagnosis requires aspiration arthrography with culture of synovial fluid.
16. Skeletal scintigraphy during the first 9 months following cemented (or longer following noncemented) total hip arthroplasty should reveal progressively decreasing abnormal activity around prosthetic

components; loosening or infection causes progressively increasing abnormal activity over the same period.

17. The scintigraphic pattern of loosening consists of focal areas of increased activity at sites of greatest movement.

18. The scintigraphic pattern of infection consists of diffuse areas of increased activity around prosthetic components.

19. Arthrography is the procedure of choice for the definitive diagnosis of component loosening, which is characterized by contrast material intravasation at cement-bone, prosthesis-bone, or cement-prosthesis interfaces; infection may exhibit abnormal bursae or sinus tracts.

20. Trochanteric nonunion is a potential complication of greater trochanteric osteotomy for placement of a femoral component that may predispose to prosthesis dislocation; contrast arthrography reveals opacification of the radiolucent gap at the nonunion site in this situation.

21. During contrast arthrography of a total hip arthroplasty, opacification of the bursa overlying the greater trochanter is diagnostic of trochanteric bursitis.

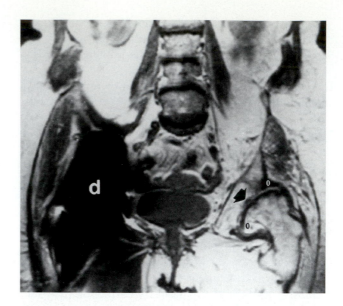

Fig. 8-13   Artifact induced by orthopedic hardware. On a coronal MR image, degradation (*d*) is noted in the immediate vicinity of a total hip prosthesis but does not preclude evaluation of the contralateral hip, which exhibits deformity of the femoral head, joint space narrowing (*arrow*), and osteophytosis (*o*) compatible with severe osteoarthritis.

Box 8-9   Total Hip Arthroplasty: Important Features

- Over 20 years, 15% to 20% documented incidence of femoral component loosening (cemented variety)
- Cup arthroplasty preferred in young patients with unilateral hip disease and following septic arthritis
- Dislocation most common in early postoperative period, associated with suboptimal cup anteversion, trochanteric nonunion, intraarticular bodies
- Good to excellent results with 85% to 90% of prostheses over 5 years
- Polymethyl methacrylate cement (nonadhesive compound) is heated during placement and subsequently contracts by about 5%, partially accounting for normal lucency at cement-bone interfaces
- Reactive histiocytosis can resemble infection radiographically and represents immune system reaction to cement or metal wear particles
- Although patients with prostheses have increased risk of hematogenous infection, most infected hardware results from iatrogenic contamination
- Acetabular cup angle should be 30 degrees to 45 degrees from horizontal with 10 to 15 degrees of anteversion
- Most radiographically apparent trochanteric complications are asymptomatic
- Heterotopic ossification occurs in 15% to 50% of cases; restricted motion, in 2% to 5%

---

22. Revision of total hip arthroplasty frequently differs from the initial surgery in the use of uncemented components, lateral cortical windows in the femoral shaft for cement extraction, bone graft, and long-stem femoral components.

23. Femoral shaft fracture is a common complication of arthroplasty revision that may require the use of circlage wiring of the femoral component.

24. Patients with subcapital fracture or ischemic necrosis of the femoral head may be treated with placement of an endoprosthesis that leaves the acetabulum intact; the two types are the Austin-Moore or Thompson prosthesis and the bipolar or Bateman component.

25. Total hip arthroplasty and other orthopedic hardware induces degratory artifacts on both CT and MRI examinations, as well as photopenic defects on radionuclide bone scans (Fig. 8-13).

Total Knee Arthroplasty (Box 8-10)

1. Asymmetry of the femorotibial joint space widths is suggestive of instability.

2. Loosening of a total knee arthroplasty is more difficult to detect both radiographically and arthrographically than that of a total hip arthroplasty.

3. Noncemented porous ingrowth tibial components may normally manifest loosened microspheres within a well-defined radiolucent interface zone of up to 2 mm in width.

4. Bone resorption along the bone-prosthesis interface at sites of stress shielding and buttressing at sites of stress may be normal radiographic features of noncemented femoral components.

5. Patellar component complications are most common and include superior migration of the patellar component and stress fracture.

Box 8-10 Total Knee Arthroplasty: Important Features

- Metallic hinge devices: inadequate freedom of movement, excessive stress at cement-bone interfaces, predisposition to loosening
- Metal component materials: stainless steel, cobalt-chromium-molybdenum alloy, cobalt-chromium-tungsten alloy, tantalum, titanium
- Wear is most pronounced on tibial side of minimal-constraint prostheses
- Postoperative pain in 15% of cases is referable to patellofemoral joint
- Insufficiency fractures are most common in the pelvis and foot
- Bone scans remain positive for 6 to 9 months following surgery
- Collateral and cruciate ligaments provide significant support to minimum constraint prostheses
- Incidence of late infection in fully constrained hinge prostheses is 10% to 15%

Miscellaneous Types of Orthopedic Hardware (Fig. 8-14)

1. Fracture is most likely to occur at the hinge of the device: Swanson arthroplasty (small joints)

2. It is used in the management of intertrochanteric femoral fractures, which normally settle with weight-bearing in the early postoperative period: Richard-type dynamic hip screw

3. They may delay fracture healing by causing devascularization and periosteal stripping: circlage wires

4. Complications include rotation and reactive synovitis with osseous erosion: carpal arthroplasty, Swanson arthroplasty

5. They are used for maintenance of length in a fractured limb with severe comminution or infection: external fixators

6. Interlocking screws are used to prevent rotation of the device: intramedullary rod

---

PEARL

Cancellous bone or compression screws are threaded only at their pointed ends, whereas cortical bone screws are threaded along their entire length.

---

7. They are used in the management of comminuted diaphyseal fractures: circlage wires, intramedullary rod

8. Flange dislocation is a potential complication, particularly in the setting of joint contracture or rheumatoid arthritis: carpal arthroplasty, Swanson arthroplasty

9. The screw head of the device should be positioned several millimeters from the articular surface of the femoral head, slightly posterior and inferior to its center: Richard-type dynamic hip screw

10. It is used for the replacement of diseased metacarpophalangeal, metatarsophalangeal, and interphalangeal joints: Swanson arthroplasty

11. The device is designed to telescope with progressive fracture impaction, thus avoiding cutout through osteoporotic bone: Richard-type dynamic hip screw

12. Shortening of the bone is a potential complication in the setting of a butterfly fragment: intramedullary rod

13. Pin tract infection is a potential complication: external fixators

14. Nonanatomic fracture reduction, including relative medial shaft displacement and valgus alignment, is stable and acceptable: Richard-type dynamic hip screw

15. They are used in the management of pelvic fractures with significant displacement: external fixators

16. It is used in the treatment of subcapital femoral fractures to prevent rotation at the fracture site: three- or four-point fixation

17. It is used to correct significant limb length discrepancy, usually secondary to previous trauma: Ilizarov device

---

BONE SCINTIGRAPHY

Radiopharmaceuticals

The radiopharmaceuticals used are technetium 99 (metastable)–labeled phosphate analogs (Box 8-11).

The current standard is methylene diphosphonate (MDP). This compound has more rapid renal clearance than others; thus, bone/background ratios are higher at the time of imaging.

A

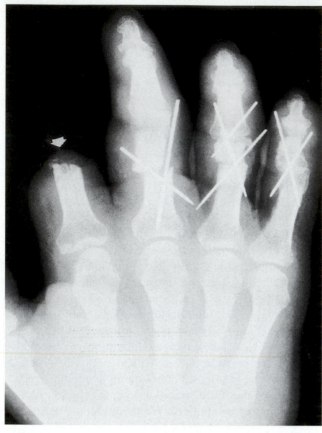

B

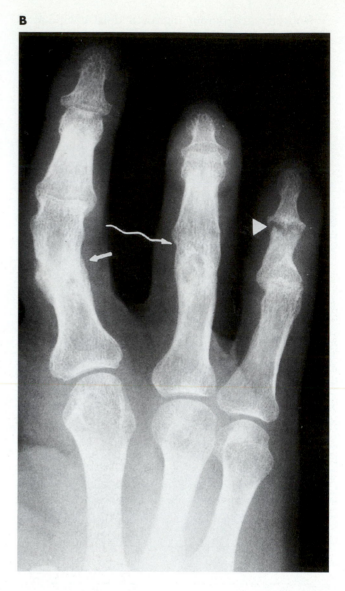

Fig. 8-14   Kirschner wire fixation for digital replantation following amputation. **A,** The third through fifth digits have been reattached, whereas the index finger could not be salvaged (*arrow*). Soft-tissue swelling is diffuse. **B,** Follow-up radiograph demonstrates fracture healing (*straight arrow*), interphalangeal joint ankylosis (*wavy arrow*), and neuroarthropathy affecting the fifth distal interphalangeal joint (*arrowhead*). This device is employed in a variety of situations that require orthopedic stabilization.

PEARL

Technetium-99m phosphate compounds:
- Localization in bone at 1 hour is higher for methylenediphosphonate (58%) than for pyrophosphate (47%) or hydroxyethylene-diphosphonate (48%)
- Renal excretion is greater for diphosphonates (68%) than for polyphosphate (46%) or pyrophosphate (50%)
- Polyphosphate and pyrophosphate are inorganic compounds
- Methylene diphosphonate and hydroxyethylene-diphosphonate are organic compounds
- Of total injected dose of pyrophosphates, 10% to 30% binds to erythrocytes
- All inhibit bone turnover (resorption and formation)

Box 8-11   Characteristics of Technetium-99m

- Physical half-life: 6 hours
- Principal emission energy: 140 keV
- Dose for skeletal scintigraphy: 15 to 20 mCi
- Parent isotope: ^{99}Mb
- Form as eluted from generator: $^{99m}TcO_4^-$
- Oxidation number of $^{99m}TcO_4^-$: +7
- Reducing agent for binding to phosphate: $Sn^{+2}Cl_2$
- Oxidation number in bound state: +4

Technique (Boxes 8-12, 8-13; Table 8-1)

No patient preparation is necessary. A dose of approximately 20 mCi of ^{99m}Tc-MDP is injected intravenously; the patient is asked to drink and void extensively, and imaging is performed at appropriate times, as follows:

1. Metastatic survey: record image of whole body at 2-3 hours after injection
2. Osteomyelitis study: radionuclide angiogram; blood pool image; images at 2-3, 5, and 24 hours
3. Ischemic necrosis, certain neoplasms: radionuclide angiogram; blood pool and 2 to 3 hour images

Bladder catheterization or sitting view may be required if area of interest is obscured by urinary residual.

Box 8-12 Protocols for Bone Single Photon Emission Computed Tomography (SPECT)

ACQUISITION

- 20 mCi ^{99m}Tc-methylene diphosphonate (MDP); image 2-3 hours later
- 64×64 matrix (400-mm field of view gamma camera)

GENERAL METHOD

- Low-energy all-purpose collimator
- 20 seconds per projection, 64 projections over 360 degrees

HIGH-RESOLUTION METHOD

- Option for lumbar spine
- High-resolution collimator
- 25 seconds per projection, 64 projections over 360 degrees

PROCESSING

- Uniformity correction
- Hanning filter (frequency cutoff = 0.8 cycle/cm preprocessing)
- Reconstruction by filtered backprojection with Ramp filter
- No attenuation correction
- Transaxial, sagittal, and coronal images of 6 mm (1 pixel) thickness

DISPLAY

- Use linear gray-scale map for temporomandibular joint, lumbar spine, and knee SPECT
- Use log gray scale when searching for femoral head ischemic necrosis

---

Box 8-13 Gamma Camera Quality Control for Bone SPECT

DAILY

- Extrinsic flood for uniformity check:
  - 3.0 million counts for 400-mm field of view camera
  - 4.5 million counts for 500-mm field of view camera

WEEKLY

- Update energy correction per manufacturer recommendation
- Intrinsic flood for uniformity check:
  - 3.0 million counts for 400-mm field of view camera
  - 4.5 million counts for 500-mm field of view camera
- Update tomographic center of rotation
- Update high-count extrinsic flood for uniformity correction:
  - 30 million counts for 64 × 64 matrix
  - 120 million counts for 128 × 128 matrix

MONTHLY

- Image bar phantom for check of planar resolution
- Image tomographic phantom (optional)

Physiology

The amount of radiopharmaceutical deposited at a given site depends on three general mechanisms:

1. Adherence to chemical surfaces (chemisorption), reflecting osteoblastic activity
2. Vascular delivery to site
3. Vascular permeability at site (Box 8-14)

Box 8-14 Factors Responsible for Radiopharmaceutical Uptake by the Skeleton

- Blood supply
- Vascular permeability
- Rate of bone turnover

From a pathophysiologic point of view, an area demonstrating increased uptake may be thought of as having increased bone turnover, whereas an area demonstrating decreased uptake has decreased bone turnover. Clearance

Table 8-1   Special patient positioning for bone SPECT

Bony structure	Special positioning	Pitfalls
Knees	Place 2- to 3-in pad between knees; secure knees with straps to prevent motion; secure feet in neutral position to prevent rotation	With obese patients both knees may not fit in field of view
Hips and pelvis	Empty bladder before exam; position hips symmetrically and secure knees and/or feet to prevent motion	Bladder filling during exam creates artifacts
Lumbar spine	Keep arms out of field of view; a pillow under the knees may relieve back pain	Patients with back pain often move during exam
Temporomandibular joint	Secure neck in comfortable hyperextension; instruct patient not to talk	Chin may not be in field of view; check lateral view to be certain it is

of ^{99m}Tc-MDP from the blood is quite rapid, and at 2 hours the distribution is roughly as follows:

Blood	10%
Bone	50%
Urine	40%

Biologic half-life is approximately 40 days.

Dosimetry

Whole body	0.01 R/mCi
Bladder wall	0.1-0.2 R/mCi
Skeleton	0.05 R/mCi
Kidneys	0.03 R/mCi
Gonads	0.03-0.05 R/mCi
Bone marrow	0.04 R/mCi
Liver	0.01 R/mCi

Bladder dose can be greatly reduced by frequent emptying of the bladder.

Indications (Box 8-15)

1. Detection of skeletal metastases
2. Early detection of osteomyelitis (Table 8-2)

PEARL

Osseous metastatic disease:
- At least 50% of cancellous bone must be destroyed before radiographic changes become evident
- False negative bone scans are common in patients with neuroblastoma
- Less than 5% of lesions are scintigraphically negative despite positive radiographic findings
- Multiple myeloma, eosinophilic granuloma, and anaplastic tumors have a high incidence of false negative scintigrams

Continued

PEARL—cont'd

- Distribution of lesions: skull (5%), long bones (15%), spine, ribs, and pelvis (80%)
- Metastatic Ewing sarcoma: approximately 15% of patients are scintigram positive at presentation, one third at follow-up
- Of all lesions, 10% to 40% are positive scintigraphically despite negative radiographs
- In stage 3 disease, scintigraphy is positive in approximately 30% of breast and 20% of prostate cancer cases

PEARL

Scintigraphy in osteomyelitis:
- Bone scans remain positive at 6 weeks in 50% of successfully treated acute infections
- Cold lesions may occur in early disease
- Combined three-phase bone and gallium or indium-white cell imaging best distinguishes osteomyelitis from cellulitis
- Combined bone and gallium or indium-white cell imaging is 75% sensitive and specific for infection following total joint replacement
- Acute infection: scintigraphy is positive within 24 hours, whereas radiography requires 10 to 14 days
- Degree of uptake is important in distinguishing between infection and sterile loosening of joint prostheses; a normal scan excludes both

3. Diagnosis of stress and occult fractures
4. Staging extent of Paget disease
5. Early diagnosis of osteonecrosis
6. Detection of injuries in suspected child abuse
7. Evaluation of fracture healing versus nonunion
8. Detection of complications in total joint arthroplasty
9. Detection of nidus of osteoid osteoma

Table 8-2 Radiopharmaceuticals for scintigraphic evaluation of musculoskeletal emergencies

Agent used	Reason	Adult dose	Pediatric dose
^{99m}Tc-MDP	Stress fracture, occult fracture	20 mCi	2 mCi minimum; Minimum dose $+ \dfrac{(\text{Adult dose} \times \text{ kg})}{70}$
^{67}Ga citrate	Infection—acute or chronic	3-5 mCi	500 μCi minimum; Minimum dose $+ \dfrac{(\text{Adult dose} \times \text{ kg})}{70}$
^{111}In white blood cells	Infection—acute	500 μCi	100 μCi minimum; 10 μCi per kg

Box 8-15 Traumatic and Postoperative Disorders in Which Scintigraphy Can Provide Useful Information

- Fractures:
  - Uncomplicated (including evaluation of physically abused children)
  - Stress
  - Occult
  - Delayed union
  - Nonunion
  - Pseudoarthrosis
  - Infection complicating fracture
  - Infection complicating joint prosthesis
- Shin-splints
- Painful tendinous and ligamentous insertions
- Traumatic arthritis
- Bone graft viability
- Trauma to skeletal muscles
- Reflex sympathetic dystrophy syndrome

Interpretation

Almost any lesion can produce increased uptake on a bone scan. Considerations should include the following:

1. Primary and metastatic tumor
2. Osteomyelitis
3. Arthritis
4. Fracture
5. Metabolic bone disease
6. Paget disease
7. Transient osteoporosis

The bone scan is highly sensitive but has poor specificity. Findings must be correlated with radiographs and clinical history.

PEARL

Skeletal scintigraphy:
- Eighty percent of bone scans are positive at 24 hours following fracture
- Ninety-five percent of bone scans are positive at 3 days following fracture
- Baseline imaging of joint prosthesis: 6 to 9 months following surgery
- With increasing age, bone scans become positive later after acute fracture and remain abnormal for longer periods
- Increased uptake can occur in bone islands over 3 cm in size
- Disuse osteoporosis without long-term paralysis and reflex sympathetic dystrophy syndrome exhibit diffusely increased scintigraphic activity
- Hyperparathyroidism and disseminated metastatic disease can result in a "superscan"
- Useful in treatment planning for osteoarthritic knee (high tibial valgus osteotomy versus total joint or unicompartmental replacement)
- Following therapeutic irradiation of bone, scintigrams initially show increased activity and are subsequently cold
- Scintigraphy is more sensitive than radiography in determining activity of rheumatoid and seronegative arthritis

Warnings

1. Examine kidneys:
   a. Presence
   b. Location
   c. Outline
   d. Obstruction

2. Watch for cold lesions:

    a. Necrotic metastasis

    b. Early osteonecrosis

    c. Early osteomyelitis

    d. Fracture nonunion

    e. Solitary bone cyst

    f. Total joint arthroplasty

3. Observe soft-tissue activity:

    a. Conditions associated with soft-tissue calcification or ossification

    b. Normal or diseased breast tissue

    c. Neuroblastoma and other neoplasms

    d. Muscular disease or inflammation

PEARL

Soft-tissue uptake of bone-scanning agents:
- Myocardial infarcts that are 1 to 7 days old
- Calcifying conditions: neuroblastoma, metastatic colon carcinoma, myositis ossificans, dermatomyositis, hyperparathyroidism
- Mitral stenosis: valvular or pulmonary calcification
- Heterotopic ossification of paralysis (20% to 30% of spinal cord injuries); scans can be positive before bone is evident radiographically
- Bone scans are less accurate in predicting maturation of heterotopic ossification (requires 12 to 18 months) than bone marrow scans or MRI
- Breast tissue: normal, lactating, or diseased
- Noncalcifying soft-tissue tumors, cerebral infarcts, soft-tissue infections, inflammation in cardiac or skeletal muscle, amyloidosis, intramuscular injection sites, healing wounds

MUSCULOSKELETAL MRI

Techniques

Magnetic resonance imaging is the most sensitive noninvasive imaging method for a broad spectrum of diseases affecting the musculoskeletal system. The advantage of MRI over previous imaging techniques resides primarily in increased contrast resolution of body tissues and pathologic processes. Thus, high resolution images can be acquired with excellent separation of body organs and disease. In the musculoskeletal system, conventional radiographs continue to be useful in detecting bony fractures, although most subtle bony injuries are not detectable with plain radiographs. Radiography may also be useful in detecting gross destruction of bone, but the increased sensitivity and accuracy of MRI

in delineating lesion extent is usually necessary for clinical characterization. Plain radiographs may be more specific than MRI in differentiating potential etiologies of lesions because of their proven ability to characterize specific calcification patterns and periosteal reaction. Although computed tomography superbly characterizes bony cortical changes, this is usually less important clinically than detecting the presence and extent of soft-tissue disease. The radionuclide bone scan is useful in localizing the extent of multifocal bone disease, but is less sensitive than MRI in detecting metastases and does not have sufficient spatial resolution to detail the extent and anatomical association of disease processes that is often necessary for optimal clinical decision making. MRI is now the primary imaging method for detailed evaluation of a broad spectrum of musculoskeletal disease processes.

PEARL

MRI of musculoskeletal disorders:
- The intense signal of normal bone marrow is caused by the short T1 relaxation time of fat.
- Cortical bone is dark on MR images because of a paucity of mobile protons.
- Osseous lesions with long T1 values have high water content.
- Inversion-recovery sequences enhance T1 tissue characteristics.
- Superconducting magnets are preferred over resistive units, since higher field strengths and thinner sections are possible with the former.
- Increasing field strength is more effective than increasing scan time in improving signal-to-noise ratio.
- Short TE yields best signal-to-noise ratio and spatial resolution.
- Long TE maximizes differences among signal intensities of various tissues.
- Spin-echo technique with shorter TR produces images that are more heavily T1 weighted because only partial T1 relaxation occurs between pulses.

In the evaluation of musculoskeletal masses, the MR scanner must be able to evaluate the entire compartment of the mass with pulse sequences that provide adequate signal-to-noise and tissue contrast to detect, characterize, and stage the disease process. This requires a combination of T1W, PDW, and T2W images with adequate signal-to-noise and fields-of-view just large enough to visualize the diseased compartment and surrounding anatomy. The use of intravenous gadolinium chelates and gradient echo images may be of additional benefit in selected cases.

In the evaluation of joints, it is necessary to use properly designed surface or local volume coils to obtain optimal contrast and spatial resolution. All joints should be evaluated in at least two planes. Patients must be positioned comfortably to minimize patient motion during the scan. For joints that cannot be placed in the center of the magnet, the scanner must be able to electronically off-center the acquisition field-of-view to allow high resolution image acquisition tailored to the size of the joint of interest. Slice thicknesses less than or equal to 5 mm and pixel dimensions less than 1 mm are necessary for adequate spatial resolution. A combination of T1W, PDW, and T2W images and adequate signal-to-noise are necessary for accurate MR evaluation of a broad spectrum of joint diseases. T2-weighted images obtained with gradient recalled echoes may be useful in evaluating anatomy and selected disease processes. Intravenous gadolinium chelates may be useful in evaluating specific musculoskeletal pathology. Before and after contrast administration, T1-weighted images should be performed in one or more planes. The use of intraarticular gadolinium shows promise in the evaluation of the knee, shoulder, and other joints. Its potential clinical roles are currently under evaluation.

Indications and Specific Techniques for MRI of the Musculoskeletal System

Soft tissues

Spectroscopic fat saturation or short inversion time inversion recovery (STIR) fat suppressed images may be helpful in evaluating soft-tissue masses.

MRI is indicated for the evaluation of traumatic muscle and tendon injuries, hematomas, compartment syndromes, entrapment syndromes (Fig. 8-15), tendi-nosis, tenosynovitis, and bursitis. MRI is useful for the evaluation of infections and abscesses, as well as myositis. It is indicated for the evaluation of masses such as simple nonneoplastic cysts, ganglion cysts, parameniscal cysts (Fig. 8-16), hematomas, muscle tears, and ligament or tendon tears. It is also useful for the detection, staging, and characterization of benign and malignant soft-tissue neoplasms, and for the follow-up evaluation of neoplastic disease after therapy. Intravenous contrast agents may be helpful in evaluating potentially malignant disease. For the evaluation of genetic muscle diseases, MR spectroscopy may also be helpful.

Bones

Whereas most bone pathology is well visualized with short T1W, PDW, and T2W images, subtle injuries, infiltrative marrow disease (Figs. 8-17, 8-18), and partially treated infections may be more sensitively detected with spectroscopic fat suppressed (FAT SAT) (Fig. 8-19) or short inversion time inversion recovery (STIR) images.

In the setting of trauma, MRI is indicated for the evaluation of suspected radiographically occult injuries of the metaphysis and epiphysis (Figs. 8-19, 8-20, 8-21), and to assess fracture union (Fig. 8-22). MRI is useful to detect and stage acute and chronic osteomyelitis, and to evaluate periprosthetic infections in selected cases. MRI is indicated to detect and stage primary bone lesions both nonneoplastic and neoplastic (Figs. 8-23, 8-24), and to detect and stage occult bony metastases. MRI is also useful to detect and determine the extent of congenital lesions such as fibrous dysplasia and C1-C2 anomalies. MRI is indicated for the evaluation of suspected early osteonecrosis and bone infarcts (Figs. 8-25, 8-26).

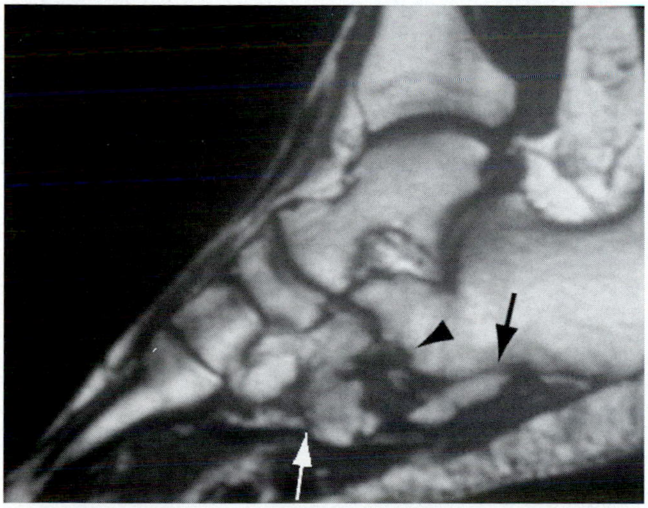

Fig. 8-15 Dysplasia epiphysealis hemimelica. Sagittal T1-weighted MR image demonstrates osteochondromatous proliferation arising from the lateral aspects of the cuboid and calcaneus (*arrows*), with impingement on the peroneal tendons (*arrowhead*). In general, MRI is extremely valuable in documenting tendon pathology such as entrapment, tenosynovitis, and partial or complete tears.

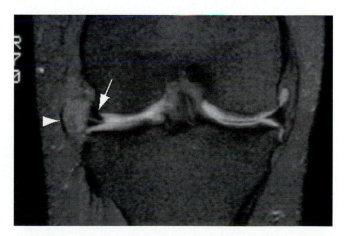

Fig. 8-16 Recurrent meniscal tear following partial arthroscopic meniscectomy. Coronal fat-suppressed T1-weighted MR image following intraarticular injection of gadolinium-DTPA documents a horizontal tear of the meniscal remnant (*arrow*) and an associated meniscal cyst (*arrowhead*). Because the accuracy of routine knee MRI is lower in the setting of previous surgery, MR arthrography may be indicated in selected patients with equivocal findings.

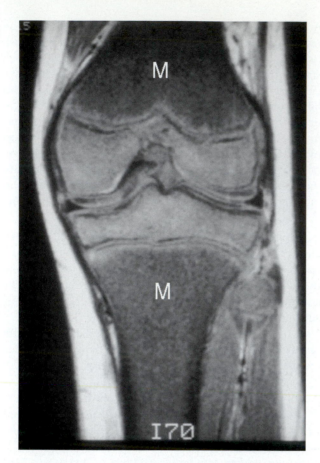

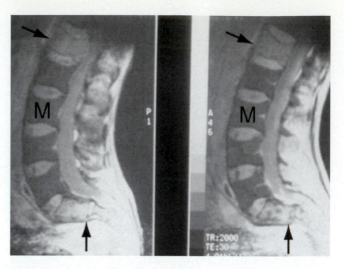

Fig. 8-18   Gaucher disease affecting the spine. Sequential sagittal proton density-weighted MR images demonstrate abnormally low marrow signal intensity (*M*) in the intact vertebral bodies; high signal intensity with pathologic fracture is present in L1 and the sacrum (*arrows*), secondary to localized accumulations of Gaucher cells simulating neoplasm. In general, MRI is the procedure of choice for staging diffuse bone marrow disease, and exhibits high sensitivity in the detection of its potential complications.

Fig. 8-17   Sickle cell anemia. Coronal T1-weighted MR image documents abnormally low signal (*M*) affecting the distal femoral and proximal tibial metaphyses (with relative epiphyseal sparing), secondary to compensatory hematopoietic marrow overgrowth and hemosiderin-laden reticuloendothelial cells. In general, MRI is the optimal imaging method for characterization of both diffuse and localized bone marrow disease.

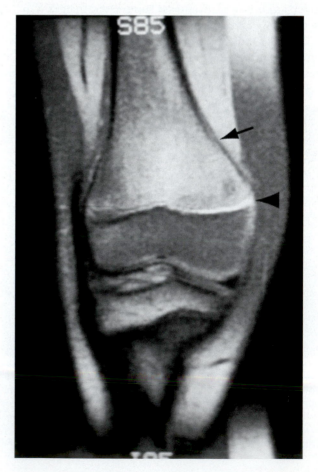

Fig. 8-19   Radiographically occult Salter-Harris type two fracture of the distal femur. Coronal fat-suppressed T2-weighted MR image reveals high signal intensity within the metaphysis (*arrow*) and medial aspect of the physis (*arrowhead*). Although MRI is extremely sensitive for the early diagnosis of such subtle injuries, a more cost-effective approach is presumptive treatment and follow-up radiography.

PEARL

- Tumor invasion of bone is manifested as diminished signal intensity from marrow-bearing areas as seen on short TR sequences.
- MRI demonstrates decreased T1-weighted signal intensity in ischemic necrosis of the femoral head, and it may be more sensitive than radionuclide scanning early in the disease.
- Contrast between normal and abnormal marrow is most pronounced on inversion-recovery sequences, which optimally demonstrate increased water content of inflamed marrow in osteomyelitis.
- Abnormalities in osteomyelitis can be seen with MRI when CT studies, conventional radiographs, and radionuclide studies are normal or equivocal.
- Soft-tissue swelling contiguous to infected bone is well visualized by MRI.
- MRI is useful in distinguishing active from inactive chronic osteomyelitis.
- Recovery of signal from bone marrow fat may occur following chemotherapy for osteogenic sarcoma and other neoplasms.
- Idiopathic aplastic anemia exhibits a generalized increase in signal intensity from bone marrow.
- Solid tumors and infiltrative disorders such as leukemia and neuroblastoma cause diminished signal intensity in bone marrow.

Joints

Diseases affecting all joints

MRI is indicated for the evaluation of traumatic injuries to joints and adjacent muscles, tendons, and ligaments; selected articular cartilage injuries (Fig. 8-27); bursitis and synovitis from overuse; fragment stability and cartilage status in osteochondritis dissecans (Fig. 8-28), posttraumatic osteonecrosis and degenerative joint disease; loose bodies; and tenosynovitis. MRI is useful in the evaluation of joint infections, noninfectious inflammatory joint disease such as rheumatoid and the seronegative arthritides, overuse synovitis, and tenosynovitis. It is useful for the evaluation of ganglion cysts, bursal cysts with bursitis, abscesses, benign neoplastic masses, and primary and metastatic masses. MRI is indicated for the evaluation of osteonecrosis including avascular necrosis, and degenerative joint disease.

Diseases affecting specific joints

SHOULDER   Imaging planes should be selected to coincide with the primary axes of the shoulder. Long TR, double echo coronal images should be obliqued electronically parallel to the supraspinatus muscle or perpendicular to the glenoid fossa of the scapula. T1 or T2W sagittal images should be obtained perpendicular to the oblique coronal images. The field of view (FOV) should be 12 to 16 cm. The slice thickness should be

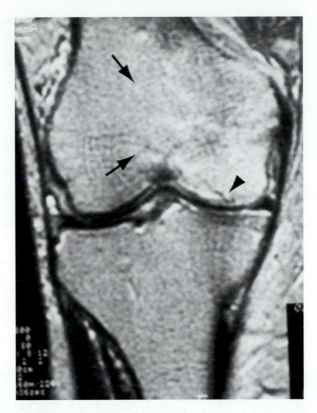

Fig. 8-20   Radiographically occult fracture of the medial femoral condyle following trauma. Coronal T2-weighted MR image reveals diffuse bone marrow edema (*arrows*) extending to the subchondral surface; the presence of a linear component (*arrowhead*) distinguishes the abnormality from a simple bone contusion. Both types of osseous abnormality are frequent incidental findings on knee MR examinations performed for clinically suspected internal derangement.

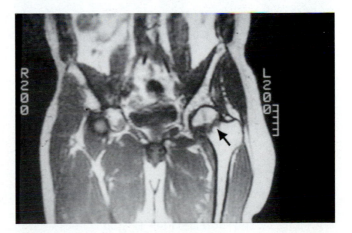

Fig. 8-21   Subcapital fatigue fracture of the proximal femur. Coronal T1-weighted MR image documents a linear zone of low signal intensity (*arrow*) traversing the head-neck junction. This portion of the skeleton is an area where MRI is perhaps preferred over radiography for detection of fatigue, insufficiency, and occult traumatic fractures, owing to its higher sensitivity and the potential morbidity associated with delayed or missed diagnosis.

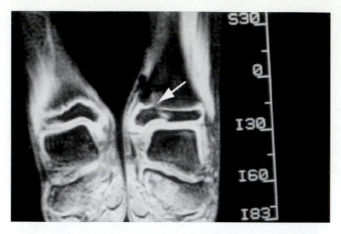

Fig. 8-22   Healing triplane fracture (Salter-Harris type four) of the ankle in a 7-year-old child. Coronal gradient-echo MR image documents an osseous bridge (*arrow*) across the distal tibial physis, with significant potential for future aberrant bone growth. In general, gradient-echo sequences (particularly SPGR, as in this case) are optimal for demonstrating abnormalities affecting physeal and hyaline cartilage.

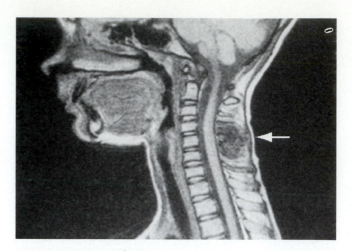

Fig. 8-24   Osteoblastoma. Sagittal T1-weighted MR image from a child demonstrates a mass of low signal intensity (*arrow*) arising from the posterior elements of the cervical spine. Although MRI is the best technique for staging of most primary bone and soft-tissue neoplasms, CT may be preferred in the setting of heavily calcified or ossified lesions such as parosteal osteosarcoma, osteoblastoma, and osteoid osteoma.

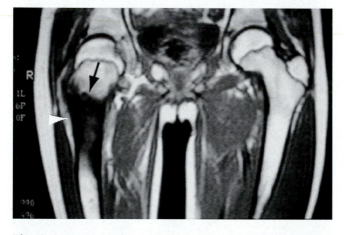

Fig. 8-23   Osteoid osteoma at capsular margin of hip joint. T1-weighted MR image reveals the nidus (*arrow*) and extracapsular reactive osseous sclerosis (*arrowhead*); the unusual location of the lesion also induced reactive synovitis with effusion in the joint, typical of intracapsular osteoid osteoma. Although T2-weighted or inversion recovery MR sequences optimally depict the marrow and soft-tissue inflammation associated with these lesions, CT is probably a preferred method for nidus localization, and can be used to guide percutaneous resection.

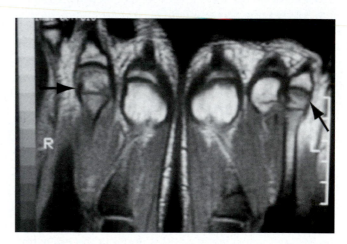

Fig. 8-25   Freiberg infraction. Coronal T1-weighted MR image reveals low signal intensity and subtle deformity affecting the right second and left third metatarsal heads (*arrows*) in an adolescent girl with a history of prolonged ambulation in high-heeled shoes. In general, MRI is the procedure of choice for the early diagnosis of clinically-suspected osteonecrosis.

less than 5 mm and spatial resolution less than 1 mm. Images in the oblique coronal plane should be acquired with long TR and short and long TE spin echoes such as with a TR of 2000 ms and TEs of 20 and 80 ms. Images in the oblique sagittal plane require contrast adequate for anatomical evaluation, such as TR 500, TE 15 ms or TR 2000 and TEs of 20 and 80 ms. The axial images require optimized anatomical information as well as sequences sensitive for fluid. Gradient-echo or long TR double spin echo images are adequate in this plane.

Acquisition matrices of 192 × 256 are desirable, with the number of excitations/averages selected to ensure adequate signal-to-noise.

MRI is indicated for the evaluation of shoulder pain of undetermined etiology, detection and staging of rotator cuff degeneration (tendinosis) and tears, impingement syndromes, labral degenerative changes and tears, bicipital tendon disease and dislocation, suprascapular notch syndrome, glenohumeral ligament injuries, coracoclavicular and acromioclavicular separations,

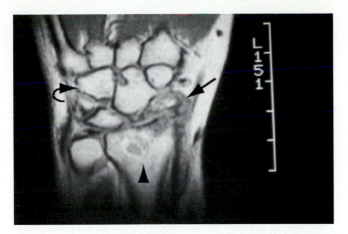

Fig. 8-26 Carpal osteonecrosis in systemic lupus erythematosus treated with corticosteroids. Although collapse of the scaphoid (*straight arrow*) and fracture of the hamate (*curved arrow*) were evident radiographically, coronal proton density-weighted MR also documents occult infarction of the distal radius (*arrowhead*). Other applications of MRI in the wrist include diagnosis of scaphoid ischemia following wrist fracture, Kienböck disease, triangular fibrocartilage and intercarpal ligament tears, carpal tunnel syndrome, and tenosynovial pathology.

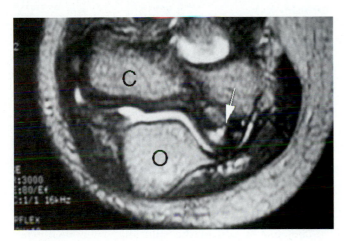

Fig. 8-27 Radiographically occult subacute chondral fracture of the trochlea in a child. Transaxial T2-weighted MR image of the elbow reveals abnormal high signal intensity (*arrow*) within the unossified portion of the ulnar aspect of the distal humerus (*C*=capitellum, *O*=olecranon). In general, MRI is the preferred technique for depicting abnormalities in the incompletely mineralized epiphyses and apophyses of skeletally immature patients.

subacromial bursitis, and articular anatomy after bony fracture. Suspected instability and labral tears may require intraarticular contrast administration for optimal detection (either MRI after gadolinium chelate or saline injection, or CT arthrography).

ELBOW Local surface coils and FOV of less than 12 cm are recommended. MRI is indicated for the evaluation of medial epicondylitis (tennis elbow), fractures in children, osteochondral defects (Fig. 8-27), and osteonecrosis, as well as other soft-tissue injuries.

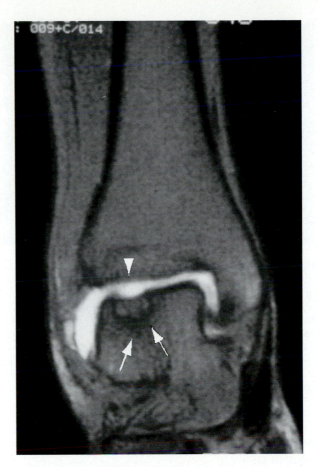

Fig. 8-28 Osteochondritis dissecans of the talar dome. Coronal fat-suppressed T1-weighted MR image following intraarticular injection of gadolinium-DTPA demonstrates a stable residual fragment (*arrows*) and a defect in the osteochondral surface (*arrowhead*) secondary to previous fragmentation with body formation. Although air CT arthrography is the preferred method for documenting intraarticular osteochondral bodies, MR-arthrography is superior for accurate staging of osteochondritis dissecans.

WRIST, HAND, AND FINGERS Local surface coils and FOV of 10 cm or less are recommended. Imaging planes should be carefully selected to be in the true coronal, sagittal, and axial planes. Electronic angulation may be necessary to ensure proper plane selections. An MR compatible brace on the wrist may be useful to ensure uniform wrist positioning.

MRI is indicated for the detection and evaluation of the cause of carpal tunnel syndrome, tendon and ligamentous injuries, triangular fibrocartilage injuries, extensor and flexor tenosynovitis, de Quervain syndrome, Kienböck disease and other causes of osteonecrosis (Fig. 8-26), injuries of the flexor and extensor tendons, tenosynovitis, and masses.

HIP Most imaging of the hips is performed without surface coils because of the difficulties of surface coil design for hip anatomy. T1-weighted coronal images are sensitive for osteonecrosis and fractures. T2-weighted images or high resolution 3D gradient-echo images may

Fig. 8-29 Developmental hip dysplasia (type two). Coronal gradient-echo MR image reveals subluxation of the left femoral head (*arrowhead*), which is smaller and exhibits delayed secondary ossification (*straight arrow*); the fibrocartilaginous acetabular labrum is inverted (*curved arrow*), preventing closed reduction. Other applications of MRI in the pediatric hip include early diagnosis of Legg-Calvé-Perthes disease and other causes of osteonecrosis, septic and transient synovitis, juvenile chronic arthritis, intracapsular osteoid osteoma, and slipped capital femoral epiphysis.

be useful for evaluating fine anatomical details, especially in children.

MRI is indicated for the evaluation of ischemic necrosis, occult hip fractures (Fig. 8-21), developmental hip dysplasia (Fig. 8-29), transient osteoporosis, and iliopsoas bursitis.

KNEE   Imaging should be performed in at least two planes utilizing at least one T2W technique. FOV should be less than 16 cm, pixel dimension less than 1 mm, and slice thickness less than 5 mm. Acquisition matrices of 192 × 256 are advantageous. Magnified, narrow window width display of short TE images containing the menisci may be of benefit in interpretation of meniscal disease.

MRI is indicated for the evaluation of knee pain of undetermined etiology, meniscal tears (Fig. 8-16), discoid menisci, suspected cruciate and collateral ligament tears, bone contusions (trabecular fractures), patellar chondromalacia, patellar tracking abnormalities, popliteal cysts and aneurysms, pes anserinus and prepatellar bursitis, pigmented villonodular synovitis, evaluation of anterior cruciate ligament reconstructions, and meniscal and ganglion cysts (Fig. 8-16).

ANKLE AND FOOT   FOV less than 12 cm is recommended with appropriately designed local coils. Images should be obtained in the coronal, sagittal, and axial planes with respect to the anatomic region of interest. Suspected disease in the tendons coursing through the ankle may be best evaluated with angled axial images, perpendicular to the tendon at the level of suspected disease. Long TR double echo images and STIR images

are advantageous in evaluating tendon and subtle bony injuries. When a tear of the Achilles tendon is suspected, a large FOV sagittal image may be warranted to evaluate the extent of possible retraction of the superior segment. When subtle findings are present, imaging of the contralateral ankle may be helpful.

MRI is indicated for the evaluation of ankle pain of undetermined etiology, tendon and ligament injuries, sinus tarsi syndrome, tarsal tunnel syndrome, plantar fasciitis, masses (Fig. 8-15), postoperative evaluation of Achilles tendon repair, diabetic foot disease, neuropathic joint disease, and healing triplane fractures in children (Fig. 8-22).

TEMPOROMANDIBULAR JOINT (TMJ)   Local surface coils are necessary with FOVs of less than 12 cm. Slice thickness should be 3 mm. Sagittal high resolution T1W images are usually performed to determine disk position. If inflammatory disease is present, then long TR, long TE (T2-weighted) images are helpful in evaluating pathology. Kinematic imaging with rapid T1 or gradient-echo images may be helpful in characterizing disk mobility and placement abnormalities. Because TMJ disease may be bilateral in up to 80% of affected patients, both joints should be imaged, ideally simultaneously.

MRI is indicated for the evaluation of joint pain and clicking, abnormal disk displacement, degenerative joint disease, osteonecrosis of the mandibular condyle, foreign body reaction following surgery, inflammatory joint disease, trauma, ganglion cysts, and synovial osteochondromatosis.

MRI OF THE SPINE

Techniques

The role of MRI in the spine has been well established by comparative studies with conventional imaging methods using surgical correlation as an objective measure of accuracy. The areas of greatest proven value include degenerative diseases involving both the cervical and lumbar spine, vertebral inflammatory lesions, congenital malformations, and intramedullary lesions such as syringomyelia and neoplasms. Equally useful are the applications of MRI in evaluating extradural neoplasms, intradural extramedullary neoplasms, trauma, and vertebral body deformities (Fig. 8-30).

In the performance of the spinal MR examination, decisions have to be made regarding the appropriate coil, imaging plane, slice thickness, imaging matrix, number of excitations, and pulse sequence parameters. These choices are influenced by the anatomic region to be investigated, desired field of view, spatial resolution, and contrast needs. The use of gadolinium chelates for enhancement permits more specific evaluation of moment nulling, a presaturation pulse, and appropriate gradient rotation. This is particularly true for T2-weighted images with relatively long TE times.

Indications for MRI of the Spine

In general, sagittal and axial views are indicated, including a T1W and a T2W spin echo, or a low flip angle, gradient recalled echo (GRE) sequence. A surface coil is mandatory to provide the resolution required for the delineation of subtle abnormalities. The surface coil is most efficient when it precisely covers the area of interest.

Cervical spine

For suggested extramedullary disease, the slice thickness should be no more than 3 to 4 mm and the pixel dimension 1 mm or less. A typical sequence, for example, might include a low flip angle gradient echo or a long TR, dual spin echo sagittal sequence and a short TR T1W spin echo or low flip angle gradient echo axial sequence, ideally with flow compensation. Transverse axial and sagittal images should be obtained with low flip angle gradient echo or T1 or T2W spin echo.

MRI is indicated for the evaluation of neoplasms (Fig. 8-24), spinal stenosis, cervical disk disease, spondylosis, and congenital anomalies.

Thoracic spine

MRI is indicated for the evaluation of thoracic disk disease, spondylosis, and congenital anomalies. For such indications, two orthogonal views are needed, including a T1W, a T2W, and a GRE sequence. A surface coil is mandatory to provide sufficient resolution in order to demonstrate the CSF-cord and CSF-extradural interfaces.

MRI is also indicated for the evaluation of primary or metastatic tumors, infections, and inflammation. In such cases evaluation should be performed with gadolinium chelates. A minimum of three sequences is required, including one precontrast T1W sequence, one precontrast T2W sequence, and at least one postcontrast T1W sequence.

Lumbar spine

MRI is indicated for the evaluation of lumbar disk disease (Fig. 8-31), spondylosis, and congenital anomalies (Fig. 8-32). In such cases MRI should be performed without contrast and should include (ideally) axial and sagittal long TR double echo sequences or low flip angle, GRE sequences. For patients with suspected or possible metastatic disease, T1W sagittal views should be added. Spatial resolution should be on the order of 1 mm and slice thickness should be 4 to 5 mm. Nonangled axial sections should minimally cover from L3 to S1.

For suspected primary or metastatic tumors, infections, inflammation, and postoperative patients (disk disease vs. epidural fibrosis, see Fig. 8-31), MRI should be performed with gadolinium chelates. A minimum of four sequences is required including one precontrast T1W sequence, one precontrast T2W sequence, and two postcontrast T1W sequences in the sagittal and axial planes.

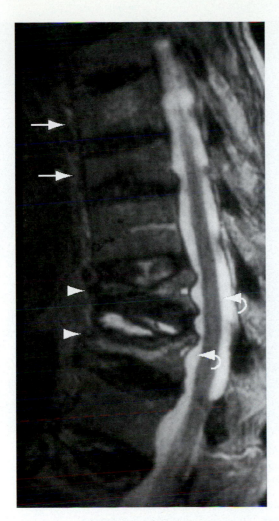

Fig. 8-30 Severe osteoporosis with multiple spinal compression fractures. Sagittal fast spin-echo T2-weighted MR image reveals numerous vertebral endplate deformities (*straight arrows*), some of which are recent owing to the presence of high signal intensity (*arrowheads*). In the setting of a traumatic vertebral body deformity, MR is useful not only in estimating fracture age but also for depicting spinal canal compromise (*curved arrows*) and distinguishing benign from malignant collapse.

intradural-extramedullary and extradural lesions. Gadolinium chelates are also routinely utilized in the evaluation of the postoperative "failed back syndrome" (Fig. 8-31).

Specific needs in the evaluation of the spine include visualization of the cerebrospinal fluid (CSF) and the CSF-extradural interface. This visualization is based on appropriate T1, T2, and proton density contrast, spatial resolution, slice thickness, and examination time. The use of flow compensation or cardiac gating to minimize artifacts from CSF pulsation further improves the examination.

Motion artifacts arising from moving structures (e.g., blood) anterior to the spine degrade the image quality. These artifacts can be suppressed using presaturation pulses and/or gradient rotations. The ideal examination includes a combination of techniques such as gradient

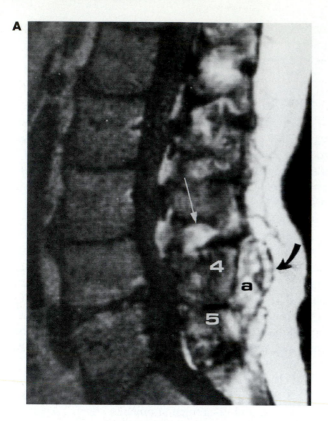

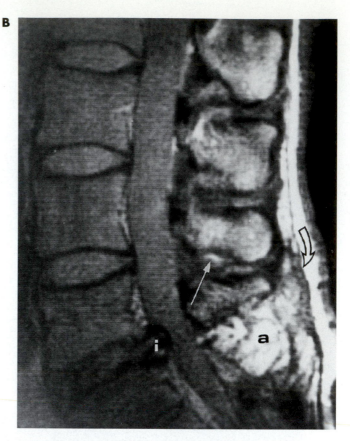

Fig. 8-31   Postoperative alterations in the lumbar spine. **A,** On a T1-weighted sagittal MR image, thinning and posterior convexity of the supraspinatus ligament (*curved arrow*) are associated with subjacent areas of low and high signal intensity (*a*) dorsal to the L4 (*4*) and L5 (*5*) spinous processes. Associated fatty replacement or bursa formation (*straight arrow*) is noted within the interspinous ligaments at the L3-4 level. The patient had undergone limited previous surgery for intervertebral disk disease. **B,** On an intermediate-weighted sagittal MR image, prominent abnormal areas of low and high signal intensity (*a*) are evident subjacent to the posteriorly displaced and disrupted supraspinous ligament (*curved arrow*) at the L5-S1 level. An associated localized zone of fatty replacement or bursa formation (*straight arrow*) can also be appreciated at the L4-5 level. The patient had undergone operative intervention for the degenerated protruding L5-S1 intervertebral disk (*i*). In general, gandolinium chelates increase the specificity of MRI following spine surgery.

PEARL

- Posterior synovial extension with extradural defects at C1-2 and atlantoaxial subluxation can be shown by use of MRI in rheumatoid and seronegative arthritis, infection, and crystal deposition disorders.
- MRI is more sensitive to disk degeneration and identification of a normal nucleus pulposus than CT, myelography, or conventional radiography.
- Spin-echo technique is preferable to inversion-recovery sequences in MR imaging of the spine, because of its better signal-to-noise ratio.
- Normal nucleus pulposus has higher signal intensity than the surrounding annulus fibrosus because of its higher water content.
- A degenerated intervertebral disk without herniation has low nuclear signal because of desiccation.

Continued

PEARL—cont'd

- On long TE sequences, increased signal intensity in an intervertebral disk and adjacent endplates occurs in spinal infection secondary to inflammation.
- Long TE and TR sequences provide optimal contrast between cerebrospinal fluid, degenerated disk, and cortical bone.
- Spondylosis, spondylolysis, and bony spinal stenosis are better evaluated by CT than by MRI.
- Following chemonucleolysis, MRI demonstrates retraction of protruding disks with loss of signal intensity and disk space narrowing.
- Contrast material–enhanced MRI allows distinction among fibrosis, disk, and thecal sac in postoperative patients.

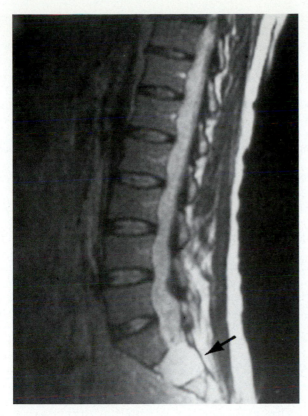

Fig. 8-32 Intrasacral meningocele. Sagittal T2-weighted MR image demonstrates a well-defined homogeneous mass of high signal intensity (*arrow*) involving the second sacral segment. In general, MRI is the procedure of choice for the diagnosis and characterization of congenital and developmental abnormalities affecting the spine.

SUGGESTED READINGS

1. Advances in musculoskeletal imaging, Radiol Clin North Am 32(2):201, 1994.
2. Bahk YW: Combined scintigraphic and radiographic diagnosis of bone and joint diseases, Berlin, 1994, Springer-Verlag.
3. Bard and Laredo JD, editors: Interventional radiology in bone and joint, New York, 1988, Springer-Verlag.
4. Bassett LW, Gold RH, and Seeger LL: MRI atlas of the musculoskeletal system, London, 1989, Martin Dunitz.
5. Beltran J: MRI: musculoskeletal system, Philadelphia, 1990, JB Lippincott Co.
6. Bloem JL and Sartoris DJ, editors: MRI and CT of the musculoskeletal system: a text-atlas, Baltimore, 1992, Williams & Wilkins.
7. Brown ML et al: Technical aspects of bone scintigraphy, Radiol Clin North Am 31:721, 1993.
8. Buckwalter KA and Braunstein EM: Digital skeletal radiography, AJR 158:1071, 1992.
9. Cann CE: Skeletal structure-function revisited, Radiology 179:607, 1991.
10. Chan WP, Lang P, and Genant HK: MRI of the musculoskeletal system, Philadelphia, 1994, WB Saunders Co.
11. Cohen MD: Magnetic resonance imaging of the pediatric musculoskeletal system, Semin US CT MR 12:506, 1991.
12 Current review of musculoskeletal radiology, Curr Opin Radiol 4:1, 1992.
13. Dalinka MK: Arthrography, New York, 1980, Springer-Verlag.
14. DeSchepper AMA, Degryse HR, and Ramon FA: Five years experience in musculoskeletal magnetic resonance imaging on a 0.5 T imager, Eur J Radiol 14:104, 1992.
15. Edeiken J, Dalinka M, and Karasick D: Edeiken's roentgen diagnosis of diseases of bone, ed 4, Baltimore, 1990, Williams & Wilkins.
16. Firooznia H et al: MRI and CT of the musculoskeletal system, St Louis, 1992, Mosby.
17. Freiberger RH and Kaye JJ: Arthrography, New York, 1979, Appleton-Century-Crofts.
18. Gabriel H et al: MR imaging of hip disorders, Radiographics 14:763, 1994.
19. Gentili A, Miron SD, and Bellon EM: Nonosseous accumulation of bone-seeking radiopharmaceuticals, Radiographics 10:871, 1990.
20. Georgy BA and Hesselink JR: MR imaging of the spine: recent advances in pulse sequences and special techniques, AJR 162:923, 1994.
21. Greenfield GB, Warren DL, and Clark RA: MR imaging of periosteal and cortical changes of bone, Radiographics 11:611, 1991.
22. Greenspan A: Orthopedic radiology: a practical approach, Philadelphia, 1988, JB Lippincott Co.
23. Greenspan A et al: Condensing osteitis of the clavicle: a rare but frequently misdiagnosed condition, AJR 156:1011, 1991.
24. Janzen DL et al: Cystic lesions around the knee joint: MR imaging findings, AJR 163:155, 1994.
25. Jaramillo D and Hoffer FA: Cartilaginous epiphysis and growth plate: normal and abnormal MR imaging findings, AJR 158:1105, 1992.
26. Kang HS and Resnick D: MRI of the extremities: an anatomic atlas, Philadelphia, WB Saunders Co.
27. Kessler JR, Wells RG, and Sty JR: Skeletal scintigraphy: radiographic artifacts, Clin Nucl Med 30:907, 1992.
28. Kier R et al: MR appearance of painful conditions of the ankle, Radiographics 11:401, 1991.
29. Kirsky DM et al: Increased long-bone periosteal cortical uptake in skeletal scintigraphy, Semin Nucl Med 22:54, 1992.
30. Kohler A et al: Borderlands of normal and early pathologic findings in skeletal radiography, ed 4, New York, 1992, Thieme.
31. Kricun ME: Imaging modalities in spinal disorders, Philadelphia, 1988, WB Saunders Co.
32. Kumar R et al: The calcaneus: normal and abnormal, Radiographics 11:415, 1991.
33. Lenchik L, Dovgan DJ, and Kier R: CT of the iliopsoas compartment: value in differentiating tumor, abscess, and hematoma, AJR 162:83, 1994.
34. Manaster BJ: Handbook of skeletal radiology, Chicago, 1989, Year Book.
35. Mettler FA Jr, editor: Radionuclide bone imaging and densitometry, New York, 1988, Churchill Livingstone, Inc.
36. Milgram JW: Radiologic and histologic pathology of nontumorous diseases of bones and joints, Northbrook, Ill, 1990, Northbrook.

37. Mink JH and Deutsch AL, editors: MRI of the musculoskeletal system: a teaching file, New York, 1990, Raven Press.

38. Mink JH et al: MRI of the knee, ed 2, New York, 1993, Raven Press.

39. Mirowitz SA et al: MR imaging of bone marrow lesions: relative conspicuousness of T1-weighted, fat-suppressed T2-weighted, and STIR images, AJR 162:215, 1994.

40. Moore SG et al: Pediatric musculoskeletal MR imaging, Radiology 179:345, 1991.

41. Murphey MD et al: Computed radiography in musculoskeletal imaging: state of the art, AJR 158:19, 1992.

42. Murphy WA Jr, editor: Musculoskeletal disease test and syllabus, Reston, Va, 1994, American College of Radiology.

43. Oestreich AE and Crawford AH: Atlas of pediatric orthopedic radiology, New York, 1985, Thieme.

44. Olson EM, Wong WHM, and Hesselink JR: Extraspinal abnormalities detected on MR imagings of the spine, AJR 162:679, 1994.

45. Orthopedics, Radiol Clin North Am 28(2):233, 1990.

46. Otake S: Sarcoidosis involving skeletal muscle: imaging findings and relative value of imaging procedures, AJR 162:369, 1994.

47. Ozonoff MB: Pediatric orthopedic radiology, ed 2, Philadelphia, 1992, WB Saunders Co.

48. Parekh JS and Teates CD: Mixed "hot" and "cold" lesions on bone scans, Semin Nucl Med 22:289, 1992.

49. Park YH et al: Patterns of vertebral ossification and pelvic abnormalities in paralysis: study of 200 patients, Radiology 188:561, 1993.

59. Patel N et al: High-resolution bone scintigraphy of the adult wrist, Clin Nucl Med 17:449, 1992.

51. Peller PJ, Ho VB, and Kransdorf MJ: Extraosseous Tc-99m MDP uptake: a pathophysiologic approach, Radiographics 13:715, 1993.

52. Poznanski AK: The hand in radiologic diagnosis with gamuts and pattern profiles, ed 2, Philadelphia, 1984, WB Saunders Co.

53. Reicher MA and Kellerhouse LE: MRI of the wrist and hand, New York, 1990, Raven Press.

54. Renton P: Orthopaedic radiology: pattern recognition and differential diagnosis, Chicago, 1990, Year Book.

55. Resnick D: Bone and joint imaging, Philadelphia, 1989, WB Saunders Co.

56. Resnick D, editor: Diagnosis of bone and joint disorders, ed 3, Philadelphia, 1995, WB Saunders Co.

57. Resnick D and Sartoris DJ, editors: Bone disease test and syllabus (fourth series), Reston, Va, 1989, American College of Radiology.

58. Sartoris DJ: Musculoskeletal imaging workbook, St Louis, 1993, Mosby.

59. Sartoris DJ, editor: Principles of shoulder imaging, New York, 1995, McGraw-Hill, Inc.

60. Seeger LL, editor: Diagnostic imaging of the shoulder, Baltimore, 1992, Williams & Wilkins.

61. Steiner GM and Sprigg A: Value of ultrasound in the assessment of bone, Br J Radiol 65:589, 1992.

62. Stoller DW: Magnetic resonance imaging in orthopaedics and sports medicine, Philadelphia, 1993, JB Lippincott Co.

63. Stoller DW et al, editors: Magnetic resonance imaging in orthopaedics and rheumatology, Philadelphia, 1989, JB Lippincott Co.

64. Taylor JAM et al: Painful conditions affecting the first metatarsal sesamoid bones, Radiographics, 13:817, 1993.

65. Tehranzadeh J and Steinbach LS: Musculoskeletal manifestations of AIDS, St Louis, 1994, Green.

66. Vanarthos WJ et al: Diagnostic uses of nuclear medicine in AIDS, Radiographics 12:731, 1992.

67. Wandtke JC and Plewes DB: Improved imaging of bone with scan equalization radiography, AJR 157:359, 1991.

68. Wardlaw JM, Best JJK, and Hughes SPF: Dynamic bone imaging in the investigation of local bone pathology: when is it useful? Clin Radiol 43:107, 1991.

69. Weissman DS: Radiology of the foot, ed 2, Baltimore, 1989, Williams & Wilkins.

70. Wong WL and Pemberton J: Musculoskeletal manifestations of epidermolysis bullosa: an analysis of 19 cases with a review of the literature, Br J Radiol 65:480, 1992.

Index

Page numbers in *italic type* refer to figures. Tables are indicated by *t* following the page number.